SOLAR SYSTEM HISTORY FROM ISOTOPIC SIGNATURES OF VOLATILE ELEMENTS

Space Sciences Series of ISSI

Volume 16

The International Space Science Institute is organized as a foundation under Swiss law. It is funded through recurrent contributions from the European Space Agency, the Swiss Confederation, the Swiss National Science Foundation, and the University of Bern. For more information, see the homepage at http://www.issi.unibe.ch/.

Cover illustration: Artist's view of the solar nebula: Gas, dust, and planetesimals orbit the early Sun in the accretion disk. Copyright by William K. Hartmann, Planetary Science Institute, Tucson, Arizona, USA.

The titles published in this series are listed at the end of this volume.

SOLAR SYSTEM HISTORY FROM ISOTOPIC SIGNATURES OF VOLATILE ELEMENTS

Volume Resulting from an ISSI Workshop
14–18 January 2002, Bern, Switzerland

Edited by

R. KALLENBACH
International Space Science Institute,
CH-3012 Bern, Switzerland

T. ENCRENAZ
DESPA, Observatoire de Paris,
F-92195 Meudon Cedex, France

J. GEISS
International Space Science Institute,
CH-3012 Bern, Switzerland

K. MAUERSBERGER
Max-Planck-Institut für Kernphysik,
D-69029 Heidelberg, Germany

T.C. OWEN
University of Hawaii at Manoa,
Honolulu, HI 96822, USA

F. ROBERT
CNRS-Muséum, Muséum d'Histoire Naturelle,
F-75005 Paris, France

Reprinted from *Space Science Reviews*, Volume 106, Nos. 1–4, 2003

INTERNATIONAL SPACE SCIENCE INSTITUTE

Space Sciences Series of ISSI

A.C.I.P. Catalogue record for this book is available from the Library of Congress

ISBN: 1-4020-1177-6

Published by Kluwer Academic Publishers,
P.O. Box 990, 3300 AZ Dordrecht, The Netherlands

Sold and distributed in North, Central and South America
by Kluwer Academic Publishers,
101 Philip Drive, Norwell, MA 02061, U.S.A.

In all other countries, sold and distributed
by Kluwer Academic Publishers,
P.O. Box 322, 3300 AH Dordrecht, The Netherlands

Printed on acid-free paper

Printed in the Netherlands

TABLE OF CONTENTS

SUN AND PROTOSOLAR NEBULA

OUTER SOLAR SYSTEM

EARTH, TERRESTRIAL PLANETS, AND MOON

ISOTOPIC FRACTIONATION PROCESSES

WORKING GROUP REPORTS

EPILOGUE

Solar System History from Isotopic Signatures of Volatile Elements
ISSI Workshop, 14–18 January 2002, Bern, Switzerland

Group Photograph

1. Konrad Mauersberger	9. Robert O. Pepin	17. Richard H. Becker
2. Eric Herbst	10. François Robert	18. Tom J. Millar
3. Kathrin Altwegg	11. Daniel Gautier	19. Ursula Pfander
4. Thérèse Encrenaz	12. Rainer Wieler	20. Jeffrey Linsky
5. Tobias Owen	13. Ulrich Ott	21. Johannes Geiss
6. Helmut Lammer	14. Evelyne Roueff	22. Ricardo Hueso
7. Scott Messenger	15. Robert N. Clayton	23. Silvia Wenger
8. Bernard Marty	16. Brigitte Fasler	24. Reinald Kallenbach

Not in this picture: Erik M. Galimov, Monica M. Grady, Alex N. Halliday

Picture taken by Urs Lauterburg, Physikalisches Institut, University of Bern

List of Participants

Kathrin Altwegg
Physikalisches Institut, University of Bern
CH-3012 Bern, Switzerland
kathrin.altwegg@phim.unibe.ch

Richard H. Becker
School of Physics and Astronomy, University of Minnesota
Minneapolis, MN 55455, USA
rhbecker@maroon.tc.umn.edu

Robert N. Clayton
Enrico Fermi Institute, University of Chicago
Chicago, IL 60637, USA
r-clayton@uchicago.edu

Thérèse Encrenaz
DESPA, Observatoire de Paris
F-92195 Meudon Cedex, France
therese.encrenaz@obspm.fr

Erik M. Galimov
V.I. Vernadsky Institute, Russian Academy of Sciences
117975 Moscow, Russia
galimov@geokhi.msk.su

Daniel Gautier
Observatoire de Paris-Meudon, CNRS
F-92195 Meudon Cedex, France
daniel.gautier@obspm.fr

Johannes Geiss
International Space Science Institute
CH-3012 Bern, Switzerland
johannes.geiss@issi.unibe.ch

Monica M. Grady
Dept. of Mineralogy, The Natural History Museum
UK-London SW7 5BD, United Kingdom
mmg@nhm.ac.uk

Alex N. Halliday
Dept. of Earth Sciences, ETH Zürich
CH-8092 Zürich, Switzerland
halliday@erdw.ethz.ch

Eric Herbst
Departments of Physics and Astronomy, The Ohio State University
Columbus, OH 43210, USA
herbst@mps.ohio-state.edu

Ricardo Hueso
Laboratorie Cassini, Observatoire de la Côte d'Azur
F-06304 Nice Cedex 4, France
hueso@obs-nice.fr

Reinald Kallenbach
International Space Science Institute
CH-3012 Bern, Switzerland
reinald.kallenbach@issi.unibe.ch

Helmut Lammer
Space Research Institute, Austrian Academy of Sciences
A-8042 Graz, Austria
helmut.lammer@oeaw.ac.at

Jeffrey Linsky
JILA, University of Colorado
Boulder, CO 80309-0440, USA
jlinsky@jila.colorado.edu

Bernard Marty
Centre de Recherches Pétrographiques et Géochimiques, CNRS
F-54501 Vandoeuvre-les-Nancy Cedex, France
bmarty@crpg.cnrs-nancy.fr

Konrad Mauersberger
Max-Planck-Institut für Kernphysik, Division of Atmospheric Physics
D-69029 Heidelberg, Germany
Konrad.Mauersberger@mpi-hd.mpg.de

Scott Messenger
Laboratory for Space Sciences, Dept. of Physics, Washington University
Saint Louis, MO 63130-4899, USA
dbunny@wuphys.wustl.edu

Tom J. Millar
Physics Department, UMIST
Manchester M60 1QD, UK
Tom.Millar@umist.ac.uk

Ulrich Ott
Max-Planck-Institut für Chemie (Otto-Hahn-Institut)
D-55020 Mainz, Germany
ott@mpch-mainz.mpg.de

Tobias Owen
Institute for Astronomy, University of Hawaii at Manoa
Honolulu, HI 96822, USA
owen@ifa.hawaii.edu

Robert O. Pepin
School of Physics and Astronomy, University of Minnesota
Minneapolis, MN 55455, USA
pepin001@maroon.tc.umn.edu

François Robert
Laboratoire de Minéralogie, CNRS-Muséum, Muséum d'Histoire Naturelle
F-75005 Paris, France
robert@cimrs1.mnhn.fr

Evelyne Roueff
LUTH and FRE2462 du CNRS, Observatoire de Paris
F-92190 Meudon Cedex, France
Evelyne.roueff@obspm.fr

Rainer Wieler
Dept. of Earth Sciences, ETH Zürich
CH-8092 Zürich, Switzerland
wieler@erdw.ethz.ch

AUTHOR INDEX

PREFACE

This volume, number 16 in the *Space Sciences Series of ISSI*, reviews the present knowledge on many aspects of the history of the solar system. It focuses on isotopic signatures of volatile elements as tracers for the evolutionary processes during the formation of the Sun and the planets from an interstellar molecular cloud and, in turn, illuminates how the isotopic compositions of present-day solar system objects had been established. This integrated collection of articles complements the information provided by previous ISSI volumes such as *From Dust to Terrestrial Planets* and *Chronology and Evolution of Mars*, expanding the time scale for the history of the solar system.

The convenors, Thérèse Encrenaz, Observatoire de Paris, Meudon; Johannes Geiss, ISSI, Bern; Reinald Kallenbach, ISSI, Bern; Konrad Mauersberger, Max-Planck-Institut für Kernphysik, Heidelberg; Tobias Owen, University of Hawaii, Honolulu; and François Robert, CNRS-Muséum d'Histoire Naturelle, Paris, invited experts in planetary science, solar and heliospheric physics, astrophysics, mineralogy, and chemistry to an interdisciplinary workshop in Bern, 14–18 January 2002. The resulting review articles are grouped into four sections 'Sun and Protosolar Nebula', 'Outer Solar System', 'Earth, Terrestrial Planets, and Moon', and 'Isotopic Fractionation Processes'.

In the first section, J. Geiss and G. Gloeckler constrain the hydrogen, helium, and neon isotopic abundances in the protosolar cloud as reference values for solar system processes. For understanding the chemical evolution of the Galaxy, precise protosolar abundances are also important, because they represent the composition of a sample of galactic matter as it existed about 5 Gy ago. Planetary samples, however, are mainly categorized by the fractionation pattern in their oxygen isotopic ratios $^{18}O/^{16}O$ and $^{17}O/^{16}O$, although the solar values of $^{18}O/^{16}O$ and $^{17}O/^{16}O$ are not known, yet. In the course of this workshop collaboration, R.N. Clayton renewed the interest in isotopic fractionation caused by self-shielding from dissociating solar radiation of the major solar nebula molecules $^{12}C^{16}O$ and $^{14}N_2$. This process may result in enrichments of ^{17}O, ^{18}O, and ^{15}N in compounds formed in the solar accretion disk described by the recent "X-model" (Shu *et al., Science* **271**, 1545–1552, 1997), but not in enrichments of ^{13}C as carbon atoms re-equilibrate rapidly with the reservoir of non-dissociated molecules. This is concordant with isotopic ratios observed in meteoritic and planetary samples, whereas pre-solar grains additionally show strong variations of the $^{13}C/^{12}C$ ratio due to stellar nucleosynthetic sources. As reviewed by U. Ott, pre-solar solids also prove their stellar origin in many isotopic signatures of noble gases.

The strongest variations among volatile element isotope ratios are observed for hydrogen. Therefore, D/H ratios are particularly characteristic for planetary processes, but they are also important for monitoring the nuclear evolution in the Galaxy and even in the Universe as a whole (see the article by J.L. Linsky). Chemical fractionation changing the D/H ratio by orders of magnitude is modeled by T.J. Millar and is observed in molecules of the interstellar medium, as summarized by E. Roueff and M. Gerin. Based on the D/H ratios measured in chondrites, F. Robert points out that these models are in principle also applicable to the (proto-) solar nebula, if they include all physical properties of a dynamical accretion disk.

In the second section, R. Hueso and T. Guillot present the current understanding of the evolution of the outer solar nebula, the formation of icy and rocky planetesimals, and the formation of giant planets. The latter study is also bound to shed some light on the formation of the newly discovered extrasolar giant planets. This provides the input parameters for potential isotopic fractionation models of the outer solar nebula. Data on volatile isotope abundance measurements in outer solar system objects are extensively reviewed by T. Owen and Th. Encrenaz, by K. Altwegg and D. Bockelée-Morvan, and by S. Messenger with co-authors (presuming that interplanetary dust particles originate from comets and asteroids). While the helium isotopic ratio in Jupiter is representative for the composition of the protosolar cloud, other isotopic ratios show the influence of some combination of interstellar chemical fractionation processes, processes that have occured in the early solar nebula, for instance during the formation of planetesimals, or special isotopic fractionation on the satellites of Jupiter and Saturn.

In the third section, B. Marty and co-authors present new intriguing data and interpretations on variations in the $^{15}N/^{14}N$ in the archive of lunar samples. Variations in the lunar nitrogen isotopic composition are explained by a secular change in the infall of meteorites, micrometeorites, and possibly comets, with the latter bodies carrying non-solar compositional signatures. The solar $^{15}N/^{14}N$ ratio has been identified to be lower than the terrestrial. This is in line with the hypothesis by R.N. Clayton that solar nitrogen is isotopically lighter than many solid planetary samples due to a self-shielding process in the solar nebula. As reported by R. Wieler and V.S. Heber, lunar samples also offer an archive of the isotopic compositions of the noble gases in the solar wind, from which important aspects of the history of the sun can be inferred over the past four billion years. The outgassing history of the Moon itself also appears to be represented in this archive. R.O. Pepin explores escape fractionation processes that have been studied in planetary atmospheres, but also may apply to planetesimals in the solar accretion disk. It remains as yet unclear, how the isotopic compositions of noble gases in the "phase Q" of meteorites have been established. M. Grady and I.P. Wright review the data on carbon and nitrogen isotopic ratios in meteorites, which provides the possibility to trace the reprocessing of presolar grains and the formation of solids in the solar nebula. Regarding the basic question on the origin of life itself, E.M. Galimov concludes that organic compounds in carbonaceous chondrites are not linked to the evolution of biological life.

In the fourth section, the principles of isotopic fractionation processes are discussed. For the example of processes related to ozone formation, K. Mauersberger and co-authors review the kinetics of chemical reactions that selectively transfer isotopes from molecular reservoirs into minor compounds. These processes may not only be relevant for the Earth's atmosphere but also for the early solar system. Particularly strong fractionation mechanisms that must have been active in the early solar system are discussed by H. Lammer and S. Bauer. Gravitational escape in a strong early solar wind appears to be responsible for a potential fractionation of nitrogen isotopes by more than a factor 4 at Saturn's moon Titan. E. Herbst outlines that such strong fractionation is by far not expected from ion-molecule reactions in cold molecular clouds or similar objects such as the solar nebula even at low temperatures of order 10-30 K. Finally, R. Kallenbach reports, as a result of the SOHO mission, that the strong plasma flow from the Sun to interplanetary space, the solar wind – not to be confused with solar energetic particles – , only fractionates heavy isotopes in the long term by about 1% per atomic mass unit or even less. This implies strong hopes for the data to be expected from the Genesis mission, which at present collects the solar wind in space.

A fifth section has been added, containing two working group articles that summarize the views of the workshop participants and address the issues left open by present research results. The working group report by R. Kallenbach and co-authors summarizes the knowledge on the fate of volatile isotopes in the pre-solar molecular cloud and in the (proto-) solar nebula. The subsequent evolution of the Sun's isotopic composition due to nuclear processes as well as the impact of the solar wind and solar energetic particles on the isotopic composition of the surrounding planetary system and the influence of the changing interstellar environment of the solar system are reviewed as well. R.H. Becker and co-authors summarize the available abundance data of volatile isotopes on terrestrial planets, and discuss the processes that have changed them during the history of the solar system.

A glossary and a list of acronyms has been added to this volume to the benefit of the reader and to ease the use of this book for students.

It is our pleasure to thank all those who have contributed to this volume and to the workshop in general. We are grateful to all authors for their contributions, and to the reviewers for their critical reports, which helped to improve the quality of the book. The referees all agreed to be identified and are listed in the following, with the number in parenthesis indicating the number of papers they have reviewed:

Kathrin Altwegg (1)
Isabelle Baraffe (1)
Richard H. Becker (1)
Claude Camy-Peyret (1)
Eric Chassefière (1)
Robert N. Clayton (1)
Jacques Crovisier (1)
Daniel Gautier (2)
George Gloeckler (1)
Ko Hashizume (1)
Eric Herbst (1)
Peter Hoppe (1)
Reinald Kallenbach (3)
Dieter Krankowsky (1)
Marty Lee (1)
Kurt Marti (1)
Glenn Mason (1)
Tom J. Millar (1)
Konrad Mauersberger (3)
Uli Ott (1)
François Robert (2)
Timothy D. Swindle (1)
Igor Tolstikhin (1)

The results and conclusions presented in this volume are based on meteorite measurements in the laboratory, on remote observations of planetary atmospheres, and on data obtained from many space missions of the US National Aeronautics and Space Administration, the European Space Agency and the Japanese and Russian Space Agencies. The editors acknowledge the work of all individuals who have contributed the enormous data base that is necessary for reconstructing the origin of the solar system and the evolution of individual solar system bodies.

Last but not least, we express our thanks to the directorate and staff of ISSI, for their support in making the workshop happen and in getting the book finalized; we thank in particular B. Fasler, D. Taylor, and S. Wenger for support in the workshop organisation, X. Schneider (until 30 April 2002) and S. Saliba (since 1 May 2002) for maintaining the computer infrastructure of ISSI, and U. Pfander for editorial assistance.

November 2002
The Editors

I. SUN AND PROTOSOLAR NEBULA

ISOTOPIC COMPOSITION OF H, HE AND NE IN THE PROTOSOLAR CLOUD

JOHANNES GEISS[1] and GEORGE GLOECKLER[2]
[1] *International Space Science Institute, Hallerstrasse 6, 3012 Bern, Switzerland* (johannes.geiss@issi.unibe.ch)
[2] *Dept. of Physics and IPST, University of Maryland, College Park, MD 20742, USA University of Michigan, Ann Arbor, MI 48109, USA* (gg10@umail.umd.edu)

Received: 19 August 2002; Accepted in final form: 1 November 2002

Abstract. Observations and measurements in the solar wind, the Jovian atmosphere and the gases trapped in lunar surface material provide the main evidence from which the isotopic composition of H, He and Ne in the Protosolar Cloud (PSC) is derived. These measurements and observations are reviewed and the corrections are discussed that are needed for obtaining from them the PSC isotopic ratios. The D/H, ^{3}He/^{4}He (D+^{3}He)/H, ^{20}Ne/^{22}Ne and ^{21}Ne/^{22}Ne ratios adopted for the PSC are presented. Protosolar abundances provide the basis for the interpretation of isotopic ratios measured in the various solar system objects. In this article we discuss constraints derived from the PSC abundances on solar mixing, the origin of atmospheric neon, and the nature of the "SEP" component of neon trapped at the lunar surface. We also discuss constraints on the galactic evolution provided by the isotopic abundances of H and He in the PSC.

1. Introduction

Isotopic abundance measurements are invaluable for tracing the origin and evolution of matter in the universe, the galaxy, and the solar system as a whole as well as the sun, planets, asteroids, moons and comets. In this paper we present the best estimates presently available of the isotopic abundances of H, He and Ne in the Protosolar Cloud (PSC) out of which the bodies in the solar system formed. In order to achieve this, the isotopic ratios observed in these bodies of the solar system have to be corrected for chemical or physical isotope fractionation and for possible alterations by nuclear reactions.

Generally, isotope fractionation by physical or chemical processes are most severe in the case of strong element depletion. From this it follows that measurements of the composition in the Outer Convective Zone (OCZ) of the Sun and in the atmosphere of Jupiter gives us the best evidence for deriving elemental and isotopic abundances of H, He and Ne in the PSC.

Space Science Reviews **106:** 3–18, 2003.

TABLE I

$^3He/^4He$ in the "Planetary Component" of Meteorites and in Jupiter.

	$^3He/^4He$ (10^{-4})
Planetary Component in Meteorites[1,2]	$\approx 1.5 \pm 0.2$
Q-Phase of the Carbonaceous Chondrite Isna[3]	1.23 ± 0.02
Jupiter[4]	1.66 ± 0.06
Protosolar Cloud[5]	$\mathbf{1.66 \pm 0.06}$

[1] Eberhardt (1974); [2] Frick and Moniot (1977); [3] Busemann *et al.* (2000, 2001); [4] Mahaffy *et al.* (1998); [5] see text.

2. Protosolar $^3He/^4He$

For several decades the $^3He/^4He$ ratio measured in the "Planetary Component" of meteorites served as a proxy for the isotopic abundance of helium in the PSC (*e.g.*, Eberhardt, 1974; Frick and Moniot, 1977), see Table I. More recently, it was realized that this Planetary Component is a mixture of components with somewhat different compositions. We have included in Table I the $^3He/^4He$ ratio measured in the "Q-phase" of the carbonaceous chondrite Isna (*e.g.*, Busemann *et al.*, 2000). This meteorite has a very short cosmic ray exposure age and, therefore, the data can be relatively well corrected for cosmic ray-produced ^3He.

The situation changed when the $^3He/^4He$ ratio in the atmosphere of Jupiter was measured by the Galileo Probe Mass Spectrometer (GPMS), see Niemann *et al.* (1996). The ratio of $(1.66 \pm 0.06) \times 10^{-4}$ obtained by Mahaffy *et al.* (1998) given in Table I is presently the best approximation to the protosolar value.

In the Jovian atmosphere helium is depleted by $\approx 18\%$ relative to the protosolar cloud (von Zahn *et al.*, 1998). The degree of helium depletion in the OCZ of the Sun is similar, but the processes that led to the helium depletion in the Solar and Jovian atmospheres are quite different, diffusive separation in the case of the Sun, and — probably — descent of helium-rich droplets in the case of Jupiter (Stevenson and Salpeter, 1976; von Zahn *et al.*, 1998). The Jovian process should be less efficient in fractionating isotopes than the solar process, and, therefore, $^3He/^4He$ fractionation in the Jovian atmosphere ought to be less than the 2% fractionation that has occurred in the OCZ of the Sun (Gautier and Morel, 1997; Vauclair, 1998). Thus, we adopt the Jovian $^3He/^4He$ ratio obtained by Mahaffy *et al.* (1998) as the protosolar value without correction.

The $^3He/^4He$ ratios in the Planetary Component of meteorites are typically lower by 10 to 20% than the Jovian ratio (cf. Table I). This is not surprising, because in many trapping and loss processes, the lighter isotope is depleted relative to the heavier one. For instance, the diffusion constants D_1 and D_2 of two

TABLE II

Direct $^{3}He/^{4}He$ Measurements in the Solar Wind.

Apollo / SWC [1]	5 Foil Collections	1969–1972	High Precision
ISEE-3 / ICI[2]	Magnetic Mass Spec	1978–1985	High Time Resolution
Ulysses / SWICS[3]	ToF Mass Spec	1990–	Data from Slow Wind & Fast Streams, High Time Resolution

Publications on $^{3}He/^{4}He$: [1] Geiss *et al.* (1970, 1972); [2] Ogilvie *et al.* (1980), Coplan *et al.* (1984), Bochsler (1984); [3] Bodmer *et al.* (1995), Gloeckler and Geiss (1998, 2000).

isotopes of an element are approximately related to the corresponding masses m_1 and m_2 by $D_1/D_2 = (m_1/m_2)^{-1/2}$, resulting in $D(^{3}\mathrm{He})/D(^{4}\mathrm{He}) = 1.15$ for the helium isotopes. Thus, the ratio between $^{3}He/^{4}He$ in Jupiter and in the Planetary Component is of the order of the ratio of the diffusion constants of the two isotopes. This is a modest reduction, considering that helium is depleted by several orders of magnitude in the host mineral of the Planetary Component.

3. The Protosolar (D + ^{3}He)/H Ratio

In the early Sun, D was converted by the (p,γ)-reaction into ^{3}He, which has not been further processed in the OCZ of the Sun, as can be surmised from the existence there of beryllium (Geiss and Reeves, 1972), cf. Section 6. Thus, $^{3}He/^{4}He$ in the OCZ basically represents the protosolar (D+^{3}He)/H ratio.

Present-day isotopic abundances of noble gases in the OCZ can best be obtained from solar wind (SW) investigations. $^{3}He/^{4}He$ in the solar wind has been measured by several space-borne instruments. Comprehensive results have been obtained from three investigations (cf. Table II): the Apollo Solar Wind Composition (SWC) experiments, using solar wind collection in foils with subsequent analysis by laboratory mass spectrometry; the Ion Composition Instrument (ICI) on the International Sun Earth Explorer 3 (ISEE 3) using an electromagnetic mass spectrometer allowing unambiguous measurement of the mass/charge ratio of the ions; and the Solar Wind Composition Spectrometer (SWICS) on Ulysses, a time-of-flight system giving the mass/charge ratio as well as the mass of the ions. The results of the three investigations show significant variations in the $^{3}He/^{4}He$ ratio. Therefore, some understanding of the causes for the changes in $^{3}He/^{4}He$ is required and a correction method needs to be developed for obtaining the best estimate of the $^{3}He/^{4}He$ ratio in the present-day OCZ.

We follow here the method introduced by Gloeckler and Geiss (2000), cf. Figure 1. Prior to the Ulysses mission, all solar wind composition data were taken in the ecliptic plane, where the low speed solar wind dominates. Thanks to the polar orbit of Ulysses and the large energy range of the SWICS instrument, it became

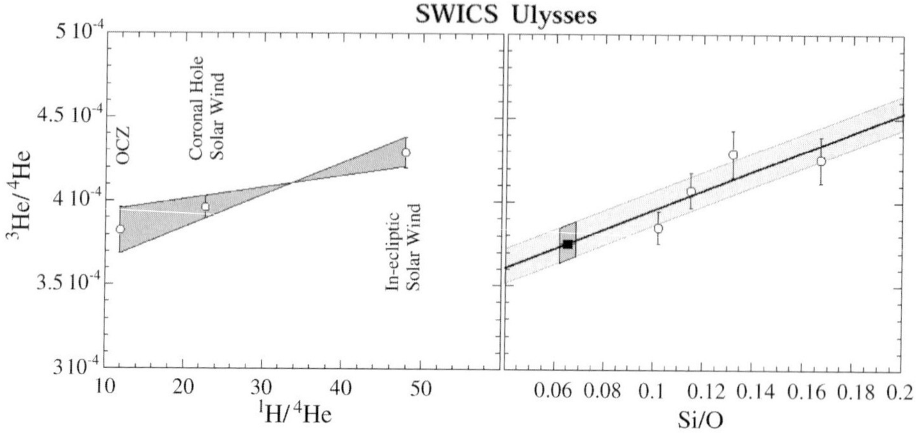

Figure 1. Determination of the $^3He/^4He$ ratio in the Protosolar Cloud by extrapolation of solar wind abundance data, updated from Gloeckler and Geiss (2000). *Left:* Correlation of $^3He/^4He$ with $^1H/^4He$ measured in high-speed streams coming out of the coronal holes and in the slow wind that dominates near the ecliptic. The OCZ value of $^3He/^4He$ is obtained by extrapolation to the $^1H/^4He$ ratio in the OCZ. *Right:* Correlation of $^3He/^4He$ with Si/O measured in two high-speed stream and in two slow solar wind periods. $^3He/^4He$ in the OCZ is obtained by extrapolation to the photospheric Si/O ratio of 0.0652, shown as a square symbol. The Si/O ratios are from von Steiger *et al.* (2000), the photospheric abundances of Si and O from Grevesse and Sauval (1998) and Holweger (2001), respectively.

possible for the first time to investigate systematically the helium isotopes not only in the slow wind but also in the high-speed streams coming out of the large polar coronal holes. Gloeckler and Geiss (2000) noticed that in the SW $^3He/^4He$ and Si/O are correlated, and, using linear regression and the photospheric value of Si/O, they estimated the $^3He/^4He$ ratio in the OCZ to be $(3.75 \pm 0.17[\text{stat.} + \text{var.}] \pm 0.20[\text{sys.}]) \cdot 10^{-4}$ (Figure 1a). The 1-σ errors include statistical and systematic instrumental uncertainties as well as the spread due to solar wind variability.

In a similar way, Gloeckler and Geiss (2000) used the correlated variations of the $^3He/^4He$ and $^4He/H$ ratios in the slow solar wind and the high-speed streams. In Figure 1b the average $^3He/^4He$ ratios measured in two high-speed stream periods and in two low-speed wind periods are plotted against the $^1H/^4He$ ratios for the corresponding periods. Using a linear extrapolation and the value of 11.9 for $(H/He)_{OCZ}$ (from Pérez Hernández and Christensen-Dalsgaard, 1994), the value $(^3He/^4He)_{OCZ} = (3.82 \pm 0.14[\text{stat.} + \text{var.}] \pm 0.20[\text{sys.}]) \cdot 10^{-4}$ is obtained (Figure 1b). The two-extrapolation methods lead to essentially the same result, giving the average of the two values $(^3He/^4He)_{OCZ} = (3.78 \pm 0.11[\text{stat.} + \text{var.}] \pm 0.20[\text{sys.}]) \cdot 10^{-4}$ — which is in remarkable agreement with earlier values of $(^3He/^4He)_{OCZ}$ obtained by other methods (cf. Geiss and Gloeckler, 1998).

The $(^3He/^4He)$ value derived from the extrapolations shown in Figure 1 refers to the present-day OCZ. Two processes, however, could have changed this ratio during the lifetime of the Sun.

TABLE III
Protosolar D/H Ratio.

	$D/H_{Protosolar}$ (10^{-5})
From Solar Wind $^3He/^4He$ and Jupiter $^3He/^4He$	1.97 ± 0.35[1]
From ISO H_2 and CH_4 spectra, corrected for D-Excess in the Ice-Phase	2.1 ± 0.4[2]
Protosolar Cloud	$\mathbf{2.0 \pm 0.35}$

[1] Updated from Gloeckler and Geiss (2000); [2] Lellouch *et al.* (2001)); cf. Table VI.

Solar seismic data and solar models show that He/H in the OCZ is 16% lower than it was in the PSC (e.g. Bahcall and Pinsonneault, 1995). The difference is interpreted as being due to settling of helium out of the OCZ into deeper layers of the Sun. ^3He settles more slowly than ^4He, resulting in an increase in present-day $(^3He/^4He)_{OCZ}$ ratio of a few percent (Gautier and Morel, 1997).

The second possible change of $(^3He/^4He)_{OCZ}$ over solar history is due to solar mixing. During the lifetime of the Sun, the pp-reaction produces additional ^3He outside the solar core at intermediate depth in the Sun. A significant increase of ^3He in the OCZ could have been caused by mixing of pp-produced ^3He into the OCZ. This increase is small and depends on solar models and solar rotation (Vauclair, 1998; Turck-Chièze *et al.*, 2001).

Addition of pp-produced ^3He to the OCZ can also be investigated by comparing solar wind helium trapped in very old and more recent lunar surface material. Using this method, Wieler and Heber (2003) concluded, that the increase of $^3He/^4He$ in the OCZ by solar mixing is at most 5%. Combining the two effects, settling of helium out of the OCZ and solar mixing, we thus adopt a correction of $-(4 \pm 2)\%$ for $(^3He/^4He)_{OCZ}$ and obtain $[(D{+}^3He)/H]_{PSC} = (3.63 \pm 0.35) \cdot 10^{-5}$.

4. Protosolar Deuterium Abundance

Two methods are available for a reliable determination of the deuterium abundance in the PSC. For the first method, the relation

$$(D/H)_{Protosolar} = [(D +^3 He)/H]_{Protosolar} - (^3He/H)_{Protosolar}$$

is used. With protosolar $(D + {}^3He)/H = (3.63 \pm 0.35) \times 10^{-5}$ and protosolar $^3He/H = (1.66 \pm 0.06) \times 10^{-5}$ (cf. Table I) we obtain $(D/H)_{Protosolar} = (1.97 \pm 0.36) \times 10^{-5}$ (Table III). The second method is based on a direct deuterium abundance measurement in Jupiter, either by remote spectroscopy (*e.g.* Beer and Taylor, 1973; Drossart *et al.*, 1982; Encrenaz *et al.*, 1996) or by *in-situ* mass spectrometry (Niemann *et al.*, 1996; Mahaffy *et al.*, 1998). The values obtained by infrared spectroscopy are more precise than those obtained by the Galileo Probe Mass Spectrometer

TABLE IV

Deuterium Abundance in Jupiter.

	D/H (10^{-5})
Galileo Probe[1]	2.6 ± 0.7
ISO[2]	
H_2	2.4 ± 0.4
CH_4	2.2 ± 0.7
H_2, CH_4 combined	2.25 ± 0.35
Corrected for D-Excess in the Ice-Phase	2.1 ± 0.4

[1] Mahaffy *et al.* (1998); [2] Lellouch *et al.* (2001).

(cf. Table IV). Lellouch *et al.* (2001) measured $D/H = (2.4 \pm 0.4) \times 10^{-5}$ in molecular hydrogen and $D/H = (2.2 \pm 0.7) \times 10^{-5}$ in methane. Recently, Bézard *et al.* (2002) obtained a D/H ratio in the methane even below 2×10^{-5}. We adopt here the results of Lellouch *et al.* (2001) who combined their H_2 and CH_4 results and obtained $D/H = (2.25 \pm 0.35) \times 10^{-5}$ as representative for Jupiter. Jupiter's D/H ratio is probably enriched above the protosolar value by the admixture of deuterium-rich ices to the nebular gas during the planets formation (see also Owen and Encrenaz, 2003). According to Guillot (1999) the enrichment is about 5–10%. Lellouch *et al.* (2001) have corrected their ISO value for this effect and obtain $(D/H)_{\mathrm{Protosolar}} = (2.1 \pm 0.4) \times 10^{-5}$. The agreement between the D/H ratio determined using solar wind data and the ratio based on Jupiter measurements is excellent. We give the solar wind derived value a somewhat higher weight and adopt $(2.0 \pm 0.35) \times 10^{-5}$ as the protosolar D/H ratio.

5. Neon Isotopes in the Protosolar Cloud

The solar wind is by far the best source of information for determining the isotopic composition of neon in the Protosolar Cloud. Very precise results on $^{20}Ne/^{22}Ne$ in the solar wind were obtained during the 1969–72 period with the Apollo SWC experiments. The solar wind was collected five times during exposure periods lasting from 70 minutes to 40 hours, and afterwards the trapped solar wind particles were analyzed in the laboratory (Table V). The SW was sampled at five different locations on the Moon for differing solar aspect angles. Varying geometrical conditions for the collection periods was considered important, because little was known in the pre-Apollo era about strength and configuration of the lunar magnetic field. The $^{20}Ne/^{22}Ne$ results of the five SWC experiments were identical within the limits of error (Table V). ^{21}Ne data with reasonably small limits of error were obtained only by the Apollo 15 and Apollo 16 missions.

TABLE V
Apollo SWC Experiments, Neon Isotopes[1].

	$^{20}Ne/^{22}Ne$	$^{22}Ne/^{21}Ne$
Apollo 11	13.5 ± 1.0	—
Apollo 12	13.1 ± 0.6	26 ± 12
Apollo 14	13.65 ± 0.50	—
Apollo 15	13.65 ± 0.30	31 ± 4
Apollo 16	13.80 ± 0.40	31 ± 4

[1] from Geiss *et al.* (1972)

TABLE VI
Isotopic Abundances in Solar System Material.

	$^{20}Ne/^{22}Ne$	$^{21}Ne/^{22}Ne$
Present-day In-ecliptic Solar Wind		
Apollo SWC[1]	13.7 ± 0.3	0.033 ± 0.003
SOHO/CELIAS[2]	13.8 ± 0.7	0.031 ± 0.008
Protosolar Cloud	$\mathbf{13.7 \pm 0.3}$	$\mathbf{0.0328 \pm 0.0005}$[3]
Solar Wind in Lunar Soil[3]	13.8 ± 0.1	0.0328 ± 0.0005
"SEP" in Lunar Soil[3]	11.2 ± 0.2	0.0295 ± 0.0005
Earth Atmosphere[4]	9.77 ± 0.03	0.0292 ± 0.0011

[1] Geiss *et al.* (1972); [2] Kallenbach *et al.* (1997); [3] Benkert *et al.* (1993); [4] Holden (1997).

$^{20}Ne/^{22}Ne$ data were also obtained by the MTOF detector of the SOHO/CELIAS experiment (Kallenbach *et al.*, 1997). The Apollo SWC and SOHO/CELIAS results are summarized in Table VI. Kallenbach *et al.* (1998) concluded from Ne, Mg and Si measurements in the solar wind that on average the lighter isotopes of elements in this mass range are enhanced in the slow solar wind by 1 or 2%. For neon this enhancement is well within the limits of error. Therefore, at this time, we do not correct for any mass discrimination of the solar wind $^{20}Ne/^{22}Ne$ ratio, but include the unknown fractionation factor in the limits of error in the adopted $(^{20}Ne/^{22}Ne)_{PSC}$ ratio.

The direct investigations of the present-day solar wind give much larger relative errors for $^{21}Ne/^{22}Ne$ than for $^{20}Ne/^{22}Ne$ (cf. Table V). Therefore, we base the protosolar $^{21}Ne/^{22}Ne$ ratio on investigations of solar wind neon trapped in lunar surface material, and we adopt in Table VI the value given by Benkert *et al.* (1993).

6. Discussion and Conclusions

In the preceding sections we have summarized the evidence from which the protosolar abundances of the isotopes of H, He and Ne are derived, and we have presented our best estimate for these abundances.

Representative protosolar abundances provide the basis for the interpretation of isotopic ratios measured in various solar system objects and samples, as is demonstrated in several articles of this volume. In this final section of our article, we discuss how the protosolar abundances of the H, He or Ne isotopes are used to constrain solar mixing, to identify excess ^{21}Ne in the Earth atmosphere, to understand the origin of neon trapped in lunar surface material, and to constrain galactic evolution.

6.1. Solar Mixing: Constraints from Isotopic Measurements

In the OCZ thermally driven convection evens out and dilutes inhomogeneities that might be produced locally at the solar surface or elsewhere. With $\approx$2 percent of the solar mass (see Gough and Scherrer, 2001), the OCZ is a huge reservoir. Therefore, neither the loss of matter through the solar wind nor nuclear reactions at the solar surface are likely to have produced significant changes in its composition during the past few Gy.

It was recognized long ago (*e.g.*, Schatzman, 1970; Schatzman *et al.*, 1981) that the abundances of the lightest nuclides at the solar surface can be used to constrain solar history. In the solar core nuclides with $Z \leq 8$ are affected by thermonuclear reactions. With increased distance from the solar center the temperature and the number of species involved in thermonuclear reactions decreases. At the surface we have definite evidence of changes in abundance only for the very lightest species, see Grevesse and Sauval (1998).

The lifetime of a nuclide determines the degree of its depletion. In Figure 2 lifetimes of several light nuclides are plotted as a function of temperature and, for simplicity, normalized to the density and composition at the bottom of the OCZ (reaction rates after Fowler *et al.*, 1975, and other compilations). The observed depletion of nuclides is consistent with the lifetime hierarchy: D and ^{6}Li are absent in the solar OCZ, ^{7}Li is depleted by a factor of $\approx$160. No depletion is observed for species heavier than ^{7}Li. Recently, Israelian *et al.* (2001) discovered ^{6}Li in the atmosphere of the metal-rich solar-type star H/D 82943, which is orbited by a giant planet. Based on the short lifetime of ^{6}Li (cf. Figure 2) the authors conclude that the observed ^{6}Li would not have survived the early epoch in the life of this star, and they interpret the presence of ^{6}Li as "evidence for a planet (or planets) having been engulfed by the parent star".

In the OCZ ^{3}He could not have been depleted, as is evident from the full preservation of ^{9}Be and the large difference in the lifetimes of ^{3}He and ^{9}Be seen in Figure 2 (Geiss and Reeves, 1972). However, ^{3}He could have been augmented by

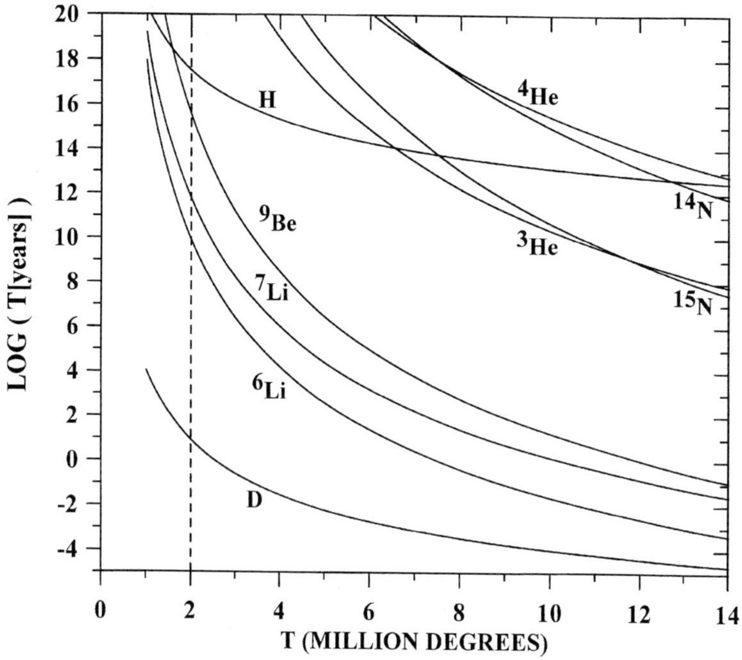

Figure 2. Life-time of light nuclides as a function of temperature. Given are the life-times due to thermonuclear reactions with density and composition normalized to the present conditions at the bottom of the OCZ. The dashed line marks the present temperature at the bottom of the OCZ.

secular mixing (Schatzman, 1970; Bochsler and Geiss, 1973; Schatzman *et al.*, 1981; Bochsler *et al.*, 1990): In the course of solar history a bulge of ^{3}He was built up by incomplete hydrogen burning at intermediate solar depth, and even a modest admixture to the OCZ of material from below could increase ^{3}He/^{4}He at the solar surface. However, the recently established absence of significant differential rotation below $r = 0.7R_\odot$ (Gough and Scherrer, 2001) has made such an increase less likely (Vauclair, 1998). In any case, the addition of surplus ^{3}He would mainly occur late in solar history, *i.e.* during and after the epoch when the older lunar samples received their solar wind irradiation (Geiss, 1973). The most recent data of solar wind helium in lunar surface material indicate that the helium isotopic composition in the solar wind has remained constant within a few percent during the past almost 4 Gy (Wieler and Heber, 2003).

6.2. Lunar Neon and Atmospheric Neon

Figure 3 shows a ^{21}Ne/^{22}Ne vs. ^{20}Ne/^{22}Ne diagram of solar wind and atmospheric neon and of the "SEP" component trapped in lunar surface material. Such three isotope plots are convenient for distinguishing isotopic differences caused by mass fractionation from those caused by mass-independent fractionation or nuclear reactions.

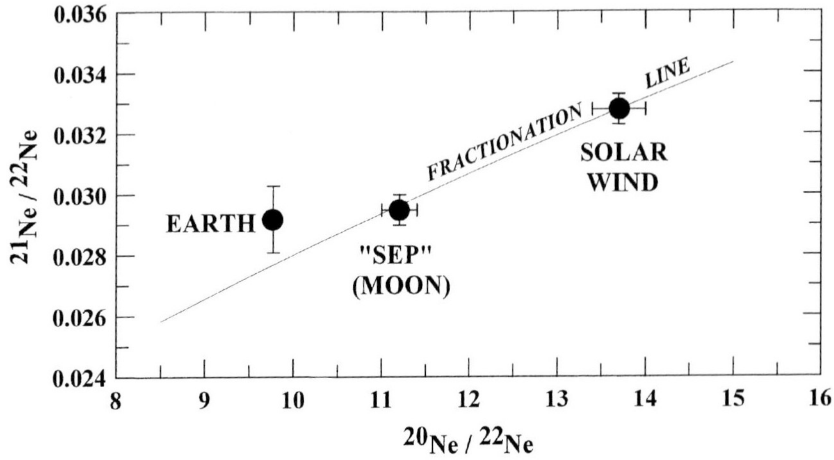

Figure 3. Three isotope plot of neon. The solar wind data and the isotopic composition of the "SEP" component in lunar soil are taken from Table VI. "SEP" neon has a composition that can be produced from solar wind neon by mass dependent isotope fractionation. On the other hand, the neon in the earth atmosphere lies above the "fractionation line". The excess ^{21}Ne is thought to be produced by nuclear reactions inside the earth (see text).

The change in the relative abundances of three isotopes A, B and C of an element is approximately given by (see Clayton, 2003)

$$d(n_A/n_C)/(n_A/n_C) = k\, d(n_B/n_C)/(n_B/n_C), \tag{1}$$

with

$$k = [(m_A - m_C)/(m_A + m_C)]/[(m_B - m_C)/(m_B + m_C)], \tag{2}$$

where n_i is the number of atoms of isotope i and m_i is its mass. For the case of neon $k = 0.49$, and integrating Equation (1) gives the "fractionation line".

$$(^{21}\mathrm{Ne}/^{22}\mathrm{Ne})/(^{21}\mathrm{Ne}/^{22}\mathrm{Ne})_0 = [(^{20}\mathrm{Ne}/^{22}\mathrm{Ne})/(^{20}\mathrm{Ne}/^{22}\mathrm{Ne})_0]^{0.49} \tag{3}$$

The fractionation line for protosolar neon is shown in Figure 3. Atmospheric neon lies slightly above the fractionation line. This is — at least in part — due to a small excess of ^{21}Ne produced by nuclear reactions inside the Earth (Wetherill, 1954; Eikenberg *et al.*, 1993).

The neon contained in lunar surface material consists of three major components: (1) Trapped solar wind neon, (2) neon produced by spallation, and (3) "SEP" neon. The acronym "SEP" was originally chosen because it was thought that this component consisted of solar energetic particles. An alternative explanation is that the "SEP" component consists of solar wind particles that are isotopically mass fractionated by diffusion or other processes.

Recently, Wimmer-Schweingruber and Bochsler (2001) proposed an interstellar origin for the "SEP" particles. Since, however, the "SEP" data point lies virtually

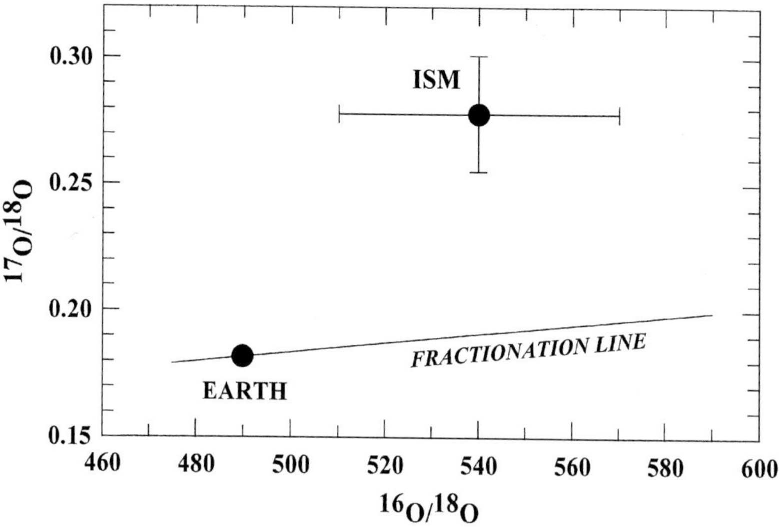

Figure 4. Three isotope plot of oxygen. The "fractionation line" defines isotopic ratios that could be obtained from terrestrial oxygen by mass dependent isotope fractionation. The ISM data point is an average of oxygen isotope measurements in the interstellar medium (after Prantzos *et al.*, 1996). Terrestrial oxygen could not be derived from ISM oxygen by mass dependent isotope fractionation, demonstrating that the nucleosynthetic origin of the two types of oxygen is significantly different.

on the fractionation line, it is much more likely that its origin is the protosolar cloud (see also Wieler and Heber, 2003).

Figure 4 shows a 3-isotope plot for oxygen. It is seen that the average isotopic composition of oxygen in the present-day interstellar medium (ISM) given by Prantzos *et al.* (1996) lies $\approx$40 percent above the fractionation line for terrestrial oxygen. Terrestrial oxygen could isotopically differ from PSC oxygen (see Clayton, 2003), but only by a few percent. Thus, in any case, ISM oxygen will lie far away from the fractionation line through PSC oxygen, proof for a significantly different nucleosynthetic origin of the oxygen in the PSC and in the ISM. Recently, Geiss *et al.* (2002) have presented evidence that the infall of matter into the galactic disc carries the compositional signature of dwarf galaxies, which is characterized by a high ratio of primary to secondary nuclides, thus explaining the low abundance of N in the LIC and of ^{18}O in the ISM.

Could ISM neon accidentally lie as close to the PSC fractionation line as lunar "SEP" neon does (see Figure 3)? The best isotopic result on interstellar neon is the ^{22}Ne/^{20}Ne = (13.0 + 5.5 − 6.8) ratio obtained by Leske *et al.* (1996) in the Anomalous Cosmic Rays (ACR). For distinguishing between different hypotheses concerning the nature of the "SEP" data, however, the error limit of this result is too large. Also, galactic evolution modeling of neon comparable to the oxygen modeling by Prantzos *et al.* (1996) has not yet been done. We note however, that ^{16}O and ^{20}Ne are both primary nuclides released by supernovae, ^{18}O and ^{22}Ne are typical secondary nuclides produced from ^{14}N during He-burning. ^{17}O and ^{21}Ne

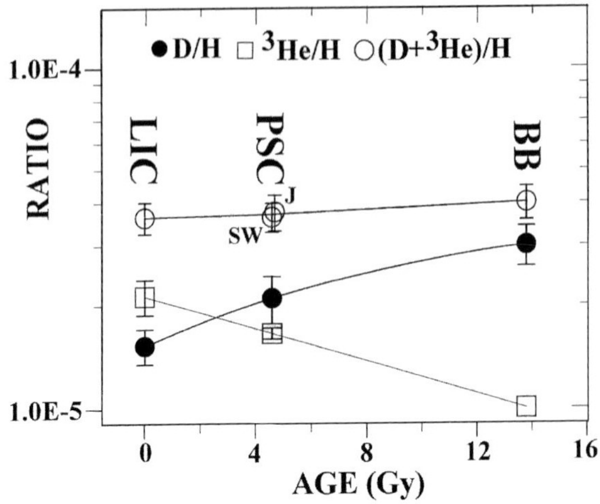

Figure 5. D/H, ^{3}He/H, and (D+^{3}He)/H in the Local Interstellar Cloud (LIC), the Protosolar Cloud (PSC) and in the baryonic matter produced in the Big Bang (BB) are plotted as a function of age. As expected from the theory of nucleosynthesis and galactic evolution, D/H decreases with time and ^{3}He/H increases. This leads to a relatively constant (D+^{3}He)/H ratio over time (see text). The PSC data points are from this paper, for (D+^{3}He)/H in the PSC there are two independent values, from solar wind (SW) data and from Jupiter (J) data. D/H in the LIC is from Linsky and Wood (2000) and ^{3}He/H in the LIC is the average of the results of Gloeckler and Geiss (1996, 1998) and Salerno *et al.* (2001). The primordial D/H ratio is from O'Meara *et al.* (2001). The ^{3}He/H ratio is a theoretical value (cf. Geiss *et al.*, 2002).

are produced from ^{16}O and ^{20}Ne, respectively, ^{17}O by core H burning, and ^{21}Ne by neutron capture mostly during He- and C-burning (Prantzos *et al.*, 1996; Prantzos, 2002). Thus, the three neon isotopes are as different in nucleosynthetic origin as the oxygen isotopes. We consider it to be unlikely therefore, that neon in the present-day ISM could lie as close to the fractionation line as does the SEP neon. In order to resolve this question definitely, it would be important to have improved neon isotope data in the interstellar gas entering the heliosphere and also to develop models for the nucleosynthesis and galactic evolution of the neon isotopes. Data on neon isotope abundances in the LIC and a better understanding of the nucleosynthesis of these isotopes would also provide a further test for the "Mixing Model" of late galactic evolution introduced by Geiss *et al.* (2002).

6.3. Galactic Evolution of Deuterium and ^{3}He

D is the only nucleus that is produced exclusively in the Big Bang. Also ^{3}He is in a category of its own. It is produced by three processes: (1) Big Bang nucleosynthesis; (2) D-burning in the early life of stars of all sizes; and (3) incomplete H-burning. The third process operates at intermediate depth in small stars (see Charbonnel and do Nascimento, 1998), and the gas relatively rich in ^{3}He is released into interstellar space when the star had left the main sequence, *i.e.* $\approx 10^{10}$ years after its birth. Therefore, this mode of ^{3}He production came into effect only in the later stages of the life of the galaxy. However, rotating stellar models predict that the net effect of this process is small (Charbonnel and do Nascimento, 1998; Tosi, 1998). From the limited importance of process (3) it follows that (D+^{3}He)/H is relatively constant. This is shown in Figure 5. The PSC and the LIC are the only representative galactic samples for which we have both D and ^{3}He determinations. It is seen that (D+^{3}He)/H has decreased by only $\approx$10 percent from the BB to the PSC and the LIC. We propose that (D+^{3}He) could become useful as an approximate invariant for modeling the chemical evolution in galaxies. Since the chemical evolution in dwarf galaxies proceeds under conditions different from those in our own galaxy, it would be important if a representative abundance of both D and ^{3}He could be measured in one of the Magellan clouds or another dwarf galaxy.

Acknowledgements

The authors wish to thank Thérèse Encrenaz, Emmanuel Lellouch, Uli Ott and Rainer Wieler for discussions and suggestions and they are indebted to Ursula Pfander and Silvia Wenger for preparing the manuscript. The conclusions given in this paper are based on several NASA and ESA space missions. We like to mention in particular the Apollo programme and the Ulysses mission, and thank the astronauts and the NASA and ESA Teams, who prepared and executed these projects, for the essential support and help they gave to our experiments. The Apollo SWC and the Ulysses/SWICS experiments were generously supported by the Swiss National Science Foundation and SWICS also by NASA/Caltech grant NAG5-6921 and NASA/JPL contract 955460.

References

Bahcall, J. N., and Pinsonneault, M. H.: 1995, *Rev. Modern Phys.* **67**, 781.

Beer, R., and Taylor, F.W.: 1973, 'The Abundance of CH_3D and the D/H Ratio in Jupiter', *Astrophys. J.* **179**, 309–328.

Benkert, J.-P., Baur, H., Signer, P., and Wieler, R.: 1993, 'He, Ne, and Ar from the Solar Wind and Solar Energetic Particles in Lunar Ilmenites and Pyroxenes', *J. Geophys. Res.* **98**, 13147–13162.

Bézard, B., Greathouse, T., Lacy, J., Richter, M., Moses, J., and Griffith, C.: Presentation at the Conference "Jupiter after Galileo and Cassini", June 2002, Lisbon, Portugal.

Bochsler, P.: 1984, *Helium and Oxygen in the Solar Wind*, University of Bern, Habilitation Thesis.

Bochsler, P., and Geiss, J.: 1973, 'Solar Abundances of Light Nuclei and Mixing of the Sun', *Sol. Phys.* **32**, 3–11.

Bochsler P., Geiss, J., and Maeder, A.: 1990, 'The Abundance of He-3 in the Solar Wind — A Constraint for Models of Solar Evolution', *Sol. Phys.* **128**, 203–215.

Bodmer, R., Bochsler, P., Geiss, J., von Steiger, R., and Gloeckler, G.: 1995, 'Solar Wind Helium Isotopic Composition from SWICS/Ulysses', *Space Sci. Rev.* **72**, 61.

Busemann, H., Baur, H., and Wieler, R.: 2000, 'Primordial Noble Gases in "Phase Q" in Carbonaceous and Ordinary Chondrites Studied by Closed-system Stepped Etching', *Met. Planet. Sci.* **35**, 949–973.

Busemann, B., Baur, H., and Wieler, R.: 2001, 'Helium Isotopic Ratios in Carbonaceous Chondrites: Significant for the Early Solar Nebula and Circumstellar Diamonds?', *Lunar Planet. Sci.* **XXXII**, abstract #1598.

Charbonnel, C., and do Nascimento, J.D.: 1998, 'How many Low-mass Stars do Destroy ^{3}He?', *Astron. Astrophys.* **336**, 915–919.

Clayton, R.N.: 2003, 'Oxygen Isotopes in the Solar System', *Space Sci. Rev.*, this volume.

Coplan, M.A., Ogilvie, K.W., Bochsler, P., and Geiss, J.: 1984, 'Interpretation of He-3 Abundance Variations in the Solar Wind', *Sol. Phys.* **93**, 415–434.

Drossart, P., Encrenaz, T., Kunde, V., Hanel, R., and Combes, M.: 1982, 'An Estimate of the PH_3, CH_3D, and GeH_4 Abundances on Jupiter from the Voyager IRIS Data at 4.5 μm', *Icarus* **49**, 416–426.

Eberhardt, P.: 1974, 'A neon-E-rich Phase in the Orgueil Carbonaceous Chondrite', *Earth Planet. Sci.* **24**, 182–187.

Eikenberg, J., Signer, P., and Wieler, R.: 1993, 'U-Xe, U-Kr, and U-Pb Systematics for Dating Uranium Minerals and Investigations of the Production of Nucleogenic Neon and Argon', *Geochim. Cosmochim. Acta* **57**, 1053–1069.

Encrenaz, T. *et al.*: 1996, 'First Results of ISO-SWS Observations of Jupiter', *Astron. Astrophys.* **315**, L397–L400.

Fowler, W.A., Caughlan, G.R., and Zimmerman, B.A.: 1975, 'Thermonuclear Reaction Rates, II', *Ann. Rev. Astron. Astrophys.* **13**, 69.

Frick, U., and Moniot, R.K.: 1977, 'Planetary Noble Gas Components in Orgueil', *Proc. 8th Lunar Planet Sci. Conf.*, p. 229.

Gautier, D., and Morel, P.: 1997, 'A Reestimate of the Protosolar $(^2H/^1H)_p$ Ratio from $(^3He/^4He)_{SW}$ Solar Wind Measurements', *Astron. Astrophys.* **323**, L9–L12.

Geiss, J.: 1973, *Conf. Pap. Int. Cosmic Ray Conf. 13th*, pp. 3375.

Geiss, J., and Reeves, H.: 1972, 'Cosmic and Solar System Abundances of Deuterium and Helium-3', *Astron. Astrophys.* **18**, 126.

Geiss, J., and Gloeckler, G.: 1998, 'Abundances of Deuterium and Helium-3 in the Protosolar Cloud', *SSSI* **4**, and *Space Sci. Rev.* **84**, 239–250.

Geiss, J., Eberhardt, P., Bühler, F., Meister, J., and Signer, P.: 1970, 'Apollo 11 and 12 Solar Wind Composition Experiments: Fluxes of ^{3}He and ^{4}He Isotopes', *J. Geophys. Res.* **75**, 5972–5979.

Geiss, J., Bühler, F., Cerutti, H., Eberhardt, P., and Filleux, Ch.: 1972, *Apollo-16 Preliminary Science Report*, NASA SP-315, Section 14.

Geiss, J., Gloeckler, G., and Charbonnel, C.: 2002, 'Chemical Evolution in our Galaxy During the Last 5 Gy', *Astrophys. J.* **578**, 862–867.

Gloeckler, G., and Geiss, J.: 1996, 'Abundance of ^{3}He in the Local Interstellar Cloud', *Nature* **381**, 210–212.

Gloeckler, G., and Geiss, J.: 1998, 'Measurement of the Abundance of Helium-3 in the Sun and in the Local Interstellar Cloud with SWICS on ULYSSES', *SSSI* **4** and *Space Sci. Rev.* **84**, 275.

Gloeckler, G., and Geiss, J.: 2000, 'Deuterium and Helium-3 in the Protosolar Cloud', *in* L. da Silva, M. Spite, and J.R. de Medeiros (eds.), *The Light Elements and Their Evolution*, IAU Symposium, Vol. 198, Publ. ASP, pp. 224–233.

Gough, D.O., and Scherrer, P.H.: 2001, 'The Solar Interior', *in* J. Bleeker, J. Geiss, M.C.E. Huber (eds.), *The Century of Space Science*, Kluwer Academic Publishers, Dordrecht, pp. 1035–1063.

Grevesse, N., and Sauval, A.J.: 1998, 'Standard Solar Composition', *SSSI* **7**, and *Space Sci. Rev.* **85**, 161–174.

Guillot, T.: 1999, 'A Comparison of the Interiors of Jupiter and Saturn', *Planet. Space Sci.* **47**, 1183–1200.

Holden, N.O.: 1997, *in* D.R. Lide (ed.), *Handbook of Chemistry and Physics*, 78th Edition, CRC Press, Boca Raton, New York, 11.41.

Holweger, H.: 2001, 'Photospheric Abundances: Problems, Updates, Implications', *in* R.F. Wimmer-Schweingruber (ed.), *Solar and Galactic Composition*, AIP Conf. Proc., Vol. 598, Melville, N.Y. pp. 23.

Israelian, G., Santos, N.C., Mayor, M., and Rebolo, R: 2001, 'Evidence for Planet Engulfment by the star HD82943', *Nature* **411**, 163–166.

Kallenbach, R., *et al.*: 1997, 'Isotopic Composition of Solar Wind Neon Measured by CELIAS/MTOF on Board SOHO', *J. Geophys. Res.* **102**, 26895–26904.

Kallenbach, R., *et al.*: 1998, 'Fractionation of Si, Ne, and Mg Isotopes in the Solar Wind as Measured by SOHO/CELIAS/MTOF', *Space Sci. Rev.* **85**, 357–370.

Lellouch, E., Bézard, B., Fouchet, T., Feuchtgruber, H., Encrenaz, T., and de Graauw, T.: 2001, 'The Deuterium Abundance in Jupiter and Saturn from ISO-SWS Observations', *Astron. Astrophys.* **370**, 610–622.

Leske, R.A., Mewaldt, R.A., Cummings, A.C., Cummings, J.R., Stone, E.C., and von Rosenvinge, T.T.: 1996, 'The Isotopic Composition of Anomalous Cosmic Rays from SAMPEX', *SSSI* **1**, and *Space Sci. Rev.* **78**, 149–154.

Linsky, J.L., and Wood, B.E.: 2000, *in* L. da Silva, M. Spite, and J.R. de Medeiros (eds.), *The Light Elements and Their Evolution*, ASP, p. 141.

Mahaffy, P.R., Donahue, T.N., Atreya, S.K., Owen, T.C., and Niemann, H.B.: 1998, 'Galileo Probe Measurements of D/H and ^{3}He/^{4}He in Jupiter's Atmosphere', *SSSI* **4**, and *Space Sci. Rev.* **84**, 251–263.

Niemann, H.B., *et al.*: 1996, 'The Galileo Probe Mass Spectrometer: Composition of Jupiter's Atmosphere', *Science* **272**, 846–849.

Ogilvie, K.W., Coplan, M.A., Bochsler, P., and Geiss, J.: 1980, 'Abundance Ratios of $^4He^{2+}/^3He^{2+}$ in the Solar Wind', *J. Geophys. Res.* **85**, 6021–6024.

O'Meara, J.M., Tytler, D., Kirkman, D., Suzuki, N., Prochaska, J.X., Lubin, D., and Wolfe, A.M.: 2001, 'The Deuterium to Hydrogen Abundance Ratio toward a Fourth QSO: HS 0105+1619', *Astrophys. J.* **552**, 718–730.

Owen, T., and Encrenaz, T.: 2003, 'Element Abundances and Isotope Ratios in the Giant Planets and Titan', *Space Sci. Rev.*, this volume.

Pérez Hernández, F., and Christensen-Dalsgaard, J.: 1994, 'The Phase Function for Stellar Acoustic Oscillations - III. The Solar Case', *Mon. Not. Roy. Astron. Soc.* **269**, 475–492.

Prantzos, N.: 2002, personal communication.

Prantzos, N., Aubert, O., and Audouze, J.: 1996, 'Evolution of the Carbon and Oxygen Isotopes in the Galaxy', *Astron. Astrophys.* **309**, 760–774.

Salerno, E., *et al.*: 2001, Direct Measurement of ^{3}He/^{4}He in the LISM with the COLLISA Experiment', *in* R.F. Wimmer-Schweingruber (ed.), *Solar and Galactic Composition*, AIP Conf. Proc., Vol. 598, Melville, N.Y. pp. 275–280.

Schatzman E.: 1970, *Physics and Astrophys. Lecture Notes*, CERN, 70–31.

Schatzman, E., Maeder, A., Angrand, F., and Glowinski, R.: 1981, 'Stellar Evolution with Turbulent Diffusion Mixing. III – The Solar Model and the Neutrino Problem', *Astron. Astrophys.* **96**, 1–16.

Stevenson, D.J., and Salpeter, E.E.: 1976, 'Interior Models of Jupiter', *in* T. Gehrels (ed.), *Jupiter*, Univ. Arizona Press, Tucson, pp. 85–112.

Tosi, M.: 1998, 'Galactic Evolution of D and ^{3}He', *SSSI* **4**, and *Space Sci. Rev.* **84**, 207–218.

Turck-Chièze, S., Nghiem, P., Couvidat, S., and Turcotte, S.: 2001, 'Solar Internal Composition and Nuclear Reaction Rates in the Light of Helioseismology', *Sol. Phys.* **200**, 323–342.

Vauclair, S.: 1998, 'Evidence of Element Diffusion Inside the Sun and the Stars and its Consequences on the Lithium Primordial Abundance', *SSSI* **4**, and *Space Sci. Rev.* **84**, 265–271.

von Steiger, R. *et al.*: 2000, 'Composition of Quasi-stationary Solar Wind Flows from Ulysses/Solar Wind Ion Composition Spectrometer', *J. Geophys. Res.* **105**, 27217–27238.

von Zahn, U., Hunten, D.M., and Lehmacher, G.: 1998, 'Helium in Jupiter's Atmosphere: Results from the Galileo Probe Helium Interferometer Experiment', *J. Geophys. Res.* **103**, 22815–22829.
Wetherill, G.: 1954, 'Variations in the Isotopic Abundances of Neon and Argon Extracted from Radioactive Minerals', *Phys. Rev.* **96**, 679–683.
Wieler, R., and Heber, V.S.: 2003, 'Noble Gas Isotopes on the Moon', *Space Sci. Rev.*, this volume.
Wimmer-Schweingruber, R.F., and Bochsler, P.: 2001, 'Lunar Soils: A Long-Term Archive for the Galactic Environment of the Heliosphere?', *in* R.F. Wimmer-Schweingruber (ed.), *Solar and Galactic Composition*, AIP Conf. Proc., Vol. 598, Melville, N.Y. pp. 399–404.

OXYGEN ISOTOPES IN THE SOLAR SYSTEM

ROBERT N. CLAYTON
Enrico Fermi Institute and Departments of Chemistry and of the Geophysical Sciences, University of Chicago, Chicago, IL 60637, USA

Received: 19 July 2002; Accepted in final form: 31 August 2002

Abstract. The oxygen three-isotope system has major advantages over the two-isotope systems of hydrogen, carbon, and nitrogen because different fractionation laws govern intraplanetary and interplanetary processes. This permits discriminating between solar nebular processes and parent-body processes. Oxygen isotopes also serve as a sensitive natural tracer for meteorite classification.

Keywords: Oxygen, meteorites, chondrites, achondrites, solar nebula

1. Isotopic Composition of the Sun and the Solar Nebula

Oxygen has three stable isotopes with mean terrestrial abundance ratios: $^{17}O/^{16}O = 1/2700$, $^{18}O/^{16}O = 1/490$. These ratios are variable in natural materials due to a variety of physical and chemical processes, ranging from stellar nucleosynthesis to ordinary mass-dependent chemical fractionation. As a natural isotopic tracer, oxygen has an advantage over other light elements, such as hydrogen, carbon, and nitrogen, in that the combined variation of two isotope ratios ($^{17}O/^{16}O$ and $^{18}O/^{16}O$) helps to identify the underlying process. Chemical and physical processes within an isolated planetary body, such as Earth, Moon, and Mars, almost always obey a simple mass-dependent relationship between variations in $^{17}O/^{16}O$ and $^{18}O/^{16}O$ ratios:

$$d\delta^{17}O/d\delta^{18}O \simeq 0.52.$$

Meteorite studies have shown that oxygen isotope abundance variations between solar system bodies are governed primarily by a different systematic relationship:

$$d\delta^{17}O/d\delta^{18}O \simeq 1.0.$$

In this delta-notation, isotope ratios are given relative to a standard reservoir. For solar system studies, the ideal standard would be the Sun itself, representing the reservoir from which planetary bodies were formed. Unfortunately the solar isotopic composition is not yet known with sufficient accuracy for that purpose. Therefore, the solar composition must be inferred from meteorite observations by a boot-strap process with inherent model-dependent ambiguities.

The cosmic abundance of oxygen is several times greater than the sum of the abundances of elements with which it can form solid compounds (mostly Mg, Si,

Space Science Reviews **106:** 19–32, 2003.

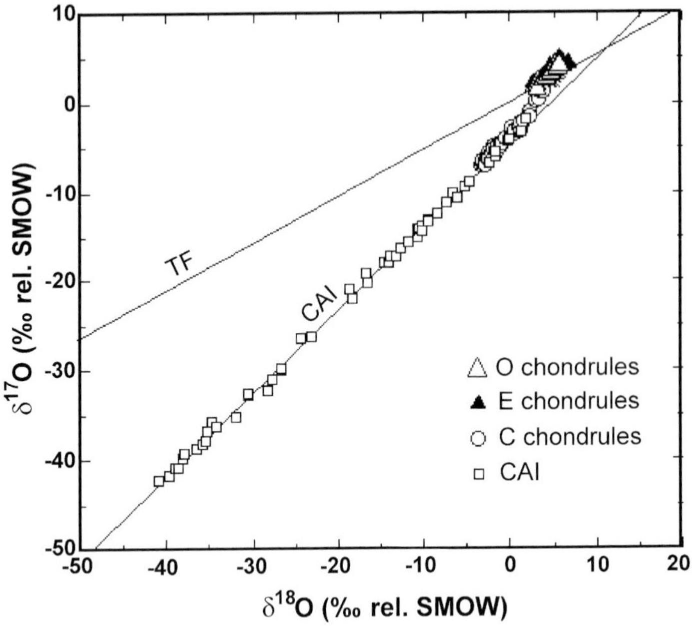

Figure 1. Oxygen three-isotope plot illustrating slope-1 variations in meteoritic chondrules and CAIs, in contrast to the terrestrial mass-independent fractionation line (TF).

and Fe). As a consequence, in a solar nebula that is cool enough for solid minerals to exist, only about 15% of the oxygen is bound in solids, and 85% is in the gas phase, primarily as H_2O and CO molecules, or O atoms, depending on temperatures and radiation fields. Most primitive stony meteorites contain abundant chondrules and less-abundant calcium-aluminum-rich refractory inclusions (CAIs) that were once molten droplets in contact with the nebular gas. Their oxygen isotopic compositions provide the most direct evidence for the isotopic composition of the gas, and hence for the nebula as a whole. Figure 1 is a three-isotope graph showing the systematic behavior of oxygen isotopic compositions of these objects. This near-linear array has been interpreted as (1) a two-component mixing line (Clayton, 1993), (2) the result of a special mass-independent chemical fractionation (MIF) process (Thiemens and Heidenreich, 1983), and (3) the result of isotope-selective photochemistry (Clayton, 2002a). The mixing model postulates a gas-dust nebula with initial isotopic heterogeneity, consisting of a ^{16}O-rich dust component and a ^{16}O-poor gas component. Since the gas component contains 85% of the oxygen, the model predicts a solar isotopic composition that is near the upper end of the line in Figure 1. The MIF model postulates an isotopically homogeneous gaseous nebula, in which gas-phase chemical reactions generate two reservoirs: one ^{16}O-rich, the other ^{16}O-poor, from which the isotopic signatures are transferred to solid phases. The photochemical model postulates an isotopically homogeneous nebula in which isotope-selective photolysis of carbon monoxide results in production of

^{16}O-depleted condensates. The model predicts a solar isotopic composition at the lower end of the line in Figure 1.

In both the first and third models, initial solid materials in the solar nebula had isotopic compositions near $\delta^{17}O = -41‰$, $\delta^{18}O = -40‰$, a composition widely observed in CAIs and amoeboid olivine aggregates from many classes of primitive meteorites: CV (Aléon *et al.*, 2002; Fagan *et al.*, 2002), CO (Itoh *et al.*, 2002), CR (Krot *et al.*, 2002). This composition has recently been observed in olivine in chondrules in a CO chondrite (Jones *et al.*, 2002). At present, most rocky material in the inner solar system has compositions near that of the Earth: $\delta^{17}O = +3‰$, $\delta^{18}O = +6‰$, implying massive isotopic processing in the solar nebula by interaction with a gas reservoir that is ^{16}O-poor, either from nucleosynthetic inheritance or from photochemical processing. An argument against the inheritance of a ^{16}O-rich presolar dust component is the absence of such isotopic effects in presolar corundum grains found in some primitive meteorites (Nittler *et al.*, 1997). These grains typically have ^{17}O excesses and ^{18}O depletions, and are probably derived from atmospheres of red giant stars, rather than from supernovae which produce ^{16}O enrichments. The MIF model could, in principle, produce a CAI-like composition in one of its products, and an Earth-like composition in the other, beginning with a solar gas composition somewhere between them. However, the only MIF reaction for which there is any theoretical understanding is the production of O_3 from O_2 (Gao and Marcus, 2002; Robert and Camy-Peyret, 2001), and it has not been shown that similarly large effects occur in reactions of the principal oxygen-bearing gases in the solar nebula, CO and H_2O. In addition there does not appear to be a plausible chemical process to trap a specific isotopic composition of residual ^{16}O-rich O_2 into CAI minerals.

Photochemical self-shielding in the photolysis of O_2 was discussed in detail by Navon and Wasserburg (1985). They showed that this was not a plausible mechanism for production of meteoritic isotopic anomalies, due to the rapid isotope exchange of $O + O_2$, which scrambles the isotopic labels before trapping into solid compounds. This limitation does not apply to self-shielding in the photolysis of CO, which is a well-known phenomenon in molecular clouds (van Dishoeck and Black, 1988).

What can be learned about solar system formation from the unusual isotopic behavior of oxygen in meteorites and planets? A first-order observation based on elemental abundances in meteorites is that volatility played a dominant role in the fractionation of chemical elements (Grossman and Larimer, 1974). This does not only apply to the gas-forming elements such as those which constitute the bulk of the giant planets, but also to the moderately volatile elements, such as the alkalies, the halogens, sulfur and manganese. These latter elements exhibit systematic depletions in the inner planets and in some asteroids (Wänke and Dreibus, 1988). These observations, coupled with the expectation of a radial temperature gradient in the solar nebula, led to a widely-accepted picture of formation of solids by condensation from a cooling hot gas, governed by chemical equilibrium (Grossman and Larimer, 1974). Condensates, and the planetesimals formed therefrom, were assumed to have formed near their present radial distance from the Sun. It is difficult to see how the observed oxygen isotope distributions would be produce in this scenario.

A radically different model of solar system formation has been proposed by Shu and co-workers, based on observations of Young Stellar Objects (YSO), characterized by magnetically-driven outflows and bipolar jets during the accretionary growth of a star (Shu *et al.*, 1997). In this picture, chondrules and CAIs are formed very near the star (a few stellar radii away), and are carried by outflowing gas into the distant planet-forming region. The chondrule-forming environment is characterized by intense fluxes of electromagnetic radiation and energetic nuclei. These radiations may have left observable effects in the irradiated materials, such as stable isotope effects (Clayton, 2002a, 2002b) and short-lived radionuclides (Gounelle *et al.*, 2001).

The chemical properties of a solar nebular gas subject to intense ultraviolet radiation differ substantially from those in equilibrium models. In particular, the gas may be more oxidizing, leading to higher condensation temperatures, and to the possibility of condensing molten chondrules containing ferrous iron (Ebel and Grossman, 2000).

A correlation has been observed between oxidation state of chondritic materials and $\Delta^{17}O$, for example (1) in whole-rock H, L, LL, and R chondrites (Clayton *et al.*, 1991; Bischoff *et al.*, 1994), (2) in chondrules and rims in a CV chondrite (Rubin *et al.*, 1990), and in whole-rock ureilites (Clayton and Mayeda, 1988). These correlations can be accounted for in the photochemical shielding model, in which atomic oxygen is enriched in the heavy (rare) isotopes.

A resolution of the various alternative models of solar system processes should be provided by oxygen isotope analysis of a returned sample of the solar wind, being collected by the Genesis spacecraft (Wiens *et al.*, 1999).

2. Carbonaceous Chondrites

The isotopic data for individual chondrules and CAI from CV chondrites conform to a single near-linear array (Figure 1), implying reservoir interactions at temperatures high enough to make the mass-dependent fractionation effects small relative to the mass-independent differences in reservoir compositions. In addition, the isotopic compositions of other carbonaceous chondrites show evidence for aqueous alteration on the parent body at low temperatures, converting high-temperature anhydrous materials (mostly olivine and pyroxene) into low-temperature alteration products (phyllosilicates, carbonates, magnetite). The simplest case is the alteration of material of CO-type carbonaceous chondrites to produce the trend seen among the CM chondrites (Figure 2) (Clayton and Mayeda, 1999). The observed trend has a slope of 0.7, indicating a low-temperature process involving large

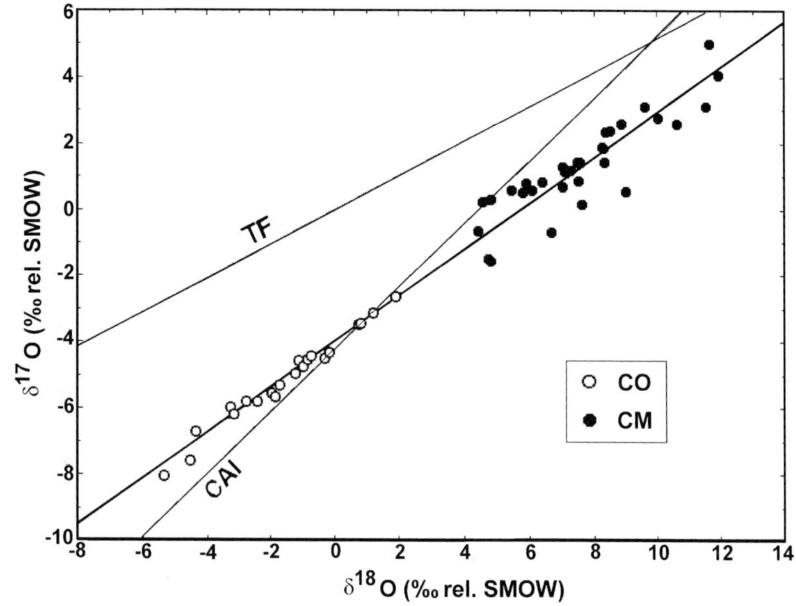

Figure 2. Oxygen isotopic compositions of whole-rock samples of CO and CM carbonaceous chondrites. CAI and TF lines from Figure 1 are shown for reference. The CO–CM line has a slope of 0.7, and reflects progressive low-temperature aqueous alteration from a CO-like precursor.

mass-dependent isotopic fractionation in the chemical alteration reactions. Material balance requires reaction with a water reservoir that was more ^{16}O-depleted than the precursor solids. It is likely that the water component was accreted to the parent body as ice, and that the aqueous alteration reactions occurred at temperatures near 0°C, when the ice melted.

The bulk oxygen isotopic compositions of all of the carbonaceous chondrites (CV, CO, CM, CI, as well as CR and CH) can be understood in terms of a combination of two processes: an initial nebular stage producing the slope-1 array of Figure 1, followed by a low-temperature parent-body stage producing a much shallower slope, as in Figure 2. Although the CV, CM, CO, CI groups form a coherent trend, consistent with progressive aqueous processing of a common precursor, the CR-CH trend, shown in Figure 3, is approximately parallel to the CO-CM trend, but is displaced toward higher values of Δ^{17}O ($= \delta^{17}\text{O} - 0.52\delta^{18}\text{O}$), implying a different anhydrous parent of the phyllosilicates (Clayton and Mayeda, 1999). The CR-CH and bencubbinites also exhibit much higher δ^{15}N values than the CV-CO-CM-CI group, perhaps as a result of solar photochemical processes (Clayton, 2002b).

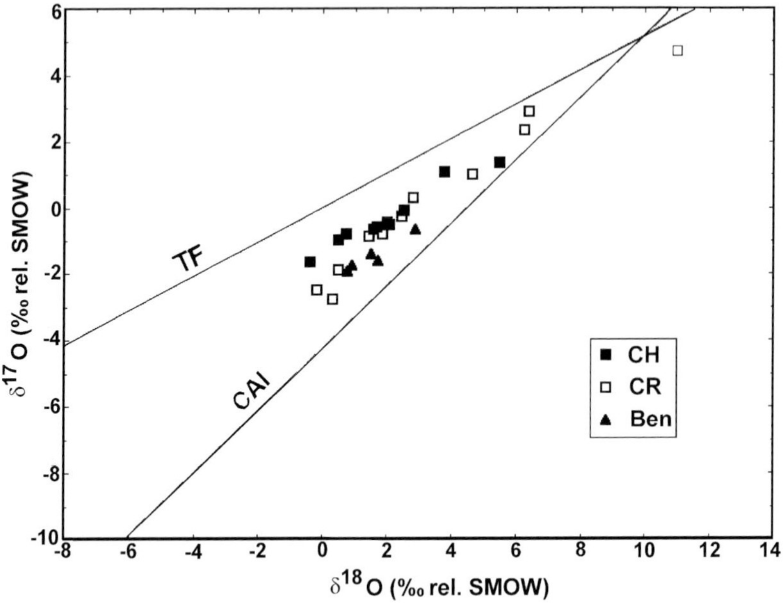

Figure 3. Oxygen isotopic compositions of CR, CH, and bencubbinite whole-rocks. CAI and TF lines from Figure 1 are shown for reference. The CR–CH line is approximately parallel to the CO–CM line of Figure 2, but is displaced to more positive $\Delta^{17}O$.

3. Ordinary, Enstatite, and Rumuruti Chondrites

In contrast to the carbonaceous chondrites, ordinary chondrites and enstatite chondrites have experienced little or no aqueous alteration on the parent bodies, although most show the effect of thermal metamorphism, which is evident by decreased internal oxygen isotopic fractionation between minerals, without changing the bulk isotopic composition of the rock. Mineral-pair isotopic thermometry yields peak metamorphic temperatures ranging from 600° to 850°C, corresponding to petrographic grades 4 to 6 (Clayton *et al.*, 1991). In individual chondrules, the oxygen isotopes of enstatite chondrites (EC), ordinary chondrites (OC), and carbonaceous chondrites form three separate groups (Figure 4). Rare ^{16}O-rich chondrules are found in EC and OC, and lie well outside the range of common chondrules.

The ordinary chondrites are subdivided into three chemical groups: H, L, and LL (for high-iron, low-iron, and low-iron, low-metal). Whole-rock meteorites from these groups have distinctive oxygen isotopic compositions (Figure 5). However, the individual chondrules from unmetamorphosed ordinary chondrites appear to form a single continuous trend, irrespective of the iron group of the parent meteorite (Clayton *et al.*, 1991). The isotopic differences among the parent meteorites appear to arise from size-sorting of chondrules, which increase in mean diameter from H to L to LL.

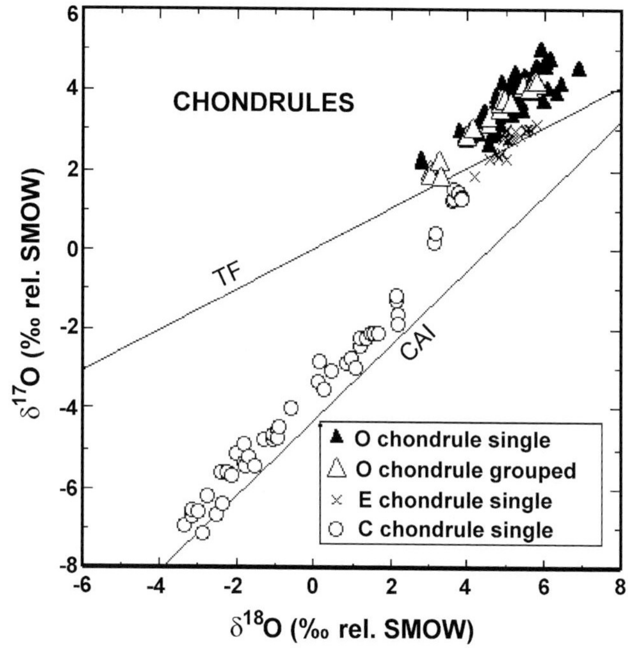

Figure 4. Oxygen isotopic compositions of chondrules from ordinary (O), enstatite (E), and carbonaceous (C) chondrites. These form three separate isotopic groups.

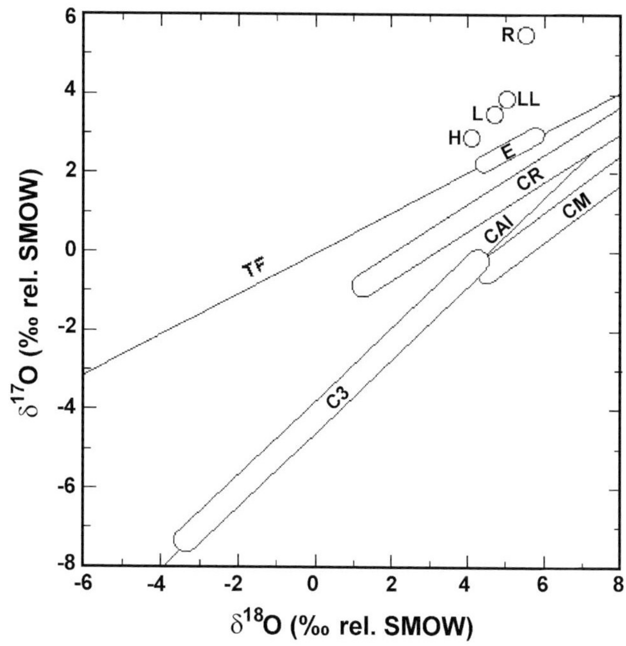

Figure 5. Schematic locations of various chondrite classes in the three-isotope graph.

The rumurutiites are a small, isotopically distinctive group of chondrites, characterized by very high ferrous iron contents, and by $\Delta^{17}O$ values that are the most positive seen in meteorites. Their chondrule isotopic compositions do not overlap the ordinary chondrite population (Bischoff *et al.*, 1994).

4. Relationships Among Chondrites, Achondrites, and Irons

Oxygen isotopic compositions may reveal genetic relationships between achondritic meteorites, which have been melted, and chondritic meteorites, which have not. A single parent body could conceivably have a metallic core, an achondritic mantle, and a chondritic surface layer. The clearest chondrite-achondrite association is that between enstatite chondrites and aubrites (enstatite achondrites). The EH and EL groups overlap one another in isotopic composition, and the aubrites overlap both (Clayton *et al.*, 1984). Thus, all of the enstatite meteorites could have been derived from a common isotopic reservoir.

Other genetic associations implied by oxygen isotopic compositions are those between H-chondrites and IIE irons, and between L-chondrites and IVA irons (Clayton and Mayeda, 1978). The former association is supported by the properties of a series of transitional stony-iron meteorites: Netschaëvo, Miles, Watson, Kodaikanal. The latter association is supported by nitrogen isotope similarities.

Another association of stony and iron meteorites is that between the winonaites (classed as primitive achondrites) and the IAB-IIICD iron meteorites (Clayton and Mayeda, 1996). These two groups also have similar mineral chemistry, and may be samples of the same parent body (Kracher, 1985). Wasson and Kallemeyn (2002) proposed that these meteorites were derived from impact melts in a carbonaceous chondrite parent body.

An especially intriguing relationship between chondrites and achondrites is seen in the ureilites. These are a relatively abundant group of meteorites with clear evidence of melting, but not following the mass-dependent oxygen isotope fractionation law associated with planetary differentiation. As shown in Figure 6, whole-rock isotopic compositions of ureilites are scattered around the slope-1 mixing line which is an extrapolation of the CAI line from Figure 1. In the context of the photochemical model of nebular fractionation, this compositional trend implies that the oxygen in the ureilites has undergone ^{16}O-depletion relative to the initial nebular composition to a greater extent than all other meteorites.

5. Differentiated Bodies

The Earth and Moon are prime examples of differentiated bodies, with chemical, isotopic, and petrographic evidence of extensive melting, homogenization, and segregation into core, mantle, and crust. No vestiges of their primary "building

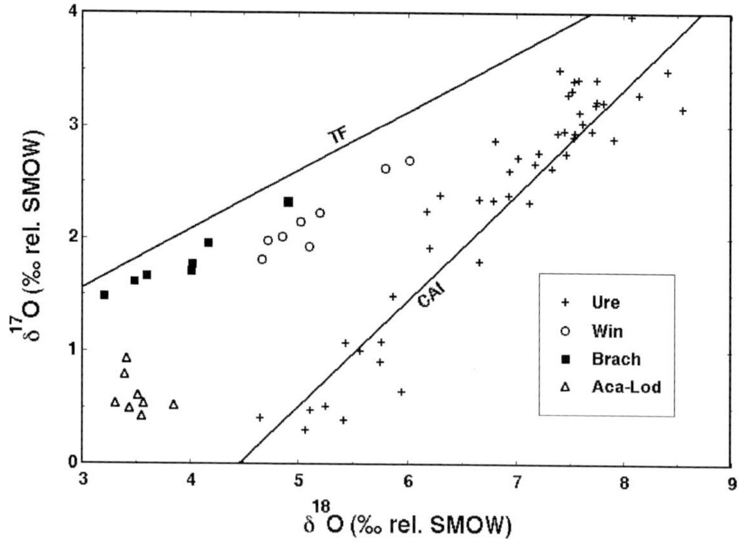

Figure 6. Oxygen isotopic compositions of primitive achondrites.

blocks" remain. The same appears to be true of the parent bodies of some achondritic meteorites. The SNC meteorites (for Shergotty, Nakhla, and Chassigny) are basalts to ultramafic igneous rocks with a uniform value of $\Delta^{17}O = +0.30$ (Clayton and Mayeda, 1996; Franchi *et al.*, 1999), indicating derivation from a common reservoir. Petrologic relations among these meteorites are consistent with derivation from a single parent body. On the basis of their young crystallization ages, and the isotopic composition of trapped gases, the parent body has been identified as the planet Mars (see Kallenbach *et al.*, 2001, for a review). Some of the SNC meteorites contain small amounts of hydrous minerals and carbonates, from which water and carbon dioxide can be extracted. These are found to have $\Delta^{17}O$ a few tenths of permil more positive than the bulk rocks, indicating derivation from an atmospheric or hydrospheric reservoir that is not isotopically equilibrated with the rock of the crust (Karlsson *et al.*, 1992; Farquhar *et al.*, 1998). The origin of the ^{17}O excess may be atmospheric reactions involving ozone, analogous to those occurring in the Earth's atmosphere (Mauersberger, 1981).

Another group of achondrites with a uniform value of $\Delta^{17} = -0.27‰$ are the HED meteorites (howardites, eucrites, and diogenites) (Figure 7). The eucrites (basalts) and diogenites (orthopyroxenites) are genetically related, and the howardites are polymict breccias containing eucritic and diogenitic clasts, so that the three groups probably have a common parent body. The asteroid 4 Vesta is the only large asteroid with a reflectance spectrum characteristic of basalt, which closely matches laboratory spectra of eucrites. Hence, it has become widely accepted that Vesta is the HED parent body. There are two other major achondrite groups that have $\Delta^{17}O$ identical to that of the HED groups: the mesosiderites and the main-group pallasites (Figure 7). Thus, all of these meteorites may have been derived

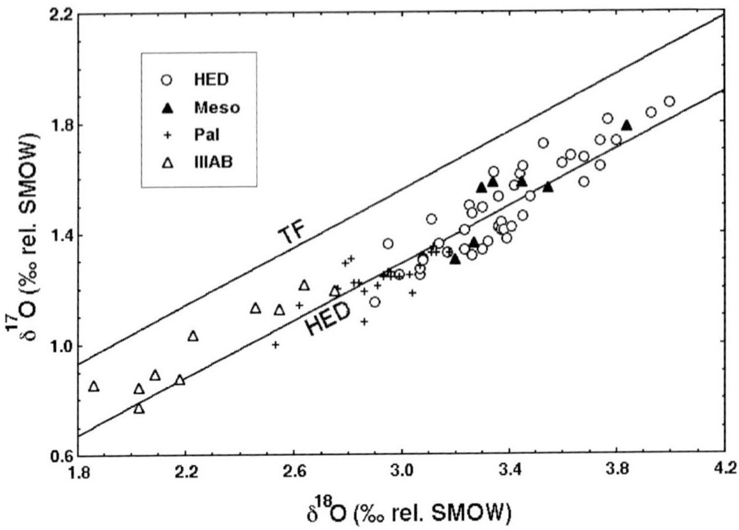

Figure 7. Oxygen isotopic compositions of differentiated achondrites: howardites, eucrites, diogenites, mesosiderites, pallasites, and IIIAB irons. These all fall on a single mass-dependent fractionation line.

from a single oxygen reservoir. The origins of mesosiderites and pallasites are still unclear, and there is disagreement concerning the possibility of producing the mesosiderites in the HED parent body (Rubin and Mittlefehldt, 1993). Oxygen isotopes in chromite, phosphate, and silicate inclusions in iron meteorites of type IIIA are also indistinguishable in $\Delta^{17}O$ (= -0.21‰, Clayton and Mayeda, 1996) from those of the HED-mesosiderite-pallasite groups. In addition, chemical compositions of metal in IIIA irons and main-group pallasites are consistent with an origin in the same parent body, representing the core and core-mantle boundary, respectively (Kracher, 1983).

The angrites are a chemically distinct group of basaltic achondrites that are barely resolvable from the eucrites in their oxygen isotopic compositions (Clayton and Mayeda, 1996).

A single eucrite-like achondrite, NWA 011, was found to have an oxygen isotopic composition far removed from the other 20 analyzed eucrites (Yamaguchi *et al.*, 2002). Its composition falls near the range of CR chondrites. The existence of NWA 011 implies that Vesta is not the only asteroid capable of producing basalts.

6. Primitive Achondrites

There are several meteorites that are achondritic in texture (i.e. lacking chondrules), but close to chondritic in chemical composition. They are thought to have been melted, but only slightly differentiated chemically by removal of small fractions of basaltic melt or metal-sulfide melt. Most of these "primitive achondrites" fall into three clusters on the oxygen-isotope diagram: the acapulcoite-lodranite, brachinite, and winonaite groups (Clayton and Mayeda, 1996) (Figure 6). Within each group, isotopic data are scattered, and do not fall along a well-defined mass-dependent fractionation line, such as the lines seen for the HED and SNC groups. Hence, melting was probably local, and not on a whole parent-body scale. McCoy *et al.* (1997) argue that acapulcoites and lodranites are derived from the same parent body, differing primarily in grain-size and texture. The coarser-grained lodranites probably underwent local partial melting, with concomitant crystal growth. The oxygen isotopic composition of this group falls within the range of the CR carbonaceous chondrites, but no genetic significance has been proposed for the coincidence. On the basis of oxygen isotopes alone, the small group of brachinites is not readily distinguished from the HED group (Clayton and Mayeda, 1996), in that they lie along the same $\Delta^{17}O$ trend, at slightly higher values of $\delta^{18}O$ and $\delta^{17}O$. As discussed earlier, the winonaites may be samples of the same body as the IAB-IIICD iron meteorites.

7. Planets

In discussion of the detailed chemical compositions of rocky planets, attention is usually focussed on four bodies from which we have samples for laboratory study: Earth, Moon, Mars, and the HED parent body (Ganapathy and Anders, 1974; Wänke, 1981; Wänke and Dreibus, 1988). These studies have led to proposals of various relationships between these "planets" and more primitive meteorites, with the goal of identifying "building blocks" of the planets, and elucidating the processes of planet formation. Since the oxygen isotopic compositions of the meteorite groups described above are not redundant with the chemical and petrographic information, they may help to understand planet formation.

Of the four bodies considered here, two have indistinguishable oxygen isotopic compositions: Earth and Moon (Clayton and Mayeda, 1996; Wiechert *et al.*, 2001). These two are significantly different in chemical composition, with the Moon being relatively depleted relative to the Earth in metal and in volatile elements. In models of an origin of the Moon by impact on the accreting Earth by a Mars-sized body, different proportions of impactor and of Earth contribute to the final Earth and Moon, so that a difference in isotopic composition would be expected unless the materials involved already were isotopically identical (Wiechert *et al.*, 2001). It has been conjectured that the oxygen isotopic composition of planet-forming materials might have exhibited a monotonic variation related to distance from the Sun, so that planetesimals at a given distance might be expected to be of uniform isotopic composition. However, numerical models of accretion of the terrestrial planets give no evidence for such narrow "feeding zones" that would allow Earth and Moon to be identical, and Mars to be different (Chambers, 1981). Thus, the reason for identical isotopic compositions of the Earth and Moon remains obscure.

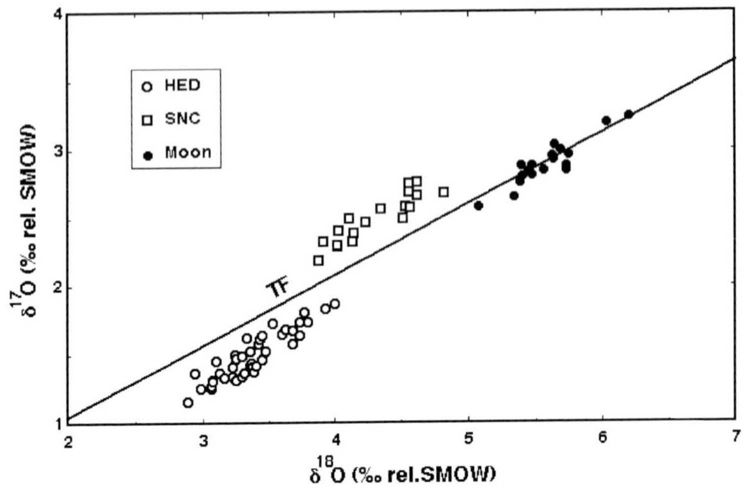

Figure 8. Oxygen isotopic comopsitions of rocks from three differentiated bodies: Mars (SNC), Moon, and Vesta (HED). These compositions cannot all be made from mixing of only two components.

Of the other two bodies under consideration, the SNC parent and the HED parent lie on opposite sides of the Earth/Moon fractionation line ($\Delta^{17}O$ is positive for SNC and negative for HED). The mean planetary compositions of Earth, Mars, and Vesta define a triangle on the three-isotope diagram (Figure 8), so that their different compositions cannot be accounted for by *any* two-component model, such as those proposed by Ringwood (1979) and Wänke (1981). It is geometrically possible to select three or more meteoritic components with compositions outside the planetary triangle, which might serve as building blocks if present in different proportions. One such selection would be: ordinary chondrites, CI chondrites, and CV chondrites (Clayton and Mayeda, 1996), with the latter two serving to provide excesses in refractory and volatile elements, respectively. However, the $\delta^{18}O$ enrichment in Earth, relative to SNC and HED, would imply a higher proportion of the volatile-rich component, contrary to observations. It has been suggested that searches for planetary construction materials among the meteorites is necessarily hopeless (Drake and Righter, 2002). Perhaps the use of meteorites to help to understand planets should focus more on processes than on materials (Ganapathy and Anders, 1974).

References

Aléon, J., Krot, A.N., McKeegan, K.D., MacPherson, G.J., and Ulyanov, A.A.: 2002, 'Oxygen Isotopic Composition of Fine-grained Ca-Al-rich Inclusions in the Reduced CV3 Chondrite Efremovka', *Lunar Planet. Sci. 33*, #1426 pdf.

Bischoff, A., Geiger, T., Palme, H., Spettel, B., Schultz L., Scherer P., Loeken T., Bland P., Clayton R.N., Mayeda T.K., Herpers U., Meltsow B., Michel R., and Dittrich-Hannen B.: 1994, 'Acfer 217 — A New Member of the Rumuruti Chondrite Group (R)', *Meteoritics* **29**, 264–274.

Chambers, J.E.: 1981, 'Making More Terrestrial Planets', *Icarus* **152**, 205–224.

Clayton, R.N.: 1993, 'Oxygen Isotopes in Meteorites', *Ann. Rev. Earth Planet. Sci.* **21**, 15–149.

Clayton, R.N.: 2002a, 'Self-shielding in the Solar Nebula', *Nature* **415**, 860–861.

Clayton, R.N.: 2002b, 'Nitrogen Isotopic Fractionation by Photochemical Self-shielding', Met. Soc. abstract #5031 pdf.

Clayton, R.N., and Mayeda, T.K.: 1978, 'Genetic Relations between Iron and Stony Meteorites', *Earth Planet. Sci. Lett.* **40**, 168–174.

Clayton, R.N., and Mayeda, T.K.: 1988, 'Formation of Ureilites by Nebular Processes', *Geochim. Cosmochim. Acta* **52**, 1313–1318.

Clayton, R.N., and Mayeda, T.K.: 1996, 'Oxygen Isotope Studies of Achondrites', *Geochim. Cosmochim. Acta* **60**, 1999–2017.

Clayton, R.N., and Mayeda, T.K.: 1999, 'Oxygen Isotope Studies of Carbonaceous Chondrites', *Geochim. Cosmochim. Acta* **63**, 2089–2104.

Clayton, R.N., Mayeda, T.K., and Olsen, E.J.: 1983, 'Oxygen Isotope Relationships in Iron Meteorites', *Earth Planet. Sci. Lett.* **65**, 229–232.

Clayton, R.N., Mayeda, T.K., and Rubin, A.E.: 1984, 'Oxygen Isotopic Compositions of Enstatite Chondrites and Aubrites', *J. Geophys. Res.* **89**, 245–249.

Clayton, R.N., Mayeda, T.K., Goswami, J.N., and Olsen, E.J.: 1991, 'Oxygen Isotope Studies of Ordinary Chondrites', *Geochim. Cosmochim. Acta* **53**, 2317–2337.

Drake, M.J., and Righter, K.: 2002, 'Determining the Composition of the Earth', *Nature* **416**, 39–44.

Ebel, D.S., and Grossman, L.: 2000, 'Condensation in Dust-enriched Systems', *Geochim. Cosmochim. Acta* **64**, 339–366.

Fagan, T.J., Yurimoto, H., Krot, A.N., and Keil, K.: 2002, 'Constraints on Oxygen Isotopic Evolution from an Amoeboid Olivine Aggregate and Ca,Al-rich Inclusion from CV3 Efremovka', *Lunar Planet. Sci. 33*, #1507 pdf.

Farquhar, J., Thiemens, M.H., and Jackson, T.: 1998, 'Atmosphere-surface Interactions on Mars: $\Delta^{17}O$ Measurements of Carbonate from ALH 84001', *Science* **280**, 1580–1583.

Franchi, I.A., Wright, I.P., Sexton, A.S., and Pillinger, C.T.: 1999, 'The Oxygen-isotopic Composition of Earth and Mars', *Met. Planet. Sci.* **34**, 657–661.

Ganapathy, R., and Anders, E.: 1974, 'Bulk Compositions of the Moon and Earth, Estimated from Meteorites', *Geochim. Cosmochim. Acta*, Suppl. 5, 1181–1206.

Gao, Y.Q., and Marcus, R.A.: 2002, 'On the Theory of the Strange and Unconventional Isotopic Effects in Ozone Formation', *J. Chem. Phys.* **116**, 137–154.

Gounelle, M., Shu, F.H., Shang, H., Glassgold, A.E., Rehm, K.E., and Lee, T.: 2001, 'Extinct Radioactivities and Protosolar Cosmic Rays: Self-shielding and Light Elements', *Astrophys. J.* **548**, 1051–1070.

Grossman, L., and Larimer, J.W.: 1974, 'Chemical History of the Solar System', *Rev. Geophys.* **12**, 71–101.

Itoh, S., Rubin, A.E., Kojima, H., Wasson, J.T., and Yurimoto, H.: 2002, 'Amoeboid Olivine Aggregates and AOA-bearing Chondrules from Y-81020 C03.0 Chondrite: Distributions of Oxygen and Magnesium Isotopes', *Lunar Planet. Sci. 33*, #1490 pdf.

Jones, R.H., Leshin, L.A., Guan, Y., Sharp, Z.D., Durakiewicz, T., and Schilk, A.J.: 2002, 'Oxygen Isotope Heterogeneity and ^{16}O-enrichments in Chondrule Olivine: Evidence for Widespread ^{16}O-rich Solids in the Solar Nebula,' *Science*, in press.

Kallenbach, R., Geiss, J., and Hartmann, W.K. (eds.): 2001, *Chronology and Evolution of Mars*, Kluwer, Dordrecht.

Karlsson, H.R., Clayton, R.N., Gibson, E.K., and Mayeda, T.K.: 1992, 'Water in SNC Meteorites: Evidence for a Martian Hydrosphere', *Science* **255**, 1409–1411.

Kracher, A.: 1983, 'Notes on the Evolution of the IIIAB/Pallasite Parent Body', *Lunar Planet. Sci. XIV*, 405–406.

Kracher, A.: 1985, 'The Evolution of Partially Differentiated Planetesimals: Evidence from Iron Meteorite Groups IAB and IIICD', *J. Geophys. Res.* **90**, 689–698.

Krot, A.N., Aléon, J., and McKeegan, K.D.: 2002, 'Mineralogy, Petrology, and Oxygen-isotopic Compositions of Ca,Al-rich Inclusions and Amoeboid Olivine Aggregates in the CR Carbonaceous Chondrites', *Lunar Planet. Sci. 33*, #1412 pdf.

Mauersberger, K.: 1981, 'Measurement of Heavy Ozone in the Stratosphere, *Geophys. Res. Lett.* **8**, 935–939.

McCoy, T.J., Keil, K., Clayton, R.N., Mayeda, T.K., Bogard, D.D., Garrison, D.H., and Wieler, R.: 1997, 'A Petrologic and Isotopic Study of Lodranites: Evidence for Early Formation as Partial Melt Residues from Heterogeneous Precursors', *Geochim. Cosmochim. Acta* **61**, 623–637.

Navon, O., and Wasserburg, G.J.: 1985, 'Self-shielding in Oxygen — a Possible Explanation for Oxygen Isotopic Anomalies in Meteorites?', *Earth Planet. Sci. Lett.* **73**, 1–16.

Nittler, L.R., Alexander, C.M.A., Gao, X., Walker, R.N., and Zinner, E.: 1997, 'Presolar Al_2O_3 Grains as Probes of Stellar Nucleosynthesis and Galactic Chemical Evolution', *Nuc. Phys. A* **621**, C113–C116.

Ringwood, A.E.: 1979, *On the Origin of the Earth and Moon*, Springer, New York.

Robert, F., and Camy-Peyret, C.: 2001, 'Ozone Isotopic Compositions: An Angular Effect in Scattering Processes?', *Annales Geophys.* **19**, 229–244.

Rubin, A.E., and Mittlefehldt, D.W.: 1993, 'Evolutionary History of the Mesosiderite Asteroid: a Chronologic and Petrologic Synthesis', *Icarus* **101**, 201–212.

Rubin, A.E., Wasson, J.T., Clayton, R.N., and Mayeda, T.K.: 1990, 'Oxygen Isotopes in Chondrules and Coarse-grained Chondrule Rims from the Allende Meteorite', *Earth Planet. Sci. Lett.* **96**, 247–255.

Shu, F.H., Shang, H., Glassgold, A.E., and Lee, T.: 1997, 'X-rays and Fluctuating X-winds from Protostars', *Science* **277**, 1475–1479.

Thiemens, M.H., and Heidenreich, J.E. III: 1983, 'The Mass-independent Fractionation of Oxygen: A Novel Isotope Effect and its Possible Cosmochemical Implications', *Science* **219**, 1073–1075.

van Dishoeck, E.F., and Black, J.H.: 1988, 'The Photodissociation and Chemistry of Interstellar CO', *Astrophys. J.* **334**, 771–802.

Wänke, H.: 1981, 'Constitution of Terrestrial Planets', *Phil. Trans. Roy. Soc. Lond.* **A303**, 287–302.

Wänke, H., and Dreibus, G.: 1988, 'Chemical Composition and Accretion History of the Terrestrial Planets', *Phil. Trans. Roy. Soc. Lond.* **A325**, 545–557.

Wasson, J.T., and Kallemeyn, G.W.: 2002, 'The IAB Iron-meteorite Complex: A Group, Five Subgroups, Numerous Grouplets, Closely Related, Mainly Formed by Crystal Segregation in Rapidly Cooling Melts', *Geochim. Cosmochim. Acta* **66**, 2445–2473.

Wiechert, U., Halliday, A.N., Lee, D.-C., Snyder, G.A., Taylor, L.A., and Rumble, D.: 2001, 'Oxygen Isotopes and the Moon-forming Giant Impact', *Science* **294**, 345–348.

Wiens, R.C., Huss, G.R., and Burnett, D.S.: 1999, 'The Solar Oxygen-isotopic Composition: Predictions and Implications for Solar Nebula Processes', *Met. Planet. Sci.* **34**, 98–107.

Yamaguchi, A., Clayton, R.N., Mayeda, T.K., Ebihara, M., Oura, Y., Miura, Y.N., Haramura, H., Misawa, K., Kojima, H., and Nagao, K.: 2002, 'A New Source of Basaltic Meteorites Inferred from Northwest Africa 011', *Science* **296**, 334–336.

Address for Offprints: Robert N. Clayton, Enrico Fermi Institute, 5640 South Ellis Avenue, University of Chicago, Chicago, IL 60637, USA; `r-clayton@uchicago.edu`

ISOTOPES OF VOLATILES IN PRE-SOLAR GRAINS

ULRICH OTT
Max-Planck-Institut für Chemie (Otto-Hahn-Institut), Postfach 3060, D-55020 Mainz, Germany
*(*ott@mpch-mainz.mpg.de*)*

Received: 21 March 2002; Accepted in final form: 6 June 2002

Abstract. Elements usually considered volatile are key constituents (C, N, O) of the pre-solar grains identified in primitive meteorites and/or key diagnostic elements (noble gases). Their isotopic compositions generally are unlike anything seen in matter of solar-system origin. The large variations are indicative of the various stellar sources and various processes of nucleosynthesis by which they were created.

1. Introduction: Overview

One of the more outstanding discoveries in cosmochemistry is that of pre-solar grains — predating the formation of the solar system — in primitive meteorites. These grains formed from material expelled in the form of stellar winds during late stages of stellar evolution or from the debris of exploding stars. Volatile elements have played a key role in their study. In fact, the very discovery is due to the fact that the most abundant ones detected (so far) are host phases of isotopically unusual noble gases. The low natural abundance of noble gases in most solid materials allowed noble gas isotope abundance anomalies to show up during stepwise degassing analyses of primitive meteorites, and, following the lead of the noble gases, the first (and still today most abundant) of the pre-solar grains were physically identified: diamond (Lewis *et al.*, 1987), silicon carbide (Bernatowicz *et al.*, 1987; Zinner *et al.*, 1987) and graphite (Amari *et al.*, 1990). Others (oxides, silicon nitride) were detected later because they followed the noble-gas-carrying grains during isolation.

An overview of the various types of pre-solar grains that have been found in meteorites is given in Table I (cf. reviews by Zinner, 1998a; Hoppe and Zinner, 2000). Concerning volatiles, the two single most important aspects are the following:

1. Known pre-solar grains are primarily composed of elements that are usually considered "volatile", but that in fact (under the right conditions of oxygen fugacity, which must have prevailed during their formation) are highly refractory.
2. Isotopic variations both in "constituent volatiles" as well as "trace volatiles" are not subtle. The enormous range of isotopic compositions reflects the variety of stellar sources where and corresponding variety of nuclesosynthesis processes by which these elements were made.

Space Science Reviews **106:** 33–48, 2003.

TABLE I

Major types of identified pre-solar dust grains in meteorites: some properties including characteristic isotopic signatures of the volatile elements.

mineral	"typical" abundance [ppm][a]	"typical" size [μm][b]	characteristic isotopic signature of volatiles[c]	contribution[c]
diamond	≈1400	≈0.0026	Xe-HL	?
silicon carbide	≈14	≈0.3 – 20	enhanced ^{13}C, ^{14}N, ^{22}Ne, s-process elements	90%
			low $^{12}C/^{13}C$, often enhanced ^{15}N	<5%
			enhanced ^{12}C, ^{15}N	1%
			low $^{12}C/^{13}C$, low $^{14}N/^{15}N$	0.1%
graphite	≈10	≈0.8 – 20	enhanced ^{12}C, ^{15}N	>80%
			Kr-S	<10%
			low $^{12}C/^{13}C$	<10%
			low $^{12}C/^{13}C$; Ne-R	2%
corundum	≈0.01	≈0.5 – 5	enhanced ^{17}O, moderately depleted ^{18}O	>70%
			enhanced ^{17}O, strongly depleted ^{18}O	20%
			enhanced ^{16}O	1%
silicon nitride	≥0.002	≈1	enhanced ^{12}C, ^{15}N	100%

[a] Listed abundances are from the compilation of Huss (1997) and refer t2o the CI chondrite Orgueil (exception: silicon nitride, where the number is for CM Murchison). Generally the abundances scale with the matrix fraction of the meteorites and decrease with increasing metamorphism of the host meteorites (Huss, 1997).
[b] Size ranges are from Zinner (1998a) and Hoppe and Zinner (2000).
[c] Isotopic signatures and relative contributions as inferred and listed in Hoppe and Zinner (2000). Enhancements and depletions are expressed relative to solar system composition.

For now, the known pre-solar grains from meteorites are all acid-resistant mineral phases and as a rule are isolated from meteorites for analysis by chemical means. Presolar silicates have just now being discovered in a class of interplanetary dust particles (IDPs; Messenger *et al.*, 2002), but so far not in meteorites.

Table I summarizes "general properties" and gives a qualitative description of isotopic features in volatile elements. Stellar sources can be inferred from the isotopic patterns. The most clear-cut case is that of the silicon carbide grains where the vast majority (the "mainstream grains" — see below) seems to derive from carbon stars, *i.e.* Red Giant stars on the asymptotic giant branch (Gallino *et al.*, 1990; Alexander, 1993; Anders and Zinner, 1993; Ott, 1993; Hoppe and Ott, 1997). The least understood case is that of the nanometer-sized diamonds — nominally the most abundant among the grains in Table I — for which it is not even clear what fraction is of pre-solar origin.

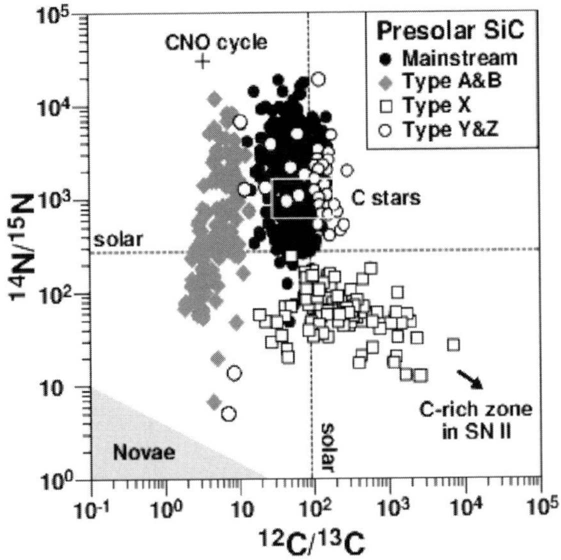

Figure 1. Carbon and nitrogen isotopic compositions measured in single pre-solar SiC grains. Solar isotopic compositions are indicated by lines. Most analyses plot in the mainstream field (cf. text and Tables I, II), displaced from normal in a direction indicating the influence of hydrogen burning via the CNO cycle. Also indicated are trends to be expected from novae and supernovae contributions. Graph courtesy of P. Hoppe (Hoppe, 2001).

2. Isotopic Compositions of CHON Elements, and Stellar Sources

2.1. Silicon Carbide

Silicon carbide has been studied in most detail, a fact that has been helped by its abundance, grain size and high content of trace elements. Isotopic analyses of carbon and nitrogen were performed on single grains >0.5 μm by secondary ion mass spectrometry (SIMS). Noble gases typically have been determined on aggregates of many grains, but larger grains have also been measured individually for He and Ne (Nichols *et al.*, 1993).

Essentially, each grain of SiC analyzed individually has turned out to be of pre-solar origin. Variations of isotopic compositions are enormous (Figure 1; Table II): $^{12}C/^{13}C$ ratios range from $\approx$2 to $\approx$7 000, i.e. from $\approx$0.02 to $\approx$80 times the solar system ratio, and $^{14}N/^{15}N$ from $\approx$7 to $\approx$20 000 ($\approx$0.03 to $\approx$70 the ratio in air). The wealth of data has allowed a classification into different categories based on C, N, and Si isotopic compositions. Most abundant ($\approx$ 90%) are the "mainstream" grains, other types have been named A, B, X, Y, Z (Table II). For most a rather clear understanding of their stellar origins exists, and for all but the X grains — of likely supernova origin — an origin from Red Giant carbon stars is highly probable (Hoppe and Ott, 1997; Zinner, 1998a; Hoppe and Zinner, 2000).

TABLE II

Distribution of single SiC grains among subgroups and diagnostic isotopic signatures of carbon and nitrogen. From Hoppe and Ott (1997), with updates according to Zinner (1998a), Hoppe and Zinner (2000) and Amari *et al.* (2001a, 2001b).

group	abundance[a] (%)	$^{12}C/^{13}C$[b]	$^{14}N/^{15}N$[b]
mainstream	87 – 94	10 – 100	50 – 20000
A	2	2 – 3.5	40 – 600
B	2 – 5	3.5 – 10	70 – 12000
X	1	18 – 7000	13 – 180
Y	1 – 2	140 – 260	400 – 5000
Z	0 – 3	8 – 180	1100 – 19000

[a] ranges refer to abundances observed in different size separates; [b] approximate ranges.

2.1.1. *Mainstream Grains*

Mainstream grains range in $^{12}C/^{13}C$ from ≈10 to ≈100, and from ≈50 to ≈20 000 in $^{14}N/^{15}N$ (Table II). Most have $^{12}C/^{13}C$ between 40 and 80 (below the solar system value of ≈89), and $^{14}N/^{15}N$ between 500 and 5 000, higher than solar, showing the input from hydrogen burning via the CNO cycle. They also contain s-process elements, with a signature similar to that of the "main component" that contributes to the solar system inventory of s-process nuclides (see noble gases). This points to an origin in low-mass (1-3 $M_{\odot}$) carbon stars in the thermally pulsing asymptotic giant branch (TP-AGB) phase. During this phase, the s-process operates in the He burning shell of the star and the products are brought to the surface by convective mixing in thermal pulses.

2.1.2. *A, B, Y, Z Grains*

A and B grains are distinguished by their very low $^{12}C/^{13}C$ ratios <10 (Table II). The dividing line is the equilibrium value for CNO cycle H burning, with $^{12}C/^{13}C$ between 3.5 and 10 for B grains, and <3.5 for grains of type A. The range in $^{14}N/^{15}N$ is from ≈40 to ≈10 000. Overall, the observed features point to an origin from specific types of carbon stars (Hoppe and Zinner, 2000; Amari *et al.*, 2001a).

Like A and B grains, Y and Z grains share many properties of the mainstream grains. They are distinguished primarily by differences in Si isotopes (Hoppe and Ott, 1997; Hoppe and Zinner, 2000). Differences exist also in the carbon isotopes. Y grains have $^{12}C/^{13}C > 100$, while Z grains have always lower than solar $^{12}C/^{13}C$. Y grains probably come from low-mass lower-than-solar metallicity AGB stars that experienced strong He shell dredge up (Hoppe and Zinner, 2000; Amari *et al.*, 2001b). For most of the Z grains the likely source are low-mass, low-metallicity AGB stars that experienced strong cool bottom processing during the Red Giant phase, but some with low $^{12}C/^{13}C$ and $^{14}N/^{15}N$ may come from novae (Hoppe and Zinner, 2000).

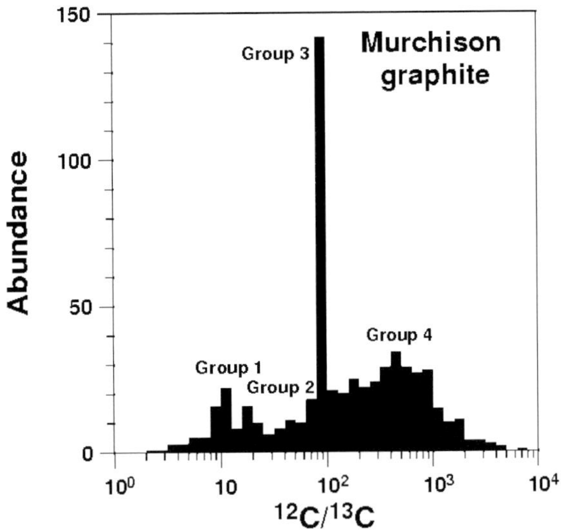

Figure 2. Distribution of $^{12}C/^{13}C$ ratios measured in single graphite grains (Hoppe and Zinner, 2000; courtesy P. Hoppe). Based on this ratio four populations can be distinguished.

2.1.3. *X Grains*

X grains (about 1% of all individually analyzed SiC grains) are clearly distinct from all other types and almost certainly derive from supernovae (Table II), as strongly indicated by, among others, huge overabundances of ^{28}Si and the former presence of ^{44}Ti ($T_{1/2} = 60$ y) showing up as excess ^{44}Ca. Ratios $^{12}C/^{13}C$ are typically higher and $^{14}N/^{15}N$ ratios lower than solar (Figure 1), which is the signature of He burning (Amari and Zinner, 1997; Hoppe and Zinner, 2000).

2.2. Graphite

Overall $^{12}C/^{13}C$ spans the same wide range of $\approx$2 to $\approx$7 000 as in SiC, the range in $^{14}N/^{15}N$ is however from $\approx$30 to 700 only, in most grains being higher than solar (Hoppe *et al.*, 1995; Hoppe and Zinner, 2000). The grains can be assigned to one out of four different groups based on the carbon isotopic ratio (Figure 2), and remarkably there seems to be a correlation with density and morphology: a) group 3 grains are characterized by normal $^{12}C/^{13}C$ and are of "non-round compact" or "spherulitic aggregate" type; they are likely to be condensates from the molecular cloud from which the solar system formed (Zinner *et al.*, 1995); b) among circumstellar grains the relative proportion assigned to each group varies with density and morphology.

Group 1 grains are characterized by $^{12}C/^{13}C$ in the range 2 – 20, similar to type A and B SiC grains, while group 2 grains with $^{12}C/^{13}C$ between 20 and 80 resemble mainstream SiC grains. Grains in group 4 are most abundant ($\approx$50% of all singly analyzed graphite grains) and isotopically light, with $^{12}C/^{13}C > 100$, up to 7 000. With these isotopic features — plus excesses of ^{18}O in trace oxygen —, it is likely that the majority of pre-solar graphite grains derives from massive stars. Type II supernovae seem the most likely sources (Zinner, 1998a; Hoppe and Zinner, 2000).

Noble gases (see below) show that other sources must have contributed as well. Krypton and xenon measured in grain aggregates show the signature of the s-process (Amari *et al.*, 1995), so some fraction must derive from AGB stars. And the presence of Ne-E(L) in a small fraction of graphite grains (Nichols *et al.*, 1994) is indicative of contributions from novae.

2.3. Nano-Diamonds

The case of the most abundant of the minerals listed in Table I, the diamonds, is not straightforward. The major problem is that we do not really know how many of them are pre-solar. A strong argument that a major part is not, comes from $^{12}C/^{13}C$. Consisting of on the average about 1000 atoms only, the diamonds cannot be analyzed individually, and "bulk analyses" of "aggregates" averaging over many grains yield a ratio of $\approx$92 (Russell *et al.*, 1991), within the range of isotopic ratios observed in solar system matter.

Nitrogen occurs in the diamonds at an abundance level close to 1% (Russell *et al.*, 1996). The isotopic ratio $^{14}N/^{15}N$ is $\approx$400, *i.e.* $\sigma^{15}N$ is ≈ -330‰ relative to the standard (air) value of 3.66×10^{-3}. While nominally this is "anomalous", new measurements by ion microprobe of surface-correlated, presumably solar wind-implanted, nitrogen in lunar grains (Hashizume *et al.*, 2000) as well as a recent re-evaluation of nitrogen in the Jovian atmosphere (Owen *et al.*, 2001) suggest that protosolar nitrogen may be similarly "light" with $^{14}N/^{15}N$ possibly as high as $\approx$400. This suggests that nitrogen in macroscopic samples of meteoritic nanodiamonds, like carbon, may be isotopically normal after all.

Diamond, alone among known pre-solar grains in meteorites, has been analyzed for hydrogen isotopic composition. Deuterium enrichments of about 200 to 300‰ have been found (Virag *et al.*, 1989), but given the history and the small size of the diamonds it is unclear if the observed signature is due to indigenous or extraneous (surface-attached or contaminant) hydrogen.

2.4. Corundum

Oxygen is the isotopically most diagnostic element (Nittler *et al.*, 1997; Zinner, 1998a; Hoppe and Zinner, 2000), with $^{16}O/^{17}O$ ranging from $\approx$70 to $\approx$30 000 (0.025 to 11$\times$ solar) and $^{16}O/^{18}O$ from 150 to 50 000 (0.3 to 100$\times$ solar), and based on oxygen four populations have been recognized (Figure 3). However, most grains analyzed in detail were selected after ion imaging indicated unusual $^{18}O/^{16}O$, so there may be a bias in the inferred distribution among the different recognized populations.

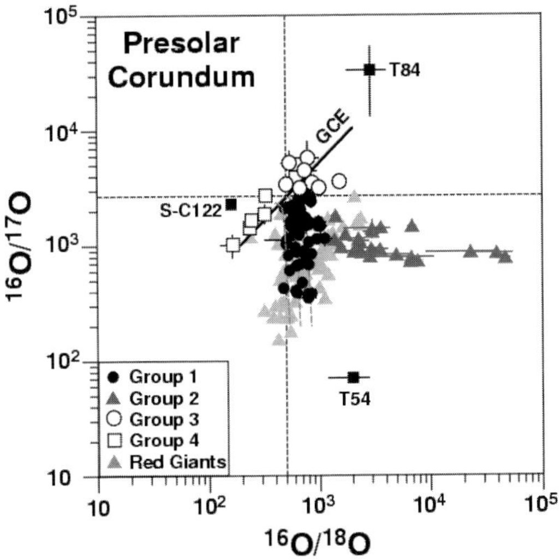

Figure 3. Oxygen isotopic ratios measured in single pre-solar corundum grains. Based on oxygen isotopes all but three grains can be assigned to one of four groups (Nittler *et al.*, 1997). Also shown are measured values for the atmospheres of red giant stars taken from the literature. The line labeled "GCE" represents the expected trend due to galactic chemical evolution (Clayton, 1988; Hoppe and Zinner, 2000). Figure courtesy P. Hoppe.

Most group 1 and 2 grains have lower than solar $^{16}O/^{17}O$ and higher than solar $^{16}O/^{18}O$ ratios. The composition of the former is similar to ratios observed in the atmospheres of red giant and AGB stars (1 to 9 $M_\odot$), which makes these the most likely stellar sources (Hoppe and Zinner, 2000). Group 3 grains may be related to group 1 grains and may — as group 4 grains — originate from Red Giant stars of low mass. Deficiencies of ^{18}O in grains of group 2 are larger than in group 1 grains; their parent stars may have been low mass stars that experienced cool bottom processing. Among the grains analyzed so far, three do not fall in any of the groups and cannot be explained by conventional red giant star models. Two of these may come from supernovae (Nittler *et al.*, 1998; Choi *et al.*, 1998).

2.5. Silicon Nitride

Silicon nitride grains have isotopic signatures similar to those of the silicon carbide X grains (Nittler *et al.*, 1995). Nitrogen is characterized by low $^{14}N/^{15}N$ (range 18 to 100). Overall, isotopic features indicate formation from supernova ejecta, as for the silicon carbide X grains (Zinner, 1998a; Hoppe and Zinner, 2000).

2.6. Other Grains

In addition to the types of grains discussed above only few other grains have been identified as being of presolar origin (Zinner, 1998a; Hoppe and Zinner, 2000). These include few grains of spinel, $MgAl_2O_4$, and hibonite, $CaAl_{12}O_{19}$, possibly a

TABLE III

Isotopic compositions of trapped He, Ne, and Ar components in diamond and SiC. Solar wind composition for comparison.

carrier	component	$^{3}He/^{4}He$ [10^{-4}]	$^{20}Ne/^{22}Ne$	$^{21}Ne/^{22}Ne$	$^{36}Ar/^{38}Ar$	ref.
diamond	P3	$\leq 1.35 \pm 0.10^{c)}$	8.910 ± 0.057	0.029 ± 0.001	5.26 ± 0.03	[1]
diamond	HL	$\leq 1.70 \pm 0.10$	8.500 ± 0.057	0.036 ± 0.001	4.41 ± 0.06	[1]
SiC	G	–	$<0.1^{a)}$	$<0.0015^{a)}$	–	[2]
	G (theory)$^{c)}$	≈ 0	*0.0827*	*0.00059*	1.52 ± 0.20	[2]
SiC	N	≤ 2.6	–	–	5.87 ± 0.07	[2]
	solar wind	4.1	13.8	0.0328	5.58	[3]

References: [1] Huss and Lewis (1994); [2] Lewis *et al.* (1994); [3] Wieler (2002).
Notes: a) In P3 a value of $(0.45 \pm 0.04) \times 10^{-4}$ has been suggested by Busemann *et al.* (2001). b) Experimental upper limits. c) Theoretical values used in the decomposition of gases measured in samples of pre-solar silicon carbide.

titanium oxide (TiO_2) grain plus carbides of Ti, Mo and Zr and Fe-Ni inside mostly grains of graphite, but also SiC. Isotopically, the oxides are similar to corundum grains and appear to come mostly from AGB stars (Zinner, 1998a; Hoppe and Zinner, 2000). As for the enclosed carbides, so far they have been too small for individual analysis but this will change with a new generation of ion microprobes coming into operation now (Zinner, 1998b).

3. Isotopic Compositions of Noble Gases

As mentioned above, noble gases have played a special role in the discovery of presolar grains in meteorites, isotopically unusual noble gases serving as “beacons” in the search for their host phases which turned out to be circumstellar condensates of diamond, silicon carbide and graphite. The most abundant of the unusual noble gases are those in the nano-diamonds. Their contribution to the total inventory of “planetary” trapped gases in primitive meteorites is substantial — for heavy noble gases such as Xe it is on the order of 10%, while for He and Ne diamond gases even dominate the “planetary” component (Ott, 2002). As in the case of the CHON elements, the case of silicon carbide is best understood. All presolar noble gas carriers — diamonds, silicon carbide, graphite — contain more than just one gas component, and the gases they carry constitute a mixture of characteristic nucleosynthetic and isotopically more normal components. Best estimates of the various components in diamond and silicon carbide are listed in Tables III–V (for a detailed discussion, see Ott, 2002). For a general overview of trapped noble gas components in meteorites see the *Glossary* and Ott (2002).

TABLE IV

Isotopic composition of Kr in pre-solar diamond (P3, HL, and P6 components) and SiC (G and N components). Solar wind for comparison. Uncertainties in the last digits are given in parentheses.

component	^{78}Kr/^{84}Kr	^{80}Kr/^{84}Kr	^{82}Kr/^{84}Kr	^{83}Kr/^{84}Kr	^{86}Kr/^{84}Kr	ref.
P3*	0.0064 (10)	0.0395 (4)	≡ 0.2023	0.2032 (5)	0.3128 (6)	[1]
HL*	0.0042 (10)	0.0305 (10)	≡ 0.1590	0.1989 (10)	0.3623 (18)	[1]
P6*	0.0059	0.0381 (8)	≡ 0.2023	0.2013 (10)	0.3147 (30)	[1]
P6 (exotic)*	0.0052	0.0354 (4)	≡ 0.1839	0.1999 (5)	0.3348 (12)	[1]
G*^{x}	–	0.0133 – 0.0183	≡ 0.4167	0.1192 (54)	0.454 – 1.176	[2]
N*	–	0.03962 (40)	0.2028 (21)	0.2018 (32)	0.2842 (36)	[2]
solar wind	0.006365	0.04088	0.20482	0.20291	0.3024	[3]

References: [1] Huss and Lewis (1994), renormalized following Busemann *et al.* (2000); [2] Lewis *et al.* (1994); [3] Pepin *et al.* (1995).

* compositions derived from mixing lines, involving assumptions about ^{84}Kr/^{82}Kr ratios in endmembers: a "typical" value from s-process nucleosynthesis of 2.40 for the G component, and for the N component a typical normal value of 4.93 ± 0.05 (corresponding to the ratio in the historical AVCC; Eugster *et al.*, 1969).

^{x} ratios ^{80}Kr/^{82}Kr and ^{86}Kr/^{82}Kr in the SiC-G component are variable. Ranges are for grain size fractions. For graphite ^{86}Kr/^{84}Kr extends to even higher values.

3.1. Silicon Carbide

Characteristic isotopic ratios in noble gases are in Ne and Kr/Xe. The isotopic signatures are indicative of the nucleosynthetic processes operating in the He-burning shell of AGB stars: He-burning and synthesis of heavy elements by slow neutron capture (the s-process). These "He-burning" gases, called "G" component, occur in silicon carbide mixed with an isotopically close to normal "N" component. "N" probably represents unmodified material from parent stars' envelopes with which the "G" gases became mixed before being trapped by SiC grains.

Best estimates for the composition of the G and N components are given in Tables III–V. Ne-G is almost monoisotopic ^{22}Ne, historically named Ne-E (Black, 1972). In detail it corresponds to the Ne-E subcomponent Ne-E(H) which had earlier been identified as being carried by a high density carrier and being released at high temperature (Eberhardt *et al.*, 1981). During He burning ^{22}Ne is produced by conversion of ^{14}N, abundantly present from previous CNO cycle H burning, via $^{14}N(\alpha,\gamma)^{18}F(e^{+}\nu)^{18}O(\alpha,\gamma)^{22}Ne$ (*e.g.*, Gallino *et al.*, 1990). Ne-G is concentrated in a small fraction (≈5%) of individually analyzed SiC grains (Nichols *et al.*, 1993).

Single grain data do not exist for Ar, Kr, Xe. S-process Kr and Xe (Kr-G, Xe-G) have been derived from the observed mixtures with Xe-N using correlation (mixing) lines and theoretical guidelines on one endmember ratio (Lewis *et al.*, 1994; Tables IV and V). They are primarily characterized by overabundances of

TABLE V

Isotopic composition of Xe in pre-solar diamond (P3, HL, P6 components), SiC (G, N components), and in the solar wind (SW). Uncertainties in the last digits are given in parentheses.

	P3*	HL*	P6*	P6 (exotic)*	G^x	N^x	SW
$^{124}Xe/^{132}Xe$	0.00446(6)	0.00839(9)	0.00433(25)	0.00679(8)	≡ 0	0.00470(13)	0.004882
$^{126}Xe/^{132}Xe$	0.00400(4)	0.00564(8)	0.00440(28)	0.00516(8)	≡ 0	0.00357(18)	0.004234
$^{128}Xe/^{132}Xe$	0.0806(2)	0.0905(6)	0.0890(20)	0.0899(5)	0.2159(23)	0.0785(11)	0.08475
$^{129}Xe/^{132}Xe$	1.042(4)	1.056(2)	1.114(8)	1.078(2)	0.1182(112)	1.000(14)	1.0420
$^{130}Xe/^{132}Xe$	0.1589(2)	0.1542(3)	0.1658(11)	0.1587(3)	0.4826(42)	0.1603(13)	0.1661
$^{131}Xe/^{132}Xe$	0.8247(19)	0.8457(13)	0.8229(47)	0.8370(13)	0.1858(117)	0.8109(130)	0.8272
$^{134}Xe/^{132}Xe$	0.3767(10)	0.6356(13)	0.3288(50)	0.5176(13)	0.0222(53)	0.4183(59)	0.3666
$^{136}Xe/^{132}Xe$	≡ 0.3096	≡ 0.6991	≡ 0.3096	≡ 0.5493	≡ 0.00343	0.4006(32)	0.2985
ref.	[1]	[1]	[1]	[1]	[2]	[2]	[3]

References: [1] Huss and Lewis (1994), renormalized following Busemann *et al.* (2000); [2] Lewis *et al.* (1994); [3] Pepin *et al.* (1995).

* compositions derived from mixing lines, involving assumptions about $^{136}Xe/^{132}Xe$ ratios in endmembers (see text). Two possible P6 compositions are listed using different assumptions.

^{x} compositions derived from mixing lines, involving assumptions about $^{136}Xe/^{130}Xe$ ratios in endmembers: endmembers: there is virtually no ^{136}Xe produced by the s-process, so the $^{136}Xe/^{130}Xe$ inferred for Xe-G from nucleosynthesis calculations is very low (0.0071; Lewis *et al.*, 1994). In deriving the N component $^{136}Xe/^{130}Xe$ was chosen so that the composition was similar to "planetary" Xe in $^{128-132}Xe/^{130}Xe$ (Lewis *et al.*, 1994). ^{124}Xe and ^{126}Xe derived in this approach may be influenced by spallation contributions. For these isotopes the theoretically expected value (≡ 0) is listed.

even-numbered isotopes of intermediate mass (^{82}Kr, ^{128}Xe, ^{130}Xe are made only by the s-process; Figures 4, 5). Noteworthy are the variations in the s-process (G-) component of the isotopic ratios $^{80}Kr/^{82}Kr$ and $^{86}Kr/^{84}Kr$. These are caused by branchings (at ^{79}Se and ^{85}Kr) in the s-process path where there is effective competition between neutron capture and β-decay (cf. Gallino *et al.*, 1990; Ott, 2002). The composition derived for the N (envelope) component is close to solar, although not exactly so, and there appears evidence for a slight overabundance in the heavy isotopes qualitatively similar to that seen in the diamonds (see below).

3.2. Graphite

Noble gases carried by graphite share similarities with those carried by silicon carbide. Some fraction must derive from AGB stars since its Kr and Xe also shows the characteristic features of the s-process. Unlike the case of SiC, where the variable Kr isotopic ratios $^{80}Kr/^{82}Kr$ and $^{86}Kr/^{84}Kr$ define almost a continuum — indicating contributions from many stars —, $^{86}Kr/^{84}Kr$ in graphite shows more extreme values (up to 2.2 as compared to 1.2 for SiC), and the distribution is rather bimodal. Neon

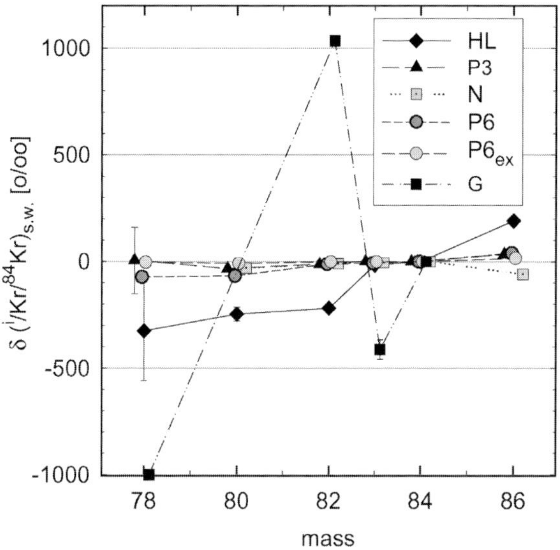

Figure 4. Kr isotopic compositions in gas components hosted by pre-solar grains: P3, P6, HL in presolar diamonds, G and N in pre-solar SiC (Table IV; Ott, 2002). Ratios are normalized to ^{84}Kr, and shown are deviations in per mill of the ^{i}Kr/^{84}Kr ratios from the corresponding ratios measured for the solar wind. Two compositions are shown for the P6 component, which involve different assumptions about ^{82}Kr/^{84}Kr in P6. For ^{78}Kr/^{84}Kr in these components no error has been given (and hence no error is shown). No experimental data have been reported for ^{78}Kr/^{84}Kr in Kr-G and Kr-N, and for the G component the theoretically expected value of -1000 ‰ is plotted. Ratios ^{80}Kr/^{84}Kr and ^{86}Kr/^{84}Kr in Kr-G are variable and not shown.

in graphite carries still another ^{22}Ne-rich component, Ne-R (Amari *et al.*, 1995). Neon-R corresponds to the former Ne-E(L) named so because of its low release temperature and association with a carrier of low density (Eberhardt *et al.*, 1981). Upper limits for ^{20}Ne/^{22}Ne and ^{21}Ne/^{22}Ne in Ne-R are 0.01 and 0.0001 (Amari *et al.*, 1995; Ott, 2002). Ne-R may represent truly monoisotopic ^{22}Ne and derive from the decay of ^{22}Na produced in novae.

3.3. Nano-Diamonds

Noble gases in diamonds play a special role since they constitute a major fraction of noble gases in the "planetary component" of primitive meteorites, the majority even of "planetary" He and Ne. Characteristic is the isotopic composition of Xe: the isotopically anomalous component Xe-HL first seen by Reynolds and Turner (1964) in stepwise heating of the Renazzo meteorite was the first nucleosynthetic (non-radiogenic) anomaly discovered in meteorites, and the hunt for its carrier phase led to the detection of the presolar nano-diamonds (Lewis *et al.*, 1987), the first of the known pre-solar phases in meteorites. Xenon-HL is characterized by about two-fold enhancement of the lightest (hence L) and the heaviest (hence

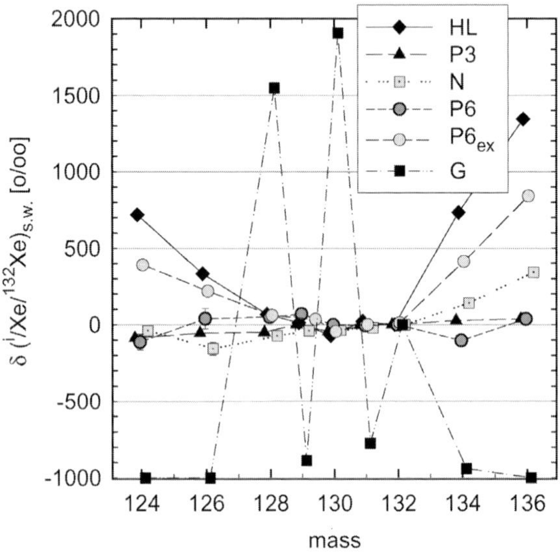

Figure 5. Xe isotopic compositions in gas components hosted by pre-solar grains: P3, P6, HL in presolar diamonds, G and N in pre-solar SiC and graphite (Table V; Ott, 2002). Ratios are normalized to ^{132}Xe, and shown are deviations in per mill of the ^{i}Xe/^{132}Xe ratios from the corresponding ratios measured for the solar wind (Table V; errors for solar wind not included). Two compositions are shown for the P6 component, which involve different assumptions about ^{136}Xe/^{132}Xe in P6. For ^{124}Xe/^{132}Xe and ^{126}Xe/^{132}Xe in the G component the theoretically expected values of -1000 ‰ are shown.

H) isotopes (made by the p- and r-processes of nucleosynthesis only) relative to intermediate ones such as s-only ^{130}Xe (Figure 5). Its origin is (still) mysterious.

Altogether 3 different noble gas components can be distinguished in the nanodiamonds. They differ in isotopic composition and can be separated according to release temperature (Huss and Lewis, 1994): an isotopically approximately normal low-temperature component (P3), anomalous HL and a less well characterized component P6. In stepwise release, at the lowest temperature steps P3 is released in essentially pure form, followed at intermediate temperatures by P3-HL mixtures, and gas released at highest temperatures is a mixture of HL and P6. Again, the compositions derived (Huss and Lewis, 1994) and listed in Table V are derived from mixing lines: the HL-composition from the intersection of P3-HL and HL-P6 mixing trends at ^{136}Xe/^{132}Xe = 0.70, P3 at the low end-side at ^{136}Xe/^{132}Xe = 0.310, where the lowest values are located. Two different choices are listed for P6: one assumes the P6 endmember to approximately "normal" with ^{136}Xe/^{132}Xe = 0.310 (as P3), the other assumes P6 to be "exotic", with a ^{136}Xe/^{132}Xe ratio of 0.55, just below the lowest ratio observed in the temperature steps where P6-HL mixtures are released (Huss and Lewis, 1994).

Each of these components has also characteristic compositions of Kr (Table IV), but for He, Ne, Ar is has proven impossible to derive a composition for the P6 component; what is listed in Table III for HL actually is a (HL+P6) mixture, but presumably the P6 contribution, as for Kr and Xe, is tiny and (HL+P6) is unlikely to be noticeably different from HL alone. Like in Xe, there is a H part of Kr-HL, but surprisingly there is no L-part — ^{78}Kr and ^{80}Kr are depleted rather than enhanced relative to s-only ^{82}Kr (Table IV). Ar-HL and Ne-HL are characterized by low ^{36}Ar/^{38}Ar and ^{20}Ne/^{22}Ne relative to other solar system reservoirs (Table III).

4. The Diamond Problems

Diamonds are (nominally at least) the most abundant, but nevertheless least understood of the pre-solar grains in meteorites, and that extends also to the isotopic compositions of their volatile elements. The problems arise from three facts:

1. carbon and possibly nitrogen in bulk are isotopically normal (see above);
2. nucleosynthetic origin and exact compositions of Xe-HL related components;
3. the surprisingly large contribution to the noble gas inventory of primitive meteorites (and maybe also the Sun).

As for the origin of Xe-HL, the enhancement of p-only and r-only isotopes (Figure 5) at first glance suggests excess contributions from just these processes. In detail, however, the case is more complex: excess r-process material, *e.g.*, would create equal relative enhancements of the two r-only isotopes ^{134}Xe and ^{136}Xe, but the observed ^{134}Xe effect is only just about half that at ^{136}Xe (Figure 5). This necessitates some new type of nucleosynthesis (*e.g.*, Howard *et al.*, 1992) or modification of the r-process composition, *e.g.* by early separating stable r-process products from their unstable precursors (Ott, 1996; Richter *et al.*, 1998). A similar situation exists for light ("p-only") ^{124,126}Xe, which do not occur in the "correct p-process ratio". Puzzling is also the close connection in which Xe-H and Xe-L are being observed in spite of the fact that they must have been produced in very different nucleosynthesis processes. Attempts of separation in the laboratory have not been very successful so far (*e.g.*, Huss and Lewis, 1994), but are ongoing (*e.g.*, Meshik *et al.*, 2001). Of course, as in the case of the isotopically normal carbon, this is made difficult by the need to analyze a large number of grains simultaneously.

In addition, experiments simulating trapping of Xe by diamonds during ion implantation indicate isotopic fractionation to take place (Koscheev *et al.*, 2001), which needs to be corrected for in order to draw conclusions about the composition at the source. Doing so might help solve another puzzle: observed deficits of light isotopes of Kr (Table IV; Figure 4) may actually be artifacts from the trapping process. The implantation experiments also suggest that there may be not much of Ne-HL and Ar-HL. Instead most of the nominal Ne-HL and Ar-HL may actually be mass-fractionated high-temperature Ne-P3 and Ar-P3. These uncertainties need to be kept in mind when interpreting the isotopic data in Tables III–V and Figures 4/5.

The final puzzle is the high apparent contribution of diamond type noble gases to solar system reservoirs. This may not only concern the "planetary component" of primitive meteorites: modeling noble gases in meteorites and the solar wind indicates that in the Sun there may be some extra ≈8% (at ^{136}Xe) of Xe similar to the Xe-H part found in diamonds (but without Xe-L), relative to a putative more primitive component (Pepin *et al.*, 1995). Solving this puzzle is definitely one of the major future tasks for xenology.

5. Summary

Elements usually considered volatile are key constituents (C, N, O) of the pre-solar grains identified in primitive meteorites and/or key diagnostic elements (noble gases). Their isotopic compositions generally are unlike anything seen in matter of solar-system origin. The large variations are indicative of the various stellar sources and the various processes of nucleosynthesis by which they were created.

Most abundant among the pre-solar grains are diamond, graphite and silicon carbide. These contain also trace elements like noble gases of unusual isotopic composition, which has been the key to their discovery. The most clear-cut case in terms of origin is that of silicon carbide where the large majority comes from carbon stars (Red Giants on the AGB). Diamonds contribute a major fraction of noble gases in primitive meteorites, but several of their features (isotopic normalcy of carbon, origin of trace noble gases) make understanding them a challenging task.

Acknowledgements

Peter Hoppe provided Figures 1–3. ISSI is thanked for hospitality and the possibility to contribute to this workshop.

References

Alexander, C. M. O'D.: 1993, 'Presolar SiC in Chondrites: How Variable and how Many Sources?', *Geochim. Cosmochim. Acta* **57**, 2869–2888.

Amari, S., and Zinner, E.: 1997, 'Supernova Grains from Meteorites', *in:* T. J. Bernatowicz and E. Zinner (eds.), *Astrophysical Implications of the Laboratory Study of Presolar Materials*, American Institute of Physics, Woodbury, N.Y., pp. 287–305.

Amari, S., Anders, E., Virag, A., and Zinner, E.: 1990, 'Interstellar Graphite in Meteorites', *Nature* **345**, 238–240.

Amari, S., Lewis, R. S., Anders, E.: 1995, 'Interstellar Grains in Meteorites: III. Graphite and its Noble Gases', *Geochim. Cosmochim. Acta* **59**, 1411–1426.

Amari, S., Nittler, L.R., Zinner, E., Lodders, K., and Lewis, R.S.: 2001a, 'Presolar SiC grains of type A and B: their Isotopic Compositions and Stellar Origins', *Astrophys. J.* **559**, 463–483.

Amari, S., Nittler, L.R., Zinner, E., Gallino, R., Lugaro, M., and Lewis, R.S.: 2001b, 'Presolar SiC Grains of Type Y: Origin from Low-metallicity Asymptotic Giant Branch Stars', *Astrophys. J.* **546**, 248–266.

Anders, E., and Zinner, E.: 1993, 'Interstellar Grains in Primitive Meteorites: Diamond, Silicon Carbide, and Graphite', *Meteoritics* **28**, 490–514.

Bernatowicz, T., Fraundorf, G., Tang, M., Anders, E., Wopenka, B., Zinner, E., and Fraundorf, P.: 1987, 'Evidence for Interstellar SiC in the Murray Carbonaceous Meteorite', *Nature* **330**, 728–730.

Black, D. C.: 1972, 'On the Origins of Trapped Helium, Neon and Argon Isotopic Variations in Meteorites - II. Carbonaceous Meteorites', *Geochim. Cosmochim. Acta* **36**, 377–394.

Busemann, H., Baur, H., and Wieler, R.: 2000, 'Primordial Noble Gases in "Phase Q" in Carbonaceous and Ordinary Chondrites Studied by Closed-system Stepped Etching', *Met. Planet. Sci.* **35**, 949–973.

Busemann, H., Baur, H., and Wieler, R.: 2001, 'Helium Isotopic Ratios in Carbonaceous Chondrites: Significant for the Early Solar Nebula and Circumstellar Diamonds?', *Lunar Planet. Sci.* **XXIII**, abstract #1598 (CD-ROM).

Choi, B.-G., Huss, G.R., Wasserburg, G.J., and Gallino, R.: 1998, 'Presolar Corundum and Spinel in Ordinary Chondrites: Origins from AGB Stars and a Supernova', *Science* **282**, 1284–1289.

Clayton, D.D.: 1988, 'Isotopic Anomalies: Chemical Memory of Galactic Evolution', *Astrophys. J.* **334**, 191–195.

Eberhardt, P., Jungck, M.H.A., Meier, F.O., and Niederer, F.R.: 1981, 'A Neon-E Rich Phase in Orgueil: Results Obtained on Density Separates', *Geochim. Cosmochim. Acta* **45**, 1515–1528.

Eugster, O., Eberhardt, P., and Geiss, J.: 1969, 'Isotopic Analyses of Krypton and Xenon in Fourteen Stone Meteorites', *J. Geophys. Res.* **74**, 3874–3896.

Gallino, R., Busso, M., Picchio, G., and Raiteri, C.M.: 1990, 'On the Astrophysical Interpretation of Isotope Anomalies in Meteoritic SiC Grains', *Nature* **348**, 298–302.

Hashizume, K., Chaussidon, M., Marty, B., and Robert, F.: 2000, 'Solar Wind Record on the Moon: Deciphering Presolar from Planetary Nitrogen', *Science* **290**, 1142–1145.

Hoppe, P.: 2001, 'Stardust in Meteorites and its Relevance to Nuclear Astrophysics', *Nucl. Phys.* **A688**, 94c–101c.

Hoppe, P., and Ott, U.: 1997, 'Mainstream Silicon Carbide Grains from Meteorites', *in:* T. J. Bernatowicz and E. Zinner (eds.), *Astrophysical Implications of the Laboratory Study of Presolar Materials*, American Institute of Physics, Woodbury, N.Y., pp. 27–58.

Hoppe, P., and Zinner, E.: 2000, 'Presolar Dust Grains from Meteorites and Their Stellar Sources', *J. Geophys. Res.* **105**, 10371–10385.

Hoppe, P., Amari, S., Zinner, E., and Lewis, R.S.: 1995, 'Isotopic Compositions of C, N, O, Mg and Si, Trace Element Abundances, and Morphologies of Single Circumstellar Graphite Grains in Four Density Fractions from the Murchison Meteorite', *Geochim. Cosmochim. Acta* **59**, 4029–4056.

Howard, W.M., Meyer, B.S., and Clayton, D.D.: 1992, 'Heavy-element Abundances from a Neutron Burst that Produces Xe-H', *Meteoritics* **27**, 404–412.

Huss, G.R.: 1997, 'The Survival of Presolar Grains in Solar System Bodies', *in:* T. J. Bernatowicz and E. Zinner (eds.), *Astrophysical Implications of the Laboratory Study of Presolar Materials*, American Institute of Physics, Woodbury, N.Y., pp. 721–748.

Huss, G.R., and Lewis, R.S.: 1994, 'Noble Gases in Presolar Diamonds I: Three Distinct Components and Their Implications for Diamond Origins', *Meteoritics* **29**, 791–810.

Koscheev, A.P., Gromov, M.D., Mohapatra, R.K., and Ott U.: 2001, 'History of Trace Gases in Presolar Diamonds as Inferred from Ion Implantation Experiments', *Nature* **412**, 615–617.

Lewis, R. S., Amari, S., Anders, E.: 1994, 'Interstellar Grains in Meteorites: II. SiC and its Noble Gases', *Geochim. Cosmochim. Acta* **58**, 471–494.

Lewis, R.S., Tang, M., Wacker, J.F., Anders, E., and Steel, E.: 1987, 'Interstellar Diamonds in Meteorites', *Nature* **326**, 160–162.

Meshik, A.P., Pravdivtseva, O.V., Hohenberg, C.M.: 2001, 'Selective Laser Extraction of Xe-H from Xe-HL in Meteoritic Nanodiamonds: Real Effect or Experimental Artifact?', *Lunar Planet Sci.* **XXXII**, abstract #2158 (CD-ROM).

Messenger, S., Keller, L.P., and Walker, R.M.: 2002, 'Discovery of Abundant Interstellar Silicates in Cluster IDPs', *Lunar Planet Sci.* **XXXIII**, abstract #1887 (CD-ROM).

Nichols, R.H. Jr., Amari, S., Hohenberg, C.M., Hoppe, P., and Lewis, R.S.: 1993, '^{20,22}Ne-E(H) and ^{4}He Measured in Single Interstellar SiC Grains of Known C-isotopic Composition', *Meteoritics* **28**, 410–411.

Nichols, R.H. Jr., Kehm, K., Brazzle, R., Amari, S., Hohenberg, C. M., and Lewis, R.S.: 1994, 'Ne, C, N, O, Mg, and Si Isotopes in Single Interstellar Graphite Grains: Multiple Stellar Sources for Neon-E(L)', *Meteoritics* **29**, 510–511.

Nittler, L.R., Alexander, C.M.O'D., Gao, X., Walker, R.M., Zinner, E.: 1997, 'Stellar Sapphires: the Properties and Origins of Presolar Al_2O_3 in Meteorites', *Astrophys. J.* **483**, 475–495.

Nittler, L.R., Alexander, C.M.O'D., Wang, J., and Gao, X.: 1998, 'Meteoritic Oxide Grain from Supernova Found', *Nature* **393**, 222.

Nittler, L.R., Hoppe, P., Alexander, C.M.O'D., Amari, S., Eberhardt, P., Gao, X., Lewis, R.S., Strebel, R., Walker, R.M., and Zinner, E.: 1995, 'Silicon Nitride from Supernovae', *Astrophys. J.* **453**, L25–L28.

Ott, U: 1993, 'Interstellar Grains in Meteorites', *Nature* **364**, 25–33.

Ott, U,: 1996, 'Interstellar Diamond Xenon and Timescales of Supernova Ejecta', *Astrophys. J.* **463**, 344–348.

Ott, U.: 2002, 'Noble Gases in Meteorites — Trapped Components', *Rev. Min. Geochem.* **46**, in press.

Owen, T., Mahaffy, P.R., Niemann, H.B., Atreya, S., and Wong, M.: 2001, 'Protosolar Nitrogen', *Astrophys. J.* **553**, L77–L79.

Pepin, R.O., Becker, R.H., and Rider, P.E.: 1995, 'Xenon and Krypton Isotopes in Extraterrestrial Regolith Soils and in the Solar Wind', *Geochim. Cosmochim. Acta* **59**, 4997–5022.

Reynolds, J.H., and Turner, G.: 1964, 'Rare Gases in the Chondrite Renazzo', *J. Geophys. Res.* **69**, 3263–3281.

Richter, S., Ott, U., and Begemann, F.: 1998, 'Tellurium in Pre-solar Diamonds as an Indicator for Rapid Separation of Supernova Ejecta', *Nature* **391**, 261–263.

Russell, S.S., Arden, J.W., and Pillinger, C.T.: 1991, 'Evidence for Multiple Sources of Diamond from Primitive Chondrites', *Science* **254**, 1188–1191.

Russell, S.S., Arden, J.W., and Pillinger, C.T.: 1996, 'A Carbon and Nitrogen Isotope Study of Diamond from Primitive Chondrites', *Met. Planet. Sci.* **31**, 343–355.

Virag, A., Zinner, E., Lewis, R.S., and Tang, M.: 1989, Isotopic Compositions of C, H, and N in Cδ Diamonds from the Allende and Murray Carbonaceous Chondrites', *Lunar Planet. Sci.* **XX**, 1158–1159.

Wieler, R.: 2002, 'Noble Gases in the Solar System', *Rev. Min. Geochem.* **46**, in press.

Zinner, E.: 1998a, 'Stellar Nucleosynthesis and the Isotopic Composition of Presolar Grains from Primitive Meteorites', *Annu. Rev. Earth Planet. Sci.* **26**, 147–188.

Zinner, E.: 1998b, 'Trends in the Study of Presolar Dust Grains from Primitive Meteorites', *Met. Planet. Sci.* **33**, 549–564.

Zinner, E., Tang, M., and Anders, E.: 1987, 'Large Isotopic Anomalies of Si, C, N and Noble Gases in Interstellar Silicon Carbide from the Murray Meteorite', *Nature* **330**, 730–732.

Zinner, E., Amari, S., Wopenka, B., and Lewis, R.S.: 1995, 'Isotopic Compositions and Structural Properties of Single Graphite Grains from Murchison', *Meteoritics* **30**, 209–226.

ATOMIC DEUTERIUM/HYDROGEN IN THE GALAXY

JEFFREY L. LINSKY
JILA, University of Colorado and NIST, Boulder, CO 80309-0440, USA

Received: 27 April 2002; Accepted in final form: 23 July 2002

Abstract. An accurate value of the D/H ratio in the local interstellar medium (LISM) and a better understanding of the D/H variations with position in the Galactic disk and halo are vitally important questions as they provide information on the primordial D/H ratio in the Galaxy at the time of the protosolar nebula, and the amount of astration and mixing in the Galaxy over time. Recent measurements have been obtained with UV spectrographs on FUSE, HST, and IMAPS using hot white dwarfs, OB stars, and late-type stars as background light sources against which to measure absorption by D and H in the interstellar medium along the lines of sight. Recent analyses of FUSE observations of seven white dwarfs and subdwarfs provide a weighted mean value of D/H = $(1.52 \pm 0.08) \times 10^{-5}$ (15.2 ± 0.8 ppm), consistent with the value of $(1.50 \pm 0.10) \times 10^{-5}$ (15.0 ± 1.0 ppm) obtained from analysis of lines of sight toward nearby late-type stars. Both numbers refer to the ISM within about 100 pc of the Sun, which samples warm clouds located within the Local Bubble. Outside of the Local Bubble at distances of 200 to 500 pc, analyses of far-UV spectra obtained with the IMAPS instrument indicate a much wider range of D/H ratios between 0.8 to 2.2 ppm. This portion of the Galactic disk provides information on inhomogeneous astration in the Galaxy.

1. Why are Accurate Measurements of the D/H Ratio Important?

Measurements of D/H, the number ratio of deuterium in all forms to hydrogen in all forms, are important for at least three reasons. First, an accurate measurement of the primordial ratio, $(D/H)_{prim}$, counts the number of baryons in the universe to determine the ratio Ω_B of the baryon density to the closure density, and tests our assumptions concerning nucleosynthesis during the first 100–1,000 seconds of the universe (e.g., Burles *et al.*, 2001). Since present theories of primordial nucleosynthesis indicate that D was formed only in the very early universe and D is the easiest isotope to be destroyed by nuclear reactions in stars (astration), D is a unique fossil for study by archeoastronomers. D is the best isotope to study because most authors agree that there are no significant sources of D after 1,000 seconds, and Ω_B is a very sensitive single-valued function of $(D/H)_{prim}$. While one expects that the best approximation to $(D/H)_{prim}$ would be an accurate measurement of D/H in an environment where there has been little chemical fractionation or star formation as measured by very low metal abundances, such measurements remain difficult.

Measurements of D/H in different locations in our Galaxy will provide an accurate test of the assumptions underlying Galactic chemical evolution models. A major problem in astrophysics is to understand how galaxies evolve with time and,

Space Science Reviews **106:** 49–60, 2003.

in particular, how the chemical element abundances evolve. In broad overview, we know that stars form out of gas clouds and over time they destroy D, create metals, and return some of this deuterium-poor and metal-rich material to the ISM by winds and supernova explosions. The detailed rates for these processes depend on the initial stellar masses. Thus with time D/H should decrease and metal abundances should increase. Theoretical models for Galactic chemical evolution rest on many assumptions that measurements of D/H in different environments can test. In particular, the temporal and spatial scales for mixing in the ISM are poorly known and likely depend on the magnetic field, which is also poorly known.

2. What is the Best Way of Measuring D/H?

While I believe that the most accurate D/H measurements are obtained from interstellar H and D Lyman line absorption in warm interstellar gas, I first summarize the various techniques that have been used to measure D/H in the Galaxy:

Deuterated molecules in cold interstellar clouds: $HDO/H_2O \geq$ **1000** ppm* and other deuterated molecules also show very high abundances. Since deuterated molecules are more tightly bound than nondeuterated molecules, the small difference in the binding energies divided by kT can be large at cold temperatures (10–20 K). For example, the reaction $HD + H_2O \leftrightarrow HDO + H_2$ at low temperatures leads to $HDO/H_2O \gg D/H$. Carbon molecule chemistry also creates huge overabundances of the deuterated molecules.

HD/H_2 in the ISM: In cold clouds nearly all D is tied up in HD molecules, so HD/H_2 measures D/H. Measurement of the HD $J = 1 \rightarrow 0$ pure rotation line (112 μm) in the Orion Bar (Wright *et al.*, 1999) by the ISO spacecraft gives D/H = **10 ± 3** ppm. This is a sensible value, but HD is not self-shielded like H_2, and thus will have a higher photodissociation rate from stellar and diffuse UV radiation fields.

Balmer-α line in the Orion Nebula: Hébrard *et al.* (2000) first detected narrow D Balmer-α and Balmer-β emission lines. Accurate measurements of the D/H ratio from the Balmer lines is difficult, however, because the D Balmer lines are fluorescent lines pumped by the hot star continuum, whereas the H Balmer lines are recombination lines (cf. O'Dell *et al.*, 2001).

Hyperfine structure line: The most recent search for the 92 cm (327 MHz) deuterium line in the ISM toward the Galactic anticenter yields a possible detection (Chengalur *et al.*, 1997) with D/H = **39 ± 10** ppm.

* I express D/H ratios in parts per million (ppm) to easily intercompare different data sets.

D/H in the Sun: A search for D Balmer-α emission at –1.785 Å relative to H Balmer-α (Beckers, 1975) gives an upper limit of D/H < **0.25** ppm. This very low value for D/H is consistent with the burning of D deep in the convective zone and the mixing of this D-depleted gas throughout the solar atmosphere.

D/H in the solar system: In their review, Robert *et al.* (2000) list for D/H in comets **300** ppm, meteorites **80–1000** ppm, Jupiter and Saturn **25** ppm, Uranus and Neptune **60** ppm. The standard explanation is that the initially highly deuterated water and other molecules become less deuterated with time by isotopic exchange with H_2 at warmer temperatures. Terrestrial water also started with a very high D/H ratio and subsequently reached its present ratio of HDO/H_2O = **150** ppm via partial isotopic re-equilibrium with warm H_2.

3. Measuring D/H with UV Spectra from HST

The Goddard High Resolution Spectrograph (GHRS) and the Space Telescope Imaging Spectrograph (STIS) instruments on HST are providing beautiful spectra of interstellar Lyman-α absorption with resolution of $\leq$ 3 km s^{-1} with which the column densities N(D I) and N(H I) and thus D/H are measurable. Several interesting surprizes have emerged from this analysis.

Virtues of this approach:

- Since no molecules are present in the warm ($T \approx 7,000$ K) ISM clouds, there is no chemical fractionation and the fractional ionization of H and D are the same. Thus N(D I)/N(H I) is the D/H ratio in these warm clouds.
- For lines of sight through the LISM, N(H I) $\approx 10^{18} - 10^{20}$ cm^{-2} and N(D I) $\approx 10^{13} - 10^{15}$ cm^{-2}. Thus for either Lyman-α or higher Lyman series lines, the D line has measurable opacity while the corresponding H line is not too optically thick to absorb completely the D line located at –82 km s^{-1}. The "horizon" set by the H I column density at which the saturated core of the interstellar H absorption is as wide as 82 km s^{-1} is 6×10^{18} cm^{-2} for Lyman-α, 4×10^{19} cm^{-2} for Lyman-β, and larger for the higher Lyman lines.

Problems with this approach:

- For many lines of sight, overlapping velocity components may permit one to measure $(D/H)_{total}$, but not D/H for each component separately.
- Low column density cloudlets of hydrogen that are Doppler shifted with respect to the main interstellar absorption feature add to the saturated H Lyman line absorption but have insufficient opacity to be detected in lines of D or any metal. When not included in the analysis, this "invisible" hydrogen can lead to large errors in N(H I) and thus the D/H ratio (Lemoine *et al.*, 2002).

As an example of the complexities in the data analysis and the serendipidous results that have emerged from measuring the D/H ratio for the lines of sight to the nearby stars, I summarize recent studies (cf. Linsky and Wood, 1996, and Wood *et al.*, 2001) of the short (1.3 pc) lines of sight to the triplet α Centauri system: A (a G2 V star like the Sun), B (a K2 dwarf), and C (Proxima Centauri, an M dwarf).

- The interstellar Fe II and Mg II resonance lines formed in the lines of sight to these stars show absorption at only one velocity, indicating that there is only one warm cloud, the so-called G (or Galactic Center) Cloud along this simple line of sight. However, the central velocity of the H Lyman-α absorption is redshifted by 2.2 km s^{-1} relative to the D Lyman α and metal line absorption, indicating the presence of a second red-shifted absorber in the H line.
- Additional absorption on the red side of the H Lyman-α absorption profile (see Figure 1) is due to the "hydrogen wall" in the heliosphere produced by the interaction and charge exchange of inflowing LISM neutral H with outflowing solar wind protons near the heliopause (e.g., Zank *et al.*, 2001). $N(\mathrm{H\ I})_{\mathrm{Hwall}} \approx 0.0004 \times N(\mathrm{H\ I})_{\mathrm{Gcloud}}$, which is sufficient to explain the additional H absorption but insufficient to provide measurable D or metal line absorption. If the H wall absorption is not included in the analysis, then the inferred N(H I) would be a factor of 2 too large and the inferred D/H ratio a factor of 2 too small.
- Additional absorption on the blue side of the H Lyman-α absorption profile of α Cen A and B (see Figure 1) is due to hydrogen wall absorption in their astrospheres produced by the interaction of LISM neutral H and their ionized stellar winds. The blue shift relative to the interstellar absorption results from viewing the decelerated H wall from the outside. The near absence of H wall absorption in the astrosphere of Proxima Centauri indicates a very low mass loss rate for this star. Studies of astrospheric absorption toward a number of nearby stars allowed Wood *et al.* (2002a) to infer stellar mass loss rates as small as $10^{-15} M_{\odot} y^{-1}$, and to estimate the mass loss rate of the young Sun, which is important for understanding the evolution of the Martian atmosphere.
- Lemoine *et al.* (2002) and Vidal-Madjar and Ferlet (2002) have argued that systematic errors in deriving N(H I) from saturated Lyman line absorption are much larger than previously assumed, leading to very uncertain D/H values. Large systematic errors can indeed be present, but in several well studied examples independent measurements of N(H I) inferred from the shape of the Lyman continuum absorption are in excellent agreement with the Lyman-α absorption results. Since the two diagnostic techniques are very different and the Lyman-α and Lyman continuum optical depths differ by a factor of 10^6, the agreement in N(H I) to better than 10% using the two techniques indicates that the systematic errors for these lines of sight are not large. Linsky *et al.* (2000) summarized the close agreement between the two different techniques for the lines of sight to the white dwarfs HZ 43 and G191-B2B, and for groups of late-type and white dwarf stars located within a few degrees of each other with lines of sight through the same clouds. Examples include the HZ 43, 31 Com, and GD 153 group, and the Capella and G191-B2B pair.

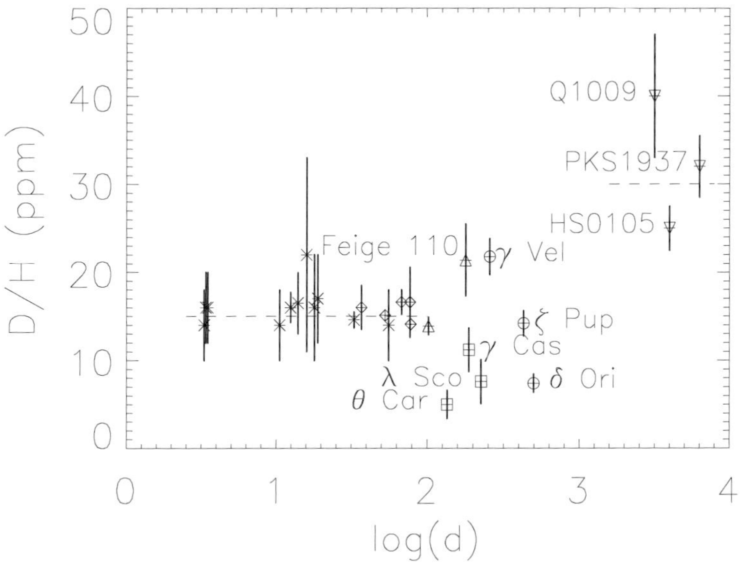

Figure 1. Comparison of the observed Lyman-α profiles toward α Cen B and Proxima Centauri. The dashed line is the interstellar absorption predicted from the observed D and metal lines. The extra absorption on the red side of the interstellar absorption (same for both stars) is due to the H wall in the heliosphere. The extra absorption on the blue side (different for the two stars and a function of the mass loss rate) is due to the H wall in the astrospheres. From Wood *et al.* (2001).

- Analysis of Lyman-α absorption for 12 sightlines through the LIC yield a mean value of D/H = **15.0± 1.0** ppm (Linsky, 1998) and no trend with distance to the target star (up to 100 pc) or Galactic longitude. Other investigators have also analyzed GHRS and STIS data using different approaches. For example, Vidal-Madjar *et al.* (1998) confirmed that the D/H ratio for the Capella line of sight through the LIC is consistent with the mean LIC value. The G191-B2B line of sight has generated more controvery, although Vidal-Madjar *et al.* (1998), Lemoine *et al.* (2002), and Sahu *et al.* (1999) agree that D/H in the LIC component is consistent with the mean value. They disagree, however, on the value of D/H in the one or two other velocity components along the line of sight to this star located only 69^{+19}_{-12} pc away.

4. Structures in the Local Interstellar Medium

The D/H ratio is unlikely to be constant throughout the Galaxy. Prime candidates for different D/H ratios are those locations where the gas has been confined for a long time and the gas composition has been altered by stellar mass loss of astrated material with limited mixing with the gas in the rest of the Galaxy. We do not know *a priori* what these structures are, but as a start we should measure the D/H ratio in ISM gas located in identifiable structures in our local region of the Galaxy.

The Galactic halo extends for many kiloparsecs (kpc) above and below the Galactic plane and is generally assumed to consist primarily of hot gas with low metal abundances. A prime goal of the FUSE mission (Far Ultraviolet Spectrograph Explorer) is to measure D/H in sightlines through the halo, but there are as yet no results available to report. The thin disk of the Galaxy, in which most giant molecular clouds are located and star formation occurs, has a vertical scale height of 325 pc and a radial scale height $\approx$ 4000 pc. The D/H ratios measured toward OB stars in the thin disk by the *Copernicus* satellite and the IMAPS instrument will be discussed below.

The Sun is located inside a region of very low density called the Local Cavity. Sfeir *et al.* (1999) have modelled the contours of the Na I absorption that likely delineates the outer edge of hot low density gas ($\log T = 6.0$–6.1) called the Local Bubble (LB), which extends outward from the Sun for 100–200 pc. It is likely that the LB fills most or all of the Local Cavity, but this is not yet demonstrated. The LB was likely formed by the winds and supernovae explosions of stars in the Scorpius-Centaurus Association as the 26 km s^{-1} flow vector is from the center of the Association. The age of the LB is a few million years and the gas within it is likely well mixed and could be D-poor and metal rich given its origin.

Within the LB are a number of small clouds consisting of warm, partially ionized gas (see Figure 2). The Sun is located within but close to the edge of the Local Interstellar Cloud (LIC). First identified from its kinematics by Lallement and Bertin (1992), the LIC was modelled by Redfield and Linsky (2000) as roughly spherical with dimensions of 5–8 pc, $T \approx 7,000$ K, and $n_{\rm total} \approx 0.2$ cm^{-3}. Within the LIC, D/H and the depletions of Mg and Fe appear to be constant. Near the LIC are at least nine other warm clouds with similar temperatures but a wide range of metal depletions, indicating that the grains in some clouds have been evaporated by shocks. The ionization fractions of H and He in the LIC are consistent with steady-state equilibrium for which the photoionization is from nearby stars (primarily ϵ CMa), the UV background, and an assumed UV radiation field formed at the boundary between warm clouds and the hot surrounding gas (cf. model 17 in Slavin and Frisch, 2002, and Wood *et al.*, 2002b).

5. FUSE Measurements of D/H Along the Lines of Sight to Nearby Hot White Dwarf Stars

The FUSE spacecraft obtains spectra of stars and extragalactic sources in the far–UV (910–1180 Å) with about 20 km s^{-1} resolution. For a description of the satellite and its capabilities see Moos *et al.* (2000) and Sahnow *et al.* (2000). A major goal of the FUSE observing program is to measure D/H in local and more distant interstellar gas. The first results of this program will be published in a series of eight papers to appear in the May 2002 issue of ApJ Supplements. Moos *et al.* (2002) summarize the results obtained from analyses of the lines of sight to five white

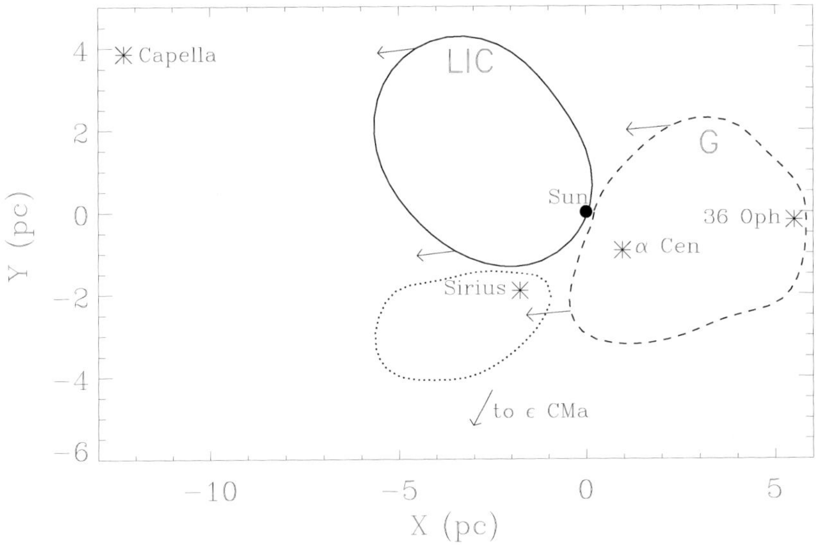

Figure 2. A schematic view of the Local Interstellar Cloud (LIC) and two other clouds as viewed from the North Galactic Pole. The Sun is located just inside the LIC toward the G cloud. Arrows designate the ISM flow direction. The star ϵ CMa is a major source of the photoionizing radiation. From Wood *et al.* (2002b).

dwarfs (HZ 43, G191-B2B, WD 0621-376, WD 1634-573, and WD 2211-495) located at distances of 37–69 pc within the LB and to two subdwarfs (Feige 110 and BD +28°4211) located at distances of 104–179 pc outside of the LB. White dwarfs are useful targets because they have a bright continuum with few stellar absorption lines against which to measure the ISM absorption, relatively simple lines of sight, and no stellar winds to complicate the analysis.

The basic approach taken in analyzing these spectra is to fit Voigt profiles to the interstellar absorption seen in the FUSE and STIS spectra to infer the number of absorption components and the total column densities for H, D, and important metals. One complication is that the FUSE spectra have insufficient spectral resolution to determine N(H I) from the shapes of the higher Lyman lines, and N(H I) is better determined from the Lyman-α profile or EUVE measurements of the Lyman continuum absorption. Uncertainties in the FUSE line spread function and velocity scale also complicate the analysis. The possible presence along the line of sight of hot hydrogen absorbers with low column densities increases the uncertainty in N(H I) and thus D/H.

As an example of this work, I summarize the analysis of the line of sight to G191-B2B by Lemoine *et al.* (2002). The line of sight to this hot DA white dwarf ($\log T_{\rm eff} = 54,000$ K, $\log g = 7.4$) has 3 ISM velocity components: 19.6 km s^{-1} (LIC), 11.5 km s^{-1}, and 7.4 km s^{-1}. The inclusion of uncertainties in N(H I) from the possible presence of hot H absorbers and uncertainties in the stellar Lyman line

TABLE I

FUSE results for D/H from D/O and O/H

Number Ratio	5 sightlines inside LB	All 7 sightlines
DI/OI (FUSE LISM)	0.076 ± 0.0020	0.0399 ± 0.0019
O/H (Sun)	$(5.2 \pm 1.1) \times 10^{-4}$	$(5.2 \pm 1.1) \times 10^{-4}$
D/H (ppm)	**19.5±4.1**	**20.7±4.3**
D/H ($O_{gas}/O_{tot} = 0.75$)	**14.6±3.1**	**15.5±3.2**
OI/HI (ISM)	$(3.43 \pm 0.15) \times 10^{-4}$	$(3.43 \pm 0.15) \times 10^{-4}$
D/H (ppm)	**11.8±0.8**	**13.7±0.9**
OI/HI (FUSE LISM)	$(3.94 \pm 0.35) \times 10^{-4}$	$(3.03 \pm 0.21) \times 10^{-4}$
D/H (ppm)	**14.8±1.5**	**12.1±0.8**

shapes against which the interstellar absorption is measured leads to log N(HI) = 18.18 ± 0.18 ($2\,\sigma$) and $(D/H)_{tot}$ = $\mathbf{16.6^{+9}_{-6}}$ ppm.

For all seven lines of sight typical uncertainties in N(D I) are ±10% (1σ), but the values of N(H I) obtained from EUVE, GHRS, STIS, or IUE spectra are typically uncertain by ±17% (1σ). The weighted mean D/H = **15.2 ± 0.8** ppm and the range in D/H values is **14–21** ppm. The line of sight with the highest D/H = **21.4±4.1** ppm is Feige 110, which is located outside of the LB. D/H for the other six lines of sight cluster closely about **15** ppm.

An alternative and perhaps more accurate way of determining D/H is from measurements of D/O and O/H. Oxygen is a good proxy for H as the ionization potentials for O I and H I are nearly the same and the ionization equilibria are closely tied to each other by charge exchange reactions. The presence of many optically thin O I lines in the FUSE spectrum lead to typical uncertainties in N(O I) of ±10%. Typical uncertainties in D/O are ±15%, and for the five white dwarfs inside the LB the weighted mean value is D/O = 0.040 ± 0.0020 (±5%). The usually cited value of OI/HI = $(3.43 \pm 0.15) \times 10^{-4}$ in the ISM at 200–1000 pc (Meyer, 2002). However, about 25% of oxygen may be tied up in grains, so $O_{gas}/O_{tot} \approx 0.75$ in the ISM and probably variable. Table 1 summarizes the D/H ratios derived using D/O and O/H under different assumptions.

6. What have We Learned about D/H in the Galaxy?

- Within the Local Bubble (out to 100 pc or more from the Sun), D/H probably has a single value (i.e., the local ISM is well mixed). The measurements of D/H summarized in Table II lead me to conclude that the best value for D/H in the Local Bubble is $(D/H)_{LB}$ = **15± 1** ppm.

TABLE II
D/H in the Local Bubble measured by different techniques

D/H Measurements	Published value	Reference
HST (solar-like stars with 100pc)	**15.0±1.0**	Linsky (1998)
FUSE (5 sightlines inside LB)	**15.2±0.8**	Moos *et al.* (2002)
FUSE (D/O and O/H in LISM)	**14.8±1.5**	Moos *et al.* (2002)
FUSE (D/O, solar O/H)	**19.5±4.1**	Moos *et al.* (2002)
FUSE (D/O, solar O/Hx0.75)	**14.6±3.1**	Moos *et al.* (2002)

TABLE III
D/H measurements beyond the Local Bubble

Star	D(pc)	D/H	Instrument	Reference
BD+28°4211	104	**13.9±1.0**	FUSE	Sonneborn *et al.* (2002)
θ Car	135	**5.0±1.6**	Copernicus	Allen *et al.* (1992)
Feige 110	180	**21.4±4.1**	FUSE	Friedman *et al.* (2002)
γ Cas	188	**13±2.5**	Copernicus	Ferlet *et al.* (1980)
λ Sco	216	**7.6±2.5**	Copernicus	York (1983)
γ^2 Vel	258	**21.8±2.0**	IMAPS	Sonneborn *et al.* (2000)
ζ Pup	430	**14.2±1.5**	IMAPS	Sonneborn *et al.* (2000)
δ Ori A	500	**7.4±1.0**	IMAPS	Jenkins *et al.* (1999)
4 QSOs		**30±4**	Keck	O'Meara *et al.* (2001)

- Table III and Figure 3 summarize the D/H measurements of gas beyond the LB, including measurements toward two hot subdwarfs by FUSE, to three O stars by Copernicus, to three O stars by the IMAPS experiment, and the mean of four quasar sightlines studied with the Keck telescope. These results show a wide range of D/H = **5–22** ppm in the Galactic disk.
- If we adopt the most recent quasar sightline value of D/H = **30**± **4** ppm (O'Meara *et al.*, 2001) as an approximate value for $(D/H)_{prim}$, then the deuterium astration in the Local Bubble, $(D/H)_{prim}/(D/H)_{LB} = (30 \pm 0.4)/(15 \pm 1) = 2.0 \pm 0.4$. The range of deuterium astration in the Galactic disk from the data in Table III is then 1.35–6.0.
- Theoretical estimates of deuterium astration over the lifetime of our Galaxy are ≤ 3 (Tosi *et al.*, 1998) and appear to be inconsistent with the wide range of observed astration values. However, the models make a number of assumptions that may not be valid. For example, the young Galaxy has primordial D/H and no metals, and the infalling gas from the halo has primordial or near-primordial abundances. Each ring of the Galaxy (several kpc wide) is assumed to be well mixed, and the gas is not mixed with gas in other rings. If the D/H results beyond the Local Bubble are valid, the Galactic chemical evolution

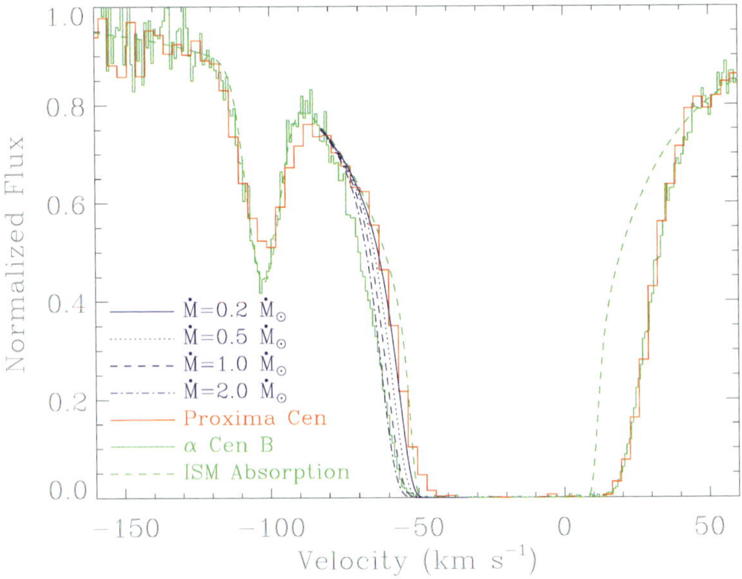

Figure 3. A summary of D/H measurements obtained with HST (*asterisks*), FUSE inside the LB (*diamonds*), FUSE outside of the LB (*triangles*), Copernicus (*squares*), IMAPS (*circles*), and Keck (*upside down triangles*). The three quasar lines of sight studied with Keck are not plotted at their correct distances. The dashed lines refer to the mean D/H values inside the Local Bubble and for the quasar lines of sight.

models are overly simplified. The next generation of Galactic chemical evolution models must include episodic star formation (star bursts) with rapid mass loss and supernovae events and more realistic mixing scenarios.

- All Galactic chemical evolution models predict that D/H and metal abundance should be anti-correlated, but the initial results from FUSE do not show this. Rather, there appears to be a weak positive correlation between D/H and O/H. Analysis of D/H and O/H for more lines of sight is needed.
- $(D/H)_{prim} \approx (D/H)_{QSO} = \mathbf{30 \pm 4}$ ppm is consistent with the primordial abundance of He and ^{7}Li according to present models of Big Bang nucleosynthesis (e.g., Burles *et al.*, 2001). The ratio of baryons to photons is $\eta = (5.5 \pm 0.5) \times 10^{-10}$, and the ratio of the baryon density to the closure density, $\Omega_B = 0.041 \pm 0.009$. Big bang nucleosynthesis theory looks basically right.

Acknowledgements

This work is supported by NASA grant S-56500-D to the University of Colorado and NIST and contract NAS5-32985 to The Johns Hopkins University for the NASA–CNES–CSA FUSE project. I would also like to thank Greg Herczeg, Seth Redfield, and Brian Wood for discussions, support, and figures.

References

Allen, M.M., Jenkins, E.B., and Snow, T.P.: 1992, 'Interstellar Absorption along the Line of Sight to θ Carinae using Copernicus Observations', *Astrophys. J. Suppl.* **83**, 261–272.

Beckers, J.M.: 1975, 'A Search for Deuterium on the Sun', *Astrophys. J.* **195**, L43–L45.

Burles, S., Nollett, K.M., and Turner, M.S.: 2001, 'Big Bang Nucleosynthesis Predictions for Precision Cosmology', *Astrophys. J.* **552**, L1–L5.

Chengalur, J.N., Braun, R., and Butler Burton, W.: 1997, 'DI in the Outer Galaxy', *Astron. Astrophys.* **318**, L35–L38.

Ferlet, R., York, D.G., Vidal-Madjar, A., and Laurent, C.: 1980, 'The Interstellar Medium on the γ Cassiopeiae Line of Sight', *Astrophys. J.* **242**, 578-583.

Friedman, S.D., Howk, J.C., Chayer, P. Tripp, T.M., Hebrard, G., Andre, M., Oliveira, C., Jenkins, E.B., Moos, H.W., Oegerle, W.R., Sonneborn, G., Lamontagne, R., Sembach, K.R., and Vidal-Madjar, A.: 2002 'Deuterium and Oxygen Toward Feige 110: Results from the Far Ultraviolet Spectroscopic Explorer (FUSE) Mission', *Astrophys. J. Suppl.* **140**, 37–49.

Hébrard, G., Péquignot, D., Vidal-Madjar, A., Walsh, J. R., and Ferlet, R.: 2000, 'Detection of Deuterium Balmer lines in the Orion Nebula', *Astron. Astrophys.* **354**, L79–L82.

Jenkins, E.B., Tripp, T.M., Wozniak, P.,W., Sofia, U.J., and Sonneborn, G.: 1999, 'Spatial Variability in the Ratio of Interstellar Atomic Deuterium to Hydrogen. I. Observations toward δ Orionis by the Interstellar Medium Absorption Profile Spectrograph', *Astrophys. J.* **520**, 182–195.

Lallement, R., and Bertin, P.: 1992, 'Northern-Hemisphere Observations of Nearby Interstellar Gas - Possible Detection of the Local Cloud', *Astron. Astrophys.* **266**, 479–485.

Lemoine, M. *et al.*: 2002, 'Deuterium Abundance toward G191-B2B: Results from the Far Ultraviolet Spectroscopic Explorer (FUSE) Mission', *Astrophys. J. Suppl.* **140**, 67–80.

Linsky, J.L.: 1998, 'Deuterium Abundance in the Local ISM and Possible Spatial Variations', *Space Sci. Rev.* **84**, 285–296.

Linsky, J.L., and Wood, B.E.: 1996, 'The α Centauri Line of Sight: D/H Ratio, Physical Properties of Local Interstellar Gas, and Measurement of Heated Hydrogen (the "Hydrogen Wall") near the Heliopause', *Astrophys. J.* **463**, 254–270.

Linsky, J.L., Redfield, S., Wood, B.E., and Piskunov, N.: 2000, 'The Three-Dimensional Structure of the Warm Local Interstellar Medium. I. Methodology', *Astrophys. J.* **528**, 756–766.

Meyer, D.: 2002, 'CNO Abundances in the Diffuse Interstellar Medium', *in* R. Ferlet *et al.* (eds.), *Gaseous Matter in the Galaxies and Intergalactic Space*, XVIIth IAP Colloquium, pp. 135–140.

Moos, H.W., *et al.*: 2000, 'Overview of the Far Ultraviolet Spectroscopic Explorer Mission', *Astrophys. J.* **538**, L1–L6.

Moos, H.W., *et al.*: 2002, 'Abundances of Deuterium, Oxygen, and Nitrogen in the Local Interstellar Medium: Overview of First Results from the Far Ultraviolet Spectroscopic Explorer Mission', *Astrophys. J. Suppl.* **140**, 3–17.

O'Dell, C.R., Ferland, G.J., and Henney, W.J.: 2001, 'Measurement and Interpretation of Deuterium-Line Emission in the Orion Nebula', *Astrophys. J.* **556**, 203–214.

O'Meara, J.M., Tytler, D., Kirkman, D., Suzuki, N., Prochaska, J., Lubin, D, Wolfe, A.M.: 2001, 'The Deuterium to Hydrogen Abundance Ratio toward a Fourth QSO: HS 0105+1619', *Astrophys. J.* **552**, 718–730.

Redfield, S., and Linsky, J.L.: 2000, 'The Three-Dimensional Structure of the Warm Local Interstellar Medium. I. The Colorado Model of the Local Interstellar Cloud', *Astrophys. J.* **534**, 825–837.

Redfield, S., and Linsky, J.L.: 2001, 'Microstructure of the Local Interstellar Cloud and the Identification of the Hyades Cloud', *Astrophys. J.* **551**, 413–428.

Robert, F., Gautier, D., and Dubrulle, B.: 2000, 'The Solar System D/H Ratio: Observations and Theories', *Space Sci. Rev.* **92**, 201–224.

Sahu, M.S., Landsman, W., Bruhweiler, F.C., Gull, T.R., Bowers, C.A., Lindler, D., Feggans, K., Barstow, M.A., Hubeny, I., and Holberg, J.B., 1999 'The D/H Ratio in Interstellar Gas toward G191-B2B', *Astrophys. J.* **523**, L159–L163.

Sfeir, D.M., Lallement, R., Crifo, F., and Welsh, B.Y.: 1999, 'Mapping the Contours of the Local Bubble: Preliminary Results', *Astron. Astrophys.* **346**, 785–797.

Sahnow, D.J., *et al.*: 2000, 'On-Orbit Performance of the Far Ultraviolet Spectroscopic Explorer Satellite', *Astrophys. J.* **538**, L7–L11.

Slavin, J.D., and Frisch, P.C.: 2002, 'The Ionization of Nearby Interstellar Gas', *Astrophys. J.* **565**, 364–379.

Sonneborn, G., Tripp, T.M., Ferlet, R., Jenkins, E.B., Sofia, U.J., Vidal-Madjar, A., and Wozniak, P.R.: 2000, 'Spatial Variability in the Ratio of Interstellar Atomic Deuterium to Hydrogen. II. Observations toward γ^2 Velorum and ζ Puppis by the Interstellar Medium Absorption Profile Spectrograph', *Astrophys. J.* **545**, 277–289.

Sonneborn, G., Andre, M., Oliveira, C., Hebrard, G., Howk, J.C., Tripp, T.M., Chayer, P., Friedman, S.D., Kruk, J.W., Jenkins, E.B., Lemoine, M., Moos, H.W., Oegerle, W.R., Sembach, K.R., and Vidal-Madjar, A.: 2002, 'Interstellar Deuterium, Nitrogen, and Oxygen Abundances Toward BD +28 4211: Results from the Far Ultraviolet Spectroscopic Explorer' Spectrograph', *Astrophys. J. Suppl.* **140**, 51–66.

Tosi, M., Steigman, G. Matteucci, F. and Chiappini, C.:1998 'Is High Primordial Deuterium Consistent with Galactic Evolution?', *Astrophys. J.* **498**, 226–240.

Vidal-Madjar, A., and Ferlet, R.: 2002, 'Hydrogen Column Density Evaluations Toward Capella: Consequences on the Interstellar Deuterium Abundance', *Astrophys. J.* **571**, L169–L172.

Vidal-Madjar, A., Lemoine, M., Ferlet, R., Hébrard, G., Koester, D., Audouze, J., Cassé, M., Vangioni-Flam, E., and Webb, J.: 1998, 'Detection of Spatial Variations in the (D/H) Ratio in the Local Interstellar Medium', *Astron. Astrophys.* **338**, 694–712.

Wood, B.E., Linsky, J.L., Müller, H.R., and Zank, G.P.: 2001, 'Observational Estimates for the Mass-loss Rates of α Centauri and Proxima Centauri using Hubble Space Telescope Lyα Spectra' *Astrophys. J.* **547**, L49–L52.

Wood, B.E., Müller, H.R., Zank, G.P., and Linsky, J.L.: 2002a, 'Measured Mass Loss Rates in Solar-like Stars as a Function of Age and Activity', *Astrophys. J.* **574**, 412–425.

Wood, B.E., Redfield, S., Linsky, J.L., and Sahu, M.S.: 2002b, 'Elemental Abundances and Ionization States within the Local Interstellar Cloud Based on HST and FUSE Observations of the Capella Line of Sight', *Astrophys. J.*, submitted.

Wright, C.M., van Dishoeck, E.F., Cox, P., Sidher, S.D., and Kessler, M.F.: 1999, 'Infrared Space Observatory-Long Wavelength Spectrometer Detection of the 112 Micron HD J = 1 –> 0 Line toward the Orion Bar', *Astrophys. J.* **515**, L29–L33.

York, D.G.: 1983, 'The Interstellar Medium near the Sun. III - Detailed Analysis of the Line of Sight to λ Scorpii', *Astrophys. J.* **264**, 172–195.

Zank, G.P., Müller, H.-R., and Wood, B.E.: 2001, 'The interaction of the Solar Wind and Stellar Winds with the Partially Ionized Interstellar Medium', *Physics of Plasmas* **8**, 2385–2393.

Address for Offprints: Jeffrey L. Linsky, JILA, University of Colorado, Boulder, CO 80309-0440, USA; jlinsky@jila.colorado.edu

DEUTERIUM IN MOLECULES OF THE INTERSTELLAR MEDIUM

E. ROUEFF[1] and M. GERIN[2]
[1] LUTH (FRE2462 of CNRS) and Observatoire de Paris, Place J. Janssen, 92190 Meudon, France
[2] LERMA and ENS (FRE2460 of CNRS), 24 rue Lhomond, 75231 Paris cedex 05, France

Received: 3 May 2002; Accepted in final form: 1 October 2002

Abstract. Deuterated molecules are detected both in interstellar translucent clouds and in cold dark clouds, as well as in star forming regions. We review the recent observational studies of deuterated molecules ranging from the VUV to the millimeter wavelength range. We outline some sources of uncertainties on the deuterium fractionation and on the subsequent derivation of the elemental deuterium to hydrogen ratio. Steady state versus time dependent models are discussed and the role of initial conditions is emphasized.

1. Introduction

The observation of deuterated molecules is in some way easier than the observation of atomic deuterium itself. In cold and relatively dense regions where the material is partially or fully molecular, it is imperative to consider deuterated molecules in the budget of the available deuterium. Table I reports the present status of the observations of interstellar deuterated molecules in a few lines of sight sampling diffuse and transluscent clouds (ζ Oph, HD 110432), warm regions (OMC1, Orion bar), cold dense cores (TMC1 (CP), L134N, L1689N) and star forming regions (IRAS16293-2422, NGC 1333, hot core). A special emphasis is put on HD in which UV electronic transitions are detected in absorption in diffuse and translucent clouds, while the infrared ro-vibrational and rotational lines have been detected either in emission and/or absorption towards a very limited number of lines of sight pertaining to different physical conditions.

Table I shows clearly a large range of deuterium fractionation values which may be analysed in different ways. In a fully molecular region, HD is the reservoir of deuterium and is probably the best indicator of the elemental D/H ratio. We present in Section 2 the corresponding observations and their limitations. Our view of the deuteration mechanisms of polyatomic molecules has been drastically renewed by the recent detection of multiply deuterated species. Triply deuterated ammonia has been found in cold dark clouds (Lis *et al.*, 2002; van der Tak *et al.*, 2001; Roueff *et al.*, 2002), following the detection of doubly deuterated ammonia and formaldehyde (Turner, 1990; Roueff *et al.*, 2000; Loinard *et al.*, 2000) in similar environments. We discuss in Section 3 the different aspects related to millimeter and submillimeter spectroscopy and give our view of the modelling aspects in Section 4. Our conclusions and perspectives are finally reported in Section 5.

Space Science Reviews **106:** 61–72, 2003.

TABLE I

Observed deuterium fractionation (XD/XH) towards several lines of sight. Numbers in parenthesis refer to powers of ten.

Molecule	ζ Oph	HD 110432	PKS 1232+082	Orion bar	OMC 1	W 49
HD/H_2	4.7(-7) [1]	6.5(-7) – 8.3(-6)[2]	2.0(-3)[3]	0.7(-5) –2.6(-5)[4]	9.0(-6)[5]	≥0.001[6]

	TMC 1 (CP)	L134N	Hot Core	NGC 1333	IRAS 16293-2422	L1689N
H_2D^+/H_3^+				0.02[7]		
DCO^+/HCO^+	0.012[8]	0.035[8]		0.01[7]	0.0086[9]	0.08[26]
N_2D^+/N_2H^+	0.0064[8]	0.12[8]				
C_2D/C_2H	0.048[8]	0.06[8]			0.18[9]	
DCN/HCN	0.011[8]	0.05[8]			0.013[9]	0.11[26]
DNC/HNC	0.016[8]	0.046[8]			0.03[9]	
HDO/H_2O			2-6(-4)[10]		≥0.01[9]	
HDS/H_2S			≤0.01[11]		0.1[9]	
NH_2D/NH_3	≤0.01[8,13]	0.05[8,12]	0.03-0.06[14,27]	0.07[13]	0.003[13]	0.08[13,17,27]
ND_2H/NH_2D		0.05[15]				0.4[17]
ND_3/NH_3				1(-3)[18]		≈1(-3)[19]
$HDCO/H_2CO$		0.068[8]	0.14[14]		0.14[9]	
$D_2CO/HDCO$			0.02[14]		0.33[16]	≥0.4[17]
$HDCS/H_2CS$	0.02[20]				0.1[9]	
C_4D/C_4H	0.0043[21]					
DC_3N/HC_3N	0.014[8]					
$c\text{-}C_3HD/c\text{-}C_3H_2$	0.05-0.08[8,22]	0.1[22]				≈0.1[28]
CH_2DOH/CH_3OH			0.04[23]			
CH_3OD/CH_3OH	≈0.02[8]	≤0.03[8]	0.01-0.06[24]		≤ 0.1[9]	
DC_5N/HC_5N	0.015[29,30]					
CH_2DCCH/CH_3CCH	0.06[25]					
CH_2DCN/CH_3CN			≈ 0.01[31]			

[1] Wright and Morton (1979), [2] Rachford *et al.* (2000), [3] Varshalovich *et al.* (2001), [4] Wright *et al.* (1999), [5] Bertoldi *et al.* (1999), [6] Caux *et al.* (2002), [7] Stark *et al.* (1999), [8] Turner (2001), [9] van Dishoeck *et al.* (1995), [10] Helmich *et al.* (1996), [11] Hatchell *et al.* (1999), [12] Tiné *et al.* (2000), [13] Shah and Wootten (2001), [14] Turner (1990), [15] Roueff *et al.* (2000), [16] Loinard *et al.* (2000), [17] Loinard *et al.* (2001), [18] van der Tak *et al.* (2001), [19] Roueff *et al.* (2002), [20] Minowa *et al.* (1997), [21] Turner (1989), [22] Bell *et al.* (1988), [23] Jacq *et al.* (1993), [24] Mauersberger *et al.* (1988), [25] Gerin *et al.* (1992a), [26] Lis *et al.* (2002), [27] Saito *et al.* (2000), [28] Gerin *et al.* (1987), [29] McLeod *et al.* (1981), [30] Schloerb *et al.* (1981), [31] Gerin *et al.* (1992b)

2. HD Observations

The deuterated hydrogen molecule has been first detected via its ultraviolet interstellar absorption transitions towards the bright star ζ Oph with the Copernicus satellite by Morton, 1975. The corresponding cloud has long been considered as a prototype of diffuse interstellar clouds where hydrogen is partially molecular ($f = 2\,N(H_2)/(2\,N(H_2) + N(H)) = 0.30$, where N stands for a column density) and where photoprocesses play a major role in the chemistry. HD 110432 is a translucent cloud observed via FUSE (Rachford *et al.*, 2000) where the molecular fraction is about 0.57. We notice the large error bar reported for this line of sight which arises both from saturation effects in the HD lines and from multiple components on the line of sight. However, the HD/H_2 fraction is seen to be small and much less than the expected value of the elemental D/H ratio of 1.5×10^{-5}. This is not surprising since a significant amount of deuterium is still atomic as emphasized by Le Petit *et al.* (2002) in an extensive study of the D – HD transition in Photon dominated regions (PDR). Ultraviolet absorption transitions are redshifted to the visible in high redshifted regions. The HD/H_2 ratio found towards the quasar PKS 1232+082 at a redshift of 2.33 from electronic absorption observations at VLT is particularly exciting. The value is as big as 2×10^{-3} whereas the molecular fraction found towards the line of sight is small ($f = 4 \times 10^{-4}$, Petitjean *et al.*, 2000). Such results should be however taken with care and confirmed by higher signal to noise observations.

The observations towards the Orion Bar, a dense PDR, have been obtained in emission via the $J = 1 - 0$ rotational transition at 112 μm with the Long-Wavelength-Spectrometer (LWS) on board the Infrared Space Observatory (ISO) (Wright *et al.*, 1999). The derivation of the D/H elemental ratio from such an observation implies the understanding of the excitation of the $J = 1$ level of HD. The density is several times 10^5 cm^{-3} and the $J = 1$ level is thermalized. A value of D/H of $(1.0 \pm 0.3) \times 10^{-5}$ is derived. Through a deep integration with the Short-Wavelength-Spectrometer (SWS) on board ISO, the pure rotational R(5) line at 19.43 μm was detected toward the Orion (OMC-1) outflow at its brightest H_2 emission region. Accounting for non-LTE HD level populations in a partially dissociated gas, the estimate for the HD/H_2 column density ratio is $(9.0 \pm 3.5) \times 10^{-6}$. The $J = 1 - 0$ transition of HD has even been observed in absorption with ISO in a cold dark cloud in front of an infrared source located in W49. The result displayed in Table I is obtained by assuming the standard CO/H_2 ratio and is about two orders of magnitude larger than the average D/H ratio. Another difficulty is showing up here: a significant fraction of CO towards the source is probably condensed on dust grains. A factor of 100 for the depletion of CO is required to reconcile the HD value with the standard D/H ratio following Caux *et al.* (2002). The HD absorption transition corresponds to the highest HI component detected by Brogan and Troland (2001) in their VLA observations towards the W49 complex.

3. Millimeter and Submillimeter Observations

Polyatomic deuterated molecules are considerably enhanced in deuterium in comparison to their hydrogenated counterparts as shown in Table I. A typical figure is [XD]/[XH] = 0.05, corresponding to $\delta D = +3000$. Chemical fractionation occurs in cold dense cores as described in Herbst (2002) via three principal ion-molecule reactions: $H_3^+ + HD \rightarrow H_2D^+ + H_2$, $CH_3^+ + HD \rightarrow CH_2D^+ + H_2$, and $C_2H_2^+ + HD \rightarrow C_2HD^+ + H_2$, allowing further reaction of the deuterated molecular ions for producing more complex deuterated molecules.

3.1. Spectroscopic Considerations

The derivation of the deuterium fractionation from radio observations is not straightforward. First, deuterium possesses a nuclear spin of 1 and the rotational spectrum, which is modified accordingly, is not always known accurately. Turner (2001) has addressed this problem and made use of his astrophysical spectra $J = 1 - 0$ and $J = 2 - 1$ to provide new or improved molecular constants on N_2H^+, N_2D^+, $HC^{15}N$, $HN^{13}C$, $H^{15}NC$ and DNC. The hyperfine structure of the N_2D^+ $J = 2 - 1$ and $3 - 2$ transition has been discussed by Gerin *et al.* (2001; see also Caselli *et al.*, 2002). This latter case shows that, contrary to frequent assumptions, the amplitude of hyperfine structure is not necessarily decreasing with the rotational quantum number. Because of the large quadrupole coupling of nitrogen and the fact that there are two nuclear spins leading to larger values of the quantum numbers, it turns out that the higher J hyperfine structure is observable and very useful (cf., Gordy and Cook, 1984). The large number of hyperfine components provides satellites and contributes to the broadening and asymmetric shapes of the N_2H^+ and N_2D^+ line profiles. It is therefore important to take into account the detailed hyperfine structure of the transitions when deducing astrophysical parameters from observed data. The intensity ratio of the various hyperfine components is used for determining the line opacity as it can be predicted in the optically thin case. When such a fit is possible, the determination of the column density from the radio data is quite accurate.

3.2. Source Structure and Radiative Transfer

Another source of uncertainty in the derivation of deuterium fractionation ratio is due to the geometrical structure of the source itself, hence a careful study of the radiative transfer must be considered. It is necessary to observe at least the main molecule and the deuterated molecule to derive an abundance ratio, which means a few lines. Systematic errors, due to, e.g., calibration or imperfect subtraction of the atmospheric absorption, may contribute differently for the hydrogenated and deuterated species if different instruments are used (telescopes, receivers), and must be taken into account in the total error budget for the XD/XH measurements.

Also, in some cases, such as for HNC, for which the opacity of the main isotope is very large, the fractionation ratio is obtained from the comparison of molecules with rare isotopes: D for DNC and ^{13}C for $HN^{13}C$ (e.g. Turner, 2001). The accuracy of the observed deuterium fractionation is therefore dependent on the accuracy of the elemental isotopic ratio. Fortunately, the relative abundance of the most commonly used isotope ^{13}C is fairly well determined. Roberts *et al.* (2002) have obtained good measurements of [HDCO]/[H_2CO] using both ^{12}C and ^{13}C isotopes. For other molecules, the discrepancy between D/H obtained using the main isotope or using ^{13}C may be large (cf., Turner, 2001), as seen from the comparison of numbers in Table I. A likely explanation is the different spatial distibution of the different isotopes. Indeed, the knowledge of the detailed source structure is important for deriving an accurate deuterium fractionation. The variation of kinetic temperature along the line of sight is of particular interest due to the high sensitivity of the deuterium fractionation to temperature. Towards star forming regions, where the very dense gas close to the protostar is warmer than the foreground gas, lines from the main isotope are probing the full extent of the gas cloud while lines from the deuterated isotopes are restricted to the coldest zones. Sophisticated models are required for accurate determinations of the molecular abundances (cf., the NGC 1333 region, Stark *et al.*, 1999). Temperature gradients are less critical in cold dense cores with no star formation where the kinetic temperature is almost constant, but density gradients along the line of sight may also affect the radiative transfer hence the determination of the column density. In all cases, it is necessary to compare the information obtained at the same positions, and with identical, or very close, spatial resolutions.

3.3. Observed Trends

Valuable information on the chemical processes is obtained when comparing many sources with similar physical conditions. We show in Figure 1 a comparison of the deuterium fractionation in dense cores. It is obvious that high fractionation regions exist where most molecules are highly enriched in deuterium. L134N and L1689N (Table I) are good examples of such regions. The typical size of the high deuterium region is quite small, of the order of 1 arcmin which is about 0.03 pc for nearby clouds (Tiné *et al.*, 2000), smaller than the typical size of dark clouds (a few pc). The observed trends can be understood as revealing the spatial distribution of deuterium fractionation, from very high values in the coldest and densest cores, to lower values in the cloud envelopes.

The isomers DCN and DNC deserve a special remark. As discussed previously by Turner (2001) the deuterium fractionation is larger for HNC than for HCN. It is well known that HNC and HCN have similar abundances in cold dense cores, while HCN is more abundant in warm regions (e.g., Hirota *et al.*, 1998). Hirota *et al.* (2001) show maps of TMC-1 for DNC and $HN^{13}C$. [DNC]/[HNC] is lower in the southern part of TMC-1 (near TMC1-D and the cyanopolyyne peak with

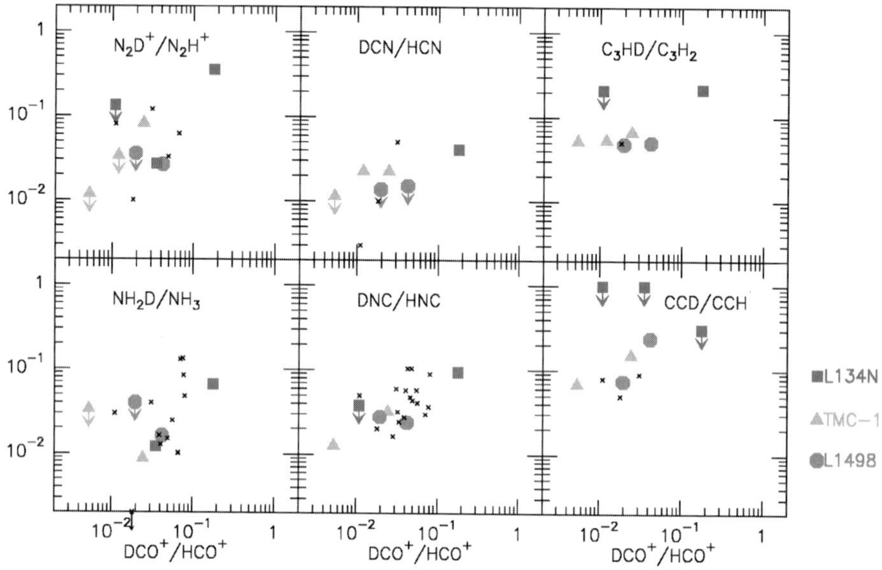

Figure 1. Observed deuterium fractionation for a sample of dense cores (*crosses*) (Butner *et al.*, 1995; Saito *et al.*, 2000; Shah and Wootten, 2001; Hirota *et al.* , 2001), and for specific positions observed at high spatial resolution (23") in L134N (*squares*), TMC1 (*triangles*) and L1498 (*dots*). The high resolution data bracket the range of values obtained from the low resolution data, and extend it to more extreme values. The mean electronic fraction (n_e/n_H) is predicted to decrease from the left (low $[DCO^+]/[HCO^+]$) to the right (high $[DCO^+]/[HCO^+]$).

low deuterium fractionation) than in the northern region (TMC1-N where the fractionation is rising). We have found the same behaviour, namely higher deuterium fractionation for both HNC and HCN in the regions where DCO^+ or N_2H^+ peaks. Also, we could study the isomeric ratio [DCN]/[DNC] and $[H^{13}CN]/[HN^{13}C]$. For all data points, the ratio of the deuterated isomers [DCN]/[DNC] is smaller than the corresponding ratio for the ^{13}C species $[H^{13}CN]/[HN^{13}C]$. Treating upper limits as detections, we obtain the following average values: $[DCN]/[DNC] = 0.29 \pm 0.16$ and $[H^{13}CN]/[HN^{13}C] = 0.42 \pm 0.15$.

4. Modeling Aspects

Chemical models aim at deriving the abundances of the various species as a function of the physical conditions present in the interstellar clouds. The low density regime implies that only binary collisions may occur. The thermal balance may in some cases be solved in parallel. A set of first-order non-linear differential equations is obtained from the balance of formation and destruction processes leading to variations of the abundances as a function of time:

$$\frac{dX_n}{dt} = \sum_{i,j} k_{ij} X_i X_j - X_n \sum_i k_{ij} X_i + \sum_i k'_i X_i - X_n \sum_i k'_i \ , \qquad (1)$$

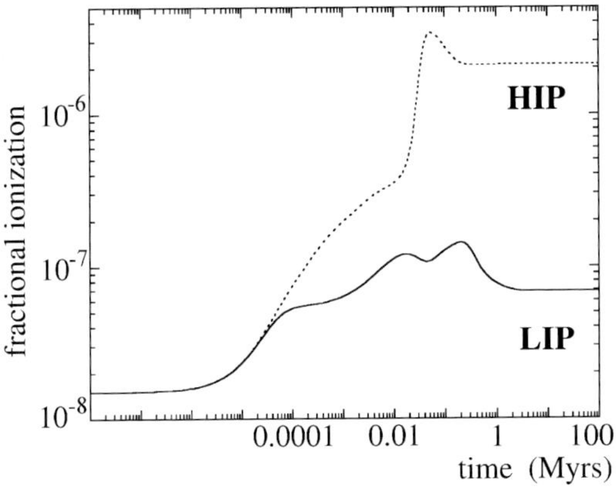

Figure 2. Time evolution solutions of the fractional ionization for a dark cloud model with $n_H = 10^4\,cm^{-3}$, $T = 10\,K$ and $\zeta = 5 \times 10^{-17}\,s^{-1}$ for 2 different initial conditions.

where X_i stands for the abundance of species i and is expressed in cm^{-3}, and k_{ij} represents the reaction rate coefficient in $cm^3\,s^{-1}$ between species i and j. It is a function of temperature, and is usually expressed as:

$$k = \gamma \cdot \left(\frac{T}{300}\right)^{\alpha} \cdot \exp\left(-\frac{\beta}{T}\right) \quad . \tag{2}$$

The corresponding values are obtained from laboratory experiments or theoretical calculations. A database often used in astrochemical modeling is maintained by the UMIST group and the corresponding values can be found following Le Teuff *et al.* (2000).

In dark clouds, k'_i, in s^{-1}, represents the ionization rate through cosmic rays or the dissociation induced by the secondary UV photons emitted by electronically excited H_2. In diffuse or translucent clouds, k'_i is the photodissociation or photoionization probability due to the penetrating interstellar ultraviolet photons.

Solving these coupled equations for a set of appropriate initial conditions allows a derivation of the time evolution of the abundances and finally the steady state values when $\frac{d}{dt}$ is equal to 0. It is well known that the corresponding equations are stiff since the chemical time scales reach very different values. Another way to obtain steady state values is to directly solve the equations

$$\frac{dX_n}{dt} = 0 \quad . \tag{3}$$

It has been recognized by Le Bourlot *et al.* (1993) that this is an open non-linear chemical system and that the steady state solutions may depend on their initial conditions. Indeed, Figure 4 displays the time dependent solutions of the fractional ionization for 2 different initial conditions in a cloud model with proton density $n_H = 10^4\,cm^{-3}$ and temperature 10 K. It is clearly seen on this figure that two different steady states may be obtained, corresponding to very different values of

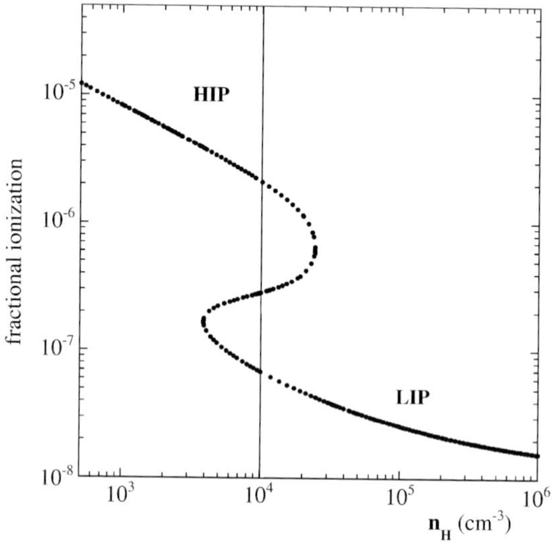

Figure 3. Calculated fractional ionization at steady state for densities between 500 and 10^6 cm^{-3}, $T = 10$ K and $\zeta = 5 \times 10^{-17}$ s^{-1}. The vertical line is drawn at $n_H = 10^4$ cm^{-3}.

the fractional ionization. All variables display the same properties. These solutions have been labeled the High Ionization Phase (HIP) and the Low Ionization Phase (LIP) as the fractional ionization property is critical to drive the chemistry. In the HIP, radicals, atomic ions and unsaturated species are favored whereas the opposite occurs in the LIP where large abundances of saturated molecular ions (H_3^+, H_3O^+, ..) and molecules such as H_2O, CO, NH_3, O_2, ... are predicted. Figure 3 displays on the other hand the ensemble of steady state values of the ionization fraction for different values of n_H. Each point in this figure corresponds to one chemical model. Three different solutions are found for proton densities between 5000 cm^{-3} and 2×10^4 cm^{-3}. The vertical line shows the steady state results for $n_H = 10^4$ cm^{-3}, the value for which time evolution is displayed (Figure 4). The two extreme values correspond respectively to the previous HIP and LIP solutions (which is mandatory from the numerical point of view), whereas the median value represents an unstable solution which is not reachable with a time dependent code. This Figure represents an hysteresis curve, outlining a quite general behaviour found in non-linear systems sometimes referred to *bistability*. The interesting feature is that this bistable behaviour occurs for density, temperature, and ionization rate conditions relevant to dark interstellar clouds. Whether this phenomenon really occurs in the ISM is not demonstrated yet. However, this possibility should be retained at least for theorists. The two chemical phases have very different behaviours concerning deuterium fractionation. In the LIP phase, the H_3^+ ion is predicted to be abundant and deuterium fractionation may proceed efficiently. This is no more the case in the HIP phase. Table II gives the predicted fractionation ratios for the two steady state solutions corresponding to the same value of the proton density. It should be

TABLE II

Deuterium fractionation (XD/XH) for 2 models at $n_H = 10^4$ cm^{-3}.

Deuterium fractionation	LIP	HIP
H_2D^+/H_3^+	2.1(-2)	5.8(-3)
DCO^+/HCO^+	9.5(-3)	1.5 (-3)
N_2D^+/N_2H^+	7.4(-3)	1.6(-3)
C_2D/C_2H	6.9(-2)	5.4(-3)
DCN/HCN	2.9(-2)	3.4(-3)
HDO/H_2O	1.2(-3)	1.7(-4)
NH_2D/NH_3	2.3(-2)	1.4(-3)

kept in mind that the steady state fractional abundances of the main isotopomers are also very different in the two phases.

The sensitivity to the initial conditions is even more drastic when one focuses on non steady state solutions. Such early time solutions are sometimes invoked to explain the abundances of specific molecules such as carbon chains (e.g., Herbst and Leung, 1989). Figure 4 displays the evolution of the NH_2D/NH_3 ratio for 4 different initial conditions. It is clear from this figure that taking the value before steady state is achieved may lead to large errors. The chemical time scale should be compared to the dynamical free-fall time, of the order of 43 / $\sqrt{n_H}$ My, where n_H is expressed in cm^{-3}, and to the accretion time on dust estimated to $3/n_H$ Gy. The presence of a magnetic field allows the cloud to resist gravitational free fall.

5. Conclusion

For many interstellar molecules detected so far, deuterated counterparts, including a triply deuterated species, have now been detected. These species bring valuable information on the chemical processes and gas evolution in cold dense cores and star forming regions. The link with the temperature and density conditions and with the amount of depletion of oxygen and other heavy elements is significant. To extract the full potential of this interesting information, care must be taken at all steps in analysing the data and modeling the sources. Steady state models, which ignore the problem of initial conditions, may be preferred for simple analysis. The use of time-dependent models should be restricted to cases for which the initial conditions are pretty well constrained. Two important processes, i) the evolution of the cloud structure itself (density, temperature) accompanying the formation of a dense core, and then a protostar, and ii) the condensation of abundant species on dust grains, act on similar time scales as the chemistry, $\approx 10^5$ years. Including

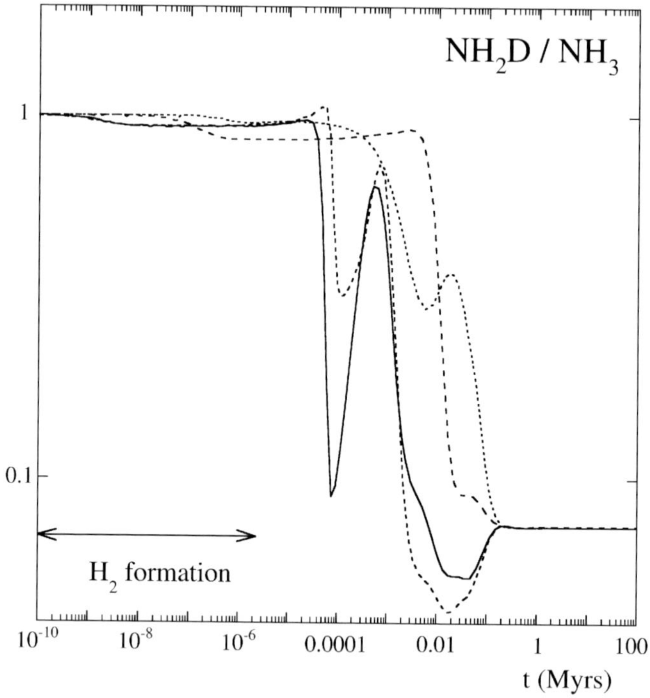

Figure 4. Time evolution solutions of the NH_2D / NH_3 ratio for a dark cloud model with $n_H = 2 \times 10^5$ cm^{-3}, $T = 10$ K and $\zeta = 5 \times 10^{-17}$ s^{-1} for 4 different initial conditions.

these processes in time-dependent models represents a major challenge for future studies.

Acknowledgements

We want to thank the referees for their careful reading and their comments. This work has benefited from the support of the Programme National "Physico-Chimie du Milieu Interstellaire" supported by INSU and CNRS.

References

Bell, M.B., Avery, L.W., Matthews, H.E., Feldman, P.A., Watson, J.K.G., Madden, S.C., and Irvine, W.M.: 1988, 'A Study of C_3HD in Cold Interstellar Clouds', *Astrophys. J.* **326**, 924–930.

Bertoldi, F., Timmermann, R., Rosenthal, D., Drapatz, S., and Wright, C.M.: 1999, 'Detection of HD in the Orion Molecular Outflow', *Astron. Astrophys.* **346**, 267–277.

Brogan, C.L., and Troland, T.H.: 2001, 'Very large array H I Zeeman observations toward the W49 complex', *Astrophys. J.* **550**, 799–816.

Butner, H.M., Lada, E.A., and Loren, R.B.: 1995, 'Physical Properties of Dense Cores: DCO^+ Observations', *Astrophys. J.* **448**, 207–225.

Caselli, P., Walmsley, C.M., Zucconi, A., Tafalla, M., Dore, L., and Myers, P.C.: 2002, 'Molecular Ions in L1544. II. The Ionization Degree', *Astrophys. J.* **565** , 344–358.

Caux, E., Ceccarelli, C., Pagani, L., Maret, S., Castets, A., and Pardo, J.R.: 2002, 'HD 112 m in Absorption and Extreme CO Depletion in a Cold Molecular Cloud', *Astron. Astrophys.* **383**, L9–L13.

Gerin, M., Wootten, H.A., Combes, F., Boulanger, F., Peters, W.L. III, Kuiper, T.B.H., Encrenaz, P.J., Bogey, M.: 1987, 'Deuterated C_3H_2 as a Clue to Deuterium Chemistry', *Astron. Astrophys.* **173**, L1–L4.

Gerin, M., Combes, F., Wlodarczak, G., Encrenaz, P., and Laurent, C.: 1992a, 'Interstellar Detection of Deuterated Methyl Acetylene', *Astron. Astrophys.* **253**, L29–L33.

Gerin, M., Combes, F., Wlodarczak, G., Jacq, T., Guélin, M., Encrenaz, P., and Laurent, C.: 1992b, 'Interstellar Detection of Deuterated Methyl Cyanide', *Astron. Astrophys.* **259**, L35–L38.

Gerin, M., Pearson, J.C., Roueff, E., Falgarone, E., and Phillips, T.G.: 2001, 'Determination of the Hyperfine Structure of N_2D^+', *Astrophys. J.* **551**, L193–L197.

Gordy, W., and Cook, R.L.:1984, *Microwave Molecular Spectra*, 3rd Edition, John Wiley and Sons publisher.

Hatchell, J., Roberts, H., and Millar, T.J.: 1999, 'Limits on HDS/H_2S Abundance Ratios in Hot Molecular Cores', *Astron. Astrophys.* **346**, 227–232.

Helmich, F.P., van Dishoeck, E.F., and Jansen, D.J.: 1996, 'The Excitation and Abundance of HDO Toward W3(OH)/(H_2O)', *Astron. Astrophys.* **313**, 657–663.

Herbst, E., and Leung, C.M.: 1989, 'Gas-phase Production of Complex Hydrocarbons, Cyanopolyynes, and Related Compounds in Dense Interstellar Clouds', *Astrophys. J. Suppl.* **69**, 271–300.

Herbst, E.: 2003, 'Isotopic Fractionation by Ion-molecule Reactions', *Space Sci. Rev.*, this volume.

Hirota, T., Yamamoto, S., Mikami, H., and Ohishi, M.: 1998, 'Abundances of HCN and HNC in Dark Cloud Cores', *Astrophys. J.* **503**, 717–728.

Hirota, T., Ikeda, M., and Yamamoto, S.: 2001, 'Observations of DNC and $HN^{13}C$ in Dark Clouds', *Astrophys. J.* **547**, 814–828.

Jacq, T., Walmsley, C.M., Mauersberger, R., Anderson, T., Herbst, E., and de Lucia, F.C.: 1993, 'Detection of Interstellar CH_2DOH', *Astron. Astrophys.* **271**, 276–281.

Le Bourlot, J., Pineau des Forêts, G., Roueff, E., and Schilke, P.: 1993, 'Bistability in Dark Cloud Chemistry', *Astrophys. J* **416**, L87–L90.

Le Petit, F., Roueff, E., and Le Bourlot, J.: 2002, 'D/HD transition in Photo-dominated Regions (PDR)', *Astron. Astrophys.* **390**, 369–381.

Le Teuff, Y.H., Millar, T.J., and Markwick, A.J.: 2000, 'The UMIST Database for Astrochemistry 1999', *Astron. Astrophys. Suppl.* **146**, 157–168.

Lis, D.C., Gerin, M., Phillips, T.G., and Motte, F.: 2002a, 'The Role of Outflows and C Shocks in the Strong Deuteration of L1689N', *Astrophys. J.* **569** , 322–333.

Lis, D., Roueff, E., Gerin, M., Phillips, T.G., Coudert, van der Tak, F.F.S., and Schilke, P.: 2002, 'Detection of Triply Deuterated Ammonia in the Barnard 1 Cloud', *Astrophys. J.* **571** , L55–L58.

Loinard, L., Castets, A., Ceccarelli, C., Caux, E., and Tielens, A.G.G.M.: 2001, 'The Enormous Abundance of D_2CO in IRAS 16293-2422', *Astrophys. J.* **552**, L163–L166.

Loinard, L., Castets, A., Ceccarelli, C., Tielens, A.G.G.M., Faure, A., Caux, E., and Duvert, G.: 2000, 'Doubly Deuterated Molecular Species in Protostellar Environments', *Astron. Astrophys.* **359**, 1169–1174.

McLeod, J.M., Avery, L.W., and Broten, N.W.: 1981, 'Detection of Deuterated Cyanodiacetylene DC_5N in Taurus Molecular Cloud 1', *Astrophys. J.* **251**, L33–L36.

Mauersberger, R., Henkel, C., Jacq, T., and Walmsley, C.M.: 1988, 'Deuterated Methanol in Orion', *Astron. Astrophys.* **194**, L1–L4.

Minowa, H., Satake, M., Hirota, T., Yamamoto, S., Ohishi, M., and Kaifu, N.: 1997, 'Laboratory Microwave Spectroscopy of HDCS and Its Astronomical Detection Toward TMC-1', *Astrophys. J.* **491**, L63–L66.

Morton, D.C.: 1975, 'Interstellar Absorption Lines in the Spectrum of Zeta Ophiuchii', *Astrophys. J.* **221**, 85–115.

Petitjean, P., Srianand, R., and Ledoux, C.: 2000, 'Molecular Hydrogen and the Nature of Damped Lyman-α Systems', *Astron. Astrophys.* **364**, L26–L30.

Rachford, B.L., Snow, T.P., Tumlinson, J., Shull, J.M., Roueff, E., Andre, M., Desert, J.-M., Ferlet, R., Vidal-Madjar, A., and York, D.G.: 2000, 'Far Ultraviolet Spectroscopic Explorer Observations of Molecular Hydrogen in Translucent Interstellar Clouds. II. The Line of Sight Toward HD 110432', *Astrophys. J.* **555**, 839–849.

Roberts, H., Fuller, G.A., Millar, T.J., Hatchell, J., and Buckle, J.V.: 2002, 'A Survey of [HDCO]/[H2CO] and [DCN]/[HCN] Ratios Towards Low-mass Protostellar Cores', *Astron. Astrophys.* **381**, 1026–1038.

Roueff, E., Tiné, S., Coudert, L.H., Pineau des Forêts, G., Falgarone, E., and Gerin, M.: 2000, 'Detection of Doubly Deuterated Ammonia in L134N', *Astron. Astrophys.* **354**, L63–L66.

Roueff, E., *et al.*: 2002, 'The Chemistry of Triply Deuterated Ammonia', *Astron. Astrophys.*, in preparation.

Saito, S., Ozeki, H., Ohishi, M., and Yamamoto, S.: 2000, 'Observations of NH_2D Toward Dark Molecular Clouds', *Astrophys. J.* **535**, 227–230.

Schloerb, F.P., Snell, R.L., Young, J.S., and Langer, W.D.: 1981, 'Detection of Deuteriocyanobutadiyne, DC_5N, in the Interstellar Cloud TMC-1', *Astrophys. J.* **251** L37–L41.

Shah, R.Y., and Wootten, A.: 2001, 'Deuterated Ammonia in Galactic Protostellar Cores', *Astrophys. J.* **554**, 933–947.

Stark, R., van der Tak, F.F.S., and van Dishoeck, E.F.: 1999, 'Detection of Interstellar H_2D^+ Emission', *Astrophys. J.* **521**, L67–L70.

Tiné, S., Roueff, E., Falgarone, E., Gerin, M., and Pineau des Forêts, G.: 2000, 'Deuterium Fractionation in Dense Ammonia Cores', *Astron. Astrophys.* **356**, 1039–1049.

Turner, B.E.: 2001, 'Deuterated Molecules in Translucent and Dark Clouds', *Astrophys. J. Suppl.* **136**, 579–629.

Turner, B.E.: 1990, 'Detection of Doubly Deuterated Interstellar Formaldehyde (D_2CO) – an Indicator of Active Grain Surface Chemistry', *Astrophys. J. Lett.* **347**, L29–L33.

Turner, B.E.: 1989, 'Detection of Interstellar C_4D - Implications for Ion-molecule Chemistry', *Astrophys. J.* **347**, L39–L42.

van Dishoeck, E.F., Blake, G.A., Jansen, J., and Groesbeck, T.D,: 1995, 'HD Molecular Lines in an Absorption System at Redshift z=2.3377', *Astrophys. J.* **136**, 579–629.

van der Tak, F.F.S, Schilke, P., Müller, H.S.P., Lis, D.C., and Phillips T.G.: 2002, 'Triply Deuterated Ammonia in NGC 1333', *Astron. Astrophys.*, in press.

Varshalovich, D.A., Ivanchik, A.V., Petitjean, P., Srianand, R., and Ledoux, C.: 2001, 'Molecular Abundances and Low-Mass Star Formation. II. Organic and Deuterated Species Toward IRAS 16293-2422', *Astronomy Let.* **27**, 683–685.

Wright, E.L., and Morton, D.C.: 1979, 'Rotationally Excited HD Toward Zeta Ophiuchii', *Astrophys. J.* **227**, 483–488.

Wright, C.M., van Dishoeck, E.F., Cox, P., Sidher, S.D., and Kessler, M.F.: 1999, 'Infrared Space Observatory-Long Wavelength Spectrometer Detection of the 112 Micron HD $J = 1 -> 0$ Line Toward the Orion Bar', *Astrophys. J.* **515**, L29–L33.

Address for Offprints: E. Roueff, LUTH and FRE2462 du CNRS, Observatoire de Paris, Place Janssen, 92190 Meudon cedex, France; Evelyne.roueff@obspm.fr

DEUTERIUM FRACTIONATION IN INTERSTELLAR CLOUDS

T. J. MILLAR (tom.millar@umist.ac.uk)
Physics Department, UMIST, PO Box 88, Manchester M60 1QD, UK

Received: 2 August 2002; Accepted in final form: 28 August 2002

Abstract. This paper reviews the chemical processes responsible for fractionating deuterium in interstellar molecules. I show that this process is intrinsically a low temperature phenomenon and discuss how the degree of enhancement of the deuterium content of molecules is related to the physical conditions, particularly abundances, in molecular clouds. If significant amounts of abundant species, such as CO, are frozen out on to interstellar dust grains, the resulting enhancement in H_2D^+ can result in its abundance being greater than that of H_3^+ at 10 K. Transfer of the deuteron from H_2D^+ can then lead to the efficient formation of multiply deuterated species, such as NHD_2 and ND_3. Fractionation can also occur in grain surface reactions and some simple models are discussed.

Keywords: deuterium, interstellar clouds

1. Introduction

In the last 30 years over 20 deuterated molecules have been detected in interstellar clouds, with abundance ratios relative to their hydrogenated analogues much larger than the cosmic D/H ratio of around 1.5×10^{-5} (Table I). It was recognised very early on (Watson, 1974), that this was due to chemical reactions extracting deuterium from its reservoir, HD, in molecular clouds. Small zero-point energy differences lead to a situation in which molecules prefer to incorporate a D atom rather than an H atom. The article by Herbst in this volume discusses these issues in much more detail. Here I will discuss the implications of these **fractionation** reactions for particular molecules and emphasise how they lead to abundance ratios which can be used to derive information on the physical conditions and chemical evolution of interstellar clouds.

TABLE I

Deuterated molecules detected in interstellar clouds

HD	HDO	HDS	DCN	DNC	C_2D
H_2D^+	DCO^+	N_2D^+	NH_2D	NHD_2	ND_3
HDCO	D_2CO	HDCS	CH_3OD	CH_2DOH	CHD_2OH
CH_2DCCH	CH_3CCD	CH_2DCN	DC_3N	DC_5N	C_4D

Space Science Reviews **106:** 73–86, 2003.

2. Fractionation Reactions

The elements H and D are mainly (likely greater than 95%) in the form of H_2 and HD in molecular clouds, with $[HD]/[H_2] = 2$ D/H, where D/H is the cosmic ratio and [X] is the abundance of species X per unit volume. In what follows the abundance ratio [XD]/[XH] will be written as $R(XD)$ and I define an enhancement ratio, $S(T)$, due to chemical reactions, as

$$R(\mathrm{XD}) = S(T)\frac{[\mathrm{HD}]}{[\mathrm{H_2}]}$$

in which, as I shall demonstrate, the enhancement ratio is a function of temperature.

There are three **primary** reactions involving HD and abundant interstellar ions which lead to fractionation

$$\mathrm{H_3^+} + \mathrm{HD} \longleftrightarrow \mathrm{H_2D^+} + \mathrm{H_2} + \Delta E_1\,, \quad (1)$$
$$\mathrm{CH_3^+} + \mathrm{HD} \longleftrightarrow \mathrm{CH_2D^+} + \mathrm{H_2} + \Delta E_2\,, \quad (2)$$
$$\mathrm{C_2H_2^+} + \mathrm{HD} \longleftrightarrow \mathrm{C_2HD^+} + \mathrm{H_2} + \Delta E_3\,, \quad (3)$$

where the reaction exoergicities, although small, in the range 220-550 K, are greater than the typical temperatures of cold interstellar clouds. However, the overwhelming abundance of H_2 compared to HD ensures that the reverse reactions in each of reactions (1)–(3) become important once the cloud temperature is greater than critical temperatures of about $\Delta E_i/10$. At temperatures larger than these critical temperatures, the reverse reactions dominate and $R(\mathrm{XD})$ approaches $R(\mathrm{HD})$. Thus, for example, H_2D^+ is inefficient in fractionation once the cloud temperature is greater than about 22 K.

The deuterated ions formed in reactions (1)–(3) are as reactive as their hydrogenated analogues and can therefore pass on, through chemical reactions, their enhanced D content to other species.

As an example, consider the interstellar chemistry of H_2D^+, formed in the forward direction of reaction (1) and destroyed by the reverse channel, as well as by dissociative recombination with electrons and by reactions with neutral species. Note that the proton (deuteron) affinity of H_3^+ (H_2D^+) is very small so that it is very reactive. Under the assumption of steady-state and equating formation and destruction of H_2D^+, one finds that

$$\frac{[\mathrm{H_2D^+}]}{[\mathrm{H_3^+}]} = \frac{k_{1f}}{k_{1r} + \alpha_1 f(e) + \Sigma k_{\mathrm{M}} f(\mathrm{M})}\frac{[\mathrm{HD}]}{[\mathrm{H_2}]}\,,$$

in which $f(\mathrm{X})$ is the fractional abundance of X relative to H_2, α_1 is the dissociative recombination coefficient of H_2D^+ with electrons, and k_{M} is the rate coefficient of H_2D^+ with species M. One can re-write this equation as

$$R(\mathrm{H_2D^+}) = S_{\mathrm{H_2D^+}}(T) f(\mathrm{HD})\,.$$

Since the rate coefficients, particularly k_{1r} and α_1, are temperature dependent – $k_{1r} = k_{1f} \exp(-\Delta E_1/T)$ and $\alpha_1 \propto 1/\sqrt{T}$ – the enhancement factor is a function of temperature.

Reaction (1) is a complex system, discussed more fully in this volume by Herbst, in which the ortho-para form of H_2 plays a role. Recent experimental studies by Gerlich have shown that k_{1f} has a value of 1/5 that of the collisional (Langevin) rate coefficient. Astrophysical implications of these new laboratory results have been discussed by Roberts *et al.* (2002). In essence, the reduction on k_{1f} leads to a reduction in the enhancement of H_2D^+, and species derived from it, by a factor of five. Since the origin of the discrepancy in rate coefficients is not understood at the present time, I present results here using the Langevin rate coefficient.

Primary fractionation can be transferred to other species producing secondary fractionation. For example,

$$H_2D^+ + N_2 \longrightarrow N_2D^+ + H_2\,, \tag{4}$$

$$H_2D^+ + N_2 \longrightarrow N_2H^+ + HD\,, \tag{5}$$

where a statistical branching ratio of 1:2 is adopted for reactions (6) and (7) respectively. Thus $R(N_2D^+) = R(H_2D^+)/3$. Because of such statistical branching ratios, secondary fractionation always leads to enhancements that are less than those arising from primary fractionation.

H_3^+ is generally more abundant than either CH_3^+ or $C_2H_2^+$, so that fractionation via H_2D^+ generally dominates at low temperature, roughly $T < 20$ K. At higher temperatures, reaction (1r) removes H_2D^+ very rapidly and CH_2D^+ and C_2HD^+ become more important in driving the fractionation. For temperatures above 60 K, the reverse reactions in (1)–(3) dominate loss of the deuterated ions and fractionation ratios tend toward the [HD]/[H_2] ratio. Thus gas-phase fractionation via the difference between forward and backward rate coefficients is intrinsically a low-temperature phenomenon.

As mentioned previously, the enhancement factor is a function of temperature as well as a function of the fractional ionisation, $f(\mathrm{e})$, and of $\Sigma f(\mathrm{M})$. The fractional ionisation is an important parameter in low-mass star formation since at low ionisation the magnetic field can decouple from the gas so that magnetic pressure cannot prevent collapse. Chemical models show that the ionisation can be related to the abundance of the metal ions, such as Mg^+, Fe^+ and Na^+. Similarly, the species M includes species such as O, C, CO, N_2, H_2O, NH_3. In fact, almost all species containing the abundant elements O, C, and N – with the exception of N and O_2 – react with H_3^+ and H_2D^+. Thus, in principle, observations of deuterated species can provide information on the abundances of these elements.

In Figure 1, I present the enhancement factors for H_2D^+, CH_2D^+ and C_2HD^+ as functions of temperature for two values of $f(\mathrm{e})$ – 10^{-6} (Figure 1, *left*) and 10^{-8} (Figure 1, *right*) – and two values of $\Sigma k_\mathrm{M} f(\mathrm{M}) = 10^{-13}$ cm^3 s^{-1} (*solid lines*) and 10^{-15} cm^3 s^{-1} (*dashed lines*). The latter value is typical of values expected in cold interstellar clouds in which the major gas-phase molecules have frozen out on to

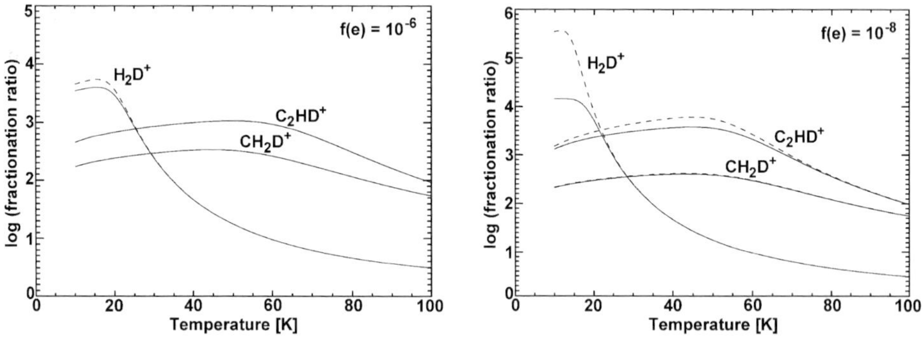

Figure 1. Left: The enhancement factors for H_2D^+, $CH_2D^+$and C_2HD^+ are shown as a function of temperature for a fixed electron fraction of 10^{-6}. Solid lines are for 'normal' molecular abundances, $\Sigma k_M f(M) = 10^{-13}$ cm^3 s^{-1} and dashed lines are for heavily depleted abundances, $\Sigma k_M f(M) = 10^{-15}$ cm^3 s^{-1}. *Right:* The same, but with a fractional electron abundance of 10^{-8}.

the surfaces of dust grains. One sees that, at low temperatures, low ionisations and low molecular abundances, the enhancement factor for H_2D^+ is greater than 10^5, that is, the $[H_2D^+]/[H_3^+]$ abundance ratio is greater than one. Clearly, in this situation, the fractionation in molecules formed from H_2D^+ can be very large, even given the dilution caused by secondary reactions as mentioned above. Figure 1 shows that the enhancements for CH_2D^+ and C_2HD^+ are not as large as that for H_2D^+ at low temperatures but are larger for T greater than about 25 K. Unlike H_2D^+, the fractionation in these two ions does not increase exponentially at low temperature. This behaviour results from the fact that both ions are destroyed in radiative association reactions with H_2, with rate coefficients which have an inverse temperature dependence. The slight decrease in the fractionation in H_2D^+ below 15 K seen in Figure 1 (*left*), is due to dissociative recombination which is important only in the high ionization case.

To date, astronomers have generally used $[DCO^+]/[HCO^+]$ to derive limits on $f(e)$ (Williams *et al.*, 1998), although $[N_2D^+]/[N_2H^+]$ has a much simpler relation to $[H_2D^+]/[H_3^+]$ since it can only be made from these species. The chemistry of DCO^+ is not simple since, in addition to deuteron transfer with H_2D^+, it can also be formed from CH_2D^+:

$$CH_2D^+ + O \longrightarrow DCO^+ + H_2 .$$

The main advantages of using DCO^+ are that CO is more abundant than N_2, collisional cross-sections for understanding the excitation of the observed lines are known, and there are no hyperfine transitions to consider. On the other hand lines are likely to be optically thick.

3. The Chemical Model

In the past ten years or so, many detailed chemical kinetic models of interstellar cloud evolution have been developed. It is generally accepted that, while H_2 is formed on dust grains, gas-phase processes dominate the formation of HD,

$$H^+ + D \longleftrightarrow D^+ + H\,, \tag{6}$$

$$D^+ + H_2 \longrightarrow HD + H^+\,, \tag{7}$$

at a rate, in diffuse clouds, about 100 times faster than that of H_2 on grains. This does not lead to an HD/H_2 ratio greater than cosmic in such clouds, since self-shielding of H_2 ensures that its photodissociation rate coefficient is about 1000 times less than that of HD.

Until recently, the received wisdom has been that gas-phase chemistry is responsible for the formation of all molecules detected in cold interstellar clouds such as Taurus Molecular Cloud 1 (TMC-1) – see Millar and Freeman (1984) and Leung *et al.* (1984). These models explicitly neglected any role for dust other than providing a source of extinction, thereby protecting molecules against photodissociation. In particular, the gas-grain collisional time-scale is approximately given by

$$t_{\rm coll} = (\pi a^2 n_d v)^{-1}\,,$$

where a is the (spherical) grain radius, n_d is the number of dust grains per unit volume and v is the velocity of the gas-phase species. Using typical values gives

$$t_{\rm coll} = 3 \times 10^9/n({\rm cm}^{-3}) \quad {\rm years}\,,$$

that is, on the same order as the time-scales for chemical evolution and star formation in these clouds. At 10 K, most collisions between neutral gas and the cold grains lead to accretion, that is, the formation of ice-type mantles. As discussed in Section 2, this accretion will lead to an enhancement of D fractionation, and depending on the degree of the depletion can lead to enhancement factors larger than 10^5 in H_2D^+. This enhancement can then be transferred to other gas-phase species leading to very large fractionation ratios.

Since chemical, accretion and dynamical time-scales are similar in cold, interstellar dust clouds, one needs to calculate fractionation as a function of time for a particular set of initial conditions using a chemical kinetic code. At UMIST we have, over a number of years, created the **UMIST Database for Astrochemistry** (see www.rate99.co.uk) which contains both rate coefficients for gas-phase reactions of importance in the interstellar clouds, but also computer codes for the creation and solution of time-dependent kinetics. The database does not implicitly include all reactions containing D but only those that are primary reactions. The generation of a deuterated analogue database is carried out by software. As an example of how the code works, consider the formation of formaldehyde, H_2CO, in the neutral-neutral reaction, contained in the database,

$$CH_3 + O \longrightarrow H_2CO + H\,,$$

with a rate coefficient whose numerical value is listed in the database but here denoted by k. The code that writes the analogous deuterated reaction replaces an H atom in a reactant by a D atom, and similarly for the products. Thus the above reaction generates **two** equivalent reactions

$$CH_2D + O \longrightarrow H_2CO + D\,, \tag{8}$$

$$CH_2D + O \longrightarrow HDCO + H\,. \tag{9}$$

Finally, the rate coefficient k has to be divided between these two channels. If experimental information is available, then that is used; otherwise the statistical branching ratios are applied. In this case the channel producing HDCO has a rate coefficient of $2k/3$. It should be noted that the adoption of statistical branching ratios may not be correct. This now appears to be the case in a number of dissociative recombination reactions of certain deuterated ions, such as H_2D^+ (Larsson *et al.*, 1996) and HD_2O^+ (Jensen *et al.*, 2000).

This above reaction is straightforward but more complex systems are not so simple and some subjective rules must be applied. Consider the following reaction which is involved in the chemistry of methanol, CH_3OH, in cold clouds:

$$CH^+ + CH_3OH \longrightarrow CH_3OH_2^+ + C\,.$$

This gives rise to four equivalent reactions when an H-atom is replaced by a D-atom in methanol:

$$CH^+ + CH_2DOH \longrightarrow CH_2DOH_2^+ + C\,, \tag{10}$$

$$CH^+ + CH_2DOH \longrightarrow CH_3OHD^+ + C\,, \tag{11}$$

$$CH^+ + CH_3OD \longrightarrow CH_2DOH_2^+ + C\,, \tag{12}$$

$$CH^+ + CH_3OD \longrightarrow CH_3OHD^+ + C\,. \tag{13}$$

Branching ratios need to be applied to these four channels. One possibility, adopted in the main in the work of the UMIST group, is to exclude channels in which the deuteron changes functional groups within the molecule during the reaction. In this case, the middle two reactions would be excluded. Finally, one needs to add the reactions for which there are no H analogues, in particular the primary fractionation reactions and reactions involving atomic D, which also lead to fractionation. The end result is a chemistry in which the number of species is about twice that in the deuterium-free case and the number of reactions is about three times larger. Typical models of D chemistry in cold clouds contain around 10,000 reactions and about 500 species. The resulting ODEs which have to be integrated to obtain the time-evolution of the abundances are extremely stiff and need to be solved using special numerical codes, such as CVODE (Cohen and Hindmarsh, 1996).

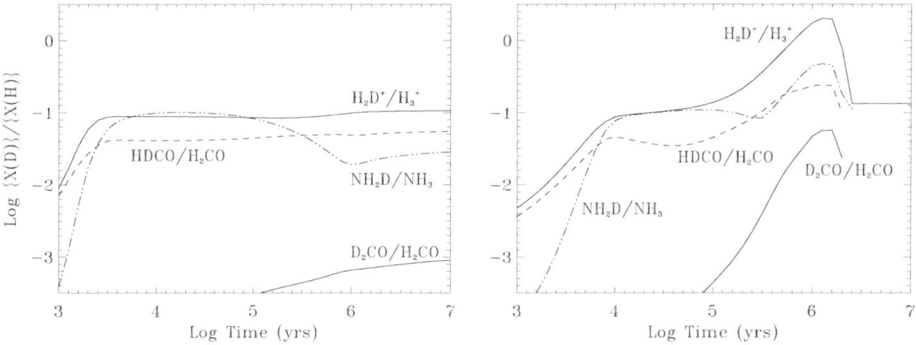

Figure 2. A comparison of selected molecular D/H ratios predicted by our gas-phase (*left*) and depletion (*right*) models. $T_{kin} = 10$ K, $n(H_2) = 10^4$ cm^{-3} (Roberts and Millar, 2000a).

4. Results

Figure 2 (*left panel*) shows the fractionation ratios of a number of species in a cold interstellar cloud. A chemical steady-state is reached in this particular case in about one million years and the fractionation ratios are fairly constant once the chemistry has evolved for about 10^4 years. However, as mentioned above, accretion onto cold grains occurs on a time-scale of less than 10^6 years. This accretion can be included by adding a loss term to the ODEs of those species that accrete. The resulting evolution, using standard grain parameters, is shown in the right hand panel of Figure 2. It is clear that, in this case, the neutral species are frozen out on to the dust in less than one million years. The ions are assumed not to stick to the dust but rather to be neutralised and returned to the gas following collision with the negatively charged grains. Once accretion begins to become important, the ratios can become very large, with $R(H_2D^+)$ larger than unity. At times longer than 10^6 y, the fractionation falls as dissociative recombination dominates. Note that at these times, the electron fraction becomes rather large, with the major positive ions being H^+ and He^+, and the decrease in the H_2D^+, and other, fractionations reflects this. Note that, beyond a few times 10^6 y, the abundances of all neutral species become vanishingly small and the fractionation ratios have no meaning.

These extremely large ratios make it possible for species to be multiply deuterated. Roberts and Millar (2000b) have presented the results of a model containing NHD_2, D_2O, D_2CO and their related species. The calculated abundances and fractionation ratios were compared with observations of the cold cloud, L134N, in which NHD_2 has been observed (Table II). Good agreement between observation and theory is obtained at a time of 10^5 years. Rodgers and Charnley (2001) presented an analytic model of multiple fractionation, including the formation of NHD_2 and showed how gas-phase and grain surface formation of these species lead to different fractionation ratios which can be used to distinguish between them. The grain surface synthesis of deuterated molecules will be discussed in Section 4.1.

TABLE II

A comparison of abundance ratios measured in L134N (Tiné *et al.*, 2000; Roueff *et al.*, 2000) with predictions from models at 10 K, $n(H_2) = 5 \times 10^4$ cm^{-3} (Roberts and Millar, 2000b).

Ratios	Observed	Modeled		
		gas-phase; 10^5 y	gas-phase; 10^7 y	accretion; 10^5 y
DCO^+/HCO^+	1.8×10^{-1}	1.6×10^{-2}	4.7×10^{-2}	1.6×10^{-1}
DCO^+/CO	7.7×10^{-6}	1.1×10^{-7}	2.5×10^{-7}	2.5×10^{-5}
NH_2D/NH_3	1.0×10^{-1}	7.5×10^{-2}	2.0×10^{-2}	1.6×10^{-1}
NH_2D/CO	1.5×10^{-4}	3.5×10^{-5}	4.4×10^{-6}	1.2×10^{-4}
N_2D^+/N_2H^+	3.5×10^{-1}	2.3×10^{-2}	3.9×10^{-2}	1.8×10^{-1}
N_2D^+/CO	3.5×10^{-6}	5.8×10^{-9}	1.1×10^{-8}	2.7×10^{-6}
NHD_2/NH_3	5.0×10^{-3}	1.3×10^{-4}	8.4×10^{-5}	4.1×10^{-3}
NHD_2/CO	7.5×10^{-6}	6.3×10^{-8}	1.8×10^{-8}	3.1×10^{-6}

Observational evidence for fractionation enhancements has been increasing in recent years, aided by the development of submillimetre cameras which can detect cold, dust clumps in interstellar clouds. Molecular line observations of some of these clumps show that CO is often highly depleted (Caselli *et al.*, 1999; Stark *et al.*, 1999; Gibb and Little, 1998). Observations of these same clumps indicate that the fractionation ratios of DCO^+/HCO^+ are several times larger than normal for many cold clouds, indicating $R(H_2D^+) \approx 1$.

For the purpose of comparing the predictions of purely gas-phase models with observation, I consider the case of TMC-1, a cold (10 K), dust cloud which contains many complex, linear chain hydrocarbon molecules. This object is nearby and has been observed extensively. Both molecular abundances and fractionation ratios vary as a function of position along this ridge-shaped molecular cloud, a point to which I shall return later in this article. Table III lists the observed fractionation ratios at the cyanopolyyne (CP) peak in this object and compares these to those calculated at steady-state (Roberts and Millar, 2000a). In general, agreement between observation and theory is reasonable, to within a factor of three or so; notable exceptions are C_3HD, DC_3N, and NH_2D, although the latter two are in good agreement at later times.

4.1. The Role of Grains

As mentioned, dust grains can affect fractionation indirectly through accretion of neutral molecules which would otherwise destroy H_2D^+. This can result in a ratio $R(H_2D^+) > 1$. This enormous fractionation can then be passed, via proton transfer reactions, to other species, before all neutral species freeze out on to the dust grains. The large enhancements enable multiply deuterated molecules to form (Section 3).

TABLE III

A comparison of abundance ratios measured in TMC-1 (CP) with predictions from gas-phase models at 10 K, $n(H_2) = 5\times10^4$ cm^{-3} at a time of 10^5 y (Roberts and Millar, 2000a).

Species	Observed	Model	Species	Observed	Model	Species	Observed	Model
C_3HD	0.08	0.006	DC_3N	0.03	0.007	HDCO	0.005-0.11	0.042
CH_2DCCH	0.11	0.083	DC_5N	0.016	0.023	N_2D^+	0.08	0.025
CCD	<0.03	0.011	DCO^+	0.013	0.019	NH_2D	0.009-0.014	0.084
C_4D	0.004	0.004	DCN	0.023	0.009			
HDCS	0.02	0.040	DNC	0.028	0.015			

Roberts and Millar (2000b) show that at 10 K, accretion of neutrals, and in particular CO, which is the most abundant molecule after H_2, leads to the formation of molecules such as NHD_2 and D_2CO. Their calculations also show that D_2O can be formed in the gas-phase. Despite sensitive searches, this molecule has not yet been detected.

Grain surface reactions may also cause fractionation in a direct manner, that is, surface production may favour deuterium-bearing molecules over their hydrogen-bearing analogues. Rodgers and Charnley (2001) have considered simple analytical models of steady-state gas phase and grain formation of multiply deuterated species, and compared their predictions to the observations of the cold, dark dust cloud L134N, in which NH_2D and NHD_2 have been detected recently (Roueff *et al.*, 2000), with fractionation ratios of 0.1 and 0.005, respectively. The latter value is equivalent to an enhancement factor of $\approx 3 \times 10^7$. Given that the kinetic temperature of this cloud is only 10 K, the observed fractionations, noting that $R(N_2D^+) = 0.45$ in this cloud, are likely to reflect the indirect effect of grains, that is, the situation in which accretion plays a role, rather than the direct route via a grain surface chemistry since it is difficult to return material from the grains to the gas at such low temperatures. Rodgers and Charnley (2001) also considered the chemistry of ND_3 and found that gas-phase and grain-surface reactions predict different fractionation for these species. A simple model for grain surface formation of multiply deuterated molecules was considered by Brown and Millar (1989) and predicts that the abundances of deuterated species scale as: $NH_2D:NHD_2:ND_3$ = 1:0.33:0.11 whereas the steady-state gas-phase values scale as: $NH_2D:NHD_2:ND_3$ = 1:0.49:0.19.

The observed values of NH_2D and NHD_2 in L134N do not allow these two mechanisms to be differentiated – future observation of ND_3 may do so – but the fractionation observed is most likely to be a gas-phase process given the difficulty of removing grain mantle material at 10 K. Rodgers and Charnley show that the magnitude of the enhancement in the ammonia species is consistent with an enhancement factor in H_2D^+ of 10^4, corresponding to a depletion of heavy molecules of a factor of two.

A further indication that accretion may be playing a role in enhancing deuteration has come from the recent detection of ND_3 in B1 (Lis *et al.*, 2002) and NGC 1333 (van der Tak *et al.*, 2002), in which the ratios of ND_3 to NH_3 are $\approx 10^{-4}$, an enhancement of 10^{11} over the cosmic ratio of $(D/H)^3 \approx 10^{-15}$. While the data is not sensitive enough to clearly differentiate between the fractionation ratios predicted by the gas and grain models, surface fractionation requires a high atomic D/H ratio on the grains – and, thus also in the gas since grain material is accreted from the gas. The model of Rodgers and Charnley (2001) requires $R(D) = 0.15$ to reproduce the observed, an enhancement larger by an order of magnitude than predicted by the numerical models (Roberts and Millar, 2000a). In NGC 1333, the depletion of CO is about a factor of 100 (Stark *et al.*, 1999).

Recent surveys have shown that multiply deuterated molecules are widespread in cold (10 K) clouds. In particular, D_2CO is seen in a number of sources (Loinard *et al.*, 2000; Ceccarelli *et al.*, 1998; Roberts and Millar, 2002). Figure 3 shows a spectrum of the 4_{04}-3_{03} transition of D_2CO, at a frequency of 231.41 GHz, toward L1448NW taken with the JCMT (Roberts and Millar, 2002). The strong line in this spectrum is the 3-2 transition of N_2D^+ at 231.32 GHz. Interestingly, these two transitions are not detected in all sources. Two of the objects observed by Roberts and Millar, L1551-IRS2 and L1527-3, show emission from D_2CO but not from N_2D^+. This may indicate that these species form in different ways. N_2D^+, as an ion, cannot be produced directly by surface chemistry, although its fractionation can be enhanced by the accretion process. Eventually, of course, accretion must remove all heavy species, including N_2, from the gas. At this point, N_2D^+ would be undetectable. D_2CO can, in principle, be formed on the grains (Charnley *et al.*, 1997), although its observation in cold clouds presents a problem since in this case it must be returned to the gas phase in order to be detectable through its rotational transitions. How this can be done, while at the same time retaining N_2, is unknown.

In regions of massive star formation, fractionation has been detected in some species at a level much larger than expected from the gas kinetic temperature, $\approx$ 100-300 K. The degree of fractionation must therefore reflect chemical processes which occurred in a previous, colder phase. These regions, called hot molecular cores, are small, $\approx$ 0.1 pc or less, and dense, $\approx 10^6$-10^9 H_2 molecules cm^{-3}. The enhancements, most easily seen in HDO and DCN, are typically a factor of 100 and are consistent with fractionation occurring at temperatures around 40-60 K. This could occur in the gas-phase but the absolute abundances of deuterated species compared to the H_2 abundance are much larger than those detected in cold clouds. This points to an active grain chemistry which produces both hydrogenated and deuterated species.

Models of surface chemistry are very uncertain. The simplest (Brown *et al.*, 1988; Brown and Millar, 1989) considered the conversion of atomic C, N and O on arrival at the grain surface. Since H atoms are as abundant as these species, and

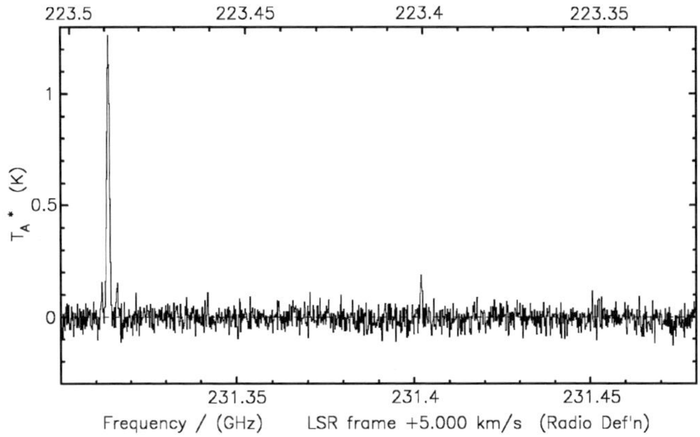

Figure 3. The 231 GHz spectrum of L1448NW taken with the JCMT. The two lines detecetd are D_2CO (at 231.41 GHz) and N_2D^+ (at 231.32 GHz). The vertical scale is T^*_A(K).

since H atoms can rapidly scan the surface by either thermal hopping or by quantum tunneling, the formation of fully hydrogenated species, CH_4, NH_3 and H_2O, is expected to dominate over the formation of species containing two heavy atoms, such as CO. Brown and Millar (1989) made some simple estimates of the formation of multiply deuterated species by incorporating the different arrival rates of H and D atoms – the former have a high velocity – and the probability of incorporating a D rather than an H atom into the final molecule. Thus, for example, the predicted ratio of NH_2D to NH_3 formed on the grain is

$$\frac{[NH_2D]}{[NH_3]} = \frac{3}{\sqrt{2}} \frac{[D]}{[H]},$$

where $\sqrt{2}$ is the ratio of the velocities of D and H atoms and the multiplier 3 arises because a D atom can be inserted into the chain of reactions forming NH_2D at three points - reaction with N, NH and NH_2. A large fractionation on the surface is therefore helped by a large D/H ratio in the gas phase.

This approach, essentially a reaction-diffusion model, may not be correct for a situation in which the number of reactive species per dust grain is less than one, as may often happen in the interstellar medium. In essence, H and D atoms may thermally evaporate before finding a reactive partner. In this case, one must follow the chemistry stochastically, using Monte Carlo, or probabilistically using a master equation approach. These approaches are difficult to implement for large networks of reactions on the grains and much effort has gone into modifying the reaction-diffusion approach to allow such systems to be studied (Caselli *et al.*, 1998; Stantcheva *et al.*, 2001), although comparisions between Monte Carlo and modified rate equations have only been made for small systems.

Caselli *et al.* (2002) have recently investigated a grain surface chemistry with D included. The chemistry is relatively simple, concentrating on H, D, C and O but includes the formation of multiply deuterated water, formaldehyde and methanol.

For surface reactions which involve activation energies, such as H + CO $\rightarrow$ HCO, the equivalent reactions with D atoms occur much more slowly, due to their lower tunneling efficiency. This has most effect on the formation of multiply deuterated species, since for singly deuterated species, there are usually a number of routes to their formation, some of which do not involve activation energy barriers. Caselli *et al.* find that, at high densities of D atoms, which corresponds to dense clouds in which accretion is occurring, multiply deuterated methanol can be produced and find fractionations in CHD_2OH and CH_2DOD larger than that in HDCO. The production of multiply deuterated species, while reasonably efficient at 10 K, falls off rapidly with increasing temperature as D atoms thermally evaporate before tunnelling and reaction. Nevertheless, the recent detection of CHD_2OH by Ceccarelli and colleagues is a significant result. It remains to be seen whether this results from active or passive dust grains.

5. Deuterium in Comets

HDO and DCN are the only deuterated molecules detected to date in comets; the former in the Comets Halley, Hale-Bopp, and Hyakutake with $R(\text{HDO}) = 6 \times 10^{-4}$, and the latter in the coma and jets of Hale-Bopp, with $R(\text{DCN}) = 0.002$ in the coma and about an order of magnitude larger in the dust jets (Blake *et al.*, 1999). The fractionation ratios in the coma agree with those observed for HDO and DCN in hot moleular cores (Gensheimer *et al.*, 1996; Hatchell *et al.*, 1998). Since the gas kinetic temperature in hot cores is larger than 100 K, this fractionation must be a memory of a colder chemistry. Comparison with detailed models show that the coma ratios are best matched at $T \approx 30 - 50$ K. Blake *et al.* suggest that the ratio observed in the jets is due to the sublimation of dust particles and, therefore, reflects the value in the nucleus, while the lower coma value arises from the re-equilibration of DCN/HCN as it passes through a layer of liquid water on the surface of the nucleus. The fractionation measured in HDO is larger than that in the Earth's oceans by a factor of two or so and indicates that the bulk of the water on Earth did not come from comets, although it is possible that comets from the inner, warmer region of the protosolar nebula, with lower fractionation, were responsible (Delsemme, 2000). D/H ratios in comets can help investigate the question as to whether or not solar system ices have been re-processed in whole or part from interstellar ices. For example, if the accretion shock which occurred as the nebula condensed is strong enough, interstellar ices, and their D/H ratios, would be completely destroyed.

Any surviving interstellar fractionation can also be altered within the solar system. The main reaction which alters the fractionation in water is

$$\text{HDO} + \text{H}_2 \longleftrightarrow \text{H}_2\text{O} + \text{HD}$$

which reduces the fractionation in HDO at high temperatures in the inner nebula. Similar reactions reduce the initial fractionation in other neutrals such as CH_3D.

Ion-molecule chemistry may reset D/H ratios in the outer, cold nebula (Aikawa and Herbst, 2001) while turbulent mixing can also be important (Mousis *et al.*, 2002).

Rodgers and Charnley (2002) have investigated D chemistry in a cometary coma using a chemical-hydrodynamical model with various initial D/H ratios in the molecular ices. They found that D/H ratios of daughter species remained constant in the coma and reflect the ratios adopted for the parent species. For example, HC_3N is formed from $CN + C_2H_2$, where C_2H_2 is a parent, so that $R(DC_3N) = R(C_2HD)/2$. In their calculations they generally assumed that the photodissociation rates for deuterated molecules were the same as those of the hydrogenated analogues and that the branching ratios for D-bearing products were statistical. Thus, photodissociation of HDO produces equal amounts of OD and OH. However, it is known that this is not necessarily correct; the channel producing OD is favoured over that producing OH by a factor of about 2.5. When the branching ratio of OD/OH was taken to be 3:1, the value of $R(OD)$ increased by about 50 per cent. Thus D/H ratios in daughters in the coma measure those of their nuclear parents to within a factor of two.

Acknowledgements

Astrophysics at UMIST is supported by a grant from PPARC.

References

Aikawa, Y., and Herbst, E.: 2001, 'Two-Dimensional Distributions and Column Densities of Gaseous Molecules in Protoplanetary Disks. II. Deuterated Species and UV Shielding by Ambient Clouds', *Astron. Astrophys.* **371**, 1107–1117.

Blake, G.A., Qi, C., Hogerheijde, M.R., Gurwell, M.A., and Muhleman, D.O.: 1999, 'Sublimation from Icy Jets as a Probe of the Interstellar Volitile Content of Comets', *Nature* **398**, 213–216.

Brown, P.D., and Millar, T.J.: 1989, 'Grain Surface Formation of Multi-Deuterated Molecules', *Month. Not. Roy. Astron. Soc.* **240**, 25P–29P.

Brown, P.D., Charnley, S.B., and Millar, T.J.: 1988, 'A Model of the Chemistry in Hot Molecular Cores', *Month. Not. Roy. Astron. Soc.* **231**, 409–417.

Caselli, P., Hasegawa, T.I., and Herbst, E.: 1998, 'A Proposed Modification of the Rate Equations for Reactions on Grain Surfaces', *Astrophys. J.* **495**, 309–316.

Caselli, P., Walmsley, C.M., Tafalla, M., Dore, L., and Myers, P. C.: 1999, 'CO Depletion in the Starless Cloud Core L1544', *Astrophys. J.* **523**, L165–L169.

Caselli, P., Stantcheva, T., Shalabiea, O., Shematovich, V.I., and Herbst, E.: 2002, 'Deuterium Fractionation on Interstellar Grains Studied with Modified Rate Equations and a Monte Carlo Approach', *Plan. Space Sci.*, in press.

Ceccarelli, C., Castets, A., Loinard, L., Caux, E., and Tielens, A.G.G.M.: 1998, 'Detection of Doubly Deuterated Formaldehyde Towards the Low-luminosity Protostar IRAS 16293-2422', *Astron. Astrophys.* **338**, L43–L46.

Charnley, S.B., Tielens, A.G.G.M., and Rodgers, S.D.: 1997, 'Deuterated Methanol in the Orion Compact Ridge', *Astrophys. J.* **482**, L203–L206.

Cohen, S.D., and Hindmarsh, A.C.: 1996, 'CVODE, a Stiff/Nonstiff ODE Solver in C', *Computers in Physics* **10**, 138–143.

Delsemme, A.H.: 2000, '1999 Kuiper Prize Lecture: Cometary Origin of the Biosphere', *Icarus* **146**, 313–325.
Gensheimer, P.D., Mauersberger, R., and Wilson, T.L.: 1996, 'Water in Galactic Hot Cores', *Astron. Astrophys.* **314**, 281–294.
Gibb, A.G., and Little, L.T.: 1998, 'Probing the Structure of Molecular Cloud Cores: Observations and Modeling of CI and $C^{18}O$', *Month. Not. Roy. Astron. Soc.* **295**, 299–311.
Hatchell, J., Millar, T.J., and Rodgers, S.D.: 1998, 'The DCN/HCN Abundance Ratio in Hot Molecular Cores', *Astron. Astrophys.* **332**, 695–702.
Jensen, M.J., Bilodeau, R.C., Safvan, C.P., Seiersen, K., Andersen, L.H., Pedersen, H.B., and Heber, O.: 2000, 'Dissociative Recombination of H_3O^+, HD_2O^+, and D_3O^+', *Astrophys. J.* **543**, 764–774.
Larsson, M., Lepp, S., *et al.*: 1996, 'Dissociative Recombination of H_2D^+ and the Cosmic Abundance of Deuterium', *Astron. Astrophys.* **309**, L1–L3.
Leung, C.M., Herbst, E., and Huebner, W.F.: 1984, 'Synthesis of Complex Molecules in Dense Interstellar Clouds via Gas-phase Chemistry – A Pseudo Time-Dependent Calculation', *Astrophys. J. Suppl.* **56**, 231–256.
Lis, D.C., Roueff, E., Gerin, M., Phillips, T.G., Coudert, L.H., van der Tak, F.F.S., and Schilke, P.:2002, 'Detection of Triply Deuterated Ammonia in the Barnard 1 Cloud', *Astrophys. J.* **571**, L55–L58.
Loinard, L., Castets, A., Ceccarelli, C., Tielens, A.G.G.M., Fauré, A., Caux, E., and Duvert, G.: 2000, 'The Enormous Abundance of D_2CO in IRAS 16293-2422', *Astron. Astrophys.* **359**, 1169–1174.
Millar, T.J., and Freeman, A.: 1984, 'Chemical Modelling of Molecular Sources I. TMC-1', *Month. Not. Roy. Astron. Soc.* **207**, 405–423.
Mousis, O., Gautier, D., and Coustenis, A.: 2002, 'The D/H Ratio in Methane in Titan: Origin and History', *Icarus* **159**, 156–165.
Roberts, H., and Millar, T.J.: 2000a, 'Modelling of Deuterium Chemistry and its Application to Molecular Clouds', *Astron. Astrophys.* **361**, 388–398.
Roberts, H., and Millar, T.J.: 2000b, 'Gas-Phase Formation of Doubly-Deuterated Species', *Astron. Astrophys.* **364**, 780–784.
Roberts, H., Herbst, E., and Millar, T.J.: 2002, 'The Importance of New Rate Coefficients for D Fractionation Reactions on Interstellar Chemistry', *Month. Not. Roy. Astron. Soc.* **336**, 283–290.
Roberts, H., and Millar, T.J.: 2002, in preparation.
Rodgers, S.D., and Charnley, S.C.: 2001, 'Gas-phase Production of NHD_2 in L134N', *Astrophys. J.* **553**, 613–617.
Rodgers, S.D., and Charnley, S.C.: 2002, 'A Model of the Chemistry in Cometary Comae: Deuterated Molecules', *Month. Not. Roy. Astron. Soc.* **330**, 660–674.
Roueff, E., Tiné, S., Coudert, L. H., Pineau des Fôrets, G., Falgarone, E., and Gerin, M.: 2000, 'Detection of Doubly Deuterated Ammonia in L134N', *Astron. Astrophys.* **354**, L63–L66.
Stantcheva, T., Caselli, P., and Herbst, E.: 2001, 'Modified Rate Equations Revisited. A Corrected Treatment for Diffusive Reactions on Grain Surfaces', *Astron. Astrophys.* **375**, 673–679.
Stark, R., van der Tak, F.F.S., and van Dishoeck, E.F.: 1999, 'Detection of Interstellar H_2D^+ Emission', *Astrophys. J.* **521**, L67–L70.
van der Tak, F.F.S., Schilke, P., Müller, H.S.P., Lis, D.C., Phillips, T.G., Gerin, M., and Roueff, E.: 2002, 'Triply Deuterated Ammonia in NGC 1333', *Astron. Astrophys.* **388**, L53–L56.
Tiné, S., Roueff, E., Falgarone, E., Gerin, M., and Pineau des Fôrets, G.: 2000, 'Deuterium Fractionation in Dense Ammonia Cores', *Astron. Astrophys.* **356**, 1039–1049.
Watson, W.D.: 1974, 'Ion-Molecule Reactions, Molecule Formation, and Hydrogen-Isotope Exchange in Dense Interstellar Clouds', *Astrophys. J.* **188**, 35–42.
Williams, J.P., Bergin, E.A., Caselli, P., Myers, P.C., and Plume, R.: 1998, 'The Ionization Fraction in Dense Cloud Cores', *Astrophys. J.* **503**, 689–699.

THE D/H RATIO IN CHONDRITES

FRANÇOIS ROBERT
Laboratoire de Minéralogie, CNRS-Muséum
Muséum d'Histoire Naturelle, 61 rue Buffon, 75005 Paris, France

Received: 20 September 2002; Accepted in final form: 31 October 2002

Abstract. The statistical distribution of all available published D/H ratios of carbonaceous chondrites is presented. The possible interpretations of this distribution are reviewed at the global scale of the solar system planetary objects and at the scale defined by the carbonaceous meteorites. New issues concerning the chemical origin of the deuterium enrichment in solar system water and organic molecules are put into light by this exercise. This distribution is a robust constraint on the origin of water on Earth.

1. Introduction

Little attention has been paid on the origin of the variations of the hydrogen isotopic composition (referred to as the D/H ratio) defined by the whole rock values of the carbonaceous chondrites. These meteorites share several common isotopic carracteristics: (1) At a bulk scale they are all enriched in deuterium by a factor 5 to 12 relative to the isotopic composition of the Universal hydrogen whose D/H ratio was $25 \pm 5 \times 10^{-6}$ at the time of the formation of the solar system (Geiss and Gloeckler, 1998; Lellouch *et al.*, 2001; Mahaffy *et al.*, 1998), (2) They contain clay minerals and organic macromolecules (hereafter referred to as OM) with OM systematically enriched in deuterium by a factor 2 to 3 relative to the minerals. Since only a few natural processes are able to cause such large isotopic enrichment, the hydrogen isotopic signature is in principle an efficient tool to reconstruct the conditions of the synthesis of hydroxylated minerals and OM. This signature is also used to constrain the origin of water on Earth. A summary of these topics can be found in Robert (2001).

A review of all the available data on meteorites is compiled in this paper (including some unpublished results) with the aim of (i) presenting some of the central ideas accounting for the deuterium enrichment in the solar system and of (ii) addressing unrecognized issues about the origin of the isotopic variations among the different classes of carbonaceous chondrites. This paper addresses only the isotopic variations in the inner solar system (i.e. for objects orbiting between 1 and 3 AU) while the outer solar system is discussed in other chapters of this book (cf. Messenger *et al.*, 2003, or Encrenaz and Owen, 2003).

Space Science Reviews **106:** 87–101, 2003.

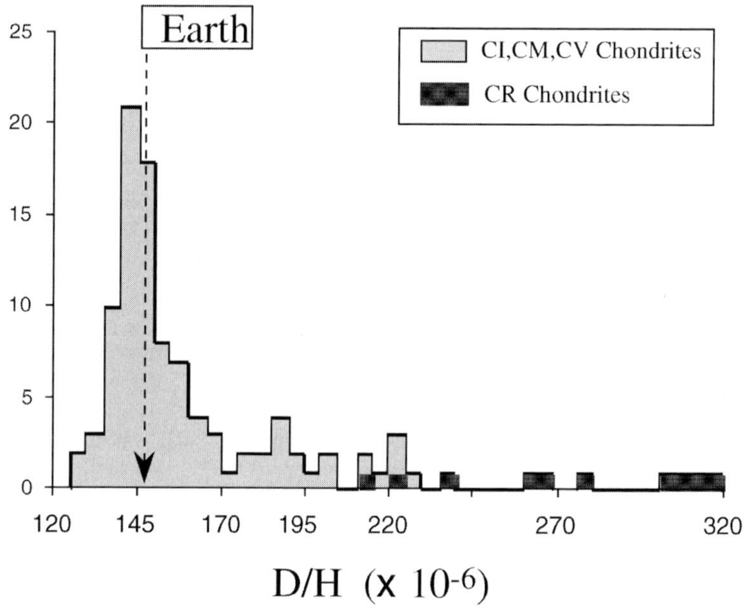

Figure 1. Distribution of the D/H ratio in the carbonaceous meteorites ($N = 106$).

2. The D/H Ratio in Chondrites

In Table I and Figure 1, the D/H ratio of the carbonaceous chondrites are reported along with the hydrogen concentration reported as water concentration and expressed as H_2O wt.-%. In meteorites H is present as a mineralogical constituent (as hydroxyls in clays, the so called phyllosilicates) and as organic hydrogen but is collected in the form of water in most experiments where the sample is outgassed by step heating. Therefore the unit H_2O wt.-% does not imply that all H is present as water in the sample.

All the data of Table I were obtained by outgassing bulk samples under vacuum or in presence of helium or molecular oxygen and by collecting and mass spectrometrically analyzing the water. Since this report is focused on the analysis of bulk samples, in situ measurements of individual minerals, such as those performed by ion microprobe, by laser desorption, or by chemical separation of OM, are not considered but only mentioned for clarity (see the review by Robert, 2002).

Concerning the data of Table I, a remnant problem annoys all the experimentalists: what fraction of water extracted from an extraterrestrial sample belong to the terrestrial contamination? Surprisingly, detailed experimental studies indicate that this fraction is small, it is a relative contribution between 0 and 15%. These studies have been performed by comparing the D/H ratio of a sample in contact with liquid deuterium-rich water with the same sample in contact with liquid water having a terrestrial D/H ratio. For example, after outgasing the Orgueil meteorite ($\approx$10 wt.-% H_20) over night at 50°C under vacuum, water extracted above 50°C shows

TABLE I

Water concentration and D/H ratio (in units of 10^{-6}) in Carbonaceous Chondrites after available literature data. The routine precision on D/H ratios is generally ±0.2.

Name	Type	H_2O wt-%	D/H ($\times 10^{-6}$)	Ref.	Name	Type	H_2O wt-%	D/H ($\times 10^{-6}$)	Ref.
Alais	CI	10.57	162.6	(6)	Murray	CM	14.04	148.4	(5)
	CI	15.39	159.0	(8)		CM	13.68	163.5	(1)
Orgueil	CI	2.90	145.3	(2)		CM	11.54	156.7	(4)
	CI	4.95	166.0	(2)		CM	8.64	143.4	(8)
	CI	7.10	189.7	(2)	Nawapali	CM	18.00	138.5	(1)
	CI	10.98	190.8	(4)	Nogoya	CM	18.47	130.8	(6)
	CI	14.58	192.4	(1)		CM	14.09	143.3	(6)
	CI	10.30	182.2	(5)		CM	14.31	140.7	(8)
	CI	4.60	175.1	(2)	Pollen	CM	16.74	148.1	(1)
	CI	14.58	151.1	(8)	Santa Cruz	CM	9.02	156.2	(6)
Ivuna	CI	14.69	183.8	(4)		CM	16.74	148.1	(6)
	CI	14.04	202.5	(1)	ALH83100	CM	14.40	131.4	(8)
	CI	11.95	188.5	(6)	Y79824	CM	12.51	135.5	(8)
	CI	15.75	164.1	(8)	EET83334	CM	13.59	126.4	(8)
LEW87232	CI	4.86	141.3	(8)	Banten	CM	12.15	151.8	(8)
Banten	CM	6.52	158.4	(6)	Pollen	CM	13.68	144.0	(8)
	CM	8.23	162.8	(3)	Y791198	CM	14.49	159.8	(8)
Bells	CM	6.19	215.7	(6)	LEW90500	CM	15.57	148.4	(8)
	CM	6.78	310.0	(6)	ALH85004	CM	3.24	159.4	(8)
Borisskino	CM	16.09	128.3	(6)	Felix	CO	0.43	146.1	(6)
Cochabamba	CM	14.13	145.9	(6)	Kainsaz	CO	0.36	179.7	(6)
Cold Bok-	CM	11.56	139.9	(7)	Lancé	CO	1.80	137.7	(6)
keveld	CM	15.84	135.0	(1)		CO	0.63	141.6	(6)
	CM	1.91	137.1	(6)		CO	2.52	142.1	(8)
	CM	6.90	137.5	(2)	Colony	CO	8.28	147.2	(8)
	CM	10.20	138.8	(2)	Kainsaz	CO	0.63	149.5	(8)
	CM	14.31	141.6	(8)	Warrenton	CO	0.72	142.8	(8)
Erakote	CM	14.44	139.4	(6)	Y791717	CO	3.06	148.0	(8)
Essebi	CM	11.29	225.9	(6)	FRO 95002	CO	1.26	144.8	(8)
Haripura	CM	14.58	144.2	(1)	ALH82101	CO	0.99	143.6	(8)
Mighei	CM	16.25	145.0	(4)	ALHA77003	CO	1.35	144.6	(8)
	CM	17.28	139.6	(1)	Y790112	CO	6.93	221.5	(8)
	CM	10.89	141.5	(8)	Ornans	CO	0.11	490.6	(6)
Murchison	CM	13.32	154.7	(5)		CO	0.54	146.7	(8)
	CM	13.93	145.6	(4)	Warrenton	CO	0.40	132.9	(6)
	CM	9.09	147.5	(8)	Allende	CV	0.12	148.0	(5)

TABLE I
(continued)

Name	Type	H_2O wt-%	D/H ($\times 10^{-6}$)	Ref.	Name	Type	H_2O wt-%	D/H ($\times 10^{-6}$)	Ref.
Allende	CV	0.03	165.7	(7)	Al Rais	CR	12.87	256.1	(8)
	CV	0.36	224.3	(6)	Kaba	CR	2.16	145.4	(8)
	CV	0.18	150.9	(8)	Renazzo	CR	8.30	301.1	(5)
Bali	CV	0.85	166.7	(6)		CR	8.12	275.7	(4)
	CV	0.45	144.8	(8)		CR	6.98	313.7	(7)
Grosnaja	CV	2.81	143.8	(6)		CR	3.87	222.5	(8)
	CV	4.77	151.9	(8)	EETB87770	CR	4.23	212.9	(8)
Kaba	CV	1.49	186.9	(6)	Karoonda	CK	0.09	142.8	(8)
Leoville	CV	1.67	148.1	(6)		CK	0.14	157.9	(6)
Mokoia	CV	1.55	186.0	(1)	Maralinga	CK	1.44	143.0	(8)
	CV	1.12	202.5	(6)	Kagangari	CH	1.71	145.0	(8)
Vigarano	CV	2.88	151.3	(8)	ALH85085	CH	2.16	172.0	(8)
	CV	1.23	155.0	(6)	ALH84025	CH	0.63	140.7	(8)
EET96286	CV	5.40	199.7	(8)	Coolidge	n.c.	3.60	143.8	(8)
ALH84028	CV	0.09	148.6	(8)	Tagish Lake	n.c.	12.33	214.0	(8)
Al Rais	CR	12.24	263.2	(6)	Adelaide	n.c.	2.09	153.4	(6)
	CR	12.06	236.8	(3)	Kaidum	n.c.	8.32	318.5	(6)

References : (1) Boato (1954); (2) Thesis by Robert (1978) and after Robert *et al.* (1979); (3) McNaughton *et al.* (1982); (4) Kolodny *et al.* (1980); (5) Robert and Epstein (1982); (6) Kerridge (1985); (7) Yang and Epstein (1983); (8) Pearson *et al.* (2001).

almost no contamination (0.04 wt.-% H_20; Engrand *et al.*, 1999). Similarly, the altered olivine in the Semarkona meteorite shows a relative contamination fraction of $\approx$15%, although water concentration in this mineral does not exceed 0.2 wt-% H_20 (Robert and Deloule, 2002). Most samples reported in Figure 1 are "falls" and, thus, have never been in contact with liquid water on Earth. Therefore the range in D/H ratios reported in Figure 1 cannot result from different degrees of terrestrial contamination.

In several CI and CM meteorites (Cold-Bokkeveld, Murray, Murchison, Orgueil) the insoluble organic matter has been chemically extracted, purified and its D/H ratio has been determined by stepwise temperature pyrolysis (Robert and Epstein, 1982; Kerridge *et al.*, 1987; Yang and Epstein, 1983). Although there are large uncertainties on these determinations because of the contamination linked to the acid extraction procedures, the OM D/H ratios lie between 310 and 370 $\times$ 10^{-6} for these meteorites. Halbout *et al.* (1990) have argued that all these extracts have a similar indigeneous D/H ratio of 370 $\times$ 10^{-6} the lower values being entirely

TABLE II

Water concentration and D/H ratio (in units of 10^{-6}) in Ordinary Chondrites after available literature data. The routine precision on D/H ratios is generally ±0.2.

Name	Type	H_2O wt-%	D/H ($\times 10^{-6}$)	Ref.	Name	Type	H_2O wt-%	D/H ($\times 10^{-6}$)	Ref.
Saint	LL6	0.0560	134.9	(1)	Hedjaz	L3-L5	0.0860	139.7	(1)
Séverin	LL6	0.0920	135.5	(1)		L3-L5	0.1120	141.3	(1)
	LL6	0.0480	123.5	(1)	Olivenza	LL5	0.0920	129.1	(1)
	LL6	0.0500	131.1	(1)		LL5	0.0670	135.8	(1)
	LL6	0.0400	133.2	(1)		LL5	0.0610	138.9	(1)
Bori	L6	0.0970	131.0	(1)	Tiechitz	L3.6		235.2	(2)
Guareña	H6	0.0110	126.2	(1)		L3.7		190.0	(2)
	H6	0.0009	128.5	(1)	Mezo Madaras	L3.7		163.5	(2)
	H6	0.0140	124.0	(1)	Prairie Dog	H3.8		143.0	(2)
Hedjaz	L3-L5	0.0660	143.5	(1)	Bremmervorde	H3.8		158.9	(2)
	L3-L5	0.0740	143.0	(1)	Carraweena	H3.9		169.0	(2)
	L3-L5	0.1030	139.4	(1)	Hvitis	E4	0.1076	143.6	(3)
	L3-L5	0.1060	140.7	(1)	Abee	E6	0.1957	145.2	(3)
	L3-L5	0.0970	141.7	(1)					

References : (1) Thesis by Robert (1978) and after Robert *et al.* (1979);
(2) McNaughton *et al.* (1982); (3) Robert, unpublished.

due to the contamination. In the case of the Renazzo meteorite, the OM D/H ratio is markedly different from CI-CM with a ratio of 550 × 10^{-6}. As observed in Figure 1 the Renazzo type meteorites (reported as CR meteorites) are systematically enriched in deuterium relative to CI-CM. Mass balance calculations using the carbon concentration in the bulk samples and the C/H ratio in the OM indicate that the clay minerals have a D/H ratio close to 300 × 10^{-6} in Renazzo (Robert and Epstein, 1982). Thus the difference between CI-CM and CR cannot be attributed to the presence of deuterium-rich OM but is truly due to an enhancement in deuterium at the scale of the bulk sample.

In Table II the D/H ratios in several types of ordinary chondrites are reported. Because of the low water concentration in these "dry" rocks, the contribution of the terrestrial water contamination can be quite large. In addition, no mineral has been identified so far which could host hydrogen in an −OH or a −CH group (expect, of course, rusty metallic minerals which are visible under naked eye and which are obviously of terrestrial origin!). Contrary to the carbonaceous meteorites, the contribution of the terrestrial water contamination has not been experimentally quantified in these rocks. The D/H variations may nevertheless reflect some indigenous signature because, as noticed by Robert *et al.* (1979) for the H-L-LL chondrites and by McNaughton *et al.* (1982) for the LL3 type chondrites, the highest the metamorphic grade, the lower the concentrations and the lower the D/H ratios.

The LL3 class of ordinary chondrites is not reported here because their bulk ratio is poorly known and varies widely in the literature (McNaughton *et al.*, 1981; Yang and Epstein, 1983; Deloule and Robert, 1995). The origin of this lack of reproducibility can be understood as follows. These samples are systematically rich in deuterium and contain a mixture of deuterium rich (720×10^{-6}) and deuterium depleted hydroxylated minerals (85×10^{-6}) mixed in various proportions with deuterium rich OM ($< 1000 \times 10^{-6}$). In such highly heterogeneous samples, the precise determination of the bulk D/H ratio should be performed on samples whose weight is probably close to 1 gram. Because these samples are quite rare in the meteorite collections, such large sample size has never been used for bulk sample analysis.

3. Interpretations and Associated Issues

The interpretations of the hydrogen isotopic variations should be considered at two scales: (1) at the global scale of the solar system by explaining the deuterium enrichment relative to the protosolar value i.e. relative to $25 \pm 5 \times 10^{-6}$ and (2) at a much smaller scale defined by the internal variations reported in Figure 1. While the interpretation (1) is relatively well understood and extensively discussed in the literature, the interpretation (2) is rarely addressed and raised numerous problems. They are presented successively in the following.

3.1. The Origin of the Chondritic Deuterium Enrichment Relative to the Protosolar Value

The common consent interpretation is based on an analogy with astrophysical observations of cold interstellar media. In these low density (10-100 H atoms per cm^3) and cold regions of the Galaxy (10-30 K) numerous carbon bearing molecules (organic molecules) have been detected. With no counter examples, they are all enriched in deuterium by a factor f up to 5000 (with a mean value around 2000) relative to the present day Universal hydrogen whose D/H ratio should be close to 16×10^{-6} (Linsky *et al.*, 1993; McCullough, 1992). The enrichment factor f is defined as

$$f = (D/H)_{OM}/(D/H)_{H} . \qquad (1)$$

Because of the UV irradiation of the surrounding stars, these interstellar media show a high degree of ionization (ions/neutrals up to 10^{-5}). Isotope exchange via ion molecule reactions are known to be fast even at low temperature and can in principle yield the observed enrichments in deuterium (Brown and Rice, 1981;

Guélin *et al.*, 1982; Rodgers and Millar, 1996; Willacy and Millar, 1998; Brown and Millar, 1989; Millar *et al.*, 1989). No other processes have been identified so far to be able to cause such a huge isotopic enrichment. In the gas phase, two classical examples of such reactions are

$$H_3^+ + HD \rightarrow H_2D^+ + H_2 \quad (\Delta E = 227\ \text{K}) , \tag{2}$$

$$CH_3^+ + HD \rightarrow CH_2D^+ + H_2 \quad (\Delta E = 370\ \text{K}) . \tag{3}$$

Under equilibrium condition, the isotopic fractionation factor α is expressed as

$$\alpha = (D/H)_{\text{ions}}/(D/H)_{H_2} = \exp(\Delta E/T) . \tag{4}$$

Contrary to thermal reactions, at low temperature the equilibrium can be reached and f can be quite large. Note, however, that the overall isotopic fractionation depends also on the degree of ionization and on the total density of the gas and, hence, is not solely a function of $\exp(\Delta E/T)$ (see for example Deloule and Robert, 1995). The first reaction represents the first step for the synthesis of water, and the second reaction starts the synthesis of organic molecules.

Interpreting the enrichment in deuterium of OM and water in the carbonaceous chondrites as resulting from a direct heritage of products synthesized in the interstellar medium, poses two problems: (1) Did the journey of the interstellar products in the protosolar nebula essentially made of H_2 cause an isotopic decrease via the thermal isotopic exchange with H_2? (2) Why do OM in the carbonaceous chondrites have an enrichment factor f close to 15 ($f = 370 \times 10^{-6}/25 \times 10^{-6}$) while interstellar molecules have an f value close to 2000?

The first question has been addressed by Drouart *et al.* (1999) (see also Mousis *et al.*, 2000; Hersant *et al.*, 2001; Robert *et al.*, 2000). In a turbulent model of the protosolar nebula where the temperature and pressure vary as a function of time and of heliocentric distances, the following isotope exchange reaction was studied (Lécluse and Robert, 1994):

$$HD + H_2O \leftrightarrow HDO + H_2 . \tag{5}$$

It has been shown that the water D/H ratio can decrease in such a nebula from an initial f value of $\approx$ 30 to a final value of $\approx$ 3. Water condensed in the outer regions of the nebula should exhibit f values $\approx$ 10 while water condensed at close distance from the Sun (1 AU) can reach an f value of $\approx$ 3. Similarly, the OM f value could have decreased by a factor of 5 (Mousis *et al.*, 2000). Recent calculations by Millar (2002) indicate that an initial interstellar f value for water equal to 30 is in agreement with the assumed value of the model by Drouart *et al.* (1999). Note that the D/H ratio of interstellar water has been measured only in the regions where the cold interstellar clouds are heated (100-200 K) by newly born stars (the so called "hot cores"; Jacq *et al.*, 1990; Gensheimer *et al.*, 1996). This is however not the case for the interstellar OM for which the D/H ratios are determined for the gaseous species down to 10 K. According to these determination, the corresponding OM f values should decrease by a factor of 100 in order to account for the meteoritic values ($f \approx 15$). Such an important decrease in the OM D/H ratio is not accounted by isotopic exchange with H2 in the protoslar nebula, as theoretically shown by Mousis *et al.* (2000). This remark yields the second question: why are carbonaceous OM different from interstellar?

This question is not yet solved, but recently two important observations have been made. It has been pointed out that the cometary water and OM (Meier *et al.*, 1998b) are both enriched in deuterium relative to carbonaceous chondrites ($D/H \approx 300 \times 10^{-6}$ and 2200×10^{-6}, respectively). This is also observed for the CR and likely for the LL3 meteorites. Thus the magnitude of the deuterium enrichment seems to be coupled in solar system objects: the higher the D/H ratio in water, the higher in OM. It has been suggested that such coupling may reflect the temperature at which the ion-molecule reactions have proceeded at the surface of the protosolar nebula when the Sun was in its T-Tauri phase (Robert, 2002).

A second observation has been made through the analysis of IDPs (Interplanetary Dust Particules) at the scale of ± 1 μm (Aléon *et al.*, 2001). In one IDP, three components intimately mixed have been identified: water with $D/H \approx 150 \times 10^{-6}$, organic molecules having a C/H ratio close to those of the carbonaceous chondrites with a $D/H \approx 500 \times 10^{-6}$ and organic molecules having a $C/H \approx 1$ (an aliphatic molecule?) with a $D/H \approx 1500 \times 10^{-6}$. This unique observation is quite important because it suggests that, under the chemical and physical conditions that prevailed at the formation of IDPs, a wide range of D/H ratios was established in the primitive components of the solar system. It can be thus proposed that the precursors of these different species corresponds to different ΔE values. The isotopic exchange reactions corresponding to these different ΔE values has not been yet identified but – as suggested by Sandford *et al.* (2001) – may reflect the different isotopic fractionation pathways taking place during homogeneous (gas-gas) and heterogeneous (gas-grain) reactions. In other words, grains and gas may have carried a wide range of different D/H ratios.

Astronomical observations provide compelling evidence that all young stars (YSO; Young Stellar Objects) go through a T-Tauri phase and emit an intense flux of X-rays that "waters" the surface of their surrounding accretion disk (André and Montmerle, 1994). Calculations show that X-rays penetrate deep into the gaseous disk due to Compton scattering and that the ionization rate of the disk greatly exceeds that of the ISM (Chiang *et al.*, 2001; Glassgold *et al.*, 2000). The degree of ionization, the thickness and the temperature of the ionized disk varies as a function of the distance to the Sun. Under such conditions, ion-molecule reactions can rapidly proceed. Once being formed at the surface of the disk – where the gas density is low enough for the ionizing X-rays to penetrate – , organic and water molecules sink towards the mid plane, in the deeper, denser and neutral regions of the disk – the so-called protosolar nebula (Cassen, 1994; Dubrulle, 1992; Drouart *et al.*, 1999) – where they are protected from ionization. However, such a detailed scenario of the evolution of the T Tauri disk of the Sun is beyond anything which can be calculated with available models of the protosolar nebula.

3.2. The Origin of the D/H Variations Among the Carbonaceous Meteorites

The most puzzling aspect of the variations reported in Figure 1 is that they greatly exceed the variations observed on Earth for clay minerals formed at temperature between 0 and 100 °C. Let us first eliminate quantitatively a series of intuitive interpretations of this problem.

An isotopic fractionation could occur during liquid water circulation in a warm parent body (Bunch and Chang, 1980; Sears *et al.*, 1995). As a general rule, clay minerals are depleted in deuterium relative to water from which they formed. For heuristic purposes, a first-order approximation of this effect can be given: at temperatures $< 100\,°C$, one should add $\approx 15 \times 10^{-6}$ to the D/H ratio of the minerals in order to obtain the D/H ratio of the water in isotopic equilibrium with the mineral. This correction is small compared to the total variations reported in Figure 1. Therefore these variations cannot be attributed to different degree of isotopic exchange under equilibrium conditions with a water reservoir having a unique D/H ratio.

A systematic rule seems to emerge when comparing solar system objects for which the D/H ratio in water and in OM are both known: OM are systematically enriched in Deuterium relative to water (Robert and Epstein, 1982; Becker and Epstein, 1982; Kerridge, 1985; Kerridge *et al.*, 1987; Epstein *et al.*, 1987; Robert *et al.*, 1987a, b; Pizzarello *et al.*, 1991; Krishnamurthy *et al.*, 1992). Thus, part of the observed variations in bulk samples (Figure 1) may be caused by different mixing ratios between clays and OM. The maximum value for the relative contribution of the organic hydrogen is $\approx 10\%$ (that is 90% of H is hosted by clays), corresponding to a shift in the D/H ratio of $\approx 20 \times 10^{-6}$. That is one order of magnitude smaller than the total D/H distribution reported in Figure 1. Note also that several authors have attempted to relate the C/H ratio to the D/H ratio of the carbonaceous chondrites in order to trace the relative contribution of the organic matter to the bulk D/H ratio. Although positive correlations have been found between these two ratios (Kerridge, 1985), these correlations suffer many counter examples. As a whole, the variations in carbonaceous chondrites cannot be caused by the differences in the mixing ratio between water and organic matter.

In Figure 1 the distribution appears clearly non-symmetrical. The possible interpretation of this asymmetry has interesting consequences. It should be first pointed out that the mixing in various proportions of OM and water having two given D/H ratios, should give a symmetrical distribution if this mixing is performed randomly. Therefore, and once again, the D/H distribution of the carbonaceous chondrites cannot be solely due to the presence of deuterium-rich organic matter distributed in variable amounts among bulk samples.

Two small peaks are present in the tail of the distribution at D/H equal to 190 and 220×10^{-6}. In the distribution reported here, all analysis methods have been used and thus, the same meteorite can be counted more than once. However, reporting an average value for each meteorite – i.e. having a unique D/H ratio per meteorite – does not alter the peaks at D/H equal to 190 and 220 $\times 10^{-6}$. This observation reinforces the statistical value of the distribution. As a firm conclusion of this discussion, the asymmetry of the distribution results from a wide range of D/H ratios both in OM and in clays. The asymmetry is likely caused by a bias in the meteorite sampling since the fragments of the disrupted asteroids – i.e. the parent bodies of the carbonaceous chondrites – have a much greater chance to intercept the Earth orbit if the asteroids were initially located at distances < 2.5 AU (Vokrouhlicky and Farinella, 2000). At higher distances, the fragments have a much greater probability to be expelled in the outer regions of the solar system. Consequently, this asymmetry suggests that the parent bodies located in the far regions of the asteroid belt have higher D/H ratios than those in the inner belts.

We therefore come back to our original question: what is the process responsible for such a wide range of isotopic variations?

A new interpretation of these variations has been recently proposed by Eiler and Kitchen (2002a, 2002b), who have shown convincingly that the least altered CM samples (e.g., Murchison) have D/H ratios in their matrix and whole rock close to 150 $\times 10^{-6}$, whereas both whole rock and matrix samples of relatively altered CM chondrites (e.g., Cold Bokkeveld) and altered chondrules in all CM samples have D/H close to 125 $\times 10^{-6}$. The degree of alteration in these samples is traced by several geochemical criteria, which correlate with D/H ratio. Among the most remarkable of these is a negative correlation between D/H and the oxygen isotope index, $\Delta^{17}O$. The protolith of the CM chondrites (i.e, pre-aqueous alteration) required to explain these relationships has abundances and isotopic compositions of many volatile elements that are indistinguishable from the Orgeuil meteorite, including a D/H ratio close to 175 $\times 10^{-6}$. However, this protolith also contained abundant ^{16}O-rich, anhydrous silicates (chondrules and CAI's), which Orgueil does not. Eiler and Kitchen conclude that the CM chondrites were derived by aqueous alteration of a volatile-rich material not represented in our meteorite collections but resembling in many respects the CI chondrites and perhaps sampling a widely distributed "primitive" and D-rich volatile reservoir in the early solar system. Such large isotopic variations associated with an alteration process have not been observed in a terrestrial environment and likely must result from a contrast in D/H ratio between water diffusing through the rocks and the initial water source; they suggest that water had a D/H ratio of 130 $\times 10^{-6}$, or ca. 250‰ depleted relative to the rock is infiltrated.

During the circulation of the fluids in the parent bodies of the carbonaceous chondrites, an isotopic exchange between water and the deuterium-rich OM may have taken place. A quantitative estimate of this effect have been proposed by Aléon *et al.* (2001) based on the experimental determinations of isotopic exchange rates between water and terrestrial kerogens, structurally similar to the macromolecular organic matter of chondrites. Clearly such an isotopic exchange could have taken place in parent bodies and is a sensitive function of the fluid temperature (at $T > 220\,^{\circ}$C, the exchange is fast at time scale of thousands of years). If this effect is at the origin of the distribution, it should be admitted that individual meteorites cannot be considered as a closed system: in a closed system, the bulk D/H ratio does not change whatever the degree of isotopic exchange between the different components of the whole rock is. On the contrary, each meteorite would represent the degree of advancement of the isotopic exchange between water and OM, a degree which should be a function of time or of the water/rock ratio or both. If correct, this alteration process should be recorded in the isotopic composition of the different organic structures of the macromolecules. Such experimental determinations have not been performed yet.

4. Implications for the Origin of Water on Earth

As extensively discussed in the literature (Balsiger *et al.*, 1995; Bockelée-Morvan *et al.*, 1998; Meier *et al.*, 1998a; Eberhardt *et al.*, 1987), detailed mass balance calculations between the D/H ratio in carbonaceous chondrites and the D/H ratio in comets, show that cometary water cannot represent more than 10% (and eventually 0%; Dauphas *et al.*, 2000) of the total budget of water on Earth (Lécuyer *et al.*, 1998; Deloule *et al.*, 1998). Consequently two scenarios (Drake and Righter, 2002; Morbidelli *et al.*, 2000) have been proposed to import water on Earth during its accretion. The two scenarios are in conflict but, considering the previous discussion of this article, each one has interesting consequences.

Morbidelli *et al.* (2000) proposed on the basis of the dynamical evolution of planetary embryos (1000 km size planetesimals) in the first 50 My of the solar system evolution that the carriers of water on Earth originated from ≈ 2.5 AU. This scenario is in agreement with the fact that the terrestrial water D/H ratio and the mean D/H of the carbonaceous chondrites (cf. Figure 1) are indistinguishable.

Drake and Righter (2002) suggested that the water-rich planetesimals at the origin of the terrestrial oceans were formed at the same heliocentric distance as the silicate material which formed the Earth. According to these authors, the chemical and isotopic composition of these planetesimals have to be different from any known type of carbonaceous chondrites in order to account for the present day bulk composition of the Earth.

The interpretation of Drake and Righter (2002) poses a new question: Did the water condensed at a terrestrial orbit – if any? – have a D/H ratio similar to the mean value defined by the carbonaceous chondrites? No answer can be proposed for this intruiguing question.

The interpretation of Morbidelli *et al.* (2000) implies that the water was carried on Earth by altered planetesimals but not by icy bodies whose D/H ratio should exceed 175×10^{-6} according to the interpretation of Eiler and Kitchen (2002a, 2002b) detailed in the previous Section 3.2.

5. Future Research Directions

Three experimental determinations should put some lights on all these issues. First by determining the D/H ratio of the icy satellites of Jupiter and Saturn in order to verify that the D/H ratio was indeed related to the heliocentric distance in the protosolar nebula. Second by determining the D/H ratio of comets from the Kuiper belt region, since these comets can also be considered as potential vectors of water on Earth. Third, by determining the true indigenous D/H ratio of the minerals in enstatite chondrites. These enstatite chondrites have been formed in reducing conditions similar to that of the proto-Earth material. Their D/H ratio is a key point in this debate.

References

Aléon, J., Engrand, C., Robert, F., and Chaussidon, M.: 2001, 'Clues to the Origin of Interplanetary Dust Particles from the Isotopic Study of their Hydrogen-bearing Phases', *Geochim. Cosmochim. Acta* **65**, 4399–4412.

André, P., and Montmerle, T.: 1994, 'From T Tauri Stars to Protostars: Circumstellar Material and Young Stellar Objects in the ρ-Ophiuchi Cloud', *Astrophys. J.* **420**, 837–862.

Balsiger, H., Altwegg, K., and Geiss, J.: 1995, 'D/H and $^{18}O/^{16}O$ Ratio in the Hydronium Ion and in Neutral Water from in Situ Ion Measurements in Comet P/Halley', *J. Geophys. Res.* **100**, 5827–5834.

Becker, R.H., and Epstein, S.: 1982, 'Carbon, Hydrogen, and Nitrogen Isotopes in Solvent-extractable Organic Matter from Carbonaceous Chondrites', *Geochim. Cosmochim. Acta* **46**, 97–103.

Boato, G.: 1954, 'The Isotopic Composition of Hydrogen and Carbon in the Carbonaceous Chondrites', *Geochim. Cosmochim. Acta* **6**, 209–220.

Bockelée-Morvan, D., Gautier, D., Lis, D.C., Young, K., Keene, J., Phillips, T., Owen, T., Crovisier, J., Goldsmith, P.F., Bergin, E.A., Despois, D., and Wootten, A.: 1998, 'Deuterated Water in Comet C/1996 B2 (Hyakutake) and its Implications for the Origin of Comets', *Icarus* **133**, 147–162.

Brown, P.D., and Millar, T.J.: 1989, 'Models of the Gaz-grain Interaction-Deuterium Chemistry', *Mon. Not. Roy. Astron. Soc.* **237**, 661–671.

Brown, R.D., and Rice, E.: 1981, 'Interstellar Deuterium Chemistry', *Phil. Trans. Roy. Soc. London* **A303**, 523–533.

Bunch, T.E., and Chang, S.: 1980, 'Carbonaceous Chondrite: II. Carbonaceous Chondrites Phyllosilicates and Light Element Geochemistry as Indicators of Parent Body Processes and Surface Conditions', *Geochim. Cosmochim. Acta* **44**, 1543–1577.

Cassen, P.: 1994, 'Utilitarian Models of the Solar Nebula', *Icarus* **112**, 405–429.

Chiang, E.I., Joung, M.K., Creech-Eakman, M.J., Qi, C., Kessler, J.E., Blake, G.A., and van Dishoeck, E.F.: 2001, 'Spectral Energy Distributions of Passive T Tauri and Herbig Ae Disks: Grain Mineralogy, Parameter Dependences and Comparison with Infrared Space Observatory LWS Observations', *Astrophys. J.* **547**, 1077–1089.

Dauphas, N., Robert, F., and Marty, B.: 2000, 'The Late Asteroidal and Cometary Bombardment of Earth as Recorded in Water Deuterium to Protium Ratio', *Icarus* **148**, 508–512.

Deloule, E., and Robert, F.: 1995, 'Interstellar Water in Meteorites?', *Geochim. Cosmochim. Acta* **59**, 4695–4706.

Deloule, E., Doukhan, J.-C., and Robert, F.: 1998, 'Interstellar Hydroxyle in Meteorite Chondrules: Implications for the Origin of Water in the Inner Solar System', *Geochim. Cosmochim. Acta* **62**, 3367–3378.

Drake, M.J., and Righter, K.: 2002, 'Determining the Composition of the Earth', *Nature* **416**, 39–44.

Drouart, A., Dubrulle, B., Gautier, D., and Robert, F.: 1999, 'Structure and Transport in the Solar Nebula from Constraints on Deuterium Enrichment and Giant Planet Formation', *Icarus* **140**, 129–155.

Dubrulle, B.: 1992, 'A Turbulent Closure Model for Thin Accretion Disks', *Astron. Astrophys.* **266**, 592–604.

Eberhardt, P., Dolder, U., Schulte, W., Krankowsky, D., Lammerzahl, P., Hoffmann, J.H., Hodges, R.R., Berthelier, J.J., and Illiano, J.M.: 1987, 'The D/H Ratio in Water from Comet p/Halley', *Astron. Astrophys.* **187**, 435–437.

Eiler, J.M., and Kitchen, N.: 2002a, 'Hydrogen-isotope Geochemistry of the CM Chondrites', Abstract presented at the 2002 Goldschmidt Conf. in Davos.

Eiler, J.M., and Kitchen, N.: 2002b, 'Hydrogen Isotope Evidence for the Origin and Evolution of the CM Carbonaceous Chondrites', *Geochim. Comochim. Acta*, in preparation.

Encrenaz, T., and Owen, T.: 2003, 'Element Abundances and Isotope Ratios in the Giant Planets and Titan', *Space Sci. Rev.*, this volume.

Engrand, C., Deloule, E., Robert, F., Maurette M., and Kurat, G.: 1999, 'Extraterrestrial Water in Micrometeorites and Cosmic Spherules from Antarctica: An Ion Microprobe Study', *Met. Planet. Sci.* **34**, 773–786.

Epstein, S., Krishnamurty, R.V., Cronin, J.R., Pizzarello, S., and Yuen, G.U.: 1987, 'Unusual Stable Isotope Ratios in Amino Acids and Carboxylic Acid Extracts from the Murchison Meteorite', *Nature* **326**, 477–479.

Geiss, J., and Gloecker, G.: 1998, 'Abundances of Deuterium and Helium in the Protosolar Cloud', *Space Sci. Rev.* **84**, 239–250.

Gensheimer, P.D., Mauersberger, R., and Wilson, T.L.: 1996, 'Water in Galactic Hot Cores', *Astron. Astrophys.* **314**, 281–294.

Glassgold, A., Feigelson, E.D., and Montmerle, T.: 2000, 'Effects on Energetic Radiation in Young Stellar Objects', *in* V. Mannings, A.P. Boss, and S.S. Russel (eds.), *Protostars and Planets IV*, University of Arizona Press, Tucson, pp. 429–455.

Guélin, M., Langer, W.D., and Wilson, R.W.: 1982, 'The State of Ionization in Dense Molecular Clouds', *Astron. Astrophys.* **107**, 107–127.

Halbout, J., Robert, F., and Javoy, M.: 1990, 'Hydrogen and Oxygen Isotope Compositions in Kerogens from the Orgueil Meteorite: Clues to Solar Origin', *Geochim. Cosmochim. Acta* **54**, 1453–1462.

Hersant, F., Gautier, D., Huré, J.-M.: 2001, 'A Two-dimensional Model for the Primordial Nebula Constrained by D/H Measurements in the Solar System: Implications for the Formation of Giant Planets', *Astrophys. J.* **554**, 391–407.

Jacq, T., Walmzey, C.M., Henkel, C., Baudry, R., Mauersberger, R., and Jewell, P.R.: 1990, 'Deuterated Water and Ammonia in Hot Cores', *Astron. Astrophys* **228**, 447–458.

Kerridge, J.F.: 1985, 'Carbon, Hydrogen, and Nitrogen in Carbonaceous Meteorites: Abundances and Isotopic Compositions in Bulk Samples', *Geochim. Cosmochim. Acta.* **49**, 1707–1714.

Kerridge, J.F., Chang, S., and Shipp, R.: 1987, 'Isotope Characterisation of Kerogen-like Material in Murchison Carbonaceous Chondrite', *Geochim. Cosmochim. Acta* **51**, 2527–2540.

Kolodny, Y., Kerridge, J.F., and Kaplan, I.R.: 1980, 'Deuterium in Carbonaceous Chondrites', *Earth Planet. Sci. Lett.* **46**, 149–158.

Krishnamurthy, R.V., Epstein, S., Cronin, J.R., Pizzarello, S., and Yuen, G.U.: 1992, 'Isotopic and Molecular Analyses of Hydrocarbons and Monocarboxylic Acids of the Murchison Meteorite', *Geochim. Cosmochim. Acta* **56**, 4045–4058.

Lécluse, C., and Robert, F.: 1994, 'Hydrogen Isotope Exchange Rates: Origin of Water in the Inner Solar System', *Geochim. Cosmochim. Acta* **58**, 2297–2939.

Lécuyer, C., Gillet, P., and Robert, F.: 1998, 'The Hydrogen Isotope Composition of Sea Water and the Global Water Cycle', *Chem. Geology* **145**, 249–261.

Lellouch, E., Bézard, B., Feuchtgruber, T., Encrenaz, T., and de Graauw, T.: 2001, 'The Deuterium Abundance in Jupiter and Saturn from ISO-SWS Observations', *Astron. Astrophys.* **370**, 610–622.

Linsky, J.L., Brown, A., Gayley, K., Diplas, A., Savage, B.D., Landsman, T.R., Shore, S.N., and Heap, S.R.: 1993, 'High-Resolution Spectrograph Observations of the Local Interstellar Medium and the D/H Ratio along the Line of Sight toward Capella', *Astrophys. J.* **402**, 694–709.

Mahaffy, P.R., Donahue, T.M., Atreya, S.K., Owen, T.C., and Niemann, H.B.: 1998, 'Galileo Probe Measurements of D/H and $^3He/^4He$ in Jupiter's Atmosphère', *Space Sci. Rev.* **84**, 251–263.

McCullough, P.R.: 1992, 'The Interstellar Deuterium to Hydrogen Ratio: A Reevaluation of Lyman Absorbtion Line Measurements', *Astrophys. J.* **390**, 213–225.

McNaughton, N.J., Borthwicks, S., Fallick, A.K., and Pillinger, C.T.: 1981, 'D/H Ratio in Unequilibrated Ordinary Chondrites', *Nature* **294**, 639–641.

McNaughton, N.J., Fallick, A.E., and Pillinger, C.T.: 1982, 'Deuterium Enrichments in Type 3 Ordinary Chondrites', *J. Geophys. Res.* **87**, 294–302.

Meier, R., Owen, T.C., Matthews, H.E., Bockelée-Morvan, D.K., Biver, N., Crovisier, J., and Gautier, D.: 1998a, 'A Determination of the HDO/H_2O Ratio in Comet C/1995 O1 (Hale-Bopp)', *Science* **279**, 842–844.

Meier, R., Owen, T.C., Matthews, H.E., Bockelée-Morvan, D., Biver, N., Crovisier, J., and Gautier, D.: 1998b, 'A Determination of the HDO/H_2O Ratio in Comet C/1995 O1 (Hale-Bopp)', *Science* **279**, 842–844.

Messenger, S., Stadermann, F.J., Floss, C., Nittler, L.R., and Mukhopadyay, S.: 2003, 'Isotopic Signatures of Presolar Materials in Interplanetary Dust', *Space Sci. Rev.*, this volume.

Millar, T.J., Bennett, A., and Herbst, E.: 1989, 'Deuterium Fractionation in Dense Interstellar Clouds', *Astrophys. J.* **340**, 906–920.

Millar, T.J.: 2002, *Proc. Meudon Symposium 2001: Deuterium in the Universe*, personal communication and *Planet. Space Sci.*, in press.

Morbidelli, A., Chambers, J., Lunine, J., Petit, J.M., Robert, F., and Valsecchi, G.B.: 2000, 'Source Regions and Timescales for the Delivery of Water on Earth', *Met. Planet. Sci.* **35**, 1309–1320.

Mousis, O., Gautier, D., Bockelée-Morvan, D., Robert, F., Dubrulle, B., and Drouart, A.: 2000, 'Constraints on the Formation of Comets from D/H Ratios Measured in H_2O and HCN', *Icarus* **148**, 513–525.

Pearson, V.K., Sephton, M.A., Gilmour, I., and Franchi, I.A.: 2001, 'Hydrogen Isotopic Composition of the Tagish Lake Meteorite: Comparison with other Carbonaceous Chondrites', *Lunar Planet. Sci.* **XXXII**, abstract # 1861.

Pizzarello, S., Krishnamurthy, R.V., Epstein, S., and Cronin, J.R.: 1991, 'Isotopic Analysis of Amino Acids from the Murchison Meteorite', *Geochim. Cosmochim. Acta* **55**, 905–910.

Robert, F.: 2001, 'The Origin of Water on Earth', *Science* **293**, 1056–1058.

Robert, F.: 2002, 'Water and Organic Matter in the Solar System as a Record of an Early Irradiation of the Protosolar Nebula', *Proc. Meudon Symposium 2001: Deuterium in the Universe, Planet. Space Sci.*, in press.

Robert, F., and Epstein, S.: 1982, 'The Concentration of Isotopic Compositions of Hydrogen, Carbon, and Nitrogen in Carbonaceous Chondrites', *Geochim. Cosmochim. Acta* **16**, 81–95.

Robert, F., and Deloule, E.: 2002, 'Using the D/H Ratio to Estimate the Terrestrial Water Contamination in Chondrites', *Lunar Planet. Sci. Conf.* **XXXIII**, 1299–1300 (pdf file).

Robert, F., Merlivat, L., and Javoy, M.: 1979, 'Deuterium Concentration in the Early Solar System: A Hydrogen and Oxygen Isotopic Study', *Nature* **282**, 785–789.

Robert, F.M., Javoy, J., Halbout, B., Dimon, B., and Merlivat, L.: 1987a, 'Hydrogen Isotopes Abundances in the Solar System, Part I: Unequilibrated Chondrites', *Geochim. Cosmochim. Acta* **51**, 1787–1806.

Robert, F.M., Javoy, J., Halbout, B., Dimon, B., and Merlivat, L.: 1987b, 'Hydrogen Isotope Abundances in the Solar System, Part II: Meteorites with Terrestrial Like D/H Ratios', *Geochim. Cosmochim. Acta* **51**, 1807–1822.

Robert, F., Gautier, D., and Dubrulle, B.: 2000, 'The Solar System D/H Ratio: Observations and Theories', *in* W. Benz, R. Kallenbach, and G.W. Lugmair (eds.), *From Dust to Terrestrial Planets*, Kluwer Acadmic Publishers, Dordrecht, *Space Sci. Rev.* **92**, 201–224.

Rodgers, S.D., and Millar, T.J.: 1996, 'The Chemistry of Deuterium in Hot Molecular Cores', *Month. Not. Roy. Astron. Soc.* **280**, 1046–1054.

Sandford, S.A., Bernstein, M.P., and Dworkin, J.P.: 2001, 'Assessment of the Interstellar Processes Leading to Deuterium Enrichment in Meteoritic Organics', *Met. Planet. Sci.* **36**, 1117–1133.

Sears, D.W.G., Morse, A.D., Hutchison, R., Guimon, K.R., Jie, L., Alexander, C.M.O'D., Benoit, P.H., Wright, I., Pillinger, C., Xie, T., and Lipschutz, M.E.: 1995, 'Metamorphism and Aqueous Alteration in Low Petrographic Type Ordinary Chondrites', *Meteoritics* **30**, 169–181.

Vokrouhlicky, D., and Farinella, P.: 2000, 'Efficient Delivery of Meteorites to the Earth from a Wide Range of Asteroid Parent Bodies', *Nature* **407**, 606–608.

Willacy, K., and Millar, T.J.: 1998, 'Adsorption Processes and the Deuterium Fractionation in Molecular Clouds', *Month. Not. Roy. Astron. Soc.* **228**, 562–568.

Yang, J., and Epstein, S.: 1983, 'Interstellar Organic Matter in Meteorites', *Geochim. Cosmochim. Acta* **47**, 2199–2216.

Address for Offprints: François Robert, Laboratoire de Minéralogie, CNRS-Muséum, Muséum d'Histoire Naturelle, 61 rue Buffon, 75005 Paris, France; `robert@cimrs1.mnhn.fr`

II. OUTER SOLAR SYSTEM

EVOLUTION OF THE PROTOSOLAR NEBULA AND FORMATION OF THE GIANT PLANETS

RICARDO HUESO (hueso@obs-nice.fr) and TRISTAN GUILLOT
Laboratorie Cassini, Observatoire de la Côte d'Azur, B.P. 4229, 06304 Nice Cedex 4, France

Received: 9 May 2002; Accepted in final form: 30 September 2002

Abstract. The formation of planetary systems is intimately tied to the question of the evolution of the gas and solid material in the early nebula. Current models of evolution of circumstellar disks are reviewed here with emphasis on the so-called "alpha models" in which angular momentum is transported outward by turbulent viscosity, parameterized by an dimensionless parameter α. A simple 1D model of protoplanetary disks that includes gas and embedded particles is used to introduce key questions on planetesimal formation. This model includes the aerodynamic properties of solid ice and rock grains to calculate their migration and growth. We show that the evolution of the nebula and migration and growth of its solids proceed on timescales that are generally not much longer than the timescale necessary to fully form the star-disk system from the molecular cloud. Contrary to a widely used approach, planet formation therefore can *neither* be studied in a static nebula *nor* in a nebula evolving from an arbitrary initial condition. We propose a simple approach to both account for sedimentation from the molecular cloud onto the disk, disk evolution and migration of solids.

Giant planets have key roles in the history of the forming Solar System: they formed relatively early, when a significant amount of hydrogen and helium were still present in the nebula, and have a mass that is a sizable fraction of the disk mass at any given time. Their composition is also of interest because when compared to the solar composition, their enrichment in elements other than hydrogen and helium is a witness of sorting processes that occured in the protosolar nebula. We review likely scenarios capable of explaining both the presence of central dense cores in Jupiter, Saturn, Uranus and Neptune and their global composition.

Keywords: Accretion, Accretion Disks, Solar System: Formation, Giant Planets

1. Introduction: Models of Protosolar Nebula Evolution

The problem of star and planet formation is one of the most fundamental of astrophysics. We will not attempt to extensively detail the numerous theories, key observations, crucial data and extended literature devoted to the subject. Instead, we will present a rather simplistic view of the problem and concentrate on simple theoretical models that hopefully include the essential physical mechanisms.

Stars are formed from giant molecular clouds, but the (tiny) amount of rotation present in these clouds is sufficient to prevent a direct collapse onto a stellar radius (about 5 to 6 orders of magnitude smaller than the initial structure), and instead yield the formation of a protostar and surrounding disk. From the "excess" infrared radiation, disks have indeed been shown to be commonly present around pre-main sequence stars (Beckwith *et al.*, 1990). Recent surveys of star formation regions show an initially high fraction of stars with disks (≈80%) that drops to almost zero on a timescale of order ≈3 My (Haisch *et al.*, 2001). Evidence for the presence of active accretion has been demonstrated (Hartmann *et al.*, 1998). Disks of 50 to 800 AU in size are also directly observed, either from visible and UV observations (e.g. Bally *et al.*, 1998) or at sub-mm wavelengths, in which case power-law mass distributions can sometimes be inferred (Guilloteau and Dutrey, 1998; Dutrey *et al.*, 1998). Finally, in the Solar System, the presence of planets and of a population of smaller bodies orbiting the Sun in the same direction and in the same plane is a testimony of the initial presence of the protosolar nebula, as first envisioned by Kant and Laplace (e.g., Lissauer, 1993).

Theoretically, the protostar+disk system can only yield the formation of a fully formed star with little surrounding material (apart from possible remaining planets) by transporting angular momentum outward. This is particularly evident in the present Solar System in which 98% of the angular momentum is in the Outer Planets orbits, and only a tiny fraction in the Sun itself (despite it holding more than 99.8% of the mass). The mechanism invoked to exchange angular momentum determines the kind of nebula evolution. For instance, in massive (very young) disks, gravitational instabilities could develop to form spiral-wave structures that efficiently transport angular momentum (Adams and Laughlin, 2000). In most cases however, disks are not massive enough for this to occur and angular momentum is transported outward by a still unknown mechanism. It is easy to show that molecular viscosity alone is much too small to produce the required transport. However, since Reynolds numbers are relatively large (10^2 to 10^9, Dubrulle, 1993) turbulence is a natural source of additional viscosity that can potentially account for the relatively rapid removal of circumstellar disks. The source of the turbulence itself is however unknown. One can cite as possible sources, shear, magnetic instabilities, convective baroclinic instabilities, etc.

A quite extended literature is devoted to the study of these sources of turbulence, either from theoretical considerations, or from intensive numerical calculations. In order to study the evolution of the protosolar nebula over several millions of years, one must use a very simplified approach in which these processes are parameterized. One such approach (which is by far the most widely used) is the alpha prescription due to (Shakura and Sunyaev, 1973) . We will use the classical alpha disk model to present the different evolutionary stages of the nebula. We will also discuss the circumstances for which more detailed modeling is required.

2. Alpha Disk Models: General Evolution of the Early Nebula

Alpha disk models are among the simplest evolutionary models that can be constructed for the study of accretion disks. In these models, turbulence alone redistributes mass and angular momentum allowing a disk to evolve and accrete onto a central star. The main advantage of such a model is that the set of equations arising can be solved in time for the whole 1-10 My of nebula evolution giving insight of the main processes occurring there. The weakest point is that results depend on an obscure parameterization for the turbulence. There is no simple way to know which values for the α parameter may be the most plausible and even if a constant and uniform value of α can be considered to study the nebula evolution.

2.1. General Description

We will assume the disk to be axisymmetric and geometrically thin. These two hypotheses are relatively reasonable, and justify a 1D approach (although clearly a 2D (r-z) approach would be more satisfactory to study a variety of problems).

It is convenient to describe the disk in terms of a Σ surface density variable obtained as the vertical integration of the volumetric density ρ at any radius. The equation that describes the evolution of this Σ surface is a diffusion equation obtained simply by the conservation of mass and angular momentum (Lynden-Bell and Pringle, 1974; Pringle, 1981)

$$\frac{\partial \Sigma}{\partial t} = \frac{3}{r}\frac{\partial}{\partial r}\left(\sqrt{r}\frac{\partial}{\partial r}\left[\nu \Sigma \sqrt{r}\right]\right). \tag{1}$$

Here, ν is the very slightly constrained turbulent kinematic viscosity. The problem of disk evolution has been reduced to determining the proper expression for ν. Shakura and Sunyaev (1973) proposed the following simple parametrization:

$$\nu = \alpha C_s H. \tag{2}$$

The parameter α controls the amount of turbulence in a medium, where its scale height H and its sound speed C_s are upper limits for its mixing length and its turbulent speed, respectively.

Observations of accretion rates for T Tauri stars indicate that $\alpha \approx 10^{-2}$ (Hartmann *et al.*, 1998). Values of $\alpha = 10^{-3}$ to 10^{-2} yield evolution timescales that are broadly consistent with the ages inferred for stars possessing gas disks. In the Solar System, the measured D/H ratio in the Earth, comets and protosolar grains can be interpreted as a temperature dependant exchange between HDO and HD reservoirs. Using this approach and an α-disk model, Drouart *et al.* (1999) suggest $\alpha \approx 0.003$, a value which is to be taken cautiously, however, because of the arbitrarily hot initial conditions chosen (see Section 4).

Note, that these values of α indicate an undefined average over the disk. Only in the case of thermal convection, α may be roughly uniform, but even in this case variability can be important. Furthermore, although values of α estimated from different methods appear to be fairly consistent, there is no consensus as to what this tells us about the source and possible mechanisms of turbulence (Cassen, 1994). Magnetorotational instabilities suggest $\alpha \approx 10^{-2}$ (Balbus and Hawley, 1991). From 3D hydrodynamical simulations due to vortexes created by baroclinic instabilities, Klahr *et al.* (1999) obtained values of α varying between 10^{-4} and 10^{-2}.

No matter which is the exact mechanism that produces the turbulence, the general solution of Equation (1) is a disk that progressively accretes mass inwards while redistributing a part of the outermost material further away, conserving angular momentum until all of the material is accreted and all of the angular momentum has been transported away.

In order to solve Equation (1), one needs to know the temperature in the nebula because the α parametrization implies that ν is a function of temperature. The inner part of the disk is optically thick and the energy escapes the disk vertically with no radial transport of energy. In a steady state, the energy generated by viscous dissipation is balanced by radiative losses from both surfaces of the disk (Ruden and Pollack, 1991). Pollack *et al.* (1985) determined Rosseland mean opacities relevant to the solar nebula and expressed these opacities as functions of temperature and density of the disk. A consistent method is required to simultaneously calculate values of ν and T (e.g., Ruden and Pollack, 1991). The outer part of the disk is optically thin and far less active being its temperature structure determined by illumination conditions from the star insolation and flaring of the disk.

Recently, another approach to the parametrization of turbulence has been proposed (Richard and Zahn, 1999), based on the observation of turbulence and angular momentum transport on experiments carried on fluid tanks with differential rotation. In this so-called *beta parametrization*, Richard and Zahn write

$$\nu = \beta \frac{\partial \Omega}{\partial R} R^3 , \tag{3}$$

where the new parameter $\beta \approx 10^{-5}$. Models using the beta prescription have been studied by Huré *et al.* (2001) who concluded that it is applicable to most disks including those around stellar objects. However a detailed comparison on model formulation and results for both turbulence prescriptions is not available yet.

2.2. Range of Applicability

Regardless of the treatment of turbulence, Equation (1) does not guaranty the stability of the disk model against e.g. gravitational perturbations. In fact, the local gravitational stability of a rotating thin disk against axisymmetric perturbations is measured by the Toomre Q-parameter, which is defined by

$$Q = \frac{k c_s}{\pi G \Sigma} = \frac{\Sigma_{\rm crit}}{\Sigma} . \tag{4}$$

Here k is the epicyclic frequency given by $k^2 = 1/R^3[\partial(R^4\Omega^2)/\partial R]$ (Toomre, 1964; Goldreich and Lynden-Bell, 1965; Goldreich and Ward, 1973). Disks are gravitationally stable if $Q > 1$ all over the disk, while they are unstable if $Q \leq 1$ anywhere. Typically, massive disks tend to produce gravitational instabilities at their outer limits. In these cases gravitational instabilities break the axial symmetry and lead to spiral waves which transport angular momentum more efficiently than turbulence alone until stability is reestablished.

2.3. Aerodynamic Properties of Particles

Up to now we have discussed models for the evolution of gaseous disks. The evolution of solid material entrained in such a disk is examined in this section.

The aerodynamic properties of solid bodies within the solar nebula were first studied by Weidenschilling (1977). He had found that the gas in the nebula rotates with slightly sub-Keplerian velocities due to the radial pressure gradient. Dust particles do not feel this pressure as much because of their larger kinetic energy. As a result, the difference in velocity between the gas and the particles increases with particle size and is maximal for large (≈km-size) particles which rotate with Keplerian velocities. The gas drag causes the dust to spiral inwards to the Sun. The effect is maximal for sizes that are such that the velocity difference between gas and particles is large and the kinetic energy of particles is small. As a result, this effect is particularly pronounced for particles that have radii ≈ 1 m. In this case, typical migration rates are $\approx 1\,\mathrm{AU}/10^4$ y. Smaller particles are better coupled to the gas and can be retained in the nebula without much migration while bigger particles with more inertia can be progressively decoupled from the gas motions.

The problem of particle migration is a crucial problem of modern theories of planet formation. In fact, the extremely rapid migration of meter-sized particles implies that either the growth of planetesimals from cm-size to km-size was extremely rapid, or most of the material that led to the formation of solid material in the Solar System initially condensed much further from the Sun than its present location. It is hence particularly important to examine the co-evolution of the gas disk and of solid particles, as pioneered by Stepinski and Valageas (1996).

In that case, the equation for the σ evolution of dust particles contains a diffusive term analogous to Equation (1) and an advective term that incorporates the inwards migration of dust particles dragged by the gas:

$$\frac{\partial \sigma}{\partial t} = \frac{3}{r}\frac{\partial}{\partial r}\left(\sqrt{r}\frac{\partial}{\partial r}\left[\frac{\nu}{S_{\mathrm{c}}}\sigma\sqrt{r}\right]\right) + \frac{1}{r}\frac{\partial}{\partial r}\left(\frac{2r\sigma\langle v_\phi\rangle_d}{\Omega_k t_{\mathrm{s}}}\right). \tag{5}$$

Here S_{c} is the Schmidt number, a measure of the coupling between gas and particles and t_{s} is the stopping time, the time a particle needs to react to velocity changes on the gas (Stepinski and Valageas, 1996). Consider the global evolution of non-evaporating dust particles of a given size in an evolving turbulent nebula. Dust is initialized in the model to have 1% of the gas density and the total abundance of gas and dust material is compared at every time. Figure 1 shows the result of one of such calculations, performed with a comparable model (Hueso and Guillot, 2002). Small dust particles remain completely coupled to the gas but particles of larger sizes have competing effects of their gas-solid coupling and inertia and they migrate inwards more and more efficiently (thus losing more material in shorter times). When particles of even larger size are considered their bigger mass decouple them from the gas and they migrate inwards at a slower rate. Particles of big enough size (1 km) do not interact with the gas and they survive in the disk without being lost to the central star.

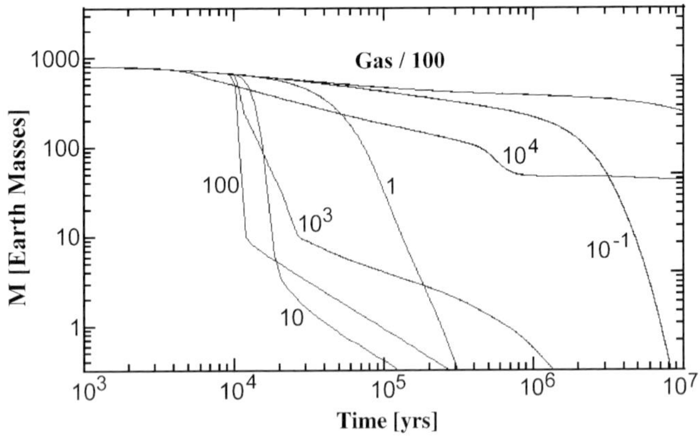

Figure 1. Evolution of Aerodynamics of non-evaporating particles as a function of size (in cm), within an evolving gas disk of initial mass $M_{\rm gas} = 0.25 M_{\odot}$, 15 AU initial radius, and assuming $\alpha = 10^{-2}$. The ratio of the surface density of solids of a given size to that of the gas was assumed to be $\sigma/\Sigma = 1/100$. Small particles (10 μm) remain coupled to the gas, particles of intermediate sizes (1 cm to 10 m) are very rapidly lost into regions where they evaportate (or the central star when the nebula is cold enough), larger planesimals (km-sized) have more inertia and are retained more effectively (Hueso and Guillot, 2002; see Stepinski and Valageas, 1996).

3. Formation of Planetesimals

As we have seen, meter-sized particles migrate extremely rapidly through the nebula. The most natural explanation for the presence of planets in the Solar System is therefore that the growth from micron-sized dust to km-sized planetesimals was extremely efficient. One possible mechanism for such an efficient growth is a gravitational instability of the solids themselves (Safronov, 1969; Goldreich and Ward, 1973). This is possible if particles settle into a layer having sufficiently high density and low velocity dispersion, density perturbations spontaneously collapse under their self-gravity, forming km-sized planetesimals. However, it was later shown that shear alone is sufficient to provide a source of turbulence that prevents dust settling into a thin layer capable of developing gravitational instabilities (Weidenschilling, 1980; Dubrulle *et al.*, 1995). Furthermore, the critical density is larger than the gas density and well before the gravitational instability could develop collective effects become important. The particles become dominant at the midplane and the gas is dragged by the particles producing a turbulent stirring that maintains the particle layer density too low for gravitational instability to develop (Cuzzi *et al.*, 1993). Therefore coagulation had to occur anyway. Weidenschilling (2000) summarizes simulations of planetesimal formation by coagulation with a full-size distribution of dust particles in a stationary gas nebula. The timescale for the formation of planetesimals from dust is a few thousand times the local orbital period at any heliocentric distance: 2000 y at 1 AU, 3×10^5 y at 30 AU.

Another possibility is to invoke turbulence in the gaseous disk, either because the presence of long-lived large-scale vortexes can concentrate dust in regions where gravitational instability could operate (Barge and Sommeria, 1995; Tanga *et al.*, 1996), or to suppress the inward drift of particles (Supulver and Lin, 2000). Anticyclonic vortexes are indeed possible stable structures in a Keplerian disk, but the question of their size and lifetime is yet unanswered.

In a sequel to their 1996 paper, Stepinski and Valageas (1997) study the combined evolution of a gas disk and of icy particles, this time allowing the particles to grow in size. They use a fixed mean particle size to represent the population of solids at any given distance from the Sun, and various simplifications for e.g. the growth rate of particles that we cannot discuss here. This approach has the advantage of being relatively feasible in terms of computer time, so that the evolution of the system can be integrated for ten million years and more. However, the monomodal distribution of particle sizes coupled to the 1D approach implies a dichotomic behavior that is not necessarily real: either particles grow beyond the km-size frontier, in which case they will survive and grow further by collecting small-size particles that orbit further from the Sun, or they do not grow that big and ineluctably end up their life into the Sun. The no-planet results of Stepinski and Valageas (1997) are nonetheless significant because an approach that would account for a distribution of particle-sizes and/or a 2D or 3D disk model would generally yield the loss of more solids (except if one invokes a mechanism capable of yielding the rapid growth of particles from cm-sizes to km-sizes).

In the case of an initially massive (0.25 $M_\odot$) and relatively small-sized (15 AU) disk, Stepinski and Valageas find that particles grow to cm-sizes and quickly migrate inwards before attaining large-enough sizes. In such a scenario the solid material is lost to the central object on timescales of a few 10^5 y. In a second scenario, they consider a less massive disk but extended up to $R_d \approx 100$ AU. In this case particles can evolve far enough from the central star and have thus enough time to grow beyond the km "barrier": they then do not share the fate of the gas which (in the α model without evaporation) ends its life in the central star.

We confirm these results using a model similar to that of Stepinski and Valageas, but more elaborate in two respects: (i) Condensation is taken into account following the equation of chemical equilibrium between the solid phase and its vapor, not simply a given critical temperature, and (ii) several condensing species can be included. Our calculations point to a faster growth when silicates are introduced. We also find a tendency to form a well-defined region at the "snow-line" in which the growth of particles is strongly enhanced by diffusion processes. This confirms the qualitative results of Stevenson and Lunine (1988). Figure 2 shows the result of one of such calculations with the significant peak of particle size at the "snowline". The peak is due to (i) the outward diffusion of gas charged with vapor that condensates on existing condensed particles as it moves from higher to lower temperatures, and (ii) the accumulation of particles that originate from further radial distances and have a faster inward migration due to their smaller sizes (<1 m).

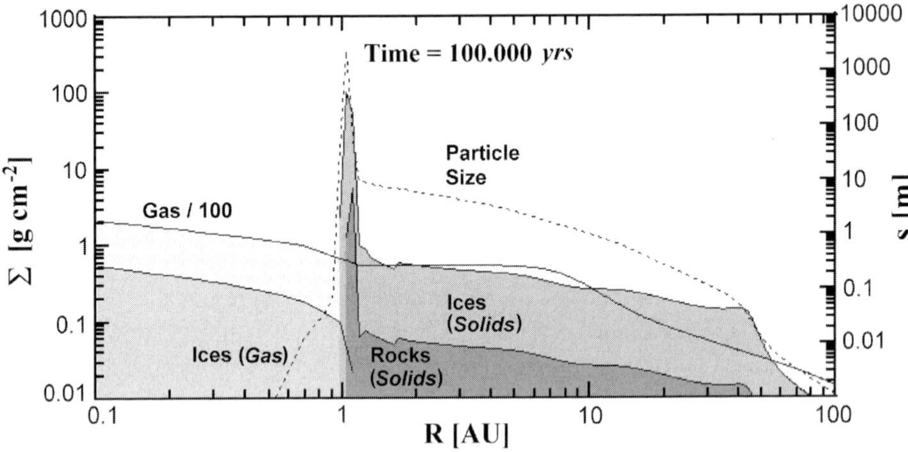

Figure 2. Planetesimals formed in the α disk model. The density of solid material is plotted for a model containing initially 99% "gas", 1% "ices" (a mixture of water, methane, ammonia, etc.), and 0.1% "rocks" (sillicates and refractory materials). Particle size is shown as a dotted line with labels on the right axis.

The presence of a well-defined snowline is difficult to ascertain because of several simplifications among which: (*i*) the condensed particles are characterized by a size which depends only on the orbital distance, not by a size distribution; (*ii*) 2D or 3D effects may critically affect the abundance of condensed particles and their growth; (*iii*) the assumption that vapor condenses on preexisting sites is valid only for grains that are at most cm-sized (when the mean free path in the gas is larger than the mean distance between condensed particles); (*iv*) the effect of the presence of grains on radiative transfer in the nebula is not included consistently. However, the important conclusions to be drawn are: (*i*) an increase of the surface density of solids is possible and could thus possibly favor the birth of a protoplanetary core (e.g. to lead to the formation of a giant planet); (*ii*) the location of such a snowline is extremely dependent on a variety of physical parameters, including time and conditions that led to the formation of a circumstellar disk.

These calculations also show that both the evolution of the protoplanetary disk and the growth of solids within that disk cannot be neglected compared to the timescale required for a collapsing giant molecular cloud core to collapse and form an isolated protostar+disk system.

4. Models Including Gravitational Collapse Processes

In order to account for mass accretion from the collapsing molecular cloud core in a simple way, we consider a uniformly rotating, isothermal, and spherical cloud of gas. Conservation of angular momentum does not allow for a direct collapse of the cloud material into a central protostar. Instead, material that is initially close to the center falls to form a low mass hot core and material further away falls into a disk whose exact size depends on mass of the inner protostar and total angular momentum. Gravitational collapse of isothermal spheres was studied by Shu (1977). In these conditions the gravitational collapse operates inside-out, i.e., the collapse starts at the center of the molecular cloud core and the collapse front propagates spherically with isothermal sound speed $a(T)$ to the outer region. In such a case, the accretion rate $\dot{M}$ over the star-disk system is a constant that depends only upon the temperature T of the molecular cloud core,

$$\dot{M} = 0.975 a^3(T)/G\,. \tag{6}$$

In such a scenario only three parameters, M_c, T_c and ω_c, describe the whole gravitational collapse phase. Typical molecular cloud cores have masses of $\approx$1-2 $M_\odot$, sizes of $\approx$0.1 pc, and rotate almost rigidly at frequencies $\omega_c \approx 2 \times 10^{-14}\ \mathrm{s}^{-1}$ (Goodman *et al.*, 1993), while temperatures may vary from 10-20 K in the cooler Taurus regions (van Dishoeck *et al.*, 1993) to 100 K in the hot Orion type regions.

Cassen and Moosman (1981) and Cassen and Summers (1983) investigated protoplanetary disk formation using a semianalytic model of disk formation and a viscous accretion disk. Nakamoto and Nakagawa (1994) studied evolutionary α-disk models including gravitational collapse from the molecular cloud over the protostar-disk system. The mass accretion rate onto the disk surface $S(r, t)$ is then

$$S(r,t) = \frac{\dot{M}}{4\pi R_d(t)}\frac{1}{r}\left(1 - \frac{r}{R_d(t)}\right)^{-1/2}, \tag{7}$$

where

$$R_d = \frac{r^4(t)\,\omega^2}{GM[r(t)]} \tag{8}$$

is the centrifugal radius, i.e. the maximum size where material can fall to the disk, and $r(t)$ is the initial position of the material that falls over the disk at time t. The Σ equation (1) can then be written with the new source term as

$$\frac{\partial \Sigma}{\partial t} = \frac{3}{r}\frac{\partial}{\partial r}\left(\sqrt{r}\frac{\partial}{\partial r}\left[\nu\Sigma\sqrt{r}\right]\right) + S(r,t). \tag{9}$$

Figure 3 shows the typical evolution of a star-disk system when including both the collapse of the molecular cloud core and the viscous evolution of the disk itself. In this case we considered an initial protostar of 0.2 M_{Star} a molecular cloud of 0.55 M_{Star} with $\omega_c = 6 \times 10^{-14}\ \mathrm{s}^{-1}$, $T_c = 14$ K and a viscosity intensity of $\alpha = 0.04$. The maximum centrifugal radius reaches 750 AU which implies most of the disk condensable material is located far from the evaporation radius. These values were chosen to fit observations of a real circumstellar disk around DM Tauri (Hueso and Guillot, 2002). Such a disk is compatible with the development of planetesimals by coagulation as discussed in Section 3.

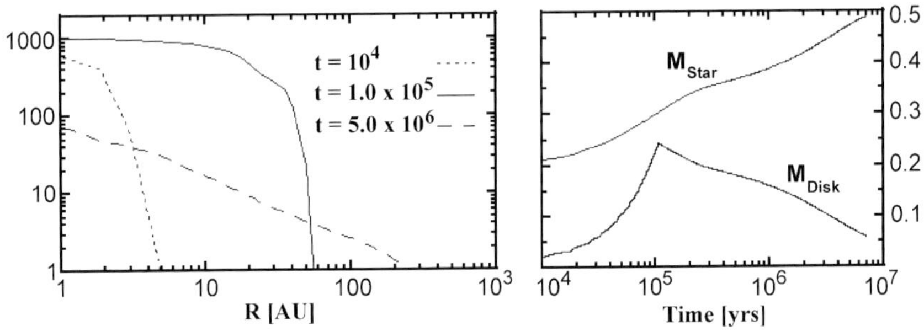

Figure 3. Left: Sigma surfaces for the gas phase for different phases of disk evolution: Early formation, maximum amount of mass and decaying by accretion to the central star. *Right:* Evolution of mass of the central star and disk in time.

It is important to stress that for certain combinations of model parameters it is possible to form highly massive disks (low values of α with respect to high values of ω_c, T_c, and M_c) that do not verify the Toomre stability criteria. This is always true for the very early phases when the inner star has a small mass. In that case, the disk is expected to form spiral arms and evolve much more rapidly than by simple turbulent diffusion (Laughlin *et al.*, 1997, 1998). Our calculations are therefore invalid during that time, but we expect that, given the relatively limited time spent with an unstable disk, our calculations should still bear the essential physical behavior of the disk for the later periods.

Our calculations that include accretion from the molecular cloud have been limited to the gas only. We find that the general structure and evolution of the disk is very sensitive to the initial condition chosen. This raises the question of the pertinence of models that are integrated from arbitrary initial conditions. This problem is particularly critical when studying the fate of planetesimals (Stepinski and Valageas, 1996, 1997), or the chemical exchange of isotopes between various reservoirs (Drouart *et al.*, 1999). Work in that direction is in progress.

5. Giant Planets Formation

Jupiter, Saturn, Uranus and Neptune are witnesses of the early formation of planets in the protosolar nebula. Further from us, the presence of a variety of companions of planetary mass (from the mass of Saturn and above) to solar-type stars is a testimony of the relatively widespread formation of giant planets around stars, but the variety of possible outcome shows that the problem is complex.

Jupiter and Saturn have masses of 318 and 95 $M_\oplus$, respectively. They are mostly formed from hydrogen and helium, but contain a higher proportion of heavy elements than the Sun (Table I). This conclusion is inferred from interior models (Guillot, 1999), but is also demonstrated in the case of several chemical species by analysis of the planets' spectra (e.g., Gautier and Owen, 1989) and *in situ* measurements of the Galileo probe in Jupiter (e.g., Owen *et al.*, 1999). On the other hand,

TABLE I

Amount of heavy elements (in $M_\oplus$) in Jupiter and Saturn

	Jupiter	Saturn
Core	0 – 10	6 – 17
Molecular region	1.6 – 6.1	2.8 – 8.8
Metallic region	0.7 – 34	0 – 17
Total (core+envelope)	11 – 42	19 – 31

Uranus (14.5 $M_\oplus$) and Neptune (17.2 $M_\oplus$) contain a relatively small amount of hydrogen and helium (less than 4 $M_\oplus$), the rest being due to an unprecised mixture of ices and rocks in which ices appear to be dominant (see Podolak *et al.*, 2000).

An important addition to the ensemble of characterized giant planets is HD 209458b, a 0.69 M_J planet orbiting at 0.047 AU from its parent star. The radius of HD209458b has been measured from transit photometry to be around 1.35 R_J (Brown *et al.*, 2001). This implies that the planet is a gas giant, but its precise composition cannot be determined because of the presently unknown atmospheric structure and the likely reduction of the cooling by stellar tides (Guillot and Showman, 2002). In a few years from now, space missions devoted to the search and characterization of transits like COROT will provide a census of extrasolar planets and their characteristics which will greatly help us to understand where this new population of astrophysical objects comes from.

Boss (2001) has proposed that all giant planets formed from a *direct* gravitational collapse of the gas in protostellar nebulae, the envelopes of Uranus and Neptune being blown away by photoevaporation due to the presence of nearby young massive stars (Boss *et al.*, 2002). This approach is interesting because it leads to quick ($\approx 10^4$ y) formation of giant planets, but it has several major problems:

- Dense clumps are obtained in simulations assuming locally isothermal conditions, but not in simulations using more elaborate thermodynamics. Even in the favorable isothermal case, it is not proved that these dense clumps lead to the presence of fully formed planets.
- The formation of central cores requires a very fast gravitational settling of heavy elements which is possible only if the structure is sufficiently cold to allow for condensation and mostly radiative (Wuchterl *et al.*, 2000).
- Without invoking any additional *ad hoc* mechanism, Jupiter and Saturn should have central cores of $\approx$ 6 $M_\oplus$ and $\approx$ 2 $M_\oplus$. Boss *et al.* (2002) proposes that photoevaporation may have allowed to have an initially massive "protosaturnian clump". However, it is extremely difficult to understand how a massive structure would have remained extended long enough ($\approx 10^7$ y) for evaporation to be possible.

- The relatively large amount of heavy elements present in Jupiter and Saturn is difficult to understand in the direct collapse scenario, since fully-formed Jupiter and Saturn see their accretion efficiency significantly lowered (Guillot and Gladman, 2001).

This way of forming giant planets must therefore be considered as largely *ad hoc* at the moment.

On the other hand, the standard formation scenario suffers no major inconsistencies (although a number of questions remain) and explains in a natural way the formation of all planets (giants and terrestrial) in the Solar System. In the case of giant planets, three phases are identified by Pollack *et al.* (1996):

1. Formation of solid cores by runaway growth up to a mass $\approx 10\,M_\oplus$ in a few 10^5 to a few 10^6 y;
2. Slow capture of the surrounding hydrogen-helium of the nebula, a process which is controlled by heat escape from the planetary envelope (a few 10^6 y);
3. Rapid accretion of the gas (on a timescale $\approx 10^5$ y).

Note, that in the 1D framework of Pollack *et al.*, the process has no limit. In reality, a giant planet is expected to open a gap which can greatly suppress accretion onto the planet, depending on the viscosity of the disk (Lin *et al.*, 2000). The final mass of the planet then depends on α and on the lifetime of the disk. The time required to fully form a giant planet strongly depends on the surface density of solids available. In their scenario, Pollack *et al.* (1996) use σ=10 g cm^{-2} at Jupiter and σ=10 g cm^{-2} at Saturn, which leads to the formation of these planets in 8 My and 10 My, respectively. The formation of Uranus and Neptune takes longer, but the authors claim that it is halted in phase (2) due to the disappearance of the protosolar nebula.

This model explains why Jupiter, Saturn, Uranus and Neptune appear to possess central dense cores, but it does not address the question of the total amount of heavy elements in the planets. Detailed dynamical simulations of the fate of massless planetesimals show that it is difficult to deliver efficiently heavy elements to the giant planets as soon as Jupiter and Saturn grow to masses close to their present ones (Guillot and Gladman, 2001). It therefore appears that the heavy elements had to be delivered *early* during the first phases of the formation of the giant planets.

Guillot *et al.* (2002) hence propose that the cores of the giant planets may have been initially larger than now and have been progressively eroded by convective activity. This scenario is quite appealing because it both explains the observed enrichments in heavy elements of the envelopes of the giant planets and allows a rapid formation of Jupiter and Saturn in the core-accretion scenario. The latter is due to the fact that the constraint of having a small primordial core is equivalent to a constraint on the value of σ. If the observed core is not primordial, the value of σ may be increased which speeds up the formation of the planet. For example, in the case of Jupiter, σ=15 g cm^{-2} leads to a formation timescale of $\approx$1.5 My.

We have sketched a coherent scenario for the formation of Jupiter and Saturn (assuming that planetesimals can indeed grow to large enough sizes for runaway growth to occur). The question of the formation of the "ice giants" Uranus and

Neptune seems more complex. Kokubo and Ida (2000) find that the time required to build protoplanetary cores beyond ≈ 10 AU becomes prohibitively long ($\approx 10^8 - 10^9$ y). A way to circumvent the problem is to assume that Uranus and Neptune initially formed in the Jupiter-Saturn region and were later expelled to the outer nebula by gravitational long-range interactions (Thommes *et al.*, 1999).

Last but not least, the question of the orbital migration of planets due to disk-planet interaction is still open. Terrestrial and giant planets are prone to a relatively rapid migration when embedded in a massive circumstellar disk (Ward and Hahn, 2000): the question of their survival is crucial. It must be stressed that migration is generally envisioned as the perturbation of a lone planet in a disk, whereas both the Solar System itself and physical processes such as runaway growth point to the presence of many "perturbers". As an example, Masset and Snellgrove (2001) show that the formation of an inner massive planet (e.g. Jupiter) and an outer lighter planet (e.g. Saturn) can halt their inward migration. However, their planets end up close to the 2:3 resonance which is not the case of Jupiter and Saturn. A better understanding of these mechanisms and their consequences requires more work.

6. Summary

In this paper we have shown that α-disk models, despite their limitations, can still be useful to understand the evolution of the protosolar nebula and the formation of the giant planets. Of course, a number of important issues remain largely undetermined and require specific, detailed studies. For instance, the mechanism responsible for angular momentum transport in the nebula is still unknown. We do not know if a model with a constant value of α can be used as representative of real diffusive processes on protoplanetary disks. To further complicate things, the early evolution of the nebula was probably dominated by gravitational instabilities able to drive nebula evolution faster and in sudden outbreaks of activity.

The question of the fate of solid material in protoplanetary disks is even less clear. In order to solve the problem, an approach that consistently solves the evolution of the gas, the motion of solids (both vertically and radially) within that gas, and their growth through a variety of mechanisms would be required. This is presently unrealistic, but we show that a simple approach using α-disks and based on the work by Stepinski and Valageas (1997) can be used to address key questions. For example, we show that a sharp concentration of planetesimals (the "snowline") appear in our simulations when vapor is assumed to condense on preexisting sites. We also verify that the formation of planets seems to be possible only in nebulae that are sufficiently extended, otherwise grains cannot grow beyond the meter-size barrier before being lost into the central star due to gas drag. Note that while this model includes a very simple prescription for grain growth, a more realistic approach should include sticking efficiencies, disruptive effects, etc.

Once planetesimals grow beyond the km-size, runaway growth is thought to efficiently drive those that are far enough from the central star to planetary size. If the process is fast enough, it can be shown that a significant amount of material can still be present in the nebula, and that therefore relatively large planetary embryos can form. It is to be stressed that within an evolving nebula, there is no reason that the planets formed when the disk itself had the so-called "minimum mass" (about 0.01 $M_\odot$). In fact, it is much more likely that Jupiter and Saturn at least, had their cores formed before that at a time when the surface density was relatively large. We show that assuming that the present cores of the giant planets are not primordial but have been progressively eroded by convection, giant planets could have formed rapidly. This limits the necessity for alternative formation scenarii.

In conclusion, while a number of important questions remain, we are confident that most of them will be addressed soon, both due to progress in computing power, but more importantly by direct observations of stars in formation and protoplanetary disks with VLT, VLTI, and ALMA, by the characterization of extrasolar planets with COROT, and by detailed studies of our own giant planets with JASSI.

Acknowledgements

This work has benefited from discussions with Pat Cassen, Anne Dutrey, Daniel Gautier, Bérangère Dubrulle, from meetings of the group *Formation et Evolution du Système Solaire* in Nice (Brett Gladman, Christiane Froeschlé, Patrick Michel, Alessandro Morbidelli, Paolo Tanga, Hans Scholl), and it has been supported by the *Programme Nationale de Planétologie*. We thank ISSI for its hospitality during the workshop. R. Hueso acknowledges a Postdoctoral fellowship from Gobierno Vasco.

References

Adams, F.C., and Laughlin, G.: 2000, 'Protostellar Disk Formation and Early Evolution', *Space Sci. Rev.* **92**, 23–38.

Balbus, S.A., and Hawley, J.F.: 1991, 'A Powerful Local Shear Instability in Weakly Magnetized Disks I - Linear analysis. II - Nonlinear evolution', *Astropys. J.* **376**, 214–233.

Bally, J., Sutherland, R.S., Devine, D., and Johnstone, D.: 1998, 'Externally Illuminated Young Stellar Environments in the Orion Nebula: Hubble Space Telescope Planetary Camera and Ultraviolet Observations', *Astronom. J.* **116**, 293–321.

Barge, P., and Sommeria, J.: 1995, 'Did Planet Formation Begin Inside Persistent Gaseous Vortices?', *Astron. Astrophys.* **295**, L1–L4.

Beckwith, S.V.W., Sargen, A.I., Chini, R.S., and Guesten, R.: 1990, 'A Survey for Circumstellar Disks Around Young Stellar Objects', *Astronom. J.* **99**, 924–950.

Brown, T.M., Charbonneau, D., Gilliland, R.L., Noyes, R.W., and Burrows, A.: 2001, 'HST Time-Series Photometry of the Transiting Planet of HD 209458', *Astrophys. J.* **552**, 699–709.

Boss, A.P.: 2001, 'Gas Giant Protoplanet Formation: Disk Instability Models with Thermodynamics and Radiative Transfer', *Astrophys. J.* **563**, 367–373.

Boss, A.P., Wetherill, G.W., and Haghighipour, N.: 2002, 'NOTE: Rapid Formation of Ice Giant Planets', *Icarus* **156**, 291–295.

Cassen, P., and Moosman, A.: 1981, 'On the Formation of Protostellar Disks', *Icarus* **48**, 353–376.

Cassen P., and Summers, A.: 1983, 'Models of the Formation of the Solar Nebula', *Icarus* **53**, 26–40.

Cassen, P.: 1994, 'Utilitarian Models of the Solar Nebula', *Icarus* **112**, 405–429.

Cuzzi, J.N., Dobrovolskis, A.R., and Champney, J.M.: 1993, 'Particle-Gas Dynamics in the Midplane of a Protoplanetary Nebula', *Icarus* **106**, 102–134.

Dubrulle, B.: 1993, 'Differential Rotation as a Source of Angular Momentum Transfer in the Solar Nebula', *Icarus* **106**, 59–76.

Dubrulle, B., Morfill, G., and Sterzik, M.: 1995, 'The Dust Subdisk in the Protoplanetary Nebula', *Icarus* **113**, 237–246.

Dutrey, A., Guilloteau, S., Prato, L., Simon, M., Duvert, G., Schuster, K., and Menard, F.: 1998, 'CO Study of the GM Aurigae Keplerian Disk', *Astron. Astrophys.* **338**, L63–L66.

Drouart, A., Dubrulle, B., Gautier, D., and Robert, F.: 1999, 'Structure and Transport in the Solar Nebula from Constraints on D Enrichment and Giant Planet Formation', *Icarus* **140**, 129–155.

Gautier, D., and Owen, T.: 1989, 'The Composition of Outer Planet Atmospheres', *in* S.K. Atreya, J.B. Pollack, and M.S. Matthews (eds.), *Origin and Evolution of Planetary and Satellite Atmospheres*, Univ. Arizona Press, Tucson, pp. 487–512.

Goldreich, P., and Lynden-Bell, D.: 1965, 'Gravitational Stability of Uniformly Rotating Disks I', *Month. Not. Roy. Astron. Soc.* **130**, 97–124.

Goldreich, P., and Ward, W.R.: 1973, 'The Formation of Planetesimals', *Astrophys. J.* **183**, 1051–1061.

Goodman, A.A., Benson, P.J., Fuller, G.A., and Myers, P.C.: 1993, 'Dense Cores in Dark Clouds. VIII – Velocity Gradients', *Astrophys. J.* **406**, 528–547.

Guillot, T.: 1999, 'Interior of Giant Planets Inside and Outside the Solar System', *Science* **286**, 72–77.

Guillot, T., and Gladman, B.: 2000, 'Late Planetesimal Delivery and the Composition of Giant Planets', *in* F. Garzon *et al.* (eds.), *Proc. Disks, Planetesimals and Planets Conf.*, ASP Conf. Ser. **219**, pp. 475–485.

Guillot, T., and Showman, A.P.: 2002, 'Evolution of "51 Pegasus b-like" Planets', *Astron. Astrophys.* **385**, 156–165.

Guillot, T., Stevenson, D.J., Hubbard, W.B., and Saumon, D.: 2002, 'The Interior of Jupiter', *in* F. Bagenal (ed.), *Jupiter*, in preparation.

Guilloteau, S., and Dutrey, A.: 1998, 'Physcial Parameters of the Keplerian Protoplanetary Disk of DM Tauri', *Astron. Astrophys.* **339**, 467–476.

Haisch, K.E., Lada, E.A., and Lada, C.J.: 2001, 'Disk Frequencies and Lifetimes in Young Clusters', *Astrophys. J.* **553**, L153–L156.

Hartmann, L., Calvet, N., Gullbring, E., and D'Alessio, P.: 1998, 'Accretion and the Evolution of T Tauri Disks', *Astrophys. J.* **495**, 385–400.

Hueso, R., and Guillot, T.: 2002, 'Formation of Planetesimals in the Solar Nebula', in preparation.

Huré, J.M., Richard, D., and Zahn, J.P.: 2001, 'Accretion Discs Models with the Beta-viscosity Prescription Derived From Laboratory Experiments', *Astron. Astrophys.* **367**, 1087–1094.

Klahr, H.H., Henning, Th., and Kley, W.: 1999, 'On the Azimuthal Structure of Thermal Convection in Circumstellar Disks', *Astrophys. J.* **514**, 325–343.

Kokubo, E., and Ida, S.: 2000, 'Formation of Protoplanets from Planetesimals in the Solar Nebula', *Icarus* **143**, 15–27.

Laughlin, G., Korchagin, V., Adams, F.C.: 1997, 'Spiral Mode Saturation in Self-gravitating Disks', *Astrophys. J.* **477**, 410–423.

Laughlin, G., Korchagin, V., Adams, F.C.: 1998, 'The Dynamics of Heavy Gaseous Disk', *Astrophys. J.* **504**, 945–966.

Lin, D.N.C., Papaloizou, J.C.B., Terquem, C., Bryden, G., and Ida, S.: 2000, 'Orbital Evolution and Planet-Star Tidal Interaction', *in* V. Mannings, A.P. Boss, and S.S. Russell (eds.), *Protostars and Planets IV*, Univ. Arizona Press, Tucson, pp. 1111–1134.

Lissauer, J.L.: 1993, 'Planet Formation', *Ann. Rev. Astron. Astrophys.* **31**, 129–174.
Lynden-Bell, D., and Pringle, J.E.: 1974, 'The Evolution of Viscous Disks and the Origin of the Nebular Variables', *Month. Not. Roy. Astron. Soc.* **168**, 603–637.
Masset, F., and Snellgrove, M.: 2001, 'Reversing Type II Migration: Resonance Trapping of a Lighter Giant Protoplanet', *Month. Not. Roy. Astron. Soc.* **320**, 55–59.
Nakamoto, T., and Nakagawa, Y.: 1994, 'Formation, Early Evolution, and Gravitational Stability of Protoplanetary Disks', *Astrophys. J.* **421**, 640–650.
Owen, T., Mahaffy, P., Niemann, H.B., Atreya, S., Donahue, T., Bar-Nun, A., and de Pater, I.: 1999, 'A Low-temperature Origin for the Planetesimals that Formed Jupiter', *Nature* **402**, 269–270.
Podolak, M., Podolak, J.I., and Marley, M.S.: 2000, 'Further Investigations of Random Models of Uranus and Neptune', *Plan. Space Sci.* **48**, 143–151.
Pollack, J.B., McKay, C.P., and Christofferson, B.M: 1985, 'A Calculation of the Rosseland Mean Opacity of Dust Grains in Primordial Solar System Nebulae', *Icarus* **64**, 471–492.
Pollack, J.B., Hubickyj, O., Bodenheimer, P., Lissauer, J.J., Podolak, M., and Greenzweig, Y.: 1996, 'Formation of the Giant Planets by Concurrent Accretion of Solids and Gas', *Icarus* **124**, 62–85.
Pringle, J.E.: 1981, 'Accretion Disks in Astrophysics', *Ann. Rev. Astrophys.* **19**, 137–162.
Richard, D., and Zahn, J.P.: 1999, 'Turbulence in Differentially Rotating Flows', *Astron. Astrophys.* **347**, 734–738.
Ruden, S.P., and Pollack, J.B.: 1991, 'The Dynamical Evolution of the Protosolar Nebula', *Astrophys. J.* **375**, 740–760.
Safronov, V.S.: 1969, 'Evolution of the Protoplanetary Cloud and the Formation of the Earth and Planets', Nauka, Moscow (Transl. 1972 NASA TT 5-677).
Shakura, N.I., and Sunyaev, R.A.: 1973, 'Black Holes in Binary Systems. Observational Appearance', *Astron. Astrophys.* **24**, 337–355.
Shu, F.H.: 1977, 'Self-similar Collapse of Isothermal Spheres and Star Formation', *Astrophys. J.* **214**, 488–497.
Stepinski, T.F., and Valageas, P.: 1996, 'Global Evolution of Solid Matter in Turbulent Protoplanetary Disks I. Aerodynamics of Solid Particles', *Astron. Astrophys.* **309**, 301–312.
Stepinski, T.F., and Valageas, P.: 1997, 'Global Evolution of Solid Matter in Turbulent Protoplanetary Disks II. Development of Icy Planetesimals', *Astron. Astrophys.* **319**, 1007–1019.
Stevenson, D.J., and Lunine, J.I.: 1988, 'Rapid Formation of Jupiter by Diffuse Redistribution of Water Vapor in the Solar Nebula', *Icarus* **75**, 146–155.
Supulver, K.D., and Lin, D.N.C.: 2000, 'Formation of Icy Planetesimals in a Turbulent Solar Nebula', *Icarus* **146**, 525–540.
Tanga, P., Babiano, A., Dubrulle, B., and Provenzale, A.: 1996, *Icarus* **121**, 158–170.
Thommes, E.W., Duncan, M.J., and Levison, H.F.: 1999, 'The Formation of Uranus and Neptune in the Jupiter-Saturn Region of the Solar System', *Nature* **402**, 635–638.
Toomre, A.: 1964, 'On the Gravitational Stability of a Disk of Stars', *Astrophys. J.* **139**, 1217–1238.
van Dishoeck, E.F., Blake, G.A., Draine, B.T., and Lunine, J.I.: 1993, *in* E.H. Levy and J. Lunine (eds.), *Protostars and Planets III*, Univ. Arizona Press, Tucson, pp. 163–241.
Ward, W.R., and Hahn, J.M.: 2000, 'Disk-Planet Interactions and the Formation of Planetary Systems', *in* V. Mannings *et al.* (eds.), *Protostars and Planets IV*, Univ. Arizona Press, p. 1135.
Weidenschilling, S.J.: 1977, 'Aerodynamics of Solid Bodies in the Solar Nebula', *Month. Not. Roy. Astron. Soc.* **180**, 57–70.
Weidenschilling, S.J.: 1980, 'Dust to Planetesimals: Settling and Coagulation in the Solar Nebula', *Icarus* **44**, 172–189.
Weidenschilling, S.J.: 2000, 'Formation of Planetesimals and Accretion of the Terrestrial Planets', *Space Sci. Rev.* **92**, 295–310.
Wuchterl, G., Guillot, T., and Lissauer, J.J.: 2000, 'Giant Planet Formation', *in* V. Mannings *et al.* (eds.), *Protostars and Planets IV*, Univ. Arizona Press, Tucson, pp. 1081–1109.

ELEMENT ABUNDANCES AND ISOTOPE RATIOS IN THE GIANT PLANETS AND TITAN

T. OWEN[1] and T. ENCRENAZ[2]

[1] *Institute for Astronomy, University of Hawaii at Manoa, Honolulu HI 96822, USA*

[2] *LESIA, Observatoire de Paris, 92195 Meudon, France;* therese.encrenaz@obspm.fr

Received: 13 May 2002; Accepted in final form: 11 November 2002

Abstract. This paper reviews our present knowledge about elemental and isotopic ratios in the Giant Planets and Titan. These parameters can provide key information about the formation and evolution of these objects. Element abundances, especially after the results of the Galileo Probe Mass Spectrometer in Jupiter, strongly support the formation model invoking an initial core formation (Mizuno, 1980; Pollack *et al.*, 1996). They also suggest that solar composition icy planetesimals (SCIPs) brought the heavy elements to Jupiter. The Jupiter value of D/H appears to be representative of the protosolar value, while the D/H enrichment observed on Uranus and Neptune is consistent with the formation scenario of these planets. The $^{15}N/^{14}N$ measurement in Jupiter seems to be representative of its protosolar value. Future measurements are expected to come from the Cassini and Herschel space mission, as well as the ALMA submillimeter observatory.

1. Introduction

Elemental and isotopic measurements can provide key information about the formation and evolution of solar-system objects, especially in the case of the giant planets (see, e.g. Gautier and Owen, 1989). The main reason is that the currently favored formation scenario for the giant planets implies an enrichment of some of these elements with respect to their solar or protosolar abundances. Such an enrichment has indeed been observed in all of the giant planets in the case of carbon, as deduced from observations of CH_4/H_2 in their atmospheres.

According to this formation scenario (Mizuno, 1980; Pollack *et al.*, 1996), all the giant planets accreted an initial core of heavy elements in the form of icy planetesimals. As these cores grew to about 10 to 15 terrestrial masses ($M_\oplus$), they became massive enough to capture gases, mostly hydrogen and helium, in a hydrodynamic collapse of the surrounding protosolar nebula. From this simple statement, we can predict the gross composition of Jupiter and other giant planets.

First, in the case of Jupiter ($318 M_\oplus$), the total mass of heavy elements should be $\approx 12 M_\oplus$, plus the contribution from the collapsing solar nebula, which would include heavy elements in solid grains and planetesimals as well as hydrogen, helium, neon and any other gases that were not trapped in the solid particles. In the simplest case, we can assume that all of the solids remained mixed with the gas collapsing from the nebula. Then the amount of heavy elements collected during

Space Science Reviews **106:** 121–138, 2003.

the collapse phase would be $\approx 6M_\oplus$, taking the solar value of the heavy element fraction as $Z = 0.02$ (Anders and Grevesse, 1989). Added to the core, this would bring the total heavy element content of the planet to $18M_\oplus$.

However, the total could be less than this if the region in the nebula from which the collapse occurred was significantly depleted in heavy elements ($Z < 0.02$) by the formation of the core. Or, if $Z > 0.02$ as a result of hydrogen (and the other gases) falling into the sun, the heavy element abundance on Jupiter could be greater than $18M_\oplus$.

Let us assume the simplest case and further assume general mixing following the infall of the surrounding protoplanetary gas. We then expect an enrichment of about 3, with respect to hydrogen, for all the heavy elements that were contained in icy planetesimals, compared to solar abundances. Since the contribution of hydrogen contained in the compounds that formed Jupiter's initial core is negligible compared to the amount captured from the solar nebula in this scenario, we expect the D/H ratio in Jupiter's outer envelope to be representative of its value in the protosolar hydrogen gas that formed the bulk of the solar nebula, to within a few percent (Guillot, 1999). The same will be true for $^3He/^4He$ – the Jovian value should be the protosolar value in solar nebula helium. In contrast, isotopic ratios of the heavy elements should have the values established in the icy planetesimals that delivered them.

In the case of the other giant planets, the mass fraction of the initial core is larger, compared to the total mass of the planet. For Uranus and Neptune it is above 60%; Saturn is intermediate at 10 to 15%. Using this same simple model and again assuming total mixing during and after the accretion phase, we thus expect for these planets an even stronger enrichment for the heavy elements: factors of 30 to 50 for Uranus and Neptune, and about 7 for Saturn, compared to solar values. Since the mass fraction of the initial core is no longer negligible, we also expect an enhancement of D/H, especially for Uranus and Neptune (Hubbard and McFarlane, 1980). This expectation is based on the observation that in the three comets from the Oort cloud in which it has been measured, D/H in cometary ice is 15 times the value deduced for protosolar hydrogen (Meier and Owen, 1999). If the icy planetesimals that formed the cores of the giant planets sampled the same reservoir of protosolar icy grains as the Oort cloud comet nuclei, we then expect the hydrogen compounds in the cores of these planets to exhibit the same high value of D/H. Even higher D/H is observed in interstellar ices (Irvine and Knacke, 1999) where it is produced by isotopic exchange in ion-molecule reactions at low temperature. As we shall see, the icy planetesimals that formed the giant planet cores may in fact not be identical to the nuclei of Oort cloud comets, so the value of D/H they brought to the planets could be somewhat higher than is commonly assumed.

Ground-based and spacecraft measurements have essentially confirmed these straightforward predictions. The measurements of the Galileo Probe Mass Spectrometer (GPMS) (Niemann *et al.*, 1998) have shown evidence for the predicted general enrichment of heavy elements in Jupiter by about a factor 3; however these results have raised other problems (Owen *et al.*, 1999a), as will be discussed below. For the other giant planets, ground-based and/or Voyager observations have shown an enrichment in C/H ranging from about 6 for Saturn (Atreya *et al.*, 1999) to 20-40 for Uranus and Neptune (Baines *et al.*, 1995), essentially what the Mizuno model suggests. Recent measurements of D/H also confirm the protosolar value for Jupiter (Mahaffy *et al.*, 1998; Lellouch *et al.*, 2001) and its enrichment in Uranus and Neptune (Feuchtgruber *et al.*, 1999). As measurements get more precise, new constraints can be derived to refine the formation models of the giant planets, and to enable a better understanding of their evolution processes.

Additional new information has come from the recent determination of $^{15}N/^{14}N$ in Jupiter, both from the ISO spacecraft (Fouchet *et al.*, 2000) and from the GPMS (Owen *et al.*, 2001). As will be discussed below, this result leads to important implications about the protosolar value of $^{15}N/^{14}N$ and about the chemical composition of the protosolar nebula. In the case of Titan, determinations of $^{15}N/^{14}N$ and $^{18}O/^{16}O$ also raise new, unanswered questions, as both ratios are strongly enhanced, compared to solar values.

This paper presents a review of the elemental and isotopic abundances of the giant planets and Titan, with special emphasis on new interpretations and open questions.

2. The Inventory

Apart from the Galileo probe results, most of the information we have on elemental and isotopic ratios comes from remote sensing spectroscopy in the infrared and millimeter range, and deals with relatively light elements: H, He, C, N, O and their isotopes. The in-situ measurements of the Galileo probe mass spectrometer (GPMS) (Niemann *et al.*, 1998), gave for the first time abundance measurements for H_2S, rare gases and their isotopes, and improved values for other isotopic ratios.

Tables I and II summarize the available data on elemental and isotopic abundances in the giant planets and Titan (see also Atreya *et al.*, 2002).

3. Abundance Ratios in the Giant Planets

Table III summarizes the abundance ratios measured in the giant planets, as compared to their solar values. It should be noted that new determinations of C/H and O/H in the Sun have been reported recently (Grevesse and Sauval, 1998; Allende Prieto *et al.*, 2001, 2002; Holweger, 2001). As these new determinations are still in conflict with one another, we have followed Atreya *et al.* (2002) and have maintained Anders and Grevesse (1989) as our reference for solar system abundances.

We can now investigate the implications of the GPMS results for Jupiter on models for the formation of the planet. First, we note that the abundance ratios

TABLE I

Elemental Ratios in the Giant Planets and Titan.

Ratio	Object	Method used
C/H	J, S, U, N, T	CH_4/H_2 GPMS (J)
		Voyager 7.7 μm (J,S,N,T)
		GB Vis-NIR (J,S,U,N,T)
N/H	J, S	NH_3/H_2 GPMS (J)
		GB + Voyager + ISO 5, 10 μm (J,S)
O/H	J, S	H_2O/H_2 GPMS (J)
		GB + Voyager + ISO 5 μm (J,S)
P/H	J, S	PH_3/H_2 GB + Voyager + NIMS 3, 5, 10 μm (J,S)
		GB 267, 800 GHz (S)
S/H	J	H_2S/H_2 GPMS (J)
Ge/H	J, S	GeH_4/H_2 GB 5 μm (J,S)
As/H	J, S	AsH_3/H_2 GB 5 μm (J,S)
He/H	J, S, U, N	GPMS (J)
		H_2-He CIA + radio-occult. (J,S,U,N)
Ne/H	J	GPMS (J)
Ar/H	J	GPMS (J)
Kr/H	J	GPMS (J)
Xe/H	J	GPMS (J)

GB = ground-based; CIA = collision-induced absorption.

measured in Jupiter exhibit an enrichment by about a factor 3 ± 1 with respect to the solar value, as would be expected from models invoking formation of a $\approx 12 M_\oplus$ core followed by collapse of $\approx 300 M_\oplus$ surrounding solar composition gas (Mizuno, 1980; Pollack *et al.*, 1996). Unless we are confronting a remarkable coincidence, it appears that the heavy elements, as gases, grains or planetesimals, were well-mixed in solar proportions with the hydrogen and helium that participated in the collapse. This is consistent with the conclusion of Guillot and Gladman (2000) that the capture of "excess" planetesimals happened very early ($1 - 10$ My), before the planet reached its present mass. Apparently there was no significant separation of gas from solids with a loss of gases to the sun during this time.

Turning to the other three giant planets, it appears that the same conclusion may apply, but the masses of heavy elements that would be associated with the hydrogen we now find in the atmospheres of these planets would be too small compared to the contribution from the $\approx 12 M_\oplus$ cores for our present precision to detect.

The second conclusion we can reach from Table III is that in all four planets, the carbon originally contained in the cores is now well mixed in the atmosphere

TABLE II
Isotopic Ratios in the Giant Planets and Titan.

Ratio	Object	Method used
D/H	J, S, U, N, T	HD/H_2 GPMS (J)
		HD/H_2 ISO (J,S,U,N)
		CH_3D/CH_4 7-9 μm Voyager + ISO (J,S,N,T)
		CH_3D/CH_4 1.6 μm GB (S,U,N,T)
$^3He/^4He$	J	GPMS (J)
$^{20}Ne/^{22}Ne$	J	GPMS (J)
$^{36}Ar/^{38}Ar$	J	GPMS (J)
$^{128\rightarrow 136}Xe$	J	GPMS (J)
$^{12}C/^{13}C$	J, S, T	GPMS (J)
		CH_4 1.1 μm GB, 7.7 μm Voyager (J, S)
		C_2H_2 13 μm GB (J)
		HCN mm GB (T)
$^{14}N/^{15}N$	J, T	GPMS (J)
		NH_3 10 μm ISO (J)
		HCN mm GB (T)
$^{16}O/^{18}O$	T	CO mm (T)

as CH_4. For Jupiter, this mixing also applies to Ar, Kr, Xe and N, but perhaps not to S, if this element was captured as H_2S in a clathrate hydrate, as discussed below.

4. A New Type of Planetesimal

There are three obvious exceptions to the factor 3 enrichment revealed in Table III for Jupiter, however: He, Ne and O. Furthermore, there is a puzzle about N and Ar. In the case of He and Ne, a differentiation process produced by condensation of He in metallic hydrogen and solution of Ne in the He droplets has been suggested to account for the depletions (Stevenson and Salpeter, 1977; Roulston and Stevenson, 1995; Niemann *et al.*, 1998). In the case of O, the last O/H measurement by the GPMS at 19 bars is still not representative of the global O/H value in the deep interior, because the Galileo Probe descended in a local "hot spot", a dry region of subsidence, a kind of gigantic downdraft, in which Jupiter's atmosphere is depleted in condensable materials (Atreya *et al.*, 1997; Owen *et al.*, 1997). Whereas the probe survived to a deep enough level to measure the well-mixed values of the condensables NH_3 and H_2S , the mixing ratio of H_2O was still increasing at the last measured data point (Niemann *et al.*, 1998; Atreya *et al.*, 2002; Wong *et al.*, 2002a).

TABLE III

Abundance Ratios in the Giant Planets.

Ratio	J/Sun	S/Sun	U/Sun	N/Sun
C/H	2.9 ± 0.5 [a)]	$6.1^{+3.3}_{-2.6}$ [b)]	23^{+10}_{-7} [c)]	31^{+7}_{-8} [c)]
N/H	3.3 ± 1.5 [d)] (P=9-12 bar)	2-4 [e)]		
O/H	2.4×10^{-4} [a)] (P=3.6 bar)	2×10^{-4} [f)] (P=3-5 bar)		
	0.033 ± 0.015 [g)] (P=12 bar)			
	$0.35^{+0.23}_{-0.16}$ [g)] (P=19 bar)			
P/H	0.75 ± 0.25 [h)] (P=1-2 bar)	3 ± 1.5 [i)]		
S/H	2.5 ± 0.15 [a)] (P>16 bar)			
He/H*	0.807 [j)]	0.56-0.85 [k)]	0.9 [j)]	1.2 [j)]
Ne/H	0.10 ± 0.01 [a)]			
Ar/H	2.5 ± 0.5 [l)]			
Kr/H	2.7 ± 0.5 [l)]			
Xe/H	2.6 ± 0.5 [l)]			

a) GPMS, Niemann *et al.* (1998); b) Voyager 7.7 μm, Courtin *et al.* (1984);
c) GB visible and NIR, Baines *et al.* (1995); d) GPMS, Atreya *et al.* (2002);
e) GB IR and radio, Marten *et al.* (1980); de Pater and Massie (1985);
f) ISO 5 μm, de Graauw *et al.* (1997); g) GPMS, Atreya *et al.* (1999); Wong *et al.* (2002a);
h) Voyager 5 μm, Drossart *et al.* (1982);
i) Voyager 10 μm, Courtin *et al.* (1984); GB submm, Weisstein and Serabyn (1994);
j) after von Zahn *et al.* (1998); k) after Conrath and Gautier (2000); l) Mahaffy *et al.* (2000);
* relative to the protosolar value of He/H.

The fact that all observed elements appear to be enriched by a constant factor of about 3 raises a question about the temperature at which these elements were trapped in the icy planetesimals that delivered them to the planet. Nitrogen is notoriously deficient in the comets in which it has been measured (Geiss, 1987; Krankowsky, 1991; Womack *et al.*, 1992), probably because most of the nitrogen in the solar nebula was in the form of N_2, which is difficult to trap in ice (Owen and Bar-Nun, 1995). As a result, the Mizuno model for giant planet formation predicts that N should be less enriched in the giant planets than C, which is present in solar abundance in comets (Pollack and Bodenheimer, 1989). Exactly the same argument applies to argon (Owen and Bar-Nun, 1995), where only upper limits that are well below solar Ar/O have been reliably determined in comets (Weaver *et al.*, 2002). According to laboratory measurements, N_2 and Ar cannot be trapped in their original abundances in amorphous ice unless the temperature at which the ice forms is < 30 K, lower than the expected formation temperature of Oort cloud comets, and much lower than expected for planetesimals at Jupiter's orbit (Owen and Bar-Nun, 1995). Yet we see from the table that N and Ar are just as enriched as carbon, sulfur, and the less volatile noble gases, Kr and Xe.

Attempting to explain this apparent paradox, Owen *et al.* (1999a) offered three working hypotheses: (a) Jupiter formed at or beyond 30 AU where temperatures were much lower and then migrated to its present position (b) models of the solar nebula are wrong, in fact the temperature reached $\approx$ 30 K at Jupiter's orbit; (c) the formation of the icy planetesimals began prior to the formation of the solar nebula in the low-temperature environment of the Interstellar Medium, perhaps during the collapse phase. All of these hypotheses seem unlikely; of the three, (c) is especially interesting for its implications for giant planet formation (Owen, 2000a).

All three of the above hypotheses are based on the assumption that the ice in the grains that trapped the gases was in the amorphous state. However, Gautier *et al.* (2001a, 2001b) have emphasized that water vapor that condensed in the solar nebula at Jupiter's orbital distance would have formed crystalline rather than amorphous ice, as suggested by Kouchi *et al.* (1994). In this case, if the ice is to capture gases, it must do so by forming clathrate hydrates. Gautier *et al.* (2001a, 2001b) propose that the gradual cooling of the solar nebula permits the successive clathration of gases in an order corresponding to the predicted stability of these clathrates at the local pressure and temperature. A temperature of 38 K is required for Ar clathrate, so again a surprisingly low temperature at Jupiter's orbit at the time that microscopic grains are available for clathration is needed to produce the icy planetesimals that bring the heavy elements to Jupiter.

This model assumes that clathrates will form at the low pressures prevailing in the solar nebula, which has not yet been demonstrated in the laboratory. Nevertheless, it is an interesting proposal that deserves elaboration. As described by Gautier *et al.* (2001a, 2001b) clathrates would deliver a greater enrichment of sulfur than is observed, requiring the sequestration of $0.22M_\oplus$ of sulfur so it does not appear in the atmosphere as H_2S. The authors propose that the missing sulfur formed FeS in the inner solar nebula. However, there is only $0.02M_\oplus$ of sulfur in the Earth and Venus presumably has a similar amount. The total mass of Mars is only $0.11M_\oplus$. Thus some other burial ground for the sulfur must be found.

Other issues that require further work are the timing and efficiency of clathrate formation. The authors implicitly assume 100% efficiency, i.e., every icy cage is occupied by an atom or molecule of gas, forming a clathrate. For this to occur, the ice must remain finely divided for at least 0.85 My in the case of CH_4 and 2.04 My, in the case of argon. Yet Weidenschilling (1997) has shown that in just 0.25 My after condensation of water in microscopic grains, icy planetesimals as large as 80 km should form by collisions in the solar nebula. Core growth is also rapid; over 80% of the core is formed in just 0.5 My (Pollack *et al.*, 1996; Hersant *et al.*, 2001). It would therefore seem that a large fraction of the ice that forms core-producing planetesimals in the nebula would not consist of clathrate hydrates. Thus the efficiency may be more like 10%.

Assuming 100% efficiency, Gautier *et al.* (2001a, 2001b) predict a mixing ratio of O/H $\approx$ 9 times solar on Jupiter to provide the extra H_2O required to make the clathrates. This leads to a heavy element abundance on Jupiter of $\approx 35 - 40M_\oplus$, close to the maximum mass of heavy elements ($42M_\oplus$) allowed by current interior models (Guillot, 1999). (They do not discuss the fate of the $\approx 18M_\oplus$ of C, N, Fe and other heavy elements corresponding to this $18M_\oplus$ of H_2O.) A more realistic value for the clathrate formation efficiency would produce a total heavy element abundance well above the limit set by Guillot (1999).

Regardless of the way in which they formed, the solar composition icy planetesimals (SCIPs) that brought the heavy elements to Jupiter differ in composition from any objects that we now find in the solar system. Assuming that SCIPs were also responsible for building the cores of the other giant planets, they must have been the most abundant solid form of matter in the early solar system. Thus it is surprising that we find no examples of these SCIPs today. Perhaps some are still present in the Oort cloud, or the Kuiper Belt, and future investigations of comets will yield examples. It is conceivable that one or more of these objects delivered the solar-like noble gases to Venus (Owen and Bar-Nun, 2000).

5. Isotopic Ratios in the Giant Planets and Titan

5.1. D/H and Noble Gases

In the Standard Big Bang Model, deuterium is entirely produced by primordial nucleosynthesis, then continuously destroyed in stars and converted into ^{3}He through the first sequence of stellar nucleosynthesis. As a result, the D/H ratio continuously decreases over the history of the Universe. Its protosolar value refers to $t = -4.6$ Gy while its value in the local interstellar medium refers to its present value. However, it is important to realize that the local interstellar medium (today) is not the ISM from which the solar system formed, owing to the motion of the sun around the galactic center.

Since Jupiter is mostly composed of protosolar gas, the D/H in Jupiter is expected to be representative of the protosolar value. As deuterium is entirely converted into ^{3}He in stars, a measurement of ^{3}He/^{4}He in Jupiter provides, by comparison with the ^{3}He/^{4}He measured in the solar wind, an independent determination of the protosolar D/H value. Using the Jovian ^{3}He/^{4}He value of $(1.66 \pm 0.05) \times 10^{-4}$ (Mahaffy *et al.*, 1998), the derived protosolar D/H is $(2.1 \pm 0.5) \times 10^{-5}$ (Geiss and Gloeckler, 1998), in very close agreement with the Jupiter measurements, as summarized in Table IV. This confirms that D/H and ^{3}He/^{4}He in Jupiter are actually protosolar.

Table IV also shows that D/H is larger in Uranus and Neptune than in Jupiter, as expected from the formation model of these planets and the equilibration of hydrogen from the compounds that formed the cores with the hydrogen gas in their atmospheres (Hubbard and McFarlane, 1980; Lecluse *et al.*, 1996). The case of Saturn is still ambiguous. One would expect D/H to be slightly larger in Saturn than in Jupiter, as the mass fraction of its initial core (10-15%) is not negligible.

TABLE IV

D/H in the Giant Planets.

Method	D/H ($\times 10^{-5}$)			
	Jupiter	Saturn	Uranus	Neptune
GPMS	2.6 ± 0.7 [a]			
ISO-HD	2.24 ± 0.04 [b] $2.3^{+1.2}_{-0.8}$ [d]	$1.85^{+0.85}_{-0.6}$ [b]	5.4^{+7}_{-3} [c]	$6.5^{+2.5}_{-1.5}$ [c]
ISO-CH_3D	2.2 ± 0.7 [b]	$2.0^{+1.4}_{-0.7}$ [b]		
5 μm-CH_3D	2.4 ± 0.7 [e]	2.1 ± 1.3 [f]		
1.6 μm-CH_3D		$1.7^{+1.7}_{-0.8}$ [g]	5.4^{+7}_{-3} [h]	9^{+12}_{-6} [i] $5.6^{+1.7}_{-1.4}$ [j] $5.0^{+4.4}_{-2.3}$ [k]
$^3He/^4He$	2.1 ± 0.5 [l]			

[a] Mahaffy *et al.* (1998); [b] Lellouch *et al.* (2001); [c] Feuchtgruber *et al.* (1999); [d] Griffin *et al.* (1986); [e] Bjoraker *et al.* (1986); [f] Noll and Larson (1991); [g] Owen *et al.* (1986); [h] de Bergh *et al.* (1986); [i] de Bergh *et al.* (1990) (revised by Feuchtgruber *et al.*, 1999); [j] Orton *et al.* (1992) (revised by Feuchtgruber *et al.*, 1999); [k] Bézard *et al.* (1997) (quoted by Feuchtgruber *et al.*, 1999); [l] Geiss and Gloeckler (1998).

There is no evidence for such an effect so far, but the error bars are still large. An accurate measurement of D/H in Saturn is a key issue for better constraining its formation scenario. The answer may come from the CIRS infrared spectrometer aboard the Cassini orbiter but it is not clear that remote sensing can achieve the required precision. If it could, a comparison of D/H on Saturn and Jupiter might then be helpful in constraining the mass of H_2O incorporated in each planet.

The D/H ratio in Titan has been measured from its thermal spectrum, by Voyager-IRIS and with ISO-SWS, using the CH_3D band at 8.6 μm and the CH_4 band at 7.7 μm. The most recent determination is $D/H = 8.8^{+3.2}_{-1.9} \times 10^{-5}$ (Coustenis *et al.*, 2001). This value is higher than the values for Uranus and Neptune but still lower than the D/H measured in comets, which is about 3×10^{-4}, (Eberhardt *et al.*, 1995; Balsiger *et al.*, 1995; Bockelée-Morvan *et al.*, 1998; Meier *et al.*, 1998; Meier and Owen, 1999). The Titan ratio is consistent with the value expected if CH_4 clathrates formed Titan's core, keeping the (calculated) solar nebula value of D/H in CH_4 (Mousis *et al.*, 2002). It also supports the current idea that Titan's atmosphere was mostly outgassed from the ices composing the satellite. However, there are opportunities for isotope exchange during the formation of Titan, e.g., with hydrogen in the Saturnian subnebula, if it was sufficiently warm, or with the hydrogen produced during the formation of the N_2 atmosphere. One should therefore keep open the idea that the present value of D/H in CH_4 was set during the formation process, rather than preserved from the solar nebula. Tests of these

TABLE V
Carbon and Nitrogen Isotopic Ratios in Jupiter.

Origin	Result	Comment	Reference
$^{12}C/^{13}C$			
CH_4 GPMS	$88 - 97$	Terrestrial	Niemann *et al.* (1998)
CH_4 1.1 μm (GB)	89^{+12}_{-10}	Terrestrial	Combes *et al.* (1977)
CH_4 7.7 μm (Voyager)	160^{+40}_{-55}	$1.8 \times$ Terrestrial	Courtin *et al.* (1983)
C_2H_2 13 μm (GB)	20^{+20}_{-10}	$0.2 \times$ Terrestrial	Drossart *et al.* (1985)
C_2H_6 12 μm (GB)	78 ± 26	Terrestrial	Orton *et al.* (1992)
$^{14}N/^{15}N$			
NH_3 GPMS	$385 - 500$	$2 \times$ Terrestrial	Owen *et al.* (2001)
NH_3 10 μm (GB)	270^{+273}_{-136}	Terrestrial	Encrenaz *et al.* (1978); Tokunaga *et al.* (1980)
NH_3 10 μm (ISO)	$357 - 1100$	$2 \times$ Terrestrial	Fouchet *et al.* (2000)

various possibilities are imminent. The model of Mousis *et al.* (2002) predicts that D/H in methane from Oort cloud comets will have the same value as that in Titan's methane, while exchange with hydrogen in the subnebula or in Titan's early atmosphere could lead to the same value of D/H in Titan's H_2O and CH_4 which the Huygens Probe may be able to measure.

In the case of Jupiter, in addition to $^3He/^4He$, isotopic ratios of other noble gases have been measured by GPMS (Mahaffy *et al.*, 2000). They appear to be consistent with solar wind measurements (Pepin *et al.*, 1999; Pepin, 2003) but the error bars are large.

6. Carbon and Nitrogen Isotopic Ratios in Jupiter

Isotopic measurements of $^{12}C/^{13}C$ and $^{14}N/^{15}N$ in Jupiter are listed in Table V. Two methods have been used, in-situ mass spectrometry measurements and infrared spectroscopy from the ground and from space (Voyager and ISO-SWS).

In addition to these results, one should mention the $^{12}C/^{13}C$, $^{14}N/^{15}N$, and $^{32}S/^{34}S$ ratios derived from HCN and CS measurements recorded at millimeter wavelengths in Jupiter's stratosphere after the SL9 collision. All ratios were found to be strongly enriched in the light isotopes compared to terrestrial values (Matthews *et al.*, 2002). However, these results are difficult to interpret, as they probably reflect the properties of the impactor, rather than those of Jupiter's atmosphere.

In the case of carbon, the in-situ measurement by GPMS has unambiguously shown that the $^{12}C/^{13}C$ ratio is close to terrestrial, which in turn agrees with the solar value (Anders and Grevesse, 1989). This also agrees with the 1.1 μm ground-based measurement that refers to the Jovian troposphere. It is interesting to note that results derived from stratospheric measurements give strong departures in both directions. In the case of the C_2H_2 measurement, the $^{13}C^{12}CH_2$ excess might be due to a fractionation effect induced by photodissociation, but the mechanism remains to be understood; in addition there is no such effect in the case of C_2H_6. There is no clear interpretation of the $^{13}CH_4$ measurement of IRIS-Voyager at 7.7 μm which remains puzzling. However, the ISO-SWS spectrum of Jupiter at 7.7 μm, which has significantly higher resolution than the IRIS data, does not confirm the high $^{12}C/^{13}C$ ratio observed by Voyager and seems more consistent with a terrestrial (solar) value of this ratio (Lellouch *et al.*, 2001). As pointed out by Fouchet (2002), the fit of this spectral range, for both the IRIS and ISO data, is probably not accurate enough for an unambiguous determination of the $^{12}C/^{13}C$ ratio.

In the case of Saturn, one measurement of $^{12}C/^{13}C$ has been obtained, from tropospheric measurements of $^{13}CH_4$ at 1.1 μm (Combes *et al.*, 1977); the result is consistent with the solar value, but has a large uncertainty.

The nitrogen isotopic ratio has only been determined on Jupiter. Two recent measurements indicate an enrichment of $^{14}N/^{15}N$ by a factor 2 compared with the terrestrial value, the first one using ISO-SWS data at 10 μm which probe the condensation region at about 0.5 bar (Fouchet *et al.*, 2000), and the second using GPMS at a pressure of several bars (Owen *et al.*, 2001). The good agreement obtained between these two determinations, which probe different atmospheric regions and use very different techniques, supports the conclusion that the $^{14}N/^{15}N$ ratio in Jupiter is indeed about twice the terrestrial value.

This result can be understood in terms of the difference in the molecular carriers of Jovian and terrestrial nitrogen (Owen and Bar-Nun, 1995; Owen *et al.*, 2001). With most of the nitrogen in the solar nebula in the form of N_2, and with N enriched by the same factor ≈ 3 as C and S, it appears that N_2 must have been the dominant source of N on Jupiter. Because it is so difficult to trap N_2 in solids, including water ice, the nitrogen in the Earth's atmosphere was probably delivered originally in the form of N-compounds. This is analogous to the situation for D/H: the major reservoir of hydrogen in the solar nebula was H_2, a highly volatile gas, which dominates the hydrogen budget on Jupiter. In contrast, the hydrogen brought to the Earth was primarily in the form of H_2O, the most abundant hydrogen compound.

As for D/H, we expect $^{15}N/^{14}N$ to be higher in compounds of nitrogen than in N_2, and calculations of isotope exchange in ion-molecule reactions in the ISM support this expectation (Terzieva and Herbst, 2000). Thus it is reasonable to conclude that the apparent enrichment of $^{14}N/^{15}N$ in Jupiter is the natural result of the dominance of N_2 over N-compounds in the solar nebula. The Jovian value of this isotope ratio should thus be the solar value as well (Owen *et al.*, 2001). In contrast, the value of $^{14}N/^{15}N = 299 \pm 30$ measured in HCN in the coma of Comet Hale-Bopp (Matthews and Jewitt, 1997) is closer to the terrestrial value. HCN is a trace constituent of comets, so to see how closely cometary nitrogen resembles

TABLE VI

Isotopic Ratios of C, N, and O in Titan.

Ratio	Origin	Result	Comment	Reference
$^{12}C/^{13}C$	$H^{13}CN$-mm	89 ± 9	Terrestrial	Hidayat *et al.* (1997)
$^{14}N/^{15}N$	$HC^{15}N$-mm	60 ± 6	$0.2 \times$ Terrestrial	Marten *et al.* (1997)
$^{16}O/^{18}O$	$C^{18}O$-mm	250	$0.5 \times$ Terrestrial	Owen *et al.* (1999b)

the nitrogen on Earth we need a measurement of the isotopes in total N, or at least in NH_3, which is the dominant nitrogen-carrying compound (Crovisier and Bockelée-Morvan, 1999).

In any case, this isotopic study supports the need for SCIPs to produce the heavy element abundances observed in Jupiter's outer envelope. To achieve the high value of $^{14}N/^{15}N$, nitrogen must have been brought in as N_2. The enhancement of N on Jupiter by the same factor of ≈ 3 as carbon and sulfur requires that nitrogen was delivered by icy planetesimals. If N_2 were captured only as a gas from the solar nebula (as was originally expected!) N/H would not be enriched over the solar value. Trapping of N_2 in amorphous ice or in clathrates would both accomplish this delivery.

The solar nebula chemistry of carbon is fundamentally different from that of hydrogen and nitrogen. Most of the carbon was originally in the form of solids, with only some 15-30% in the form of the highly volatile CO plus CH_4 (Encrenaz *et al.*, 1988). As a result, the value of $^{12}C/^{13}C$ is essentially the same (within the accuracy of remote sensing) everywhere it is measured in the solar system.

7. Isotopic Ratios of C, N, and O in the Atmosphere of Titan

Isotopic measurements of carbon, nitrogen and oxygen have been obtained on Titan using rotational transitions of HCN, CO and their isotopes in the sub-millimeter range. Results are shown in Table VI.

The depletion observed in light isotopes for nitrogen and oxygen offers a major puzzle, especially for the huge effect in nitrogen. By analogy with the ^{15}N enrichment observed on Mars (Nier and McElroy, 1977), one might consider this to be evidence for early, massive nitrogen escape. This interpretation would imply an initial N_2 pressure of at least 40 bars at Titan's surface (Hidayat, 1997), and this amount of nitrogen could have been outgassed as NH_3 if ammonia was originally trapped in Titan's ices in roughly the same portions as it is found in comets (Owen, 2000b). It appears that non-thermal escape mechanisms are needed to provide an explanation for the observed fractionation (Lunine *et al.*, 1999; Lammer *et al.*, 2000; Lammer and Bauer, 2003).

The interpretation of the ^{18}O enrichment in CO is more ambiguous. CO in Titan's stratosphere might come from the interior, but it could also result from an external oxygen source (Samuelson *et al.*, 1983), whose presence is demonstrated by the detection of H_2O and CO_2 in the stratosphere (Samuelson *et al.*, 1983; Coustenis *et al.*, 1998). The observation that the depletion of ^{16}O is so much lower than the depletion of ^{14}N suggests that both sources of CO are in play. The infall of unfractionated oxygen in the form of H_2O moderates the depletion of ^{16}O, whereas no external source of nitrogen is available.

The photochemistry producing exchanges between CO and CH_4 has been investigated by Wong *et al.* (2002b). The fact that $^{12}C/^{13}C$ does not depart from the solar value in CO despite the oxygen fractionation must be a consequence of the exchange of carbon with a large reservoir of CH_4. In fact, continuous degassing of CH_4 is required to keep this latter gas from disappearing from the atmosphere (Strobel, 1982; Yung *et al.*, 1984). This again is analogous to Mars, where $^{12}C/^{13}C$ is not factionated, despite the fractionation of $^{15}N/^{14}N$, presumably because of the large reservoir of carbon in Martian CO_2.

8. Conclusions

Recent determinations of element and isotope abundances have brought a significant contribution to our understanding of formation and evolution processes in the outer planets and their satellites. The following is an attempt to adress open questions and expected developments in the future.

The Jupiter value of D/H now appears to be the most precise way of measuring the protosolar value of this ratio, unless the planet is excessively enriched in H_2O. The Herschel submillimeter Earth-orbit observatory, to be launched by ESA in 2007, should allow us, with the PACS instrument, to get a more precise determination of this ratio on all giant planets and Titan. Prior to this, an improved measurement of D/H on Saturn and Titan is expected from the CIRS infrared spectrometer aboard the Cassini orbiter in 2005, together with the in situ measurements of Titan's atmosphere from the GCMS aboard the Cassini-Huygens probe. All these new measurements will allow us to better constrain the value of D/H in the protoplanetary ices and/or clathrates that formed the giant planets and Titan at distances ranging from 5 to 30 AU from the Sun. Measuring $^{14}N/^{15}N$ in Jupiter seems to be good way of determining this ratio in the protosolar nebula. A direct measurement in the solar wind will be accomplished by the Genesis Mission when that spacecraft returns to Earth in 2004. The present determinations on Jupiter may be improved in the future using ground-based spectroscopy at 5 and 10 microns with high spatial and spectral resolution. If $^{14}N/^{15}N$ is found to be identical on Saturn and Jupiter, it would show that SCIPs were the major providers of heavy elements to both planets. Measuring the isotopic ratios of C, N and O on Neptune (and improving them on Titan) will be possible using the millimeter and submillimeter transitions of HCN and CO with the ALMA interferometer, as well as the HIFI heterodyne spectroscopy instrument aboard Herschel. The interpretation of the resulting data may not be straightforward; however, comparison with the in-situ measurements on Titan with the GCMS aboard Cassini-Huygens will surely help. Finally, the GCMS will provide the first direct determination of minor element abundances on Titan, including the noble gases and their isotopes, which will provide key diagnostics for models of Titan's formation.

For example, if 40 bars of nitrogen were really outgassed by Titan as NH_3 one would expect a commensurate amount of ^{36}Ar to be present in the atmosphere. We will know how much argon to expect once we have determined the ratio of $^{36}Ar/NH_3$ in a few comets. The Rosetta mission may provide the first determination of this ratio in Comet Wirtanen in 2012. On the other hand, if the NH_3 and CH_4 were brought to Titan in the form of clathrates, the amount of ^{36}Ar in the satellite's atmosphere today should be extremely small (Mousis *et al.*, 2002). Any factionation of $^{36}Ar/^{38}Ar$ from the solar value will provide another constraint on models for atmospheric escape, as will the value of $^{14}N/^{15}N$ in the atmospheric N_2. If the probe survives the landing on Titan's surface, the GCMS may be able to measure D/H in exposed surface ice, if its heated inlet tube is in contact with a clean ice surface.

We await all these new results with great interest.

Acknowledgements

We thank Eric Herbst and David Lambert for respectively organizing the necessary research into nitrogen isotope exchange in the ISM and the solar carbon abundance, and Sushil Atreya for illuminating discussions about clathrates. Daniel Gautier supplied helpful comments as a referee. We are grateful to Johannes Geiss and his able staff for their hospitality at ISSI. This research was supported in part (TO) by NASA through the Galileo Project and the Planetary Astronomy and Exobiology Programs. TE acknowledges support from the Centre National de la Recherche Scientifique (CNRS).

References

Allende Prieto, C., Lambert, D.L., and Asplund, M.: 2001, 'The Forbidden Abundance of Oxygen in the Sun', *Astrophys. J.* **556**, L63–L66.

Allende Prieto, C., Lambert, D.L., and Asplund, M.:2002, 'A Reappraisal of the Solar Photospheric C/O Ratio', *Astrophys. J.* **573**, L137–141.

Anders, E., and Grevesse, N.: 1989, 'Abundances of the Elements – Meteorites and Solar', *Geochim. Cosmochim. Acta* **53**, 97–214.

Atreya, S.K., Wong, M.H., Owen, T.C., Niemann, H.B., and Mahaffy, P.R.: 1997, 'Chemistry and Clouds of Jupiter's Atmosphere: A Galileo Perspective', *in* C. Barbieri *et al.* (eds.), *The Three Galileos: the Man, the Spacecraft, the Telescope*, Kluwer Acad. Publ., p. 249.

Atreya, S.K., Wong, M.H., Owen, T.C., Mahaffy, P.R., Niemann, H.B., de Pater, I., Drossart, P., and Encrenaz, T.: 1999, 'A Comparison of the Atmospheres of Jupiter and Saturn: Deep Atmospheric Composition, Cloud Structure, Vertical Mixing, and Origin', *Planet. Space Sci.* **47**, 1243–1262.

Atreya, S.K., Mahaffy, P.R., Niemann, H.B., Wong, M.H., and Owen, T.C.: 2002, 'Composition of the Atmosphere of Jupiter – An Update', *Planet. Space Sci.*, in press.

Baines, K.H., Mickelson, M.E., Larson, L.E., and Ferguson, D.W.: 1995, 'The Abundances of Methane and Ortho/para Hydrogen on Uranus and Neptune: Implications of New Laboratory 4-0 H_2 Quadrupole Line Parameters', *Icarus* **114**, 328–340.

Balsiger, H., Altwegg, K., and Geiss, J.: 1995, 'D/H and $^{16}O/^{18}O$ in the Hydronium Ion and in Neutral Water from in Situ Ion Measurements in Comet Halley', *J. Geophys. Res.* **100**, 5827–5834.

Bézard, B., Encrenaz, T., Feuchtgruber, H., Lellouch, E., de Graauw, T., Griffin, M., and Atreya, S.K.: 1997, 'ISO–SWS Observations of Neptune', *Bull. Amer. Astron. Soc.* **29**, 993.

Bjoraker, G.L., Larson, H.P., and Kunde, V.G.: 1986, 'The Gas Composition of Jupiter Derived from 5-micron Airborne Spectroscopic Measurements', *Icarus* **66**, 579–609.

Bockelée-Morvan, D., *et al.*: 1998, 'Deuterated Water in Comet C/1996 B2 (Hyakutake) and its Implication for the Origin of Comets', *Icarus* **133**, 147–162.

Combes, M., Maillard, J.P., and de Bergh, C.: 1977, 'Evidence for a Telluric Value of the $^{12}C/^{13}C$ Ratio in the Atmospheres of Jupiter and Saturn', *Astron. Astrophys.* **61**, 531–537.

Conrath, B.J., and Gautier, D.: 2000, 'Saturn Helium Abundance: A Reanalysis of Voyager Measurements', *Icarus* **144**, 124–134.

Courtin, R., Gautier, D., Marten, A., and Kunde, V.: 1983, 'The $^{12}C/^{13}C$ Ratio in Jupiter from the Voyager Infrared Investigation', *Icarus* **53**, 121–132.

Courtin, R., Gautier, D., Marten, A., Bézard, B., and Hanel, R.A.: 1984, 'The Composition of Saturn's Atmosphere at Northern Temperate Latitudes from Voyager IRIS Spectra: NH_3, PH_3, C_2H_2, C_2H_6, CH_3D and the Saturnian D/H Isotopic Ratio', *Astrophys. J.* **287**, 899–916.

Coustenis, A., Salama, A., Lellouch, E., Encrenaz, T., Bjoraker, G.L., Samuelson, R.E., de Graauw, T., Feuchtgruber, H., and Kessler, M.F.: 1998, 'Evidence for Water Vapor in Titan's Atmosphere from ISO/SWS Data', *Astron. Astrophys.* **336**, L85–L89.

Coustenis, A., Bézard, B., Encrenaz, T., Gautier, D., Lellouch, E., Salama, A., Lacy, J., and Orton, G.S.: 2001, 'The D/H Ratio on Titan from ISO and IRSHELL Data', *Bull. Amer. Astron. Soc.* **33**, 1108.

Crovisier, J., and Bockelée-Morvan, D.: 1999, 'Remote Observations of the Composition of Cometary Volatiles', *Space Sci. Rev.* **90**, 19–32.

Drossart, P., Encrenaz, T., Kunde, V., Hanel, R., and Combes, M.: 1982, 'An Estimate of the PH_3, CH_3D, and GeH_4 Abundances on Jupiter from the Voyager IRIS Data at 4.5 Microns', *Icarus* **49**, 416–426.

Drossart, P., Lacy, J., Serabyn, E., Tokunaga, A., Bézard, B., and Encrenaz, Th.: 1985, 'Detection of $^{12}C^{13}CH_2$ on Jupiter at 13 Microns', *Astron. Astrophys.* **149**, L10–L12.

de Bergh, C., Chauville, J., Lutz, B.L., Owen T., and Brault, J.: 1986, 'Monodeuterated Methane in the Outer Solar System. II. Its Detection on Uranus at 1.6 Microns', *Astrophys. J.* **311**, 501–510.

de Bergh, C., Lutz, B.L., Owen, T., and Maillard, J.P.: 1990, 'Monodeuterated Methane in the Outer Solar System. IV. Its Detection and Abundance on Neptune', *Astrophys. J.* **355**, 661–666.

de Graauw, T. *et al.*: 1997, 'First Results of ISO-SWS Observations of Saturn: Detection of CO_2, CH_3C_2H, C_4H_2 and Tropospheric H_2O', *Astron. Astrophys.* **321**, L13–L16.

de Pater, I., and Massie, S.T.: 1985, 'Models of the Millimeter-centimeter Spectra of the Giant Planets', *Icarus* **62**, 143–171.

Eberhardt, P., Reber, M., Krankowsky, D., and Hodges, R.R.: 1995, 'The D/H and $^{16}O/^{18}O$ Ratios in Water from Comet Halley', *Astron. Astrophys.* **302**, 301–316.

Encrenaz, T., Combes, M., and Zéau, Y.: 1978, 'The Spectrum of Jupiter between 10 and 13 Microns: an Estimate of the $^{14}N/^{15}N$ Ratio', *Astron. Astrophys.* **70**, 29–36.

Encrenaz, T., d'Hendecourt, L., and Puget, J.-L.: 1988: 'The Interpretation of the 3.2–3.5 Micron Emisison Feature in the Spectrum of Comet P/Halley: Abundances in the Comet and in Interstellar Matter', *Astron. Astrophys.* **207**, 162–173.

Fegley, B., Jr., and Lodders, K.: 1994, 'Chemical Models of the Deep Atmospheres of Jupiter and Saturn', *Icarus* **110**, 117–154.

Feuchtgruber, H., Lellouch, E., Bézard, B., Encrenaz, T., de Graauw, T., Davis, G.R.: 1999, 'Detection of HD in the Atmospheres of Uranus and Neptune: a new Determination of the D/H Ratio', *Astron. Astrophys.* **341**, L17–L21.

Fouchet, T.: 2002, private communication.

Fouchet, T., Lellouch, E., Bézard, B., Encrenaz, T., Drossart, P., Feuchtgruber, H., de Graauw, T.: 2000, 'ISO-SWS Observations of Jupiter: Measurement of the Ammonia Tropospheric Profile and of the $^{15}N/^{14}N$ Ratio', *Icarus* **143**, 223–243.

Gautier, D., and Owen, T.: 1989, 'The Composition of Outer Planet Atmospheres', *in* S.K. Atreya, J.B. Pollack, and M.S. Matthews (eds.), *Origin and Evolution of Planetary and Satellite Atmospheres*, Univ. Arizona Press, Tucson, pp. 487–512.

Gautier, D., Hersant, F., Mousis, O., and Lunine, J.: 2001a, 'Enrichments in Volatiles: a new Interpretation of the Galileo Measurements', *Astrophys. J.* **550**, L227–L230.

Gautier, D., Hersant, F., Mousis, O., and Lunine, J.: 2001b, 'Erratum: Enrichments in Volatiles: a new Interpretation of the Galileo Measurements', *Astrophys. J.* **559**, L183.

Geiss, J.: 1987, 'Composition Measurements and the History of Cometary Matter', *Astron. Astrophys.* **187**, 859–866.

Geiss, J., and Gloeckler, G.: 1998, 'Abundances of Deuterium and Helium-3 in the Protosolar Cloud', *Adv. Space Res.* **84**, 239–250.

Grevesse, N., and Sauval, A.J.: 1998, 'Standard Solar Composition', *Space Sci. Rev.* **85**, 161–174.

Griffin, M.J., *et al.*: 1986, 'First Detection of the 56-Micron Rotational Line of D in Saturn's Atmosphere', *Astron. Astrophys.* **315**, L389–L392.

Guillot, T.: 1999, 'A Comparison of the Interiors of Jupiter and Saturn', *Planet. Space Sci.* **47**, 1183–1200.

Guillot, T., and Gladman, B.: 2000, 'Late Planetesimal Delivery and the Composition of Giant Planets', *in* F. Garzon, C. Eiroa, D. de Winter, and T.J. Mahoney (eds.), *Disks, Planetesimals and Planets*, ASP Conf. Ser. **219**, 475–485.

Hersant, F., Gautier, D., and Huré, J.-M.:2001, 'A Two-dimensional Model for the Primordial Nebula Constrained by D/H Measurements in the Solar System: Implications for the Formation of Giant Planets', *Astrophys. J.* **554**, 391–407.

Hidayat, T.: 1997, 'Observations hétérodynes millimétriques et submillimétriques de Titan: étude de la composition chimique de son atmosphère', Ph.D. Thesis, Université Paris VII.

Hidayat, T., Marten, A., Bézard, B., Gautier, D., Owen, T., Matthews, H.E., and Paubert, G.: 1997, 'Millimeter and Submillimeter Heterodyne Observations of Titan: Retrieval of the Vertical Profile of HCN and the $^{12}C/^{13}C$ Ratio', *Icarus* **126**, 170–182.

Holweger, H.: 2001, 'Photospheric Abundances: Problems, Updates, Implications', *in* R.F. Wimmer-Schweingruber (ed.), *Solar and Galactic Composition*, AIP Conf. Ser. **598**, pp. 24–30.

Hubbard, W.B., and McFarlane, J.J.: 1980, 'Theoretical Preditions of Deuterium Abundances in the Jovian Planets', *Icarus* **44**, 676–682.

Hunten, D.M., Tomasko, M.G., Flasar, M.F., Samuelson, R.E., Strobel, D.F., and Stevenson, D.J.: 1983, 'Titan', *in* T. Gehrels and M.S. Matthews (eds.), *Saturn*, Univ. Ariz. Press, pp. 671–759.

Irvine, W.M., and Knacke, R.F.: 1999, 'The Chemistry of Interstellar Gas and Grains', *in* S.K. Atreya, J.B. Pollack, and M.S. Matthews (eds.), *Origin and Evolution of Planetary and Satellite Atmospheres*, Univ. Arizona Press, Tucson, pp. 3–34.

Kouchi, A., Yamamoto, T., Kozasa, T., Kuroda, T., and Greenberg, J.M.: 1994, 'Conditions for Condensation and Preservation of Amorphous Ice and Crystallinity of Astrophysical Ices', *Astron. Astrophys.* **290**, 1009–1018.

Krankowsky, D.: 1991, 'The Composition of Comets', *in* R.L. Newburn, M. Neugebauer, and J. Rahe (eds.), *Comets in the Post-Halley Era*, Kluwer Acad. Publ., pp. 855–877.

Lammer, H., Stumptner, W., Molina-Cuberos, G.J., Bauer, S.J., and Owen, T.: 2000, 'Nitrogen Isotope Fractionation and its Consequence for Titan's Atmospheric Evolution', *Planet. Space Sci.* **48**, 529–543.

Lammer, H., and Bauer, S.J.: 2003, 'Isotopic Fractionation by Gravitational Escape', *Space Sci. Rev.*, this volume.

Lecluse, C., Robert, F., Gautier, D., and Guiraud, M.: 1996, 'Deuterium Enrichment in the Giant Planets', *Planet. Space Sci.* **44**, 1579–1592.

Lellouch, E., Bézard, B., Fouchet, T., Feuchtgruber, H., Encrenaz, T., and de Graauw, T.: 2001, 'The Deuterium Abundance in Jupiter and Saturn from ISO-SWS Observations', *Astron. Astrophys.* **370**, 610–622.

Lunine, J.I., Yung, Y.L., and Lorentz, R.D.: 1999, 'On the Volatile Inventory of Titan from Isotopic Abundances in Nitrogen and Methane', *Planet. Space Sci.* **47**, 1291–1303.

Mahaffy, P.R., Donahue, T.M., Atreya, S.K., Owen, T.C., and Niemann, H.B.: 1998, 'Galileo Probe Measurements of D/H and ^{3}He/^{4}He in Jupiter's Atmosphere', *Space Sci. Rev.* **84**, 251–263.

Mahaffy, P.R., Niemann, H.B., Alpert, A., Atreya, S.K., Demick, J., Donahue, T.M., Harpold, D.N., and Owen, T.C.: 2000, 'Noble Gas Abundance and Isotope Ratios in the Atmosphere of Jupiter from the Galileo Probe Mass Spectrometer', *J. Geophys. Res.* **105**, 15061–15071.

Marten, A., Courtin, R., Gautier, D., and Lacombe, A.: 1980, 'Ammonia Vertical Density Profiles on Jupiter and Saturn from their Radioelectric and Infrared Emissivities', *Icarus* **41**, 410–422.

Marten, A., Hidayat, T., Moreno, R., Paubert, G., Bézard, B., Gautier, D., Matthews, H.E., and Owen, T: 1997, IAU Circ. 6702, 1997 July 19.

Matthews, H., and Jewitt, D.: 1997, IAUC 6567, 1997 Feb. 24.

Matthews, H.E., Marten, A., Moreno, R., and Owen, T.: 2002, 'The Jovian Stratosphere after Comet Shoemaker-Levy 9: Unusual Isotopic Ratios in Molecular Trace Constituents', *Astrophys. J.* **580**, 598–605.

Meier, R., Owen, T., Matthews, H.E., Jewitt, D.C., Bockelée-Morvan, D., Biver, N., Crovisier, J., and Gautier, D.: 1998, 'A Determination of the HDO/H_2O Ratio in Comet C/1995 O1 (Hale-Bopp)', *Science* **279**, 842–844.

Meier, R., and Owen, T.: 1999, 'Cometary Deuterium', *in* K. Altwegg, P. Ehrenfreund, J. Geiss, and R. Huebner (eds.), *Composition and Origin of Cometary Materials*, *Space Sci. Rev.* **90**, 33–44.

Mizuno, H.: 1980, 'Formation of the Giant Planets', *Prog. Theor. Phys.* **64**, 544–557.

Mousis, O., Gautier, D., and Bockelée-Morvan, D.: 2002, 'An Evolutionary Turbulent Model of Saturn's Subnebula: Implications for the Origin of the Atmophere of Titan', *Icarus* **156**, 162–175.

Nier, A., and McElroy, M.B.: 1977, 'Composition and Structure of Mars' Upper Atmosphere: Results from the Neutral Mass Spectrometers at Viking 1 and 2', *J. Geophys. Res.* **82**, 4341–4350.

Niemann, H.B., *et al.*: 1998, 'The Composition of the Jovian Atmosphere as Determined by the Galileo Probe Mass Spectrometer', *J. Geophys. Res.* **103**, 22831–22845.

Noll, K.S., and Larson, H.P.: 1991, 'The Spectrum of Saturn from 1990 to 2230 cm^{-1}: Abundances of ASH_3, CH_3D, CO, GeH_4, NH_3 and PH_3', *Icarus* **89**, 168–189.

Orton, G.S., Lacy, J.H., Achtermann, J.M., Parmar, P., and Blass, W.E.: 1992, 'Thermal Spectroscopy of Neptune: the Stratospheric Temperature, Hydrocarbon Abundances, and Isotopic Ratios', *Icarus* **100**, 541–555.

Owen, T.: 2000a, 'Planets and Life', *in* B. Schuerman (ed.), *Darwin and Astronomy*, pp. 163–172.

Owen, T.: 2000b, 'On the Origin of Titan's Atmosphere', *Planet. Space Sci.* **48**, 747–752.

Owen, T., and Bar-Nun, A.: 1995, 'Comets, Impacts and Atmospheres', *Icarus* **116**, 215–226.

Owen, T., and Bar-Nun, A.: 2000, 'From the Interstellar Medium to Planetary Atmospheres via Comets', *in: Bioastronomy* **99**: A new Era in the Search for Life.

Owen, T., Lutz, B.L., and de Bergh, C.: 1986, 'Deuterium in the Outer Solar System: Evidence for Two Distinct Reservoirs', *Nature* **320**, 244–246.

Owen, T.C., Atreya, S.K., Mahaffy, P.R., Niemann, H.B., and Wong, M.H.: 1997, 'On the Origin of Jupiter's Atmosphere and the Volatiles on the Medicean Stars', *in* C. Barbieri *et al.* (eds.), *The Three Galileos: the Man,, the Spacecraft, the Telescope*, Kluwer Acad. Publ., p. 289.

Owen, T., Mahaffy, P., Niemann, H.B., Atreya, S., Donahue, T., Bar-Nun, A., and de Pater, I.: 1999a, 'A Low-temperature Origin for the Planetesimals that Formed Jupiter', *Nature* **402**, 269–270.

Owen, T., Biver, N., Marten, A., Matthews, H.E., and Meier, R.: 1999, IAU Circ. 7306, 1999 Nov. 11.

Owen, T., Mahaffy, P.R., Niemann, H.B., Atreya, S.K., and Wong, M.: 2001, 'Protosolar Nitrogen', *Astrophys. J.* **553**, L77–L79.

Pepin, R.O.: 2003, 'On Noble Gas Processing in the Solar Accretion Disk', *Space Sci. Rev.*, this volume.

Pepin, R.O., Becker, R.H., and Schlutter, D.J.: 1999, 'Irradiation Records in Regolith Materials, I: Isotopic Compositions of Solar-wind Neon and Argon in Single Lunar Mineral Grains', *Geochim. Cosmochim. Acta* **63**, 2145–2162.

Pollack, J.B., and Bodenheimer, P.: 1989, 'Theories of the Origin and Evolution of the Giant Planets', *in* S.K. Atreya, J.B. Pollack, and M.S. Matthews (eds.), *Origin and Evolution of Planetary and Satellite Atmospheres*, Univ. Arizona Press, Tucson, pp. 564–602.

Pollack, J.B., *et al.*: 1996, 'Formation of the Giant Planets by Concurrent Accretion of Solids and Gas', *Icarus* **124**, 62–85.

Roulston, M.S., and Stevenson, D.J.: 1995, 'Predictions of Neon Depletion in Jupiter's Atmosphere' (abstract), *Eos Trans. AGU* **76 (46)**, Fall Meet. Suppl. F, p. 343.

Samuelson, R.E., Maguire, W.C., Hanel, R.A., Kunde, V.G., Jennings, D.E., Yung, Y.L., and Aikin, A.C.: 1983, 'CO_2 on Titan', *J. Geophys. Res.* **88**, 8709–8715.

Stevenson, D.J., and Salpeter, E.E.: 1977, 'The Dynamics and Helium Distribution Properties for Hydrogen-helium Fluid Planets', *Astrophys. J. Suppl.* **35**, 239–261.

Strobel, D.F.: 1982, 'Chemistry and Evolution of Titan's Atmosphere', *Origins of Life* **12**, 244.

Terzieva, R., and Herbst, E.: 2000, 'The Possibility of Nitrogen Isotoic Fractionation in Interstellar Clouds', *Month. Not. Royal Astron. Soc.* **317**, 563–568.

Tokunaga, A.T., Ridgway, S.T., and Knacke, R.F.: 1980, 'High Spatial and Spectral Resolution 10-microns Observations of Jupiter', *Icarus* **44**, 93–101.

von Zahn, U., Hunten, D.M., and Lehmacher, G.: 1998, 'Helium in Jupiter's Atmosphere: Results from the Galileo Probe Helium Interferometer Experiment', *J. Geophys. Res.* **103**, 22815–22829.

Weaver, H.A., Feldman, P.D., Combi, M.R., Krasnopolsky, V., Lisse, C.M., and Shemansky, D.E.: 2002, 'A Search for Argon and OVI in three Comets using the Far Ultraviolet Spectroscopic Explorer', *Astrophys. J.* **576**, L95–L98.

Weidenschilling, S.J.: 1997, 'The Origin of Comets in the Solar Nebula; A Unified Model', *Icarus* **127**, 290–306.

Weisstein, E.W., and Serabyn, E.: 1994, 'Detection of the 267 GHz J=1-0 Rotational Transition of PH_3 in Saturn with a new Fourier-transform Spectrometer', *Icarus* **109**, 367–381.

Womack, M., Wykoff, S., and Ziurys, L.M.: 1992, 'Observational Constraints on Solar Nebula Nitrogen Chemistry', *Astrophys. J.* **401**, 728–735.

Wong, M.H., *et al.*: 2002a, 'Condensible Volatiles in the Galileo Probe Entry Site', in preparation.

Wong, A.S., Morgan, C.G., Yung, Y. L., and Owen, T.: 2002b, 'Evolution of CO in Titan', *Icarus* **155**, 382–392.

Yung, Y.L., Allen, M., and Pinto, J.P.: 1984, 'Photochemistry of the Atmosphere of Titan – Comparison between Models and Observations', *Astrophys. J. Suppl.* **55**, 465–506.

Address for Offprints: T. Owen, Institute for Astronomy, University of Hawaii at Manoa, 2680 Woodlawn Ave., Honolulu HI 96822, USA; owen@ifa.hawaii.edu

ISOTOPIC ABUNDANCES IN COMETS

KATHRIN ALTWEGG[1] and DOMINIQUE BOCKELÉE-MORVAN[2]
[1] *Physikalisches Institut, University of Bern, Sidlerstr. 5, CH-3012 Bern, Switzerland*
[2] *LESIA, Observatoire de Paris, 5 place Jules Janssen, F-92195 Meudon, France*

Received: 1 June 2002, Accepted in final form: 31 October 2002

Abstract. Isotopic ratios in comets provide keys for the understanding of the origin of cometary material, and the physical and chemical conditions in the early Solar Nebula. We review here measurements acquired on the D/H, $^{12}C/^{13}C$, $^{16}O/^{18}O$, $^{14}N/^{15}N$, $^{32}S/^{34}S$ ratios in dust and gases, and discuss their cosmogonic implications. The prospects for future measurements from cometary space missions and remote sensing observations at millimeter and submillimeter wavelengths are presented.

Keywords: Comets, Isotopic Ratios, Deuterium

1. Introduction

The most primitive material in our Solar System can be found in comets. Comets were formed at large distances from the primordial Sun and have remained for most of their lifetime outside of the orbit of Pluto. Their size is small so that internal heating is negligible. They have therefore preserved most of their original material. Their orbits are rather easily perturbed over time, often resulting in perihelia within the inner Solar System. This makes them accessible for in situ measurements by spacecraft as well as remote sensing from ground or from spacecraft. Measurement of isotope abundances, identification of complex organic molecules, discovery of the CHON-grains and low nitrogen abundance all suggest that cometary material is closely related to what is observed in the interstellar medium with perhaps only minor processing in the primordial Solar Nebula. There remain a number of open questions about cometary constituents which will be addressed by future space missions like Stardust and Rosetta.

2. Review of Results

2.1. Deuterium

Isotopic ratios provide clues to the physical conditions that prevailed during the formation of cometary volatiles. The most important one is the D/H ratio in different molecules. Table I summarizes all deuterium measurements in comets; it updates the previous review on cometary deuterium by Meier and Owen (1999).

Space Science Reviews **106:** 139–154, 2003.

TABLE I

Deuterium measurements in different comets and different molecules.

Molecule	Comet	D/H	Method	Ref.[a]
H_2O	1P/Halley	$(3.08\,^{+0.38}_{-0.53}) \times 10^{-4}$	mass spectro	[1]
	1P/Halley	$(3.06 \pm 0.34) \times 10^{-4}$	mass spectro	[2]
	C/1996 B2 (Hyakutake)	$(2.9 \pm 1.0) \times 10^{-4}$	radio spectro	[3]
	C/1995 O1 (Hale-Bopp)	$(3.3 \pm 0.8) \times 10^{-4}$	radio spectro	[4]
HCN	C/1995 O1 (Hale-Bopp)	$(2.3 \pm 0.4) \times 10^{-3}$	radio spectro	[5]
H_2CO	1P/Halley	$< 2 \times 10^{-2}$	mass spectro	[1]
	C/1995 O1 (Hale-Bopp)	$< 5 \times 10^{-2}$	radio spectro	[6]
CH_3OH[b]	1P/Halley	$< \approx 1 \times 10^{-2}$	mass spectro	[7]
CH_3OD	C/1995 O1 (Hale-Bopp)	$< 3 \times 10^{-2}$	radio spectro	[6]
CH_2DOH	C/1995 O1 (Hale-Bopp)	$< 1.4 \times 10^{-3}$	radio spectro	[6]
NH_3 (from NH)	C/1996 B2 (Hyakutake)	$< 6 \times 10^{-2}$	visible spectro	[8]
NH_3 (from NH_2D)	C/1995 O1 (Hale-Bopp)	$< 4 \times 10^{-2}$	radio spectro	[6]
H_2S	C/1995 O1 (Hale-Bopp)	$< 2.0 \times 10^{-1}$	radio spectro	[6]
CH	C/1996 B2 (Hyakutake)	$< 3 \times 10^{-2}$	visible spectro	[8]

[a] References: [1] Balsiger *et al.* (1995); [2] Eberhardt *et al.* (1995); [3] Bockelée-Morvan *et al.* (1998); [4] Meier *et al.* (1998a); [5] Meier *et al.* (1998b); [6] Crovisier *et al.* (2002); [7] Eberhardt *et al.* (1994); [8] Meier *et al.* (1998c).

[b] CH_2DOH and CH_3OD averaged.

Mass spectrometers aboard the European spacecraft Giotto obtained the first accurate D/H ratios in the water of comet 1P/Halley (Balsiger *et al.*, 1995; Eberhardt *et al.*, 1995). HDO was then detected in comets C/1996 B2 (Hyakutake) (Bockelée-Morvan *et al.*, 1998) and C/1995 O1 (Hale-Bopp) (Meier *et al.*, 1998a), from ground-based observations of its 1_{01}–0_{00} rotational line at submillimeter wavelengths. The detection of DCN in comet Hale-Bopp (Meier *et al.*, 1998b), and upper limits for several other D-bearing molecules, mainly by visible or radio spectroscopy (Table I), comprise the limited sample of D/H measurements.

All of these comets exhibit a similar D/H ratio in H_2O ($\approx 3 \times 10^{-4}$), about twice higher than the D/H value in terrestrial water, and approximately one order of magnitude larger relative to the protosolar value in H_2 ($\approx 2.5 \times 10^{-5}$). The D/H ratio in cometary HCN is 7 times higher than the value in cometary H_2O. So far, all cometary D/H measurements given in Table I refer to bulk compositions, and it is conceivable that significant departures from the mean value could occur at the grain-size level. Blake *et al.* (1999) inferred higher D/H ratios in jets compared to the surrounding coma, from radio interferometric maps of HDO and DCN in comet Hale-Bopp. These inhomogeneities are to be confirmed by observations of better quality. Note that strong isotope effects as a result of coma chemistry can be excluded for H_2O and HCN (Rodgers and Charnley, 2002).

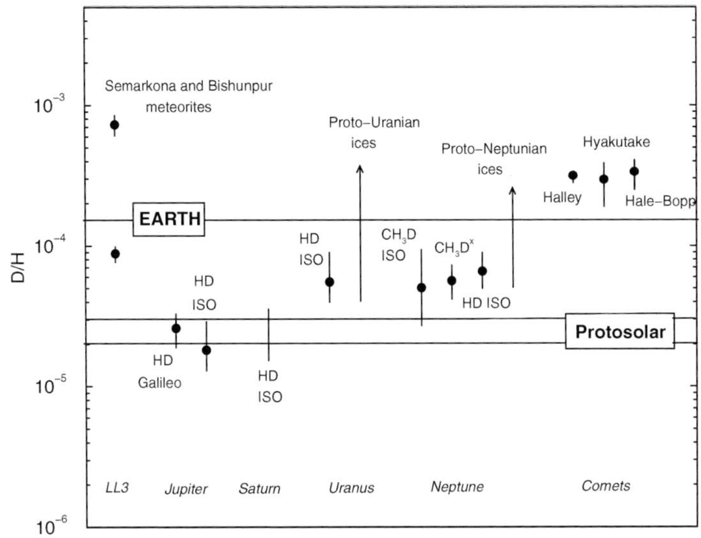

Figure 1. D/H ratios in the Solar System. From Hersant *et al.* (2001). For comets, only measurements of D/H in water are shown.

The D/H ratio of $\approx 3 \times 10^{-4}$ measured in water cannot be only explained by the isotopic exchanges between H_2O and H_2 (the main deuterium reservoir) which occurred in the dense Solar Nebula. Calculations show that, starting with a $(D/H)_{H_2O}$ value equal to the protosolar D/H value in H_2, only a moderate enrichment, by a factor of 3 at most with respect to the protosolar value, can be obtained. In contrast, deuterium fractionation is very efficient in the low temperature regions of the interstellar medium where ion-molecule and grain-surface reactions operate (e.g. Millar et al., 1989; Brown and Millar, 1989). Aikawa and Herbst (1999) show that, despite higher densities, deuterium fractionation is occurring in a way similar to the interstellar medium while matter is collapsing to form protoplanetary disks. These authors also predict significant deuterium enrichment in the outer disk, during the disk phase, providing high enough ionization rate and low temperatures. Therefore, the high D/H values measured in comets reflect, at least partly, fractionation effects which took place through ion-molecule or grain-surface processes prior to the Solar Nebula phase.

We present in Section 3.1 a detailed comparison between cometary and interstellar D/H ratios, which possibly suggests that there was some reprocessing in the Solar Nebula (Section 4). Assuming that this reprocessing did not occur, Meier *et al.* (1998b) inferred an effective formation temperature of HCN and H_2O cometary molecules of ≈ 30 K, on the basis of a simple ion-molecule gas phase model. Aikawa and Herbst (1999) were able to reproduce the D/H ratios in HCN and H_2O measured in comets with their chemical model of protoplanetary disks.

Figure 1 shows deuterium measurements in the Solar System. A comparison of the cometary $(D/H)_{H_2O}$ ratio with values found in the atmospheres of the outer planets (see the review of Owen and Encrenaz, 2003, in this volume) is consistent with the long-held idea that the giant planets formed around icy cores with a high cometary D/H ratio, and subsequently accumulated significant amounts of H_2 from the Solar Nebula with a low protosolar D/H (Hersant *et al.*, 2001).

The D/H ratio in Earth's oceans is twice lower than in cometary water, but similar to the average D/H in carbonaceous chondrites. From this, it has been estimated that impacting comets would have contributed to at most 10% of Earth's water inventory (Dauphas *et al.*, 2000). At early stages, the Earth formed from the accretion of water-poor planetesimals. The bulk of terrestrial water was likely carried at the final stage of Earth formation, by a few planetary embryos originally formed in the Asteroidal Belt with chondritic average D/H (Morbidelli *et al.*, 2000).

2.2. Other Isotopes

The measurements of isotopes other than deuterium in the volatile part of cometary coma are limited (Table II). So far, most isotope ratios reported for comets (except D/H) are compatible with the terrestrial values $^{12}C/^{13}C = 89$, $^{16}O/^{18}O = 490$, $^{14}N/^{15}N = 270$, $^{32}S/^{34}S = 24$. Differences in the isotopic composition among different components in a comet would provide important diagnostics for the origin of such material. An unequivocal detection of isotopically light carbon was reported by Jessberger and Kissel (1991): $^{12}C/^{13}C$ ratios as high as 5000 in particles composed of almost pure carbon link cometary matter to certain circumstellar grains (Zinner *et al.*, 1990; Anders and Zinner, 1993; see also the review of Ott, 2003, in this volume). These grains are almost certainly presolar (Eberhardt, 1998), although they are thought to originate from a specific nucleosynthetic (stellar) source, rather than to result from isotope fractionation. The presence of the wide range of isotopic compositions of carbon excludes any equilibration processes affecting the carbon carrier during comet formation or later in its history.

3. Comparison with the Interstellar Medium

3.1. Deuterium

Measured cometary D/H values can be compared to those measured or expected in cold molecular clouds (see the review of Roueff and Gerin, 2003, in this volume). In such regions, as the TMC-1 cloud, the DCN/HCN value is as high as 2×10^{-2}, about a factor of 10 higher than in comets. Similar high DCN/HCN values are observed in the young (Class I) low-mass stellar object IRAS 16293-2422 in Ophiuchus (van Dishoeck *et al.*, 1995), as well as around T Tauri stars (Kessler *et al.*, 2000). In IRAS 16293-2422, which presents striking similarities in deuterium fractionation with TMC-1 for several molecular species, the HDO/H_2O ratio has been estimated to > 0.01 (i.e. $D/H > 0.005$) (van Dishoeck *et al.*, 1995).

TABLE II
Other isotope measurements in different comets.

Isotopic Ratio	Value	Comet	References
$[^{12}C^{12}C/^{13}C^{12}C]$	93 ± 10	4 comets[a]	Wyckoff *et al.* (2000)
$[^{12}CN/^{13}CN]$	95 ± 12	1P/Halley	Kleine *et al.* (1995)
	90 ± 10	5 comets[b]	Wyckoff *et al.* (2000)
$[H^{12}CN/H^{13}CN]$	34 ± 12[c]	Hyakutake	Lis *et al.* (1997)
	111 ± 12	Hale-Bopp	Jewitt *et al.* (1997)
	109 ± 22	Hale-Bopp	Ziurys *et al.* (1999)
	90 ± 15	Hale-Bopp	Lis *et al.* (1997)
$[HC^{14}N/HC^{15}N]$	323 ± 46	Hale-Bopp	Jewitt *et al.* (1997)
	330 ± 98	Hale-Bopp	Ziurys *et al.* (1999)
$[H_2^{16}O/H_2^{18}O]$	518 ± 45	1P/Halley	Balsiger *et al.* (1995)
	470 ± 40	1P/Halley	Eberhardt *et al.* (1995)
$[C^{32}S/C^{34}S]$	27 ± 3	1P/Halley	Jewitt *et al.* (1997)
$[^{32}S/^{34}S]$	23 ± 6	1P/Halley	Altwegg (1996)
$[H_2^{32}S/H_2^{34}S]$	15 ± 3	Hale-Bopp	Crovisier *et al.* (2002)

[a] mean ratio in C_2 from observations in 4 comets quoted in Wyckoff *et al.* (2000): Ikeya 1963 I, Tago-Sato-Kosaka 1969 IX, Kohoutek 1973 XII and Kobayashi-Berger-Milon 1975 IX;
[b] mean ratio in CN from five comets: 1P/Halley, C/1990 K1 (Levy), C/1989 X1 (Austin), C/1989 XIX (Okazaki-Levy-Rudenko) and C/1995 O1 (Hale-Bopp);
[c] ratio possibly affected by line blending (Lis *et al.*, 1997).

There is also evidence that cometary D/H values are somewhat lower than in interstellar grains. Measurements of the HDO/H_2O ratio in interstellar grain mantles indicate D/H ratios between 4 and 15×10^{-4} (Teixeira *et al.*, 1999). However, more recent investigations of HDO ice along the line of sight of young stellar objects would suggest lower deuterium fractionation (L. d'Hendecourt, private communication).

A large amount of deuterium data has been collected towards galactic hot cores. Comparison between D/H in hot cores and comets is valuable since there are striking similarities between the gas composition of hot cores and comets (Bockelée-Morvan *et al.*, 2000; see Section 5 and Figure 2). Hot cores are parts of giant molecular clouds where high-mass stars are forming. In such warm regions (150–200 K), the deuterium fractionation observed in several species (including HDO, DCN, NH_2D, HDCO, CH_3OD) is strong evidence that the gas-phase composition reflects a previous cold phase. It is thus believed that some species observed in the gas phase result from the sublimation of icy grain mantles, while others could reflect chemistry in the hot core. For species such as NH_2D, HDCO, CH_3OD and CH_2DOH, the level of deuterium fractionation ([XD]/[XH]) in hot cores is high and exceeds a few %. The values for HDCO and CH_2DOH (D/H of 7% and 1.3%, respectively) are higher than the upper limits obtained in comet Hale-Bopp (Table I). In contrast, most $(D/H)_{H_2O}$ values in hot cores are in the range $1\text{–}3 \times 10^{-4}$ (e.g. Helmich *et al.*, 1996; Gensheimer *et al.*, 1996), i.e., comparable to the cometary value. The D-enrichment in HCN is low and the range of measured values ($1\text{–}4 \times 10^{-3}$, Hatchell *et al.*, 1998; Schilke *et al.*, 1992) comprises the cometary value. A low D/H value is also observed for H_2S in hot cores ($[HDS]/[H_2S] < 10^{-3}$, Hatchell *et al.*, 1999). These low deuterium fractionations could either be explained by ice formation at relatively high temperatures (30 to 80 K, depending on the species) or by alteration in the hot core phase. These two alternatives are proposed in the literature (Hatchell *et al.*, 1998). In the latter case, the similarity with cometary D/H values would be fortuitous.

3.2. Other Isotopes

Unlike deuterium, chemical fractionation of $^{13}C/^{12}C$, $^{18}O/^{16}O$, $^{15}N/^{14}N$, and $^{34}S/^{32}S$ is predicted to be of small importance in the interstellar medium (Langer *et al.*, 1984). Species-dependent isotopic ratios have been observed for carbon isotopes in CO and H_2CO. Although not observed yet and of small extent, species-dependent fractionation is also expected for $^{15}N/^{14}N$ (Terzieva and Herbst, 2000). Differences in the isotopic composition among different molecules in comets would provide keys for the origin of cometary material. So far, most $^{12}C/^{13}C$, $^{16}O/^{18}O$, $^{14}N/^{15}N$, and $^{32}S/^{34}S$ ratios measured in cometary volatiles are compatible with the terrestrial values. Comparing $^{13}C/^{12}C$ ratios in C_2 and HCN and CN species, Wyckoff *et al.* (2000) conclude to minimal isotopic fractionation in the protosolar precursor molecular cloud. The low $^{32}S/^{34}S$ ratio measured in H_2S in comet Hale-Bopp (Table II) is puzzling since chemical fractionation is expected to be lower for sulfur than for carbon.

Solar System isotopic ratios are significantly different from those characteristic of the local interstellar medium today. This is because C, N, O isotope ratios in our Galaxy have significantly evolved since the birth of our Solar System due to nucleosynthesis and reprocessing in stars (Wielen and Wilson, 1997).

4. Reprocessing of D/H in the Solar Nebula

The low cometary D/H ratios in H_2O and HCN, when compared to those measured in the interstellar medium, may suggest that there was some reprocessing in the Solar Nebula before these two molecules were incorporated into comets. Still more questioning, very high D-enrichments are measured in some phases of primitive meteorites (Robert *et al.*, 2000). In particular, the D-enrichment in cometary water is by a factor of 2.5 lower than in the highly D-enriched component of meteorites of LL3 class, Semarkona and Bishunpur (Figure 1). This led Drouart *et al.* (1999) to suggest that these high meteoritical D/H values might be more representative of the grains of the presolar cloud. The lower value measured in comets would then result from isotopic exchanges between vapor originating from ices in the presolar cloud and H_2, a process that would have occurred in the Solar Nebula.

Isotopic exchanges with H_2 are only possible in gaseous phase. Turbulent models of the Solar Nebula indicate that ices infalling from the presolar cloud were likely volatilized at distances up to a few ten astronomical units, owing to the high temperatures they experienced in the Solar Nebula at early times. However, for kinetic reasons, the isotopic exchange could have been substantial only in the inner hotter regions of the Solar Nebula. The D/H value measured in cometary water could then be explained if there was some mixing between water vapor reprocessed in the inner regions and the less processed component present in the outer regions. Subsequently, ices condensed as the Nebula was cooling down, they agglomerated and were incorporated in cometesimals. This scenario was examined quantitatively for HDO by Drouart *et al.* (1999). These authors used a turbulent 1-D evolutionary model of the Solar Nebula and solved diffusion equations which take into account inward mass transport into the Sun and diffusive mixing by turbulence. Mousis *et al.* (2000) extended this model to the DCN case. Hersant *et al.* (2001) used a 2-D evolutionary model which provides an improved description of the nebula. Since the rate of isotopic exchange and the turbulent diffusivity both depend on the temperature and density radial profiles of the Solar Nebula, the D/H enrichments measured in comets and meteorites were used to constrain the Solar Nebula models (Drouart *et al.*, 1999; Mousis *et al.*, 2000; Hersant *et al.*, 2001). Because isotopic exchanges stopped when H_2O and HCN vapors condensed, information on the location and time of formation of cometary ices was also obtained. From the D/H ratios in H_2O and HCN measured in comet Hale-Bopp, Mousis *et al.* (2000) and Hersant *et al.* (2001) show that the microscopic icy grains that subsequently formed the cometesimals were produced in the Uranus-Neptune region and no later than a few 10^5 years after the formation of the Sun.

The idea of turbulent radial mixing of material across wide ranges of the protoplanetary disk was first introduced by Morfill (1983, and subsequent papers). Bockelée-Morvan *et al.* (2002) and Wehrstedt and Gail (2002) suggest that the presence of crystalline silicates coexisting with amorphous silicates in comets can be attributed to large scale mixing in the Solar Nebula. Crystalline silicates would have been formed in the inner regions of the Solar Nebula from the thermal annealing of amorphous silicates issued from the presolar cloud. The D/H ratios and mass fraction of crystalline silicates measured in comet Hale-Bopp can be consistently explained by the same models of Solar Nebula and turbulent diffusion (Bockelée-Morvan *et al.*, 2002). As discussed by Mousis *et al.* (2000), the fact that the D/H ratios acquired in the "interstellar" phase are not preserved in cometary ices is not necessarily in conflict with an interstellar origin of most cometary species. This is

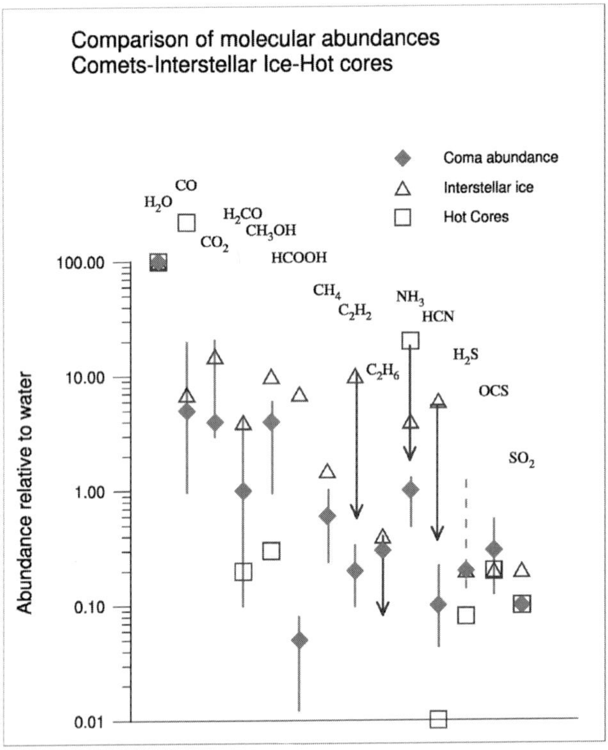

Figure 2. Comparison of molecular abundances in comets, interstellar ices and molecular hot cores.

because the activation energies for isotopic exchanges are generally much lower than those for chemical reactions or molecular pyrolysis.

5. Other Indicators for an Interstellar Origin of Cometary Material

The detection of CHON organic grains in comets, and the identification of interstellar-like organic and other exotic gases sublimated from cometary nuclei, give the definite impression that comets preserve a record of the interstellar composition. Figure 2, based on Crovisier (1998), illustrates a quantitative similarity between interstellar ice and cometary ice. The interstellar data are from infrared spectra of icy grains around protostars, and radio spectra of gaseous molecules in hot cores (e.g. see the review of Ehrenfreund and Charnley, 2000). The cometary data concern coma gases sublimed from the nucleus and dust (see the reviews of Altwegg *et al.*, 1999; Crovisier and Bockelée-Morvan, 1999; Irvine *et al.*, 2000).

The gas phase composition in the interstellar medium depends on many variables, but almost every parent species observed in comets has also been detected in the interstellar medium, either in cold dark molecular clouds or in dense cloud cores where star formation is taking place. The molecular composition of comets, interstellar ices, and hot cores are all characterized by the presence of large quantities of saturated/hydrogenated species (e.g. H_2O, CH_4, CH_3OH, NH_3, H_2S), coexisting with oxidized molecules (e.g. CO_2, SO_2). For interstellar ices, this composition presumably traces a combination of ion-molecule and grain-surface processes.

It has been pointed out a long time ago that the presence of CH_4 and NH_3 in comets is not easily explained by Solar Nebula processes. Their formation requires a high temperature/pressure environment, only found in the inner Solar Nebula or in the giant planets subnebulae. Recently, Mousis *et al.* (2002) investigated $CO \rightarrow CH_4$ and $N_2 \rightarrow NH_3$ thermochemical conversion in the inner Solar Nebula and subsequent turbulent diffusion towards the outer regions. They showed that this mechanism cannot produce significant amounts of methane and ammonia in the formation region of comets. On the basis of experimental studies, Fegley (1999, and references therein) argues for Fe-grain catalysed chemistry in the Jovian protoplanetary subnebula. Theoretical work is needed to quantify the efficiency of mixing between the planetary subnebulae and Solar Nebula. Alternatively, Gail (2001, 2002) studied carbon combustion and its conversion into CO and hydrocarbons. Using a stationary disk model, he showed that significant amounts of CH_4 and C_2H_2 can be produced in the inner nebula and transported towards outer colder regions by turbulent mixing. A quantitative evaluation of whether this process could explain cometary CH_4 and C_2H_2 abundances requires time-dependent models taking into account Solar Nebula evolution, that are actually under study by this author. Nevertheless, carbon combustion cannot account for the presence of C_2H_6 detected in comets Hyakutake, Hale-Bopp and other comets (Gail, 2001). Alkanes have not yet been detected in the interstellar medium. Cometary C_2H_6 holds a clue to later processing of interstellar material in the Solar Nebula, or to the composition of the interstellar medium itself.

Note that N_2 has not been observed directly in the interstellar medium, but its presence is inferred from its detection as protonated ion N_2H^+. N_2 is believed to be the main N-bearing species in the interstellar medium. The low N_2 abundance in comets is likely due to the fact that N_2 does not easily condense as pure ice, and is not easily trapped in water ice as a clathrate hydrate (Iro *et al.*, 2002).

Evidence for interstellar composition in the refractory component of comets comes from observations of cometary dust. In addition to the presence of organic "CHON" grains in comet Halley, which are likely presolar given their variety of $^{12}C/^{13}C$ ratios (Section 2.2), the dust appears similar in infrared spectra to interstellar grains. In some comets, the strong 3.3-3.4 μm band indicative of C-H stretch transitions has features that can be explained by very small particles similar to the polycyclic aromatic hydrocarbons (PAHs) observed in the interstellar medium. The IR continuum of cometary dust also indicates the presence of small carbon grains. As noted in Section 4, some high temperature processing is indicated by the detection of spectral features indicative of crystalline silicates, in addition to amorphous silicates which are likely presolar.

The ortho/para ratios of cometary water and ammonia both indicate a spin temperature of about 30 K (see Irvine *et al.*, 2000; Kawakita *et al.*, 2001). The meaning of this temperature is unclear, although it has been often argued that it could indicate that these species formed or were last processed at about 30 K.

The dust and gas measurements are therefore consistent with an interstellar origin, but with some measure of material that has been processed at higher temperature in the early Solar Nebula. Some cometary material must therefore have survived passage through the accretion shock in the Solar Nebula and then been partially processed in the Solar Nebula and/or partially diluted with more highly processed Solar Nebula material.

6. Future Work

6.1. In Situ Measurements

Several cometary missions which will enhance our knowledge of the isotopic composition of cometary material are either well under way or about to be launched. All of their targets are Jupiter family comets and will therefore complement the measurements done on comets Halley, Hyakutake and Hale-Bopp, which are all Oort cloud comets.

The NASA Stardust spacecraft was launched on February 7, 1999. The primary goal of Stardust is to collect dust and carbon-based samples during its closest encounter with comet 81P/Wild 2 with a flyby scheduled to take place in January 2004 (Brownlee *et al.*, 2000). Not until January 2006, will Stardust return by parachuting a reentry capsule to the Earth's surface. The $\approx$1000 dust grains collected during the comet encounter will then be analyzed in different laboratories. From these measurements we can expect insight into the non-volatile composition of comets. From isotope ratios the non-equilibrated status of dust grains can be assessed, e.g. by measuring $^{13}C/^{12}C$ in the different grains.

The goal of the International Rosetta Mission, to be launched by ESA in January 2003, is a rendez-vous with comet 46P/Wirtanen. Rosetta will study the nucleus of this comet and its environment in great detail for a period of nearly two years, the near-nucleus phase starting at a heliocentric distance of about 3.25 AU in 2011, with far-observation activities leading ultimately to close observation (from about one km distance). The prime scientific objective of the Rosetta mission is to study the origin of comets, the relationship between cometary and interstellar material and its implications with regard to the origin of the Solar System. There are several sensors on board Rosetta, which will address the question of isotopes in a comet:

– The two sensors of ROSINA (Rosetta Orbiter Spectrometer for Ion and Neutral Analysis) (Balsiger *et al.*, 2002) will determine the composition of the comet's atmosphere and ionosphere. ROSINA will have unprecedented capabilities, including: 1) very wide mass range from 1 amu (hydrogen) to more

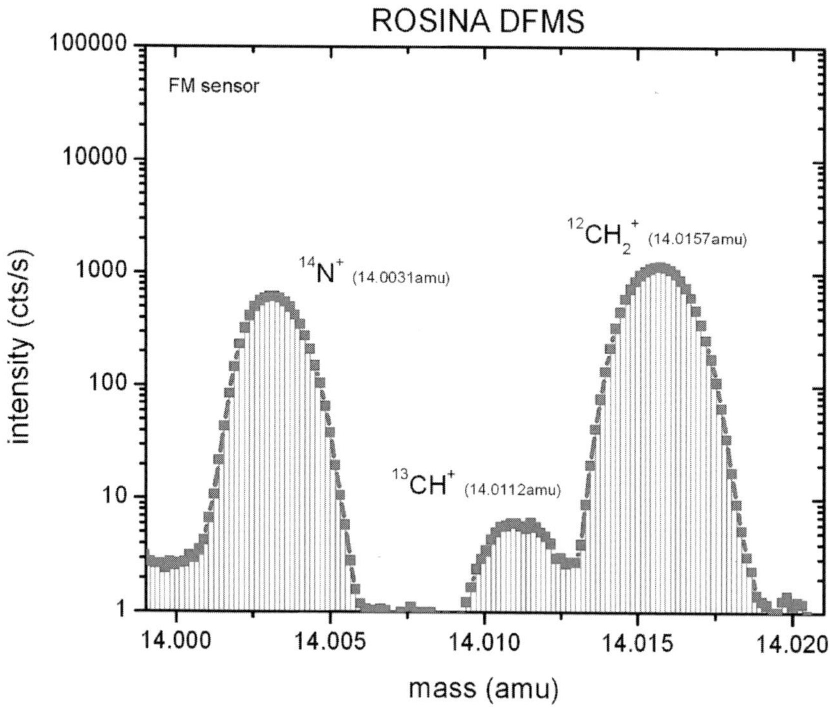

Figure 3. High resolution spectrum of mass 14, where the ability of separating isotopic species, i.e. ^{13}CH and ^{12}CH, is well demonstrated.

than 300 amu (organic molecules); 2) very high mass resolution ($m/\Delta m$ = 3000 at the 1% level, ability to resolve CO from N_2 and ^{13}C from ^{12}CH); 3) very wide dynamic range and high sensitivity to accommodate very large differences in ion and neutral gas concentrations and large changes in the ion and gas flux as the comet approaches its perihelion. An example of the capabilities of the ROSINA DFMS sensor with regard to isotopic measurements is given in Figure 3.

- COSIMA (Cometary Secondary Ion Mass Analyzer) (Kissel *et al.*, 2002) will analyze the characteristics of dust grains emitted by the comet, including their composition and whether they are organic or inorganic.
- PTOLEMY MODULUS (Wright *et al.*, 2002), which will be delivered to the comet surface on board the Rosetta lander, is a gas analysis instrument capable of performing isotopic measurements on individual components from solid samples of the comet's surface. The concept is termed "MODULUS" which is taken to mean Methods Of Determining and Understanding Light elements from Unequivocal Stable isotope compositions. The scientific goal of using MODULUS is to understand the geochemistry of light elements, such as hydrogen, carbon, nitrogen and oxygen, by determining their nature, distribution and stable isotopic compositions.

– MIRO (Microwave Instrument for the Rosetta Orbiter) consists of two heterodyne radiometers, one operating at millimeter wavelengths (190 GHz, ≈1.6 mm), and one operating at submillimeter wavelengths ≈0.5 mm (Gulkis *et al.*, 2002). This experiment will allow to measure the near-surface temperature of the comet, and to study how water, CO, NH_3 and CH_3OH sublimate in time and distance to the Sun. The submillimeter receiver, which is configurated with a high resolution spectrometer to observe molecular lines, is pretuned to cover the 1_{10}–1_{01} rotational lines of $H_2^{16}O$, $H_2^{17}O$, and $H_2^{18}O$ at 556.9, 552.0, and 547.7 GHz, respectively.

6.2. Remote Sensing Measurements

Most measurements of isotopic ratios in cometary volatile species have been made using millimeter or submillimeter spectroscopy with single-dish radio telescopes. However, the capabilities of present ground-based instrumentation limit the investigations to either very active comets, as comet Hale-Bopp (water production rate $Q(H_2O)$ of 10^{31} mol. s^{-1} at perihelion) or active comets making close approach to Earth (e.g. Hyakutake with $Q(H_2O) \approx 2 \times 10^{29}$ mol. s^{-1} at 0.1 AU from Earth).

New, very sensitive radio telescopes are currently under construction. In addition, space-based submillimeter instrumentation will be available in the next decade.

– The Large Millimeter Telescope (LMT) is a joint project of the University of Massachusetts in the USA, and the Instituto Nacional de Astrofísica, Óptica y Electronı́ca in Mexico (Irvine *et al.*, 2002). The LMT will have a diameter of 50 m, work at frequencies up to 275 GHz, and be located at 4600 m altitude in Mexico. The telescope is scheduled for completion in 2004. It will be the largest millimeter telescope of its category in the world, superseding the IRAM 30-m telescope (the Nobeyama 45-m telescope is optimized for working at frequencies lower than 100 GHz). With respect to the IRAM 30-m telescope, we might expect an improvement in sensitivity by a factor of about 2, up to ≈3 for short-lived species. This telescope will allow the investigation of the 242 GHz HDO line in comets a few times less active than comets Hale-Bopp and Hyakutake, assuming similar observational conditions (i.e., distances to Earth). This telescope promises to be very useful for measurements of high D/H fractionation in species other than H_2O (e.g., H_2S), though bright comets as comets Hale-Bopp and Hyakutake are required.
– The Herschel Space Observatory (formerly known as FIRST) is the fourth cornerstone mission in the ESA science "Horizon 2000" plan and will be in operation in early 2007 (see Pilbratt, 2001, for a review). This "observatory type" space mission will carry a 3.5 m diameter telescope with instruments working in the far infrared and submillimeter wavelength ranges (60–670 μm). The three instruments on the payload (SPIRE, PACS, HIFI) provide complementary tools for imaging and spectroscopic studies. The high spectral

resolution of the heterodyne instrument HIFI, which covers approximately the 500–1250 GHz range, is well adapted for spectroscopic investigations in comets. This instrument will provide the opportunity to detect the 1_{10}–1_{01} HDO line at 509 GHz in comets of moderate activity [$Q(H_2O) \approx 5 \times 10^{28}$ mol. s^{-1}] (Bockelée-Morvan and Crovisier, 2001), and to observe simultaneously several H_2O lines for accurate determinations of the D/H ratio.

– The Atacama Large Millimeter Array (ALMA) will be the largest ground-based astronomy project of the next decade. ALMA will comprise some 64 12-meter antennas with baselines extending up to 10 km, and be located in the Atacama desert in Chile at 5000 m altitude. Its receivers will cover the range from 70 to 900 GHz. With ALMA, it will be possible to detect HDO in faint comets with $Q(H_2O) \approx 3 \times 10^{28}$ mol. s^{-1}, as well as to search for other deuterated molecules and isotopic species. The imaging capabilities of ALMA will allow to investigate the gas distribution in the coma and possible gradients in coma composition, including variations in the D/H ratios.

In summary, thanks to the sensitivity of the Herschel Space Observatory and ALMA, it will be possible to investigate the D/H ratios in moderately active comets, including Jupiter family comets presumably formed in the Kuiper Belt.

7. Conclusion

Much progress has been made about the molecular composition of comets since the flyby of comet Halley by an armada of spacecraft. Although a few new isotopic measurements have been acquired, they all concern long-period comets formed in the Oort cloud. Thanks to a number of space cometary missions, and the availability of new telescopes working in the submillimeter domain, the forthcoming years promise D/H determinations in a large sample of comets, including Jupiter family comets. These new measurements, together with theoretical developments and new observational constraints on proto-stellar clouds and accretion disks, should unravel the question of the origin of cometary material.

References

Aikawa, Y., and Herbst, E.: 1999, 'Deuterium Fractionation in Protoplanetary Disks', *Astrophys. J.* **526**, 314–326.

Altwegg, K.: 1996, 'Sulfur in Comet Halley', Habilitationsschrift, University of Bern.

Altwegg, K., Balsiger, H., and Geiss, J: 1999, 'Composition of the Volatile Material in Halley's Coma from *in Situ* Measurements', *Space Sci. Rev.* **90**, 3–18.

Anders, E., and Zinner, E.: 1993, 'Interstellar Grains in Primitive Meteorites: Diamond, Silicon Carbide and Graphite', *Meteoritics* **28**, 490–514.

Balsiger, H., Altwegg, K., and Geiss, J.: 1995, 'D/H and $^{18}O/^{16}O$ Ratio in the Hydronium Ion and in Neutral Water from in Situ Ion Measurements in Comet P/Halley', *J. Geophys. Res.* **100**, 5827–5834.

Balsiger, H., Altwegg, K., Gliem, F., Fieth, B., *et al.*: 2002, 'Rosetta Orbiter Spectrometer for Ion and Neutral Analysis ROSINA', *ESA SP* **1165**, in press.

Blake, G.A., Qi, C., Hogerheijde, M.R., Gurwell, M.A., and Muhleman, D.O.: 1999, 'Sublimation from Icy Jets as a Probe of the Interstellar Content of Comets', *Nature* **398**, 213–216.

Bockelée-Morvan D., and Crovisier, J.: 2001, 'Comets and Asteroids with FIRST', *in: The promise of the Herschel Space Observatory, ESA SP* **460**, 279–286.

Bockelée-Morvan, D., Gautier, D., Hersant, F., and Robert, F.: 2002, 'Turbulent Radial Mixing in the Solar Nebula as the Source of Crystalline Silicates in Comets', *Astron. Astrophys.* **384**, 1107–1118.

Bockelée-Morvan, D., *et al.*: 1998, 'Deuterated Water in Comet C/1996 B2 (Hyakutake) and its Implications for the Origin of Comets', *Icarus* **133**, 147–162.

Bockelée-Morvan, D., *et al.*: 2000, 'New Molecules Found in Comet C/1995 O1 (Hale-Bopp). Investigating the Link Between Cometary and Interstellar Material', *Astron. Astrophys.* **353**, 1101–1114.

Brown, P.D., and Millar, T.J.: 1989, 'Grain-surface Formation of Multi-deuterated Molecules', *Month. Not. Roy. Astron. Soc.* **240**, 25–29.

Brownlee, D.E., *et al.*: 2000, 'Stardust: A Comet Sample Return', *Met. Planet. Sci.* **35**, A35.

Crovisier, J.: 1998, 'Physics and Chemistry in Comets: Recent Results from Comets Hyakutake and Hale-Bopp. Answers to Old Questions and New Enigmas', *Faraday Discussions* **109**, 437–452.

Crovisier, J., and Bockelée-Morvan D.: 1999, 'Remote Observations of the Composition of Cometary Volatiles', *Space Sci. Rev.* **90**, 19–32.

Crovisier, J., Bockelée-Morvan D., Colom, P., Biver, N., Despois, D., Lis, D.C., Benford, D.J., Mehringer, D., and the Team for Target-of-opportunity Radio Observations of Comets: 2002, 'The Composition of Ices in Comet C/1995 O1 (Hale-Bopp): Upper Limits on Undetected Species from Radio Spectroscopy', *Astron. Astrophys.*, to be submitted.

Dauphas, N., Robert, F., and Marty, B.: 2000, 'The Late Asteroidal and Cometary Bombardment on Earth as Recorded in Water Deuterium to Protium Ratio', *Icarus* **148**, 508–512.

Drouart, A., Dubrulle, B., Gautier, D., and Robert, F.: 1999, 'Structure and Transport in the Solar Nebula from Constraints on Deuterium Enrichment and Giant Planets Formation', *Icarus* **140**, 129–155.

Eberhardt, P.: 1998, 'The in Situ View', IAU Colloq. 168, Cometary Nuclei in Space and Time, *in* M. A'Hearn (ed.), *Astron. Soc. of the Pacific Conf. Series*, in press.

Eberhardt, P., Meier, R., Krankowsky, D., and Hodges, R.R.: 1994, 'Methanol and Hydrogen Sulfide in Comet 1P/Halley', *Astron. Astrophys.* **288**, 315–329.

Eberhardt, P., Reber, M., Krankowsky, D., and Hodges, R.R.: 1995, 'The D/H and $^{18}O/^{16}O$ Ratios in Water from Comet P/Halley', *Astron. Astrophys.* **302**, 301–316.

Ehrenfreund, P., and Charnley, S.B.: 2000, 'Organic Molecules in the Interstellar Medium, Comets, and Meteorites: a Voyage from Dark Clouds to the Early Earth', *Ann. Rev. Astron. Astrophys.* **38**, 427–483.

Fegley, B., Jr.: 1999, 'Chemical and Physical Reprocessing of Presolar Materials in the Solar Nebula and the Implications for Preservation of Presolar Materials in Comets', *Space Sci. Rev.* **90**, 239–252.

Gail, H.-P.: 2001, 'Radial Mixing in Protoplanetary Accretion Disks. I. Stationary Disk Models with Annealing and Carbon Combustion', *Astron. Astrophys.* **378**, 192–213.

Gail, H.-P.: 2002, 'Radial Mixing in Protoplanetary Accretion Disks. III Carbon Dust Oxidation and Abundance of Hydrocarbons in Comets', *Astron. Astrophys.*, in press.

Gensheimer, P.D., Mauersberger, R., and Wilson, T.L.: 1996, 'Water in Galactic Hot Cores', *Astron. Astrophys.* **314**, 281–294.

Gulkis, S., Frerking, M., Janssen, M., Hartogh, P., Beaudin, G., Koch, T., Salinas, Y., and Kahn, C.: 2002, 'MIRO: Microwave Instrument for the Rosetta Orbiter', *ESA SP* **1165**, in press.

Hatchell, J., Millar, T.J., and Rodgers, S.D.: 1998, 'The DCN/HCN Abundance Ratio in Hot Molecular Cores', *Astron. Astrophys.* **332**, 695–702.

Hatchell, J., Roberts, H., and Millar, T.J.: 1999, 'Limits on HDS/H_2S Abundance Ratios in Hot Molecular Cores', *Astron. Astrophys.* **346**, 227–232.

Helmich, E.P., van Dishoeck, E.F., and Jansen, D.J.: 1996, 'The Excitation and Abundance of HDO towards W3(OH)/(H_2O)', *Astron. Astrophys.* **313**, 657–663.

Hersant, F., Gautier, D., and Huré, J.-M.: 2001, 'A Two-dimensional Model for the Primordial Nebula Constrained by D/H Measurements in the Solar System: Implications for the Formation of Giant Planets', *Astrophys. J.* **554**, 391–407.

Iro, N., Gautier, D., Hersant, F., and Bockelée-Morvan, D.: 2002, 'An Interpretation of the Nitrogen Deficiency in Comets', *Icarus*, submitted.

Irvine, W.M., and Schloerb, F.P.: 2002, 'Observations of Deuterated Molecules with the Large Millimeter Telescope', *Planet. Space Sci.*, in press.

Irvine, W.M., Schloerb, F.P., Crovisier, J., Fegley, B., Jr., and Mumma, M.J.: 2000, 'Comets: A Link between Interstellar and Nebular Chemistry', *in* V. Mannings, A.P. Boss, and S.S. Russell (eds.), *Protostars and Planets IV*, Univ. Arizona Press, Tucson, pp. 1159–1200.

Jessberger, E.K., and Kissel, J.: 1991, 'Chemical Properties of Cometary Dust and a Note on Carbon Isotopes', *in* R.L. Newburn, Jr., M. Neugebauer, and J. Rahe (eds.), *Comets in the Post-Halley Era*, Kluwer Academic Publisher, Dordrecht, pp. 1075–1092.

Jewitt, D.C., Matthews, H.E., Owen, T.C., and Meier, R.: 1997, 'Measurements of $^{12}C/^{13}C$, $^{14}N/^{15}N$, and $^{32}S/^{34}S$ Ratios in Comet Hale-Bopp (C/1995 O1)', *Science* **278**, 90–93.

Kawakita, H., Watanabe, J., Ando. H., *et al.*: 2001, 'The Spin Temperature of NH_3 in Comet C/1999 S4 (LINEAR)', *Science* **294**, 1089–1091.

Kessler, J., Qi, C., and Blake, G.A.: 2000, 'Deuterium Fractionation in Circumstellar Disks', *BAAS* **32**, 6530.

Kissel, J., *et al.*: 2002, 'Cometary Secondary Ion Mass Analyzer – COSIMA', *ESA SP* **1165**, in press.

Kleine, M., Wyckoff, S., Wehinger, P.A., and Peterson, B.A.: 1995, 'The Carbon Isotope Abundance Ratio in Comet Halley', *Astrophys. J.* **439**, 1021–1033.

Langer, W.D., Graedel, T.E., Frerking, M.A., and Armentrout, P.B.: 1984, 'Carbon and Oxygen Isotope Fractionation in Dense Interstellar Clouds', *Astrophys. J.* **277**, 581–604.

Lis, D.C., Keene, J., Young, K., Phillips, T.G., Bockelée–Morvan, D., Crovisier, J., Schilke, P., Goldsmith, P.F., and Bergin, E.A.: 1997, 'Spectroscopic Observations of Comet C/1996 B2 (Hyakutake) with the Caltech Submillimeter Observatory', *Icarus* **130**, 355–372.

Meier, R., and Owen, T.C.: 'Cometary Deuterium', *Space Sci. Rev.* **90**, 33–43.

Meier, R., Owen, T.C., Matthews, H.E., Jewitt, D.C., Bockelée–Morvan, D., Biver, N., Crovisier, J., and Gautier, D.: 1998a, 'Deuterium in Comet C/1995 O1 (Hale-Bopp): A Determination of HDO/H_2O', *Science* **279**, 842–844.

Meier, R., Owen, T.C., Jewitt, D.C., Matthews, H., Senay, M., Biver, N. Bockelée–Morvan, D., Crovisier, J., and Gautier, D.: 1998b, 'Deuterium in Comet C/1995 O1 (Hale-Bopp): II Detection of DCN', *Science* **279**, 1707–1710.

Meier, R., Wellnitz, D., Kim, S.J., and A'Hearn, M.F.: 1998c, 'The NH and CH Bands of Comet C/1996 B2 (Hyakutake)', *Icarus* **136**, 268–279.

Millar, T.J., Bennett, A., and Herbst, E.: 1989, 'Deuterium Fractionation in Dense Interstellar Clouds', *Astrophys. J.* **340**, 906–920.

Morbidelli, A., Chambers, J., Lunine, J.L., Petit, J.M., Robert, F., Valsecchi, G.B., and Cyr, K.E.: 2000, 'Source Regions and Time Scales for the Delivery of Water to Earth', *Met. Planet. Sci.* **35**, 1309–1320.

Morfill, G.E.: 1983, 'Some Cosmochemical Consequences of a Turbulent Protoplanetary Cloud', *Icarus* **53**, 41–54.

Mousis, O., Gautier, D., Bockelée-Morvan, D., Robert, F., Dubrulle, B., and Drouart, A.: 2000, 'Constraints on the Formation of Comets from D/H Ratios Measured in H_2O and HCN', *Icarus* **148**, 513–525.

Mousis, O., Gautier, D., and Bockelée-Morvan, D.: 2002, 'An Evolutionary Turbulent Model of Saturn's Subnebula: Implications for the Origin of the Atmosphere of Titan', *Icarus* **156**, 162–175.

Ott, U.: 2003, 'Isotopes of Volatiles in Pre-solar Grains', *Space Sci. Rev.*, this volume.

Owen, T., and Encrenaz, T.: 2003, 'Element Abundances and Isotope Ratios in the Giant Planets and Titan', *Space Sci. Rev.*, this volume.

Pilbratt, G.L.: 2001, 'The Herschel Mission, Scientific Objectives, and this Meeting', *in: The Promise of the Herschel Space Observatory, ESA SP* **460**, 13–20.

Robert, F., Gautier, D., and Dubrulle, B.: 2000, 'The Solar System D/H Ratio: Observations and Theories', *Space Sci. Rev.* **92**, 201–224.

Rodgers, S.D., and Charnley, S.B.: 2002, 'A Model of the Chemistry in Cometary Molecules: Deuterated Molecules", *Month. Not. Roy. Astron. Soc.* **330**, 660–674.

Roueff, E., and Gerin, M.: 2003, 'Deuterium in Molecules in the Interstellar Medium', *Space Sci. Rev.*, this volume.

Schilke, P., Walmsley, C.M., Pineau Des Forêts, G.P., Roueff, E., Flower, D.R., and Guilloteau, S.: 1992, 'A Study of HCN, HNC and their Isotopomers in OMC-1. I - Abundances and Chemistry', *Astron. Astrophys.* **256**, 595–612.

Teixeira, T.C., Devlin, J.P., Buch, V., and Emerson, J.P.: 1999, 'Discovery of Solid HDO in Grain Mantles', *Astron. Astrophys.* **347**, L19–L22.

Terzieva, R., and Herbst, E.: 2000, 'The Possibility of Nitrogen Isotopic Fractionation in Interstellar Clouds', *Month. Not. Roy. Astron. Soc.* **317**, 563–568.

van Dishoeck, E.F., Blake, G.A., Jansen, D.J., and Groesbeck, T.D.: 1995, 'Molecular Abundances and Low-mass Star Formation. II Organic and Deuterated Species toward IRAS 16293-2422', *Astrophys. J.* **447**, 760–782.

Wehrstedt, M., and Gail, H.-P.: 2002, 'Radial Mixing in Protoplanetary Accretion Disks. II. Time Dependent Disks Models with Annealing and Carbon Combustion', *Astron. Astrophys.* **385**, 181–204.

Wielen, R., and Wilson, T.L.: 1997, 'The Evolution of the C, N, and O Isotope Ratios from an Improved Comparison of the Interstellar Medium with the Sun', *Astron. Astrophys.* **326**, 139–142.

Wright, P., *et al.*. 2002 , 'Isotopic Ratios of Light Elements by Gas Chromatography – MODULUS Berenice', *ESA SP* **1165**, in press.

Wyckoff, S., Kleine, M., Peterson, B.A., Wehinger, P.A., and Ziurys, L.M.: 2000, 'Carbon Isotope Abundances in Comets', *Astrophys. J.* **535**, 991–999.

Zinner, E., Wopenka, B., Amari, S., and Anders, E.: 1990, 'Interstellar Graphite and other Carbonaceous Grains from the Murchison Meteorite: Structure, Composition and Isotopes of C, N, and Ne', *Lunar Planet. Sci. Conf.* **21**, 1379–1380.

Ziurys, L.M., Savage, C., Brewster, M.A., Apponi, A.J., Pesch, T.C., and Wyckoff, S.: 1999, 'Cyanide Chemistry in Comet Hale-Bopp (C/1995 O1)', *Astrophys. J.* **527**, L67–L71.

Address for Offprints: Kathrin Altwegg, Physikalisches Institut, University of Bern, Sidlerstr. 5, CH-3012 Bern, Switzerland; altwegg@phim.unibe.ch

ISOTOPIC SIGNATURES OF PRESOLAR MATERIALS IN INTERPLANETARY DUST

S. MESSENGER[1], F.J. STADERMANN[1], C. FLOSS[1], L.R. NITTLER[2] and S. MUKHOPADHYAY[2]

[1] *Laboratory for Space Sciences, Department of Physics, Washington University, Saint Louis, MO 63130, USA*

[2] *Department of Terrestrial Magnetism, Carnegie Institute of Washington, 5241 Broad Branch Rd. NW, Washington D.C., 20015*

Received: 6 May 2002; Accepted in final form: 26 September 2002

Abstract. Interplanetary dust particles collected in the stratosphere frequently exhibit enrichments in deuterium (D) and ^{15}N relative to terrestrial materials. These effects are most likely due to the preservation of presolar interstellar materials. While the elevated D/H ratios probably resulted from mass fractionation during chemical reactions at very low < 100 K temperatures, the origin of the N isotopic anomalies remains unresolved. The bulk of the N-bearing material may have obtained its isotopic signatures from low temperature chemistry, but a nucleosynthetic origin is also possible.

1. Introduction

Interplanetary dust particles (IDPs) have been routinely collected in the stratosphere by NASA high-altitude research aircraft for more than two decades (Sandford, 1987). These particles are samples of both comets and asteroids, but the specific source of any given particle has not yet been determined. Nevertheless, it is clear that many IDPs are samples of *primitive* objects – parent bodies which have not been significantly affected by post-accretional alteration. Indeed, many anhydrous IDPs have escaped the hydrothermal processing that has extensively modified the matrixes of even the most primitive meteorites (Bradley *et al.*, 1988).

The view that IDPs are composed of primordial solar system materials is borne out by large, common, and highly variable excesses in deuterium (D) and ^{15}N relative to solar composition, and the recent discovery that some IDPs contain abundant grains of silicate stardust (Messenger *et al.*, 2002). The stardust is marked by extreme variations in oxygen isotopic ratios, reflecting its recent nucleosynthetic origin in the parent stars. Here we will focus on the evidence for preserved molecular cloud material, which obtained its isotopic signatures from chemical processes, not nucleosynthetic processes. The elevated D/H ratios are thought to have originated from mass fractionation during very low temperature (10–100 K) chemical reactions in a cold interstellar molecular cloud environment. Radio astronomical observations of the coldest clouds find some molecules with D/H ratios enriched by as much as 10^4 times the local D/H (Millar *et al.*, 1989). The origin of the N isotopic

Space Science Reviews **106:** 155–172, 2003.

anomalies is less certain because N isotopic fractionation has not yet been observed in the interstellar medium (ISM). The case for nitrogen is further complicated by the possible contribution of nucleosynthetically-derived anomalous N.

While similar isotopic effects are also observed in a wide variety of primitive meteorites, the anomalies are generally smaller, especially for H. This suggests that most molecular cloud material in meteorites has been diluted, chemically evolved, and/or isotopically equilibrated since it was incorporated into the meteoritic parent bodies. The carriers of H and N isotopic anomalies in meteorites are well characterized in some cases, including a wide variety of organic compounds and water of hydration in some meteorites (Messenger and Walker, 1997; Deloule and Robert, 1995). In contrast, the coordinated chemical and isotopic characterization of IDPs is still at an early stage, though significant strides have been made in recent years (Keller *et al.*, 2002).

While it remains a considerable technical challenge, the molecule-specific isotopic analysis of molecular cloud material in IDPs would provide a direct probe of some poorly constrained interstellar chemical processes, such as grain surface chemistry and photolysis of condensed materials in the ISM. Furthermore, this would offer a direct test of models of the origin of meteoritic organic compounds, and provide insight into the relative importance of chemical processes in the solar nebula. Here we review the H, C, and N isotopic measurements performed on IDPs over the past 20 years. We concentrate on presenting a comprehensive review of H, C, and N isotopic data, as this has not been previously compiled. More detailed discussions of the origin and chemical characterization of molecular cloud materials in IDPs are reviewed elsewhere (Messenger and Walker, 1997; Keller *et al.*, 2002; Messenger, 2000).

2. Experimental Techniques

Interplanetary dust particles are collected by NASA high altitude research aircraft by inertial impact onto silicone oil-coated pylons. The particles are picked from the collectors using a micromanipulator and washed of the oil in a hexane rinse. Most particles on the collectors are well separated from any other debris, but some IDPs are found as dense clusters, or sprays of material (cluster IDPs), which must have been fragile particles that disrupted upon impact.

Virtually all IDPs discussed here have been subjected to the following protocol. Each particle picked from the collector is measured for its elemental abundances by energy dispersive X-ray analysis in a scanning electron microscope in order to distinguish extraterrestrial (approximately chondritic elemental abundances) particles from terrestrial contaminants. IDPs of interest are transferred to an ultrapure Au substrate into which they are pressed with a clean spectroscopic grade quartz or sapphire disc for subsequent isotopic measurements.

Isotopic measurements are performed by SIMS (secondary ion mass spectrometry). In this technique, ions are generated from the sample by bombardment with a high energy ion beam, most commonly Cs^+, but occasionally O^- or Ga^+. These instruments are extremely sensitive, enabling isotopic measurements on samples < 1 μm, but high precision (1 ‰) isotopic measurements are difficult to achieve because of matrix-dependant instrumental mass fractionation. The degree of instrumental mass fractionation is corrected for by measuring standards of similar composition under the same conditions, immediately prior to and following the sample analysis. The typical precision for hydrogen isotopic measurements is of order 10–100 ‰. However, for the samples discussed here, sensitivity is usually more important than precision. Bulk isotopic measurements of IDPs reported here have closely followed the techniques developed by McKeegan *et al.* (1985).

As discussed below, many particles have strongly heterogeneous H and N isotopic compositions on a μm scale. Two isotopic imaging techniques have been employed to investigate isotopic variations on this scale. The first takes advantage of the fact that the spatial distribution of secondary (sample-derived) ions is retained, and the SIMS instrument is used as an ion *microscope* where ion images are detected with a multichannel plate coupled to a fluorescent screen and CCD camera (McKeegan *et al.*, 1987; Messenger, 2000). Alternatively, the primary ion beam is rastered across the sample and secondary ions are synchronously detected with one or more electron multipliers (Nittler and Messenger, 1998; Aléon *et al.*, 2001; Floss and Stadermann, 2002). In the latter technique, isotopic images are more easily quantified.

3. Hydrogen Isotopic Measurements

The H isotopic measurements of IDPs are summarized in Tables I (cluster IDPs) and II (individual IDPs). H isotopic anomalies in IDPs are common, occurring most often as enhanced D/H ratios relative to terrestrial samples. As shown in Figure 1, D enrichments are far larger, more common, and are more variable among cluster IDPs than individual IDPs. In some cases the D/H ratios observed in cluster IDPs exceed those measured in any other solar system materials, reaching 50 times the terrestrial value (1.5576×10^{-4}) in the cluster IDP Dragonfly. This exceeds the maximum D/H ratio observed in meteorites by nearly an order of magnitude. It is remarkable that this IDP also has fragments strongly depleted in D, resulting in nearly two orders of magnitude variation in D/H within this one dust particle. Most other cluster IDPs also have strong H isotopic variations among different fragments. Isotopic imaging studies have shown that significant variations in D/H often occur on a micrometer scale (Messenger, 2000; McKeegan *et al.*, 1987; Nittler and Messenger, 1998; Aléon *et al.*, 2001).

There is strong support for an interstellar origin of the high D/H ratios observed in IDPs and meteorites. Deuterium is quickly consumed as nucleosynthetic fuel in

TABLE I

Isotopic mesurements in cluster IDPs.

Ref.	Nichname	Particle ID	C1#	δD	$\sigma(\delta D)$	$\delta^{13}C$	$\sigma(\delta^{13}C)$	$\delta^{15}N$	$\sigma(\delta^{15}N)$
8	Dragonfly	L2005 A2a	31	603	76	−4	23	479	40
8		L2005 A2b	31	675	44				
8		L2005 F	31	20000	1400	−26	14	390	28
8		L2005 4	31	24800	1500	−49	15	338	78
8		L2005 3	31	8000	500	−31	13	261	24
8		L2005 *A3	31	−417	24	−1	20	396	41
8	Penelope	L2006 A1a	4	211	96				
8		L2006 A2a	4	118	29				
8		L2006 A2b	4	613	77				
8	Speedy	L2006 A9	14	3503	276				
8		L2006 A10	14	110	46				
8	Gossamer	L2006 A8	10	−14	51				
8	Claude	L2006 A4a	6	748	52				
8		L2006 A3	6	−69	49				
8		L2006 A6	6	−89	85				
8	Tweety	L2008 B2a	4	1147	92	−45	11	237	15
8		L2008 B1	4	970	56				
8	Elmer	L2008 310x	5	71	75	−4	7	89	27
8		L2008 310d	5	66	75	−11	7	41	29
8		L2008 310e	5	17	79	−33	9	63	37
8		L2008 310f	5	−14	72				
8		L2008 310g	5	−35	6	−7	6	73	23
8		L2008 310i	5	−55	18	−23	27	139	43
8		L2008 313	5	322	67				
8		L2008 19a	5	664	84	−16	26	260	34
8		L2008 19b	5	424	71	−33	5	202	16
8		L2008 110	5	822	91	−18	10	180	18
8		L2008 37a	5	−24	17				
8		L2008 37b	5	25	22				
8		L2008 37c	5	−14	36				
8	Marvin	L2009 D1	3	2145	120	−52	11	320	33
8		L2009 D2a	3	1906	80				
8		L2009 D2b	3	−34	11	140	13		
8	Pepe	L2009 D4	4	3	50				
8	Roadrunner	L2009 D5	7	11035	969				
8		L2009 D6	7	1155	115	−43	17	142	23
8		L2009 E2	7	392	117				
8	Foghorn	L2009 D7	8	1174	203	−1	20	149	27

TABLE I

Isotopic mesurements in cluster IDPs (continued).

8	Wiley	L2009 D9	10	12007	653	−52	17	241	31
8		L2009 D10	10	1080	133	−42	11	191	15
8	Taz	L2009 D11	13	5585	770				
8		L2009 D12	13	4122	571	−117	42	105	35
12		L2009 I3	13	−84	76	70	65	62	17
12*	Russell	L2011 AD2	3	3593	495	0	22	75	21
12		L2011 AD2	3	385	254				
8	Leghorn	L2011 2	5	708	72				
8		L2011 4	5	−345	25				
12		L2011 AD6	5	−173	69	0	6	−4	7
12		L2011 AD6	5	−106	52				
12		L2011 AD6	5	120	66				
12		L2011 AD7	5	87	59				
12		L2011 AD7	5	−70	63				
8	Yosemite	L2011 A1	6	1236	126	−38	15	176	12
8		L2011 A2	6	1376	154	−14	87	103	32
12		L2011 AD8	6	862	107	−10	1	600	38
12		L2011 AD8	6	220	93			295	37
8	Porky	L2011 A3	7	5570	222	−53	2	−93	4
8		L2011 A4	7	929	89	−55	60	−10	48
12*	Pasternak	L2011 AD13	8	−253	160	10	46	−7	25
12	-Mauriac	L2011 AD13	8	−64	211				
12*		L2011 AD14	8	535	336	0	11	19	11
8	Daffy	L2011 A5	11	2865	226	−27	14	94	17
8		L2011 A6	11	2914	654	33	26	290	90
8		L2011 B5	11	−100					
8	Petunia	L2011 A7	15	127	129	−76	23	17	35
8		L2011 A8	15	3579	367	−67	21	47	24
12*	Hemingway	L2011 AD19			17	−30	10	70	2
12*	-Jimenez	L2011 AD20	17	192	84	−10	11	470	159
8	Bugs	L2011 A9	22	80	80				
8		L2011 A10	22	678	61	−41	15	18	16
11		L2021 *C4	4	365	60	−40	6	445	30
11		L2021 *C4	4	780	97			334	55
11		L2021 *C6	5	100	66	−54	12	445	30
11		L2021 *C6	5	423	206			267	83
11		L2021 *C6	5	−120	152			−128	85
9		L2021 A6	5	1880	40				
9		L2021 A6	5	840	25				

TABLE I

Isotopic mesurements in cluster IDPs (continued).

11		L2036 *C10	2	984	82	−20	5	472	23
11		L2036 *C10	2	2590	200			377	107
11		L2036 *C10	2	565	124			543	129
11		L2036 *C10	2	1360	150			828	130
12*	Sienkiewicz	L2036 D3	9	278	100	−40	51	34	20
12		L2036 D3	9	58	80				
12*	Carducci	L2036 G3	14	−7	96			15	11
12*	Eucken	L2036 E2	17	−512	39			42	24
12		L2036 E2	17	−58	79				
11		L2036 *C7	19	572	85	5	50	334	95
11		L2036 *C7	19	1380	274			590	300
11		L2036 *C7	19	1670	175			363	146
2,4*	SP-56E	u2-24sp56		−340	50	−53	34	104	24
2,4		u2-24sp56		252	28				
2,4		u2-24sp56		−125	32				
2,4*	SP-88a	u2-19c3a		1345	126	−21	34	406	32
2,4		u2-19c3a		713	54				
1,4*	Calrissian	W7021 r21-m1-9a		373	107	−18	14	54	9
1,4*		W7021 r21-m1-9a		428	112	−28	39	63	16
1,4*		W7021 r21-m1-9a		776	149	−23	18	57	10
1,4*		W7021 r21-m1-9a		2191	272	−43	14	44	11
1,4*		W7021 r21-m1-9a		1078	170	2	20	42	14
1,4		W7021 r21-m1-9a		1633	224				
1,4		W7021 r21-m1-9a		515	120				
2,5*	Essex	W7021 r21-m4-3a		−242	35	−20	13	10	10
2,5*		W7021 r21-m4-3a		−269	44	−20	7	15	8
2,5*		W7021 r21-m4-3a		−225	43	−33	6	14	6
2,5*		W7021 r21-m4-3a		−235	52	−23	13	15	11
2,5*		W7021 r21-m4-3a		−256	52	−25	11	−2	8
2,5*	Ptaar/Tyson	W7021 r21-m3-4a		−169	80	−25	9	14	7
2,5*	Tyson	W7021 r21-m3-4b		−299	25				
1,5*	Skywalker-Solo	W7021 r21-m3-5a		696	96	−7	7	47	6
1,5*		W7021 r21-m3-5a		487	86	−36	7	54	9
1,5*		W7021 r21-m3-5a		297	71	−32	8	52	5
1,5*		W7021 r21-m3-5a		393	75	−41	7	46	5
1,5*		W7021 r21-m3-5a		58	59	−35	7	66	5
1,5		W7021 r21-m3-5a		56	56				
1,5		W7021 r21-m3-5a		659	89				
1,5		W7021 r21-m3-5a		266	71				
1,5		W7021 r21-m3-5a		393	84				

TABLE I

Isotopic mesurements in cluster IDPs (continued).

1,5	Solo	W7021 r21-m3-5a	742	105				
1,5		W7021 r21-m3-5a	506	65				
1,5		W7021 r21-m3-5a	270	46				
2	Butterfly	W7021 r21-m4-8a	340	54				
2		W7021 r21-m4-8a	177	88				
2		W7021 r21-m4-8a	2705	170				
2		W7021 r21-m4-8a	−322	34				
2	Viburnum	r72-m2-3a	−218					
2		r72-m2-3a	−238					
2		r72-m2-3a	−30					
1	Spray-2	u23-m4-4b	389	90				
1		u23-m4-4b	961	130				
1		u23-m4-4b	370	116				
1		u23-m4-4b	775	115				
7	Dood	W9019-3	1734	119	N/A	N/A	−15	30
7		W9019-3	95	77	N/A	N/A	−58	62
7		W9019-3	149	95	N/A	N/A	−131	70
7		W9019-3	833	73	N/A	N/A	25	26

* For these measurements H and C/N data for fragments are not correlated.
References: 1. McKeegan *et al.* (1985); 2. McKeegan *et al.* (1987); 3. McKeegan (1987); 4. Stadermann *et al.* (1989); 5. Stadermann (1991); 6. Keller and Messenger (1997); 7. Nittler and Messenger (1998); 8. Messenger (2000); 9. Aléon *et al.* (2001); 10. Floss and Stadermann (2002); 11. Nittler *et al.*, unpublished; 12. Floss and Stadermann, unpublished.

stars and its spallogenic generation is not significant. Only chemical fractionation is capable of producing significant deuterium enrichments. This process is known to be efficient at very low temperatures, but selective photodissociation in strong radiation fields may also play a role. Extremely high D/H ratios are directly observed for a variety of simple gas phase molecules in molecular clouds, reaching the highest values (D/H > 0.1) in the coldest regions. The D-rich molecules observed by radio astronomy demonstrate the efficacy of gas-phase deuteration, but grain surface reactions may be far more significant in deuterating the bulk of the solid materials in the ISM (Tielens, 1997). Sandford (2001) has also argued that selective 'unimolecular photodissociation' of PAH molecules could lead to very high D/H ratios in the diffuse ISM. While this has not yet been observationally confirmed, such a process would leave a unique fingerprint in its size-dependent effect on the D/H ratios of different PAH molecules.

TABLE II
Isotopic mesurements in individual IDPs.

Ref.	**IDP**	**IDP ID**	**C1#**	δD	$\sigma(\delta D)$	$\delta^{13}C$	$\sigma(\delta^{13}C)$	$\delta^{15}N$	$\sigma(\delta^{15}N)$
8	Dragonfly	L2005 A2a	31	603	76	−4	23	479	40
1	Attila	r21-m2-8	−171	49					
1		r21-m2-8	−176	50					
1		r21-m2-8	−82	47					
1		r21-m2-8	−129	54					
1		r21-m2-8	−193	45					
1		r21-m2-8	−226	50					
1		r21-m2-8	12	72					
1		r21-m2-8	−88	54					
2,5*	Cannonball	u23-m4-9	−59	35	−73	32	−11	39	
2,5*		u23-m4-9	−88	47	−2	31	−63	25	
2,5		u23-m4-9	−134	67					
2	Pb-1	u14-m2-8	52	38					
2		u14-m2-8	104	17					
2		u14-m2-8	−5	14					
2,4*	Cedarcreek	u014-m1-9	69	39			94	32	
2,4		u014-m1-9	494	56			10	21	
2,5*	Chicago-2	r29-j10	−49	34	−40	15	79	28	
2,5		r29-j10	−34	40					
2,5		r29-j10	111	35					
2	Chicago-3	u2001-c6	−81	35					
2		u2001-c6	−64	67					
2,5*	Chicago-5	u2015-c25	33	42	−51		46	31	
2,5		u2015-c25	4	43					
2,5		u2015-c25	29	41					
2,5*	Jefferson City	u23-m7-8	−102	61	−19	24	24	17	
2,5*		u23-m7-8	3	66	−38	31	17	28	
2,5		u23-m7-8	−168	40	−4	34	−4	17	
2,5		u23-m7-8	−96	39					
3	Nugget	u23-m1-6	−230	35					
3		u23-m1-6	−186	32					
2,5*	Pattonsburg	u014-m3-3	−132	38	−64	38	6	28	
2,5		u014-m3-3	−182	60					
2,5	Rocheport	r18-m7-16	643	52			−16	29	
2,5		r18-m7-16	474	237					
2,5*	Speckles	u21-m4-8	−154	41	−39	40	3	29	
2,5		u21-m4-8	45	60					
2,5		u21-m4-8	−266	52					

TABLE II

Isotopic mesurements in individual IDPs (continued).

2,5*	St. Elizabeth	u014-m2-2	15	47	−62	31	25	15
3,5*		u014-m2-2	173	54			185	23
3,5		u014-m2-2	11	54				
3,5		u014-m2-2	−166	77				
2,5*	Summersville	u014-m2-11	24	39	−35	23	8	7
2,5*		u014-m2-11	76	45	−21	37	28	32
2,5		u014-m2-11	47	48	13	26	13	30
2,5		u014-m2-11	−1	38				
2,5	Pb-2	u014-m2-9	−86	28				
2,5		u014-m2-9	−69	28				
2,5*	The Clown	u21-m9-4	−135	44			66	35
2,5*		u21-m9-4	−24	42			55	36
2,5		u21-m9-4	−121	42				
2,5		u21-m9-4	−145	106				
2,5*	Verona	u014-m4-7a	−237	118	−27	20	26	16
2,5*		u014-m4-7a	−131	34	2	25	−34	34
2,5*		u014-m4-7a	−74	35	−16	19	−18	13
2,5		u014-m4-7a	−119	60				
2,5*	Xavier	r21-m4-1	151	59			−85	25
2,5		r21-m4-1	−274	111				
2,5		r21-m4-1	−167	138				
2,5		r21-m4-1	106	92				
2,5*	Yoda	r21-m3-6	−224	72	−90	31	39	16
2,5*		r21-m3-6	−257	219	−18	12	25	11
2,5		r21-m3-6	−76	41				
2,5		r21-m3-6	30	34				
2,5		r21-m3-6	−113	41				
2,5		r21-m3-6	−259	35				
1	Mosquito	r21-m4-7	411	110				
1		r21-m4-7	125	85				
1		r21-m4-7	1346	170				
1		r21-m4-7	1774	207				
1		r21-m4-7	1224	160				
1		r21-m4-7	2534	260				
1		r21-m4-7	663	138				
1		r21-m4-7	963	146				
5*	Augustus	u44-m1-2	96	77	−16	31	22	14
5*		u44-m1-2	272	71	−80	34	119	60
5		u44-m1-2	152	68				
5		u44-m1-2	143	68				

TABLE II
Isotopic mesurements in individual IDPs (continued).

1,5	Essex	r21-m4-3a	−242	35					
1,5		r21-m4-3a	−269	44					
1,5		r21-m4-3a	−225	43					
1,5		r21-m4-3a	−235	52					
1,5		r21-m4-3a	−256	38					
1,5*	Lea	r21-m2-4	−386	54	−27	12	2	7	
1,5*		r21-m2-4	−381	38	−23	18	0	15	
1,5*		r21-m2-4	−31	37	−20	26	−1	12	
1,5		r21-m2-4	−350	29					
1,5		r21-m2-4	−60	43					
1,5		r21-m2-4	−66	40					
1,5		r21-m2-4	−177	42					
1,5		r21-m2-4	−39	50					
3	Shannondale-B	u014-m2-13	−209	76					
3		u014-m2-13	−101	79					
3		u014-m2-13	−247	152					
3		u014-m2-13	−70	135					
5*	Aurelian	u47-m1-2a	700	73	−20	12	39	27	
5*		u47-m1-2a	170	91			813	25	
5		u47-m1-2a	604	78					
5		u47-m1-2a	580	66					
5		u47-m1-2a	667	67					
5		u47-m1-2a	69	62					
5		u47-m1-2a	−26	44					
5		u47-m1-2a	31	55					
5		u47-m1-2a	291	59					
5		u47-m1-2a	259	56					
5		u47-m1-2a	304	67					
5		u47-m1-2a	199	76					
5		u47-m1-2a	196	82					
5		u47-m1-2a	98	69					
5		u47-m1-2a	771	85					
2	Carrollton A	u014-m2-12a	−8	50					
2		u014-m2-12a	−55	52					
5*	Florianus	u47-m1-3a	927	88	−34	11	411	20	
5*		u47-m1-3a 904	76	−23	14	409	11		
5*		u47-m1-3a	185	70	−22	20	−373	21	
5		u47-m1-3a	1120	101					
5		u47-m1-3a	1085	103					
5		u47-m1-3a	233	89					

TABLE II

Isotopic mesurements in individual IDPs (continued).

5		u47-m1-3a	766	81				
5		u47-m1-3a	788	109				
5		u47-m1-3a	933	74				
5	Galba	u44-m1-1	−176	54				
5		u44-m1-1	65	73				
5		u44-m1-1	−97	49				
5*	Gratian	u44-m4-1c	−153	35	−20	7	39	20
5		u44-m4-1c	−65	40				
5		u44-m4-1c	−112	124				
5*	Jovian	u44-m4-9a	−177	39	−43	18	28	14
5		u44-m4-9a	−252	36				
5		u44-m4-9a	−153	35				
5*	Nero	u44-m1-7	244	68	−43	13	22	9
5		u44-m1-7	−13	125				
5*	Petronius	u47-m2-12	−111	46	−15	10	7	15
5		u47-m2-12	−40	45				
5		u47-m2-12	68	64				
5*	Romulus	u44-m2-3b	−10	90	−4	22	22	96
5		u44-m2-3b	−76	54				
5		u44-m2-3b	−161	66				
5		u44-m2-3b	−20	58				
5*	Tacitus	u44-m1-11a	−64	36	−28	15	−1	20
5*	u44-m1-11a	−50	45	−37	8	42	12	
5		u44-m1-11a	−75	41				
5		u44-m1-11a	−72	39				
5*	Tiberius	u44-m1-5	−15	50	−41	12	53	6
5		u44-m1-5	30	54				
5	Titus	u44-m1-12	−84	63				
5		u44-m1-12	−257	76				
5*	Trajan	u47-m2-5	−242	31	−30	31	120	17
5		u47-m2-5	−128	45				
5		u47-m2-5	−139	66				
5		u47-m2-5	56	82				
5*	Vespasian	u47-m3-5a	−206	46	−34	8	10	7
5		u47-m3-5a	−88	27				
5		u47-m3-5a	−138	34				
5		u47-m3-5a	−97	44				
5		u47-m3-5a	−102	40				
4*	Santa Fe	u33-m3-11	−60	48	−52	10	233	11
4*		u33-m3-11	477	71	−48	15	61	11
4*		u33-m3-11			−50	32	442	29
5*	Majorian	u44-m6-1a	108	50	−23	20	53	6

TABLE II

Isotopic mesurements in individual IDPs (continued).

5*		u44-m6-1a	−55	43	−54	17	36	27
5		u44-m6-1a	−184	36				
5		u44-m6-1a	−54	39				
5*	Pupienus	u47-m2-3	336	92	−40	13	306	21
5*		u47-m2-3	172	75	−43	9	214	12
5		u47-m2-3	−38	66				
5		u47-m2-3	144	79				
5		u47-m2-3	48	66				
12*	Prudhomme	L2009 N2	305	101	−22	21	415	59
12*	L2009 N2	248	90			55	29	
12	Bjornson	L2011 Q2	125	104				
12		L2011 Q2	29	89				
6	Captain Nitrano	L2011 R11	200				250	
12*	Kipling	L2011 R12	467	122	70	9	510	3
12*		L2011 R12	945	142			1090	80
12*		L2011 R12	722	126			1250	93
12		L2011 R12	752	127				
12		L2011 R12	748	140				
12*	Mistral F.	L2036 E6	−197	66	0	45	84	35
12		L2036 E6	−315	56				
12		L2036 E6	−239	55				
12*	Echegaray	L2036 F18	460	169	−50	20	34	4
12		L2036 F18	112	86				
12		L2036 F18	173	108				
7		W9019-m4-1	345	69			49	26
7		W9019-m4-1	1077	185				
7		W9019-M4-2	−150	42			34	17
7		W9019-M4-4	−60	82			−116	76
7		W9019-M4-5	93	46				
7		W9019-M4-6	145	70				
7		W9019-M4-8	848	74			195	92
7		W9019-M4-8					598	176
9		L2021 K1	888	25				
9		L2021 K1	117	50				
9		L2021 K1	8700	800				
9		L2036 E22	5050	450				
9		L2036 E22	11070	700				
9		L2036 E22	1770	300				
9		L2036 R5	−75	5				
9		L2036 R6	−197	20				

H, C, and N isotopic measurements of individual IDPs reported in ‰ deviations from terrestrial standards. For references see Table I.

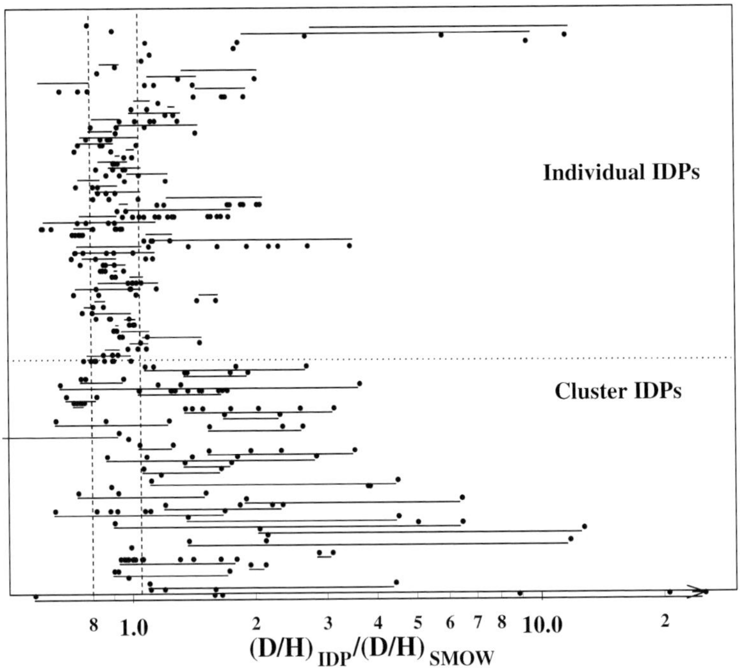

Figure 1. Comparison of D/H ratios of cluster and individual IDPs relative to the terrestrial standard mean ocean water (SMOW). The D/H ratios of different fragments of each IDP are shown as points, where the lines denote the range of values observed in each particle. The particle with the highest D/H ratio in the figure (Dragonfly) reaches a D/H ratio of 50 times solar, estimated by ion imaging (not shown). The range of D/H values observed among terrestrial rocks is shown by the vertical dashed lines.

In principle, there were regions of the solar nebula that had low enough temperatures to enable significant H isotopic fractionation during chemical reactions. However, Geiss and Reeves (1981) argued that ion molecule chemistry was required at such low temperatures, and the highly opaque solar nebula probably inhibited sufficient radiation to support such processes. Recent models by Aikawa and Herbst (1999, 2001) suggest that some regions of the outer solar nebula may have experienced sufficient ionization by cosmic rays and interstellar UV and X-ray fields to drive ion molecule reactions. In any event, the physical conditions and resulting chemistry of these models are similar to the analogous processes in molecular clouds.

4. Nitrogen Isotopic Measurements

The N isotopic measurements of IDPs are summarized in Tables I and II. Nitrogen isotopic anomalies are somewhat more common among cluster IDPs than individual IDPs, though the distinction is not as strong as that observed for D/H ratios.

Subfragments of both types of particles often show significant variations in their N isotopic ratios. While H and N anomalies are equally common, the N isotopic anomalies observed in IDPs are typically much smaller. However this is not unexpected if the ^{15}N excesses originated from chemical fractionation because the relative difference in the masses of ^{15}N and ^{14}N is much smaller than the difference between D and H.

The enrichments in ^{15}N have been attributed to chemical fractionation because: (1) they are usually associated with materials enriched in D (including meteorites), (2) as described below, there are no accompanying anomalies in C, excluding any known nucleosynthetic source, and (3) spallation reactions have negligible effects compared to the sizes of these anomalies (Geiss and Bochsler, 1982). While a correlation between the D/H and $^{15}N/^{14}N$ ratios in IDPs would constitute clear support for the chemical fractionation origin, no such correspondence is observed (see Figure 2). However, there is no reason to expect correlated anomalies in interstellar molecules as the isotopic composition of a given species is determined by a complex interplay of the reaction pathways and isotopic compositions of the precursor species (e.g., Millar *et al.*, 1989). Unfortunately, unlike the case for deuterium, low temperature N isotopic fractionation has not been definitively observed in the ISM. In fact, until recently there was little direct theoretical support for significant N isotopic fractionation in the ISM, although Adams and Smith (1981) suggested that proton exchange with N_2H could result in 1,000 ‰ enhancement in ^{15}N at 10 K. Two recent models have offered new insight into this issue. Terzieva and Herbst (2000) calculated the N isotopic fractionation in a variety of relevant gas phase reactions, finding a maximum of 250 ‰ enhancement. More recently, Charnley and Rodgers (2002) have proposed that NH_3, enriched in ^{15}N by 800 ‰, can be efficiently produced in the late stages of cold molecular cloud evolution. For comparison, the largest fractionation observed in an IDP to date is 1,250 ‰ (Floss and Stadermann, 2002). It is not yet clear whether abundant ^{15}N-rich ammonia could pass on its anomalous nitrogen to the abundant organic hosts that have been found in meteorites and IDPs.

We note that two recent studies have provided evidence that the Sun might have a bulk $^{15}N/^{14}N$ ratio considerably lower than the Earth's atmosphere. First, Hashizume *et al.* (2000) reported isotopically light N of putative solar wind origin implanted in the outer surface of lunar soil grains. Owen *et al.* (2001) reported a new value for the solar $^{15}N/^{14}N$ ratio, based on Galileo measurements from the Jovian atmosphere, as equal to $2.3 \pm 0.3 \times 10^{-3}$. Since the delta values reported in Tables I and II are calculated relative to the terrestrial value of 3.7×10^{-3}, the ^{15}N enrichments in IDPs relative to solar composition may be significantly higher, in many cases significantly greater than can be explained by even the latest theoretical isotopic fractionation models.

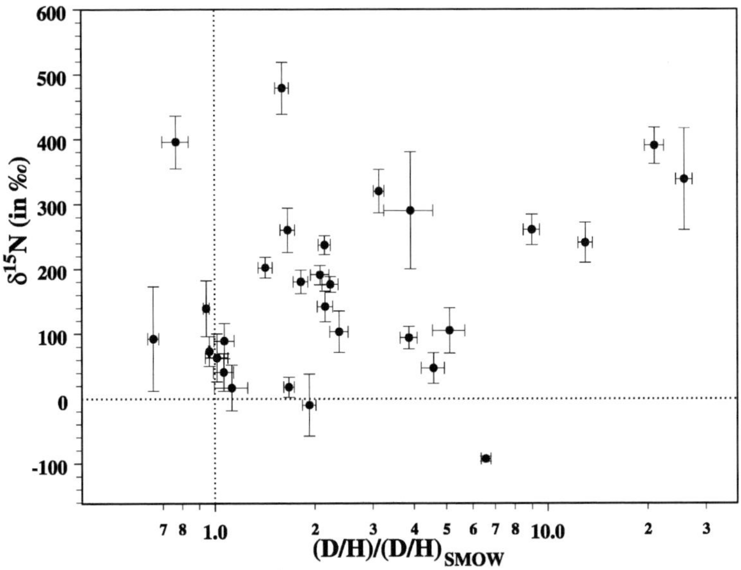

Figure 2. Comparison of H and N isotopic measurements of cluster IDPs. Despite the fact that H and N isotopic anomalies are both common and large in these particles, no clear correlation is observed. The data are taken from Messenger (2000).

5. Carbon Isotopic Measurements

The large and highly variable H and N isotopic anomalies are not accompanied by any detectable anomalies in carbon. All C isotopic measurements of IDPs so far have fallen within the terrestrial range of −70 ‰ to +10 ‰ (see Tables I and II). However, the average C isotopic composition of cluster IDPs (−45 ‰) is marginally distinct from that of individual IDPs (−30 ‰).

Given the significant and common anomalies in N, even larger anomalies in C might be expected from the larger relative difference in the masses of C isotopes. The lack of accompanying C isotopic fractionation may reflect the fact that the major reservoirs of H, C, and N (H_2, CO, and N_2) have different volatilities. CO is thought to condense onto grain surfaces and participate in grain chemistry, potentially erasing any significant isotopic fractionation between CO and other organic species. In contrast H_2 and N_2 are not significantly condensed onto grain surfaces, even at the lowest temperatures.

6. Summary and Conclusions

The common, large, and highly variable H and N isotopic anomalies of IDPs demonstrate that these materials have remained relatively unaltered since they accreted into parent bodies 4.5 billion years ago. The fact that IDPs often have D/H ratios far larger than those of meteorites suggests that they contain a better preserved record of presolar low-temperature chemistry. However, the D/H ratios of IDPs are still significantly lower than those observed in interstellar molecules. It is unclear whether this means IDPs also consist of (less) altered material, or is simply due to the fact that the observable interstellar molecules represent a very small fraction of the H-bearing material in the ISM. Indeed, as the D/H ratios of solids in the ISM are generally impossible to determine spectroscopically, our only source of such information may be preserved molecular cloud materials in IDPs and their parent bodies. By better understanding the nature of the anomalous phases in IDPs, we hope to learn more about the genesis of organic compounds in meteorites and chemical processes in the interstellar medium.

There are a number of unsolved problems to be addressed in future work. First, as with most meteorites, the specific parent bodies of IDPs are unknown, although they must come from both comets and asteroids. The presently active STARDUST mission, which will collect dust from comet 81P/Wild-2 should offer important insight into which IDPs are of probable cometary origin. It may also be possible to collect dust from specific comets in the stratosphere with appropriately timed collections (Messenger, 2002). Most of the important issues about the origins of the isotopic anomalies would be clarified by identifying their specific hosts. For instance, some molecules (*e.g.* α-amino acids) in meteorites are thought to have formed during Strecker-cyanohydrin synthesis during aqueous processing of interstellar precursor molecules (Cronin and Chang, 1993). Since some D-rich IDPs have apparently escaped aqueous alteration, they should be devoid of α-amino acids, but may contain the presumed precursor species. Additionally, there is still a great deal of uncertainty regarding the origin of the ^{15}N enrichments in IDPs. Although the majority of the ^{15}N-excesses may have been derived from chemical fractionation, the specific processes and likely hosts of the anomalies are still poorly constrained. The fact that both H and N isotopic ratios strongly vary within most IDPs implies that there are multiple, isotopically distinct phases in IDPs. Recently, Aléon *et al.* (2001) have used ion imaging to infer characteristics of these carriers, proposing distinct D/H values of water of hydration, macromolecular material, and chained aliphatics. Analytical advances now enable the direct measurement of specific aromatic hydrocarbons (Clemett *et al.*, 1993), the spectroscopic detection of aliphatic hydrocarbons (Keller *et al.*, 2002), and the detection of amino acids by fluorescent tagging (Clemett *et al.*, 2002). Ultimately, the coordinated isotopic, chemical and petrographic studies of the same IDPs on a μm scale may provide new insight into chemical processes in molecular clouds and the solar nebula.

Acknowledgements

We thank the International Space Science Institute for their hospitality during the workshop. This work was supported by NASA under grants NAG5-9801 and NAG5-10510.

References

Adams, N.G., and Smith, D.: 1981, 'N-14/N-15 Isotope Fractionation in the Reaction N_2H^++N_2 – Interstellar Significance', *Astrophys. J.* **247**, L123–L125.

Aikawa, Y., and Herbst, E.: 1999, 'Deuterium Fractionation in Protoplanetary Discs', *Astrophys. J.* **526**, 314–326.

Aikawa, Y., and Herbst, E.: 2001, 'Two-dimensional Distributions and Column Densities of Gaseous Molecules in Protoplanetary Discs II. Deuterated Species and UV Shielding by Ambient Clouds', *Astron. Astrophys.* **371**, 1101–1117.

Aléon, J., Engrand, C., Robert, F., and Chaussidon, M.: 2001, 'Clues to the Origin of Interplanetary Dust Particles From the Isotopic Study of Their Hydrogen-bearing Phases', *Geochim. Cosmochim. Acta* **65**, 4399–4412.

Bradley, J.P., Sandford, S.A., and Walker, R.M.: 1988, 'Interplanetary Dust Particles', *in* J. Kerridge and M.S. Matthews (eds.), *Meteorites and the Early Solar System*, Univ. Arizona Press, Tucson, pp. 861–895.

Charnley, S.B., and Rodgers, S.D.: 2002, 'The End of Interstellar Chemistry as the Origin of Nitrogen in Comets and Meteorites', *Astrophys. J.* **569**, L133–L137.

Clemett, S.J., Maechling, C.R., Zare, R.N., Swan, P.D., and Walker, R.M.: 1993, 'Identification of Complex Aromatic Molecules in Individual Interplanetary Dust Particles', *Science* **262**, 721–725.

Clemett, S.J., Messenger, S., Keller, L.P., Thomas-Keprta, K.L., and McKay, D.S.: 2002, 'Spatially Resolved Analysis of Amines in Interplanetary Dust Particles Using Fluorescent Molecular Probes', *Met. Planet. Sci.* **37**, A36.

Cronin, J.R., and Chang, S.: 1993, 'Organic Matter in Meteorites: Molecular and Isotopic Analyses of the Murchison Meteorite', *in* J.M. Greenberg *et al.* (eds.), *The Chemistry of Life's Origins*, Kluwer, Netherlands, pp. 209–258.

Deloule, E., and Robert, F.: 1995, 'Interstellar Water in Meteorites?', *Geochim. Cosmochim. Acta* **59**, 4695–4706.

Floss, C., and Stadermann, F.J.: 2002, 'NanoSIMS Measurements of Nitrogen Isotopic Distributions in IDPs and Renazzo: Uniform ^{15}N Enrichment in a Chondritic IDP', *LPSC* **33**, Abstract # 1350.

Geiss, J., and Reeves, H.: 1981, 'Deuterium in the Solar System', *Astron. Astrophys.* **93**, 189–199.

Geiss, J., and Bochsler, P.: 1982, 'Nitrogen Isotopes in the Solar System', *Geochim. Cosmochim. Acta* **46**, 529–548.

Hashizume, K., Chaussidon, M., Marty, B., and Robert, F.: 2000, 'Solar Wind Record on the Moon: Deciphering Presolar from Planetary Nitrogen', *Science* **290**, 1142–1145.

Keller, L.P., and Messenger, S.: 1997, 'Nitrogen Speciation in a ^{15}N-rich Interplanetary Dust Particle', *Lunar Planet. Sci.* **28**, 707–708.

Keller, L.P., Messenger, S., Flynn, G.J., and Clemett, S.J.: 2002, 'The Nature of Molecular Cloud Material in Interplanetary Dust', *Geochim. Cosmochim. Acta*, in preparation.

McKeegan, K.D.: 1987, 'Ion Microprobe Measurements of H, C, O, Mg, and Si Isotopic Abundances in Individual Interplanetary Dust Particles', Ph.D. Thesis, Washington Univ.

McKeegan, K.D., Walker, R.M., and Zinner, E.: 1985, 'Ion Microprobe Measurements of Individual Interplanetary Dust Particles', *Geochim. Cosmochim. Acta* **49**, 1971–1987.

McKeegan, K.D., Swan, P.D., Walker, R.M., Wopenka, B., and Zinner, E.: 1987, 'Hydrogen Isotopic Variations in Interplanetary Dust Particles', *Lunar Planet. Sci.* **18**, 627–628.

Messenger, S.: 2000, 'Identification of Molecular Cloud Material in Interplanetary Dust', *Nature* **404**, 968–971.

Messenger, S.: 2002, 'Opportunites for the Stratospheric Collection of Dust From Short-period Comets', *Met. Planet. Sci.* **37**, in press.

Messenger, S., Keller, L.P., and Walker R.M.: 2002, 'Discovery of Abundant Interstellar Silicates in Cluster IDPs', *Lunar Planet. Sci.* **33**, Abstract # 1887.

Messenger, S., and Walker R.M.: 1997, 'Evidence for Molecular Cloud Material in Meteorites and Interplanetary Dust', *Astrophysical Implications of the Laboratory Study of Presolar Materials*, AIP Conf. Proc. **402**, pp. 545–564.

Millar, T.J., Bennett, A., and Herbst, E.: 1989, 'Deuterium Fractionation in Dense Interstellar Clouds', *Astrophys. J.* **340**, 906–920.

Nittler, L.R., and Messenger, S.: 1998, 'Hydrogen and Nitrogen Isotopic Imaging of Interplanetary Dust', *Lunar Planet. Sci.* **29**, Abstract # 1380.

Owen, T., Mahaffy, P.R., Niemann, H. B., Atreya, S., and Wong, M.: 2001, 'Protosolar Nitrogen', *Astrophys. J.* **553**, L77–L79.

Sandford, S.A.: 1987, 'The Collection and Analysis of Extraterrestrial Dust Particles', *Fund. Cosmic Phys.* **12**, 1–73.

Sandford, S.A.: 2001, 'Assessment of the Interstellar Processes Leading to Deuterium Enrichments in Meteoritic Organics', *Met. Planet. Sci.* **36**, 1117–1133.

Stadermann, F.J.: 1991, 'Messung von Isotopen- und Elementhäufigkeiten in einzelnen interplanetaren Staubteilchen mittels Sekundärionen-Massenspektrometrie', PhD Thesis, Universität Heidelberg.

Stadermann, F.J.: 2001, 'Hydrogen, Carbon and Nitrogen Isotopic Imaging of Sub-micron Components From Interplanetary Dust Particels', *LPSC* **32**, Abstract # 1792.

Stadermann, F.J., Walker, R.M., and Zinner, E.: 1989, 'Ion Microprobe Measurements of Nitrogen Isotopic Variations in Individual IDPs', *Met. Soc. meeting 52; LPI Contributions* **712**, 228.

Terzieva, R., and Herbst, E.: 2000, 'The Possibility of Nitrogen Isotopic Fractionation in the Interstellar Medium', *Month. Not. R. Astron. Soc.* **317**, 563–568.

Tielens, X.: 1997, 'Deuterium in Interstellar Chemical Processes', *Astrophysical Implications of the Laboratory Study of Presolar Materials*, AIP Conf. Proc. **402**, pp. 523–544.

Address for Offprints: Scott Messenger; Laboratory for Space Sciences, Department of Physics, Washington University, Saint Louis, MO 63130-4899, USA; dbunny@wuphys.wustl.edu

III. EARTH, TERRESTRIAL PLANETS AND MOON

NITROGEN ISOTOPES ON THE MOON: ARCHIVES OF THE SOLAR AND PLANETARY CONTRIBUTIONS TO THE INNER SOLAR SYSTEM

B. MARTY[1], K. HASHIZUME[2], M. CHAUSSIDON[3] and R. WIELER[4]
[1] *CRPG-CNRS, BP 20, 54501 Vandoeuvre-lès-Nancy Cedex, France*
and
Ecole Nationale Supérieure de Géologie, 54501 Vandoeuvre-lès-Nancy Cedex, France
[2] *Department of Earth & Space Sciences, Osaka University, Toyonaka, Osaka 560-0043, Japan*
[3] *CRPG-CNRS, BP 20, 54501 Vandoeuvre-lès-Nancy Cedex, France*
[4] *Institute for Isotope Geology and Mineral Resources, ETH Zürich, Zürich, Switzerland*

Received: 31 May 2002; Accepted in final form: 26 July 2002

Abstract. The two isotopes of nitrogen, ^{14}N and ^{15}N, have relative abundances extremely variable among solar system reservoirs such as planets and their atmospheres, primitive and differentiated meteorites, and comets. Expressed in the delta notation ($\delta^{15}N = \{[^{15}N/^{14}N]_{sample}/[^{15}N/^{14}N]_{standard} - 1\} \times 1000$, in parts per mil, or ‰, where the standard is atmospheric N having $^{15}N/^{14}N = 0.003676$), δ^{15}N ranges from −250‰ (the lower limit of lunar soil values) up to 1600‰ (measured in the meteorites benccubinites). The lunar surface constitutes a unique archive of the past corpuscular (solar and meteoritic) contributions to planetary surfaces. Nitrogen trapped in the lunar regolith presents a highly variable isotopic composition, which represents either secular variation of the solar wind composition although this possibility conflicts with the apparent isotope stability over time of other solar wind volatile elements, or more likely different contributions from solar corpuscular radiation and non-solar sources. In this case, the solar nitrogen component is depleted by more than 24% in ^{15}N, whereas non-solar, planetary sources (meteorites, micrometeorites, possibly comets) are enriched in the heavy isotope of nitrogen by ≥10% on average. Variations in the nitrogen isotopic composition of lunar soils are explained by a secular change in the strength of the planetary flux, and a correlation between N isotopic compositions and surface exposure age for different soils suggest that the planetary contribution to the inner solar system might have increased in the last 0.4 Gy. The variability of the N isotope composition among solar system objects might be due to incomplete equilibration of nitrogen isotopes from different host phases of pre-solar origin. Alternatively, it could result from mixing between ^{15}N-depleted protosolar nitrogen originally present in the gas and presolar solid (organic?) compounds enriched in ^{15}N.

1. Introduction

Nitrogen is somewhat unique among elements in that its numerous states of valence as a function of redox conditions allow this element to be present in solid inorganic and organic compounds actually found in primitive meteorites and in gaseous forms such as NH_3, N_2 and NO_x in planetary atmospheres. Some of the N isotope variations are due to planetary atmospheric processes. This is presumably the case of Martian atmospheric N for which Viking probe data suggest a δ^{15}N value ≥ +400‰ (Nier and McElroy, 1977). However, primitive objects such undifferen-

Space Science Reviews **106:** 175–196, 2003.

tiated meteorites, comets, and interplanetary dust particles present large isotopic variations that require contributions from isotopically distinct sources. An overall interpretation of these variations has been hampered for long by the uncertainty on the value of the nitrogen isotopic composition of the Sun.

The lunar surface, which has been sampled in several locations during the Apollo and Luna missions, constitutes a unique archive of the past corpuscular radiation. The surface of the Moon has been exposed over billions of years to radiation by solar ions, as evidenced by implanted solar rare gases in the lunar regolith (e.g., Wieler and Heber, 2003), and to contributions from various exotic sources, as shown by the occurrence of 1–2% chondrite-like material in the lunar regolith (Morgan *et al.*, 1972). Nitrogen is particularly sensitive to these contributions, because, as a volatile element, it might have been extremely depleted during the Moon-forming processes. In this contribution, we review the present knowledge on different nitrogen components that are possibly present on the Moon's surface. We briefly address the production of cosmogenic N on the Moon's surface and the possible occurrence of indigenous N, both components being either easy to correct for, or present in small amount. We then review recent findings obtained with new extraction/analytical techniques that shed light on the long-standing problem of the isotopic composition of nitrogen in the solar wind. Finally, we attempt to use the lunar surface as an archive of past contributions to planetary surfaces by asteroidal/cometary material in the terrestrial region.

2. Cosmogenic and Indigenous Lunar Nitrogen

2.1. Cosmogenic Production Rate

Lunar samples have been analyzed for their volatile composition by making incremental heating steps that release sequentially different trapped components. Nitrogen released at high-temperature ($\geq$1200°C) steps is extremely enriched in ^{15}N (e.g. Becker and Clayton, 1975, 1978; Becker *et al.*, 1976; Brilliant *et al.*, 1994; Frick *et al.*, 1988; Kerridge, 1975; Thiemens and Clayton, 1980). Noble gases extracted at similarly high temperatures are characterized by enrichments in cosmic-ray produced noble gas isotopes (Eberhardt *et al.*, 1970). The ^{15}N enrichments have been attributed to the release of cosmogenic nitrogen isotopes produced by spallation reaction on ^{16}O directly via $^{16}O(p,2p)^{15}N$ and indirectly via $^{16}O(p,pn)^{15}O$ (about 55% of the direct production, Reedy, 1981). Interestingly, the dependency of the production rate upon the composition of the target is not drastic as oxygen is a major element of rocks and minerals, and its abundance is rather constant among lunar samples. Becker *et al.* (1976) estimated the production rate to be 3.6 ± 0.8 pg $^{15}N\,g^{-1}\,My^{-1}$ from bulk analysis of three rocks having well-defined exposure ages. Mathew and Marti (2001a) derived a significantly higher ^{15}N production rate of 5.8 ± 0.6 pg $^{15}N\,g^{-1}\,My^{-1}$ from analysis of lunar rocks (basalt and anorthosite) having well-characterized exposure ages and containing small amounts of nitrogen. This estimate was found to be consistent with the ^{15}N production rate of meteorites after correction for differences in the geometry of radiation (Mathew and Marti, 2001a). From the combined N-Ar isotope analysis of single regolith grains, Hashizume *et al.* (2002) found an excellent correlation between cosmogenic ^{15}N and cosmogenic ^{38}Ar and derived a ^{15}N production rate of 3.8–5.6 pg $^{15}N\,g^{-1}\,My^{-1}$, the large range of values being the result of uncertainties in the $^{38}Ar_c$ production rate of the analyzed soils, in particular its dependency with depth which could significantly vary from sample to sample during sample exposure history.

2.2. Indigenous Lunar Nitrogen

In contrast to lunar soils which contain typically 10 to 100 ppm N, lunar rocks (lavas, anorthosite) having a short exposure time contain 1 ppm or less nitrogen. From the analysis of a recently (2 My) exposed rock as well as of stepwise etching experiments on other igneous rocks, Becker *et al.* (1976) proposed that there exists an indigenous N component of the order of 2 ppm N having an isotopic signature of +10‰, but these authors acknowledged that this occurrence could be an artefact due to terrestrial contamination. Murty and Goswami (1992) proposed a $\delta^{15}N$ value of +17‰ for indigenous N in a clast of a lunar meteorite. Kerridge *et al.* (1991) measured nitrogen in 74001/74002 double-drive core samples from Shorty Crater, Apollo 17. Soils from this site contain glassy spherules some of which are orange, as a possible result of fumarolic vapor deposition. Kerridge *et al.* (1991) derived a value of around +14‰ for nitrogen after correction for spallation effects. From the stepwise heating nitrogen analysis of lunar rocks having short exposure duration, Mathew and Marti (2001a) identified a well-resolved medium ($\leq$1100°C) temperature nitrogen component having a $\delta^{15}N$ value of $+13 \pm 1$‰ and N contents between 0.7 ppm in basalt 75075, 1.4 ppm in anorthosite and 3.1 ppm in volcanic glasses, that they attributed to the presence of indigenous lunar nitrogen. For comparison, terrestrial mid-ocean ridge basalt glasses, which have been preserved to a large extent from degassing and atmospheric contamination as they were erupted under great hydrostatic pressure, contain typically less than 1 ppm (Marty and Zimmermann, 1999). The somewhat large N content of lunar rocks is surprising if one considers that (i) the Moon is drastically depleted in volatile elements compared to Earth, and (ii) lunar basalts were erupted at extremely low pressure and should have been degassed thoroughly. Notably, no indigenous noble gas component has been unambiguously identified in lunar rocks. One possibility is that indigenous nitrogen has been preferentially retained relative to noble gases during lunar crust formation and lunar basalt eruption. Nitrogen solubility experiments under variable f_{O_2} conditions have shown that the N solubility in basaltic melt is low (0.08 ppmN/atm) and comparable to that of argon for f_{O_2} between air and the iron-wustite (IW) buffer which corresponds to metal-silicate f_{O_2} equilibrium f_{O_2}

(Humbert *et al.*, 1999). However, it reaches about 1 ppm N/atm for $f_{O_2}\approx$IW-2 and increases dramatically for lower oxygen fugacity as nitrogen becomes chemically bound to the silicates (Humbert *et al.*, 1999). Therefore, trapping of about 1–2 ppm N in the lunar crust would require a significantly low f_{O_2}, which is consistent with estimates derived from mineral assemblages in lunar basalts. Indeed most lunar basalts have crystallized in the f_{O_2} range of IW-2 to IW-4 at 1125°C (Haggerty, 1978).

An alternative explanation for the relatively high nitrogen content of lunar rocks could be contamination during sample processing and storage in pure nitrogen atmosphere. Stepwise heating experiments of bulk terrestrial basalt glasses have shown that nitrogen is released continuously at temperatures up to 1600°C (Marty and Zimmermann, 1999). Excluding nitrogen released under vacuum at low to medium temperature ($\leq$600°C), the total N content varies between 0.1 ppm and several ppm, with δ^{15}N values in the range +5 to +15‰ (Marty and Zimmermann, 1999, and unpublished data). Combining crushing experiments with stepwise heating experiments, Marty and Zimmermann (1999) have shown that mantle-derived nitrogen resides almost exclusively in CO_2-rich fluid inclusions and form a minute fraction of total nitrogen. The origin of the non-indigenous nitrogen in terrestrial basalts is unclear and this component could represent either atmospheric and/or organic contamination, in which case it would need to be more tightly bound to the silicates than adsorbed atmospheric noble gases, or could be derived from sedimentary contamination by magma assimilation of sedimentary N present in the basaltic crust. Although the parallel between these experiments and those aimed to characterize the lunar indigenous N is not straightforward, we suggest that it will be very interesting to document the trapping processes and reactions of nitrogen on surfaces representative of lunar samples using different chemical treatments or isotopically labeled atmosphere to decipher if the so-called indigenous N represents a pristine lunar component or is due to terrestrial contamination that follows paths different from those of noble gases.

3. Nitrogen in Bulk Lunar Soils

Noble gases trapped in lunar soils are mostly derived from the solar corpuscular radiation (Eberhardt *et al.*, 1970). Solar noble gases are implanted at the grains surfaces and are for this reason surface-correlated, *i.e.*, their amount increases with decreasing grain size of soil grains (Eberhardt *et al.*, 1970). In lunar soil samples, the amount of trapped nitrogen correlates to a first order with that of solar noble gases (*e.g.*, Kerridge, 1993), indicating that lunar soil nitrogen is also surface-correlated. However, nitrogen is overabundant with respect to noble gases as the ^{14}N/^{36}Ar ratio of lunar soils is on average a factor of 10 higher than the solar abundance ratio of 37 (Anders and Grevesse, 1989). This overabundance is apparently not due to elemental fractionation during transfer from the solar photosphere to the

Moon's region. Fractionation in the solar wind relative to the solar composition has been observed for several elements, and may depend on their first ionization potential (FIP), but N and Ar have very similar FIPs. Furthermore, a solar $^{14}N/^{36}Ar$ ratio has been directly measured in the case of solar energetic particles (Brenemann and Stone, 1985). Therefore, the enhancement of N relative to Ar in lunar soils has been interpreted either as being due to elemental fractionation between N and Ar during lunar processes, (e.g., Kerridge, 1993), or as resulting from contribution of other surface-correlated nitrogen component(s) of non-solar origin (Geiss and Bochsler, 1982; Wieler *et al.*, 1999).

The nitrogen isotopes add further insight into this long-lasting problem but also raise specific questions. The isotopic composition of nitrogen trapped in lunar samples, once corrected for spallation effects, has been found to be extremely variable compared to other elements of comparable volatility (e.g., C, S). The $^{14}N/^{15}N$ ratio of nitrogen trapped in the lunar regolith varies by more than 30% among different bulk samples (e.g. Becker and Clayton, 1975; Kerridge, 1975) as well as in different extraction steps of a given sample (Brilliant *et al.*, 1994; Frick *et al.*, 1988). In order to explain the isotopic variability of lunar soils, two contrasted views have been proposed. In the first type of explanation, the N isotopic composition of the solar corpuscular radiation has changed with time. As reviewed in this volume by Wieler and Heber, the bulk $^{40}Ar/^{36}Ar$ ratios of different soils vary largely from about 0.4 to 20. Assuming that the ^{36}Ar flux (from solar radiation) has remained approximately constant over most of the Moon history and that the ^{40}Ar contribution (from the decay of lunar 40 K) has decreased with time, the $^{40}Ar/^{36}Ar$ ratio has been regarded as an indicator of the epoch (the "antiquity") at which a given soil sample was exposed at the Moon's surface (e.g., Eugster *et al.*, 2001, and references therein). It has been found that the N isotope composition of different soils follows different trends depending on the respective $^{40}Ar/^{36}Ar$ ratios (Clayton and Thiemens, 1980; Kerridge, 1975). Apollo soils with $^{40}Ar/^{36}Ar$ ratios lower than 1 are enriched in ^{15}N, whereas soils having $^{40}Ar/^{36}Ar$ ratios higher than 1 have variable and often low $^{15}N/^{14}N$ with $\delta^{15}N$ values reaching −250‰ (Figure 1). Based on a $^{40}Ar/^{36}Ar$-antiquity calibration curve, samples having $^{40}Ar/^{36}Ar$ ratios lower than 1 were exposed in the last Gy, whereas samples with higher ratios were exposed between 1 and 4 Gy ago.

Kerridge (1975) proposed that "young" soils have been exposed to the "recent" solar wind, which is characterized by enrichments in ^{15}N, whereas ancient soils contain a mixture of "recent" solar wind N and "ancient" solar wind N depleted in ^{15}N. This author concluded that the isotopic composition of nitrogen transported in the solar corpuscular radiation presents a significant secular change. The 10 fold enrichment of nitrogen relative to solar argon was regarded as resulting from preferential retention of nitrogen relative to noble gases (Kerridge, 1993). This model, which proposes a heavy N composition for recent solar wind, is compatible with the heavy N isotope composition of +40‰ measured in recently exposed lunar samples (a soil excavated 50 My ago during formation of North ray crater, and a

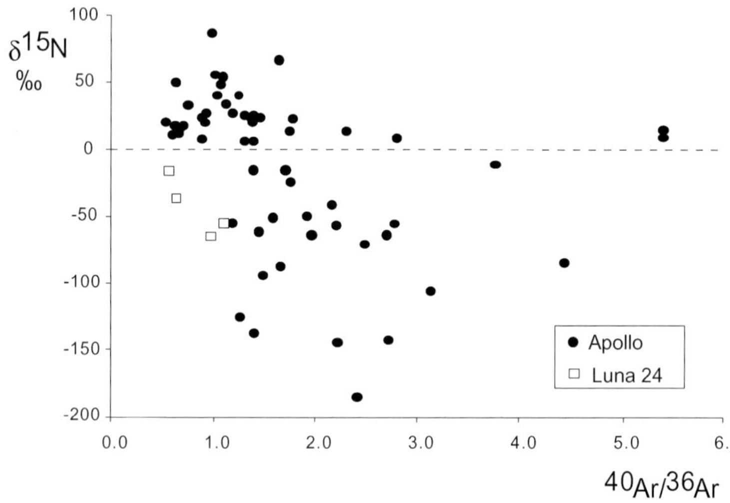

Figure 1. Variation of the nitrogen isotopic composition as a function of the $^{40}Ar/^{36}Ar$ ratio for bulk lunar soils. The black dots refer to the Apollo samples and the open squares represent the Luna 24 soil samples analyzed by Assonov *et al.* (2002).

rock excavated during formation of South ray crater 2 My ago, Kim *et al.*, 1995), and with the stepwise heating extraction of heavy N at low temperature (recently implanted N) followed by higher temperature extraction of light N (ancient SW nitrogen implanted long ago). However, it also faces difficulties. Recent N-Ar analyses of samples recovered by the Luna 16 and 24 probes have shown that bulk Luna soils as well as individual grains from these soils have $^{40}Ar/^{36}Ar$ ratios generally lower than 1 and are depleted in ^{15}N with by up to 100‰ (Assonov *et al.*, 2001, 2002). Furthermore, based on single grain analysis of Luna soils, Assonov *et al.* (2001) have shown that the N-Ar association was not an effect of mixing between old and recent soils, and that there exist soils not sampled by the Apollo missions having been exposed recently and containing light N (Figure 1).

A view opposite to that of Kerridge (1975) was proposed by Geiss and Bochsler (1982) who noted that no known solar process could produce the required ≈300‰ secular variation of the N isotope composition in the solar radiation. The spallation rates required to produce the ^{15}N enrichments observed in the recently exposed Apollo samples are not recorded in the isotopic composition of other volatile elements such as noble gases or carbon. The layered convection of the Sun does presumably not allow transfer of nuclides from inner nuclear reactions zones towards the surface, and recent observations on the isotope composition of solar-derived lithium (Chaussidon and Robert, 1999) and helium implanted in lunar soils (Wieler and Heber, 2003) strongly suggest that the outer convective zone of the Sun has been well isolated from the inner Sun for at least the last 3 Gy, precluding injection of isotopically distinct N into the solar-wind feeding zone. Furthermore, isotopic fractionation inside the Sun remained below one percent per amu over the

Sun's lifetime, except for the D/H and $^{3}He/^{4}He$ ratios (Geiss and Gloeckler, 2003). Geiss and Bochsler (1982) instead proposed that nitrogen trapped in lunar soils consists in a mixture of light N of planetary derivation (comets, meteorites) and of heavy N of solar origin. These authors also mentioned several other potential nitrogen sources, apart from atmospheric contamination, such as ionized nitrogen escaping from the top of the terrestrial atmosphere, subsequently focused by the terrestrial magnetic field and implanted in lunar soils. This nitrogen component should be enriched in ^{14}N due to preferential escape of the lighter N isotope and could account for the light N component found in lunar soils.

A predominantly non-solar origin of nitrogen in the lunar regolith was argued by Wieler *et al.* (1999) from the analysis of nitrogen and argon in single ilmenite grains from a recently exposed soil (Apollo 17 sample 71501). These authors noted that, although the average $^{14}N/^{36}Ar$ ratio of the analyzed grains was identical to the bulk soil value (Frick *et al.*, 1988), the $^{14}N/^{36}Ar$ ratio of different grains varied by two orders of magnitude and that the variation was mainly due to variable ($\times 20$) solar ^{36}Ar contents whereas the N content varied by only a factor of 3. The variability of the $^{14}N/^{36}Ar$ ratio is in striking contrast with the uniform relative abundances of Ar, Kr and Xe among different grains analyzed from the same soils (Wieler *et al.*, 1996). Wieler *et al.* (1999) concluded that on average $\approx$90% of nitrogen trapped in lunar soils is of non-solar origin and that the $\approx$300‰ variation of the solar N isotope composition is highly questionable. Clearly, lunar soil grains have had a complex history of solar wind radiation, meteoritic/cometary contribution and re-working that needs to be addressed.

4. Lunar Nitrogen: Microscopic Studies

Transmission electron microscopy (TEM) studies have shown that lunar grains are a complex material having heterogeneous rims towards their outer surface (Keller and McKay, 1997). Some grains present outer 50–100 nm amorphous rims that were produced by solar wind radiation. "Saturation", which characterizes a state for which the density of implanted solar wind ions does not vary any more with time because their amount removed by sputtering matches the amount of implanted ions, is a geologically rapid process that has been successfully reproduced in the laboratory (Futagami *et al.*, 1990). When saturation is reached, the medium has its crystalline structure destroyed by solar radiation and appears as an amorphous layer in TEM examination, and also contains numerous nuggets of iron that were produced by reduction of iron oxide contained in the grains (Keller and McKay, 1997). Other grains show a complex radiation history with amorphous rims alternating with layers enriched in volatile (S, K) elements. The latter are attributed to vapor deposition resulting from condensation of local vapor clouds formed by micrometeorite impacts (Keller and McKay, 1997). Few grains also present submi-

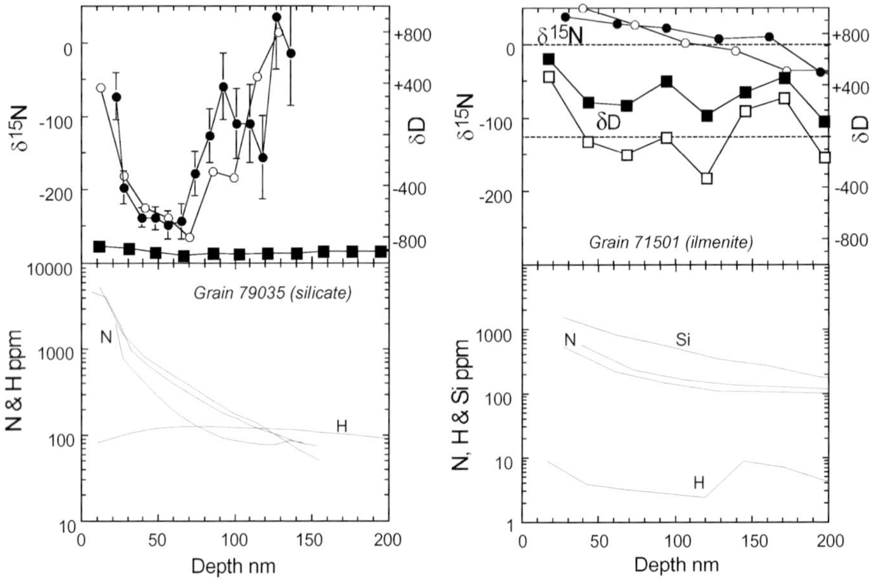

Figure 2. Ion probe depth profiling of N and H isotopic ratios for grains from two Apollo 17 soils (adapted from Hashizume *et al.*, 2000). For grains from soil 79035 (*left*), δ^{15}N values are negative and reach a minimum of −240‰ at a typical depth (distance from the grain surface) of 50 nm. The light N isotope composition is accompanied by hydrogen virtually free of deuterium of solar derivation. Note that the H content is low and presents a near-flat gradient, as a result of diffusive loss of hydrogen from regolith grains. In contrast, hydrogen in grains from soil 71501 (*right*) is rich in deuterium, indicating a planetary origin, and is associated with heavy N and a silicon coating onto the grain surface, attributed to contribution of planetary material to the analyzed ilmenite (Si-free) lunar soil grains (see text for further explanation).

crometric "splash"-like features thought to represent remnants of micrometeorite melts projected onto the grain surfaces.

Significant advance in the understanding of the origin of volatile elements has been achieved using a new technique of in-situ ion probe depth profiling of isotopes in regolith grains (Hashizume *et al.*, 2000). It consists of bombarding relatively flat surfaces of unpolished lunar grains coated with gold (to render it electrically conductive) with monoenergetic Cs^+ ions. Ions from the substratum (the sample), called secondary ions, are extracted and subsequently filtered in energy and mass. Nitrogen is analyzed as CN^- ions at masses 26 and 27 (the carbon isotopic composition is also analyzed as CC^- ions). During sputtering, the surface is progressively carved and the isotopic ratios of nitrogen, carbon and hydrogen, together with the amounts of these elements and others of interest (Si in this case) are simultaneously measured. This technique allows to establish N and H elemental and isotopic profiles from the grain surface towards its interior with a depth resolution of about 10 nm.

Two soils having different exposure histories were analyzed. Based mainly on their $^{40}Ar/^{36}Ar$ ratios, Soil 71501 is thought to have been exposed to the lunar surface within the last ≈100 My whereas soil 79035 has been exposed at least 1 Gy ago. Hence these two soils, which have been extensively documented for their noble gas and N compositions (Frick *et al.*, 1988; Wieler and Baur, 1994) are somewhat representative of the N isotopic evolution recorded among bulk soils in the sense that the "recently" exposed soil 71501 contains heavy N and sample 79035 thought to have been exposed ≥1 Gy (from a $^{40}Ar/^{36}Ar$ ratio of about 1.0–1.5) contains light N. The results, an example of which is given in Figure 2, show that the N content of some of the grains decreases exponentially from the surface towards the grain interior within a depth of about 100–200 nm from the surface. Grains from soil 79035 show a strong decrease of the $\delta^{15}N$ value having a minimum of −240‰ at a depth of about 50 nm. Such depth is typical of the thickness of amorphous rims thought to represent the grain crystalline network destroyed by solar wind ions (Keller and McKay, 1997). Its thickness is comparable to the depth at which low speed solar wind ions having energy in the range ≈ 1 keV/amu (which contribute the largest part to the flux of solar corpuscular radiation) are implanted. D/H measurements of $\delta D = -950‰$ show the near absence of deuterium and indicate without ambiguity the presence of solar hydrogen. It must be noted that the hydrogen content and H/N ratios are much lower than those expected from closed system contribution from solar corpuscular radiation, evidencing diffusion loss of hydrogen in these grains. Diffusion of light elements is also evidenced by (i) the He/Ar ratios of lunar soils which are lower than solar in comparable proportion relative to H/N, and (ii) the concentration profile of hydrogen (Figure 2) compared to that of nitrogen, fully consistent with diffusive loss of the former element.

In contrast, grains from soil 71501 do not contain such a light N component. They instead show higher $\delta^{15}N$ values, typically varying with depth from about +50‰ at the grain surface to down to ≈ −100‰ at depths exceeding 200 nm (Figure 2). Hydrogen associated with this heavy N component is rich in deuterium (Figure 2), evidencing the contribution of a "planetary" (e.g., non-solar) component. Furthermore, the grains analyzed from this soil were ilmenite but small amounts of Si were detected at their surface. The ion probe depth profiles showed that heavy N (and D-rich hydrogen) is associated with a silicon-rich layer, leading Hashizume *et al.* (2000) to propose that this Si-rich layer represents a deposit of a planetary component added to the grain surface (Figure 3). During one analytical run, a "bright" C- and N-rich spot was observed, characterized by a $\delta^{15}N$ value of > +100‰, interpreted as being due to a "splash"-like structure resulting from addition of a micrometeoritic impact.

From these observations, Hashizume *et al.* (2000) proposed that the light N component associated with D-free hydrogen was implanted from the solar wind, leading these authors to propose an upper limit for SW $\delta^{15}N$ value of −240‰. The heavier N isotope composition associated with D-rich hydrogen and Si-rich layering was interpreted as representing the contribution from a non-solar N component associated with vapor and material depositions from meteoritic, micrometeoritic, or cometary impact.

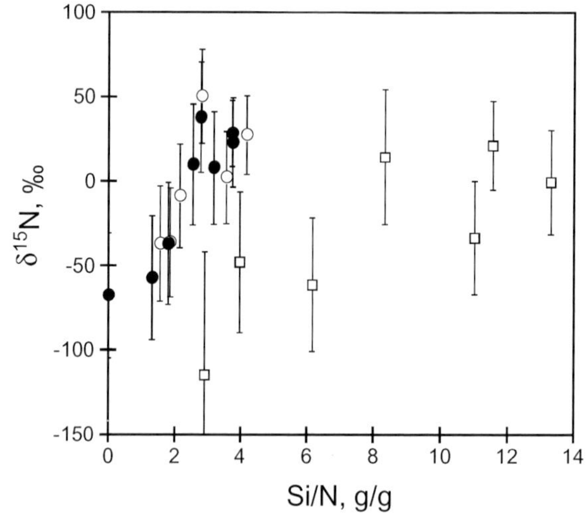

Figure 3. Nitrogen isotopic composition versus the Si/N ratio for 3 different spots on the same grain from soil 71501 (data from Hashizume *et al.*, 2000, and unpublished data). The observed correlation suggests that there exists a nitrogen component characterized by a δ^{15}N value of < -70‰ located below the Si coating covering the ilmenite grains, which could correspond to the solar energetic particle component of nitrogen.

5. In Search of the Solar Energetic Particle (SEP) Nitrogen Component

Mathew *et al.* (1998), using bulk and fractional data obtained by stepwise heating of soil 71501, obtained a correlation between solar wind Ne (characterized by a ^{20}Ne/^{22}Ne of 13.8) and heavy N (δ^{15}N $> +68$‰) on the one hand, and between solar energetic (SEP) Ne (characterized by a ^{20}Ne/^{22}Ne of 11.2) and light N (δ^{15}N < -80‰) on the other hand. The noble gas SEP component is implanted at greater depth (several hundreds of nm, from ion impact simulation and from acid etching experiment) than the solar wind (Wieler and Baur, 1994). Mathew *et al.* (1998) concluded that the light N component associated with SEP Ne is the SEP N component. The SEP N component might have been also identified during the ion probe experiments of the Nancy group (Figure 3). For sample 71501, there is a good correlation between the Si/N ratios and the δ^{15}N values extending from one end-member free of Si and having δ^{15}N ≤ -70‰, and another end-member enriched in ^{15}N and characterized by high Si/N ratios. Because the Si abundance decreases from the surface towards the grain interior at a rate greater than that of nitrogen (Figure 2), the isotopically light N component has to be deeper than the heavy N component and could correspond to the SEP N. The surface heavy N component appears to be characterized by a δ^{15}N value of about +50‰ for the analyzed grains.

Interestingly, Mathew *et al.* (1998) ascribed a modern solar wind origin for the heavy N component associated with solar wind neon, supporting therefore the view originally proposed by Kerridge (1975) (Subsection 3) that modern solar wind N is rich in ^{15}N. However, a problem with this interpretation is that in this case the SEP-N component would be more depleted in ^{15}N than the SW-N component, a trend opposite to that of noble gases where the heavy isotopes are enriched in the SEP component relative to SW. In impulsive flares, gyro-resonant plasma waves seem to accelerate ions in a single stage according to the gyro-frequencies of ions in the solar corona which are function of their respective charge-to-mass (Q/A) ratios (Fisk, 1978; Paesold *et al.*, 2002) (see the review by Kallenbach, 2003). SEPs from gradual events are fractionated following power laws in Q/A over wide ranges of Q/A ratios, and "implantation" fractionation modeled by Mewaldt *et al.* (2001) may scale with A or A^2 depending on the energy range. Therefore, assuming that the SEP-N component is ^{15}N-depleted, then the SW-N component should be even more depleted in ^{15}N to be compatible with the trend observed for noble gases. This would imply that the heavy nitrogen component observed at the very surface of 71501 is non-solar, in agreement with the view of Hashizume *et al.* (2000).

6. Nature of the Non-solar Nitrogen Contribution to the Moon's Surface

Analyses of single grains from soils 71501, 79035, 79135 (Figure 4) (Hashizume *et al.*, 2002) and Luna 24 (Assonov *et al.*, 2001) show highly variable δ^{15}N values and N/^{36}Ar ratios for a given soil, confirming that soils are extremely heterogeneous at a grain scale and represent a population of sub-millimeter objects having had contrasted histories in terms of radiation and processing. The results confirm that there exists a pole enriched in solar ^{36}Ar ($[^{36}Ar/N]_{sample}/[^{36}Ar/N]_{solar} > 0.2$) and presenting a light N isotope composition ($\delta^{15}N < -200‰$) (Figure 4) solar nitrogen. For samples depleted in solar ^{36}Ar, the δ^{15}N values tend to be higher and may define an end-member with a positive δ^{15}N value up to $\approx$+100 ‰, although negative δ^{15}N values down to −100‰ also exist. Hence the single grain data suggest that non-solar N may not be always characterized by a ^{15}N-rich composition and that there may exist several non-solar N components having variable δ^{15}N values. In the δ^{15}N vs. $[^{36}Ar/N]_{sample}/[^{36}Ar/N]_{solar}$ format displayed in Figure 4, component mixings result in straight lines joining the component end-members. The data plots can be delineated by mixing lines between two non-solar poles having δ^{15}N values of +100‰ and −100‰, and a solar end-member whose value is fixed at either −240‰, the upper limit for lunar soils (Hashizume *et al.*, 2000), or at −374‰, the Jupiter atmospheric value (Owen *et al.*, 2001). The corresponding $[^{36}Ar/N]_{sample}/[^{36}Ar/N]_{solar}$ values are between 0.2 and 0.4. In other words, the single grain analysis experiments suggest that the solar component as defined by the δ^{15}N value is depleted in solar ^{36}Ar by 60–80%. A relative depletion of solar argon implies that the enhanced N/^{36}Ar ratio relative to solar is not uniquely due to a $\approx$90% overabundance of non-solar nitrogen as previously proposed by Wieler *et al.* (1999), but is rather the result of two processes acting in the same way, i.e. contribution of non-solar nitrogen accounting for a factor of $\approx$2 (see below) *and*

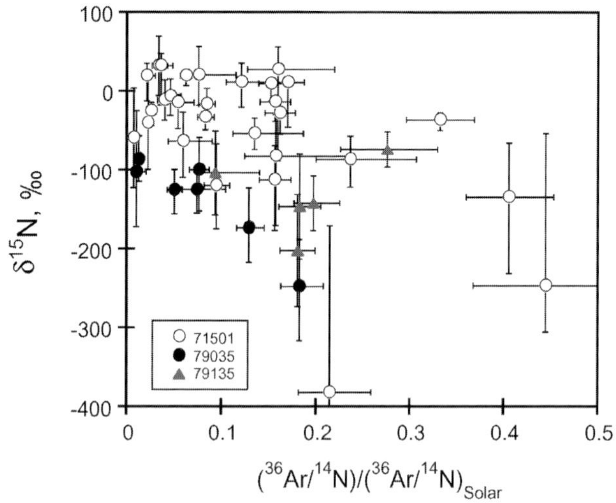

Figure 4. δ^{15}N value (corrected for spallation) versus the ^{36}Ar/^{14}N ratio (normalized to the solar value of 0.0272, Anders and Grevesse, 1989) for individual grains from "recent" soil 71501, and "old" 79035, 79135 soils, see text for definitions (Hashizume *et al.*, 2002). Samples containing solar-derived ^{36}Ar tend to contain isotopically light nitrogen whereas samples with low ^{36}Ar/^{14}N present δ^{15}N between ≈ -100‰ and +100‰. Extrapolation of data to solar δ^{15}N values (< -240‰, Hashizume *et al.*, 2000; -374‰, Owen *et al.*, 2001) points towards ^{36}Ar/^{14}N ratios from 0.2 to 1 times the solar ratio, suggesting preferential trapping of nitrogen relative to argon.

depletion by a factor of $\approx$3 of solar ^{36}Ar. Hashizume *et al.* (2002) have recently proposed a model in which noble gases and nitrogen are sputtered off the lunar soils, and re-implanted following a process that favors retention of chemically reactive nitrogen implantation relative to that of noble gases, the latter being lost from the transient lunar atmosphere.

All known planetary reservoirs host nitrogen enriched in ^{15}N relative to the solar wind value proposed by Hashizume *et al.* (2000) (δ^{15}N < -240‰) and to the Jovian atmospheric value (δ^{15}N $= -374 \pm 130$‰, Owen *et al.*, 2001) (see Section 8 and Figure 8). A possible and notable exception is cometary N for which infrared measurement in HCN in Hale-Bopp have led Jewitt *et al.* (1997) to propose a δ^{15}N value of -160 ± 170‰. Hence the δ^{15}N range from -100‰ to $+150$‰ of the non-solar nitrogen end-member present in lunar soils may represent a mixture of contributions of various bodies falling on the Moon.

Assuming that the solar N flux has been constant over the time interval recorded in the different lunar soils recovered by the Apollo mission, the N/Ar ratios, which are comparable within a factor of $\approx$2 for bulk soils (but very different on a grain to grain basis) would require that non-solar N has been supplied continuously to the different soils samples. Hashizume *et al.* (2001) noted that the flux of micrometeorites on the Moon, as inferred from terrestrial studies, can be assimilated to a continuous flux of non-solar nitrogen. At present, the flux of extraterrestrial matter on Earth is dominated by IDPs and micrometeorites and is on average

40000 ± 20000 tons/y (Love and Brownlee, 1993; Maurette, 1998). Among these objects, micrometeorites in the size range 50–200 μm make most of the infalling mass (Love and Brownlee, 1993). Micrometeorites share similarities with two classes of carbonaceous chondrites (Engrand and Maurette, 1998; Kurat *et al.*, 1994): CM-type (N≈1000 ppm, δ^{15}N = +40‰, Kerridge, 1985) and CR-type (N ≈ 600 ppm, δ^{15}N = +150‰, Kerridge, 1985). Hence this class of bodies make a candidate of choice for contributing non-solar N at the surface of terrestrial planets including the Moon.

A possible contribution of cometary material is more difficult to assess since its flux is unknown. It is possible that some of the IDPs are cometary grains, and therefore represent a continuous flux of cometary material as micrometeorites do so in the case of the meteoritic matter. Ion probe analyses of various classes of IDPs have shown that nitrogen is rich in ^{15}N (Messenger, 2000), with δ^{15}N values up to +810‰ (Stadermann and Floss, 2001). Three different nitrogen and hydrogen isotope end-members have been identified in organic compounds of IDPs (Aléon *et al.*, 2002). One is probably from a material sharing analogies with carbonaceous chondrites and is moderately rich in ^{15}N and D. Another one presents moderate enrichments in D and in ^{15}N and is interpreted as representing some material having been reprocessed in the solar system. The third one is extremely rich in D and ^{15}N and may represent a pristine interstellar phase showing high D/H ratio $\approx 2 \times 10^{-3}$, high δ^{15}N ≈ +400‰, high aromaticity (C/H ≈ 3.0) and high N content. Thus, if IDPs and comets have incorporated material from common reservoirs, the light N composition reported for HCN in Hale-Bopp (Jewitt *et al.*, 1997) would indicate that HCN represents proto-solar N from cometary ices and the ^{15}N enrichments seen in IDPs would be related to organics that have been left unprocessed.

7. Fluxes of Solar and Non-solar Contributions to the Moon's Surface

Cosmic-ray exposure ages are somewhat related to the duration at which a given soil was close to the surface since the average thickness of the lunar regolith is about 10 m, and the accumulation of cosmogenic noble gas isotopes occurs mostly within the first meter (Becker and Clayton, 1975; Kerridge *et al.*, 1975). There exists a very good correlation between δ^{15}N and cosmic-ray exposure ages (Figure 5) computed from noble gas isotopes (either $^{21}Ne_c$ or $^{38}Ar_c$, for example, in Figure 5 we used $^{21}Ne_c$ because its identification from trapped Ne is easier than in the case of Ar and the data base is therefore larger). Remarkably, individual grains from three soils from the Apollo 17 site display a very similar correlation between δ^{15}N values and $^{38}Ar_c$ exposure ages (Hashizume *et al.*, 2002), strongly suggesting that a common process all over the Moon's surface sampled to date is governing contributions of different nitrogen components over time. Such a correlation, already seen in Apollo 16 and 17 bulk soils (Becker and Clayton, 1975; Kerridge *et al.*, 1975), cannot be due to addition of spallation ^{15}N which is negligible compared to the

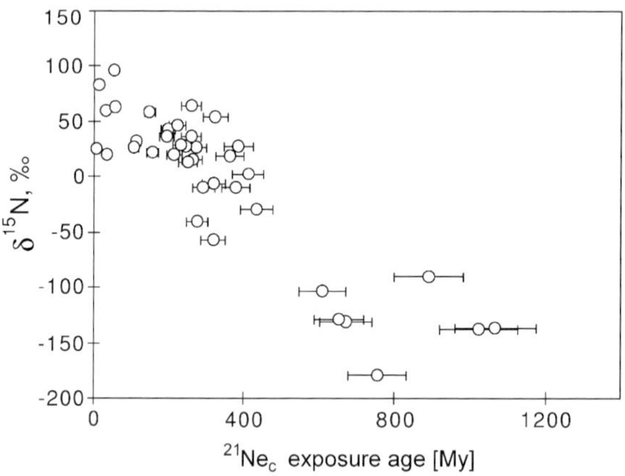

Figure 5. $\delta^{15}N$ values versus $^{21}Ne_c$ cosmic ray exposure (CRE) ages for bulk Apollo soils. CRE ages of lunar soils represent average time intervals spent by a given soil fraction close to the surface (mostly within the first meter of depth) in order to accumulate significantly CR-produced isotopes.

amount of bulk nitrogen in most cases. Samples exposed for "short" periods of time (of the order of few 100 My) are enriched in ^{15}N and correspond to samples having low $^{40}Ar/^{36}Ar$ ratios (Figure 5). Soils having $^{21}Ne_c$ exposure ages between 600 and 1200 My contain predominantly light N of solar origin. Because the antiquity-$^{40}Ar/^{36}Ar$ calibration is imprecise, especially for $^{40}Ar/^{36}Ar$ values around 0.6–2 which form the vast majority of samples used to construct Figure 5, we tentatively consider that cosmic-ray exposure ages are better proxies for the chronology of a given soil accumulating solar and non-solar nitrogen components. Therefore, the tendency of having heavy N associated with short exposure age and light N associated with larger exposure age requires to vary the respective fluxes of the N end-member components.

The $\delta^{15}N$-cosmic ray exposure age relationship suggests that the mixing ratio between solar light nitrogen and planetary ^{15}N-rich nitrogen has changed with time. One possibility would be that the strength of the solar radiation flux has changed within the last 1 Gy, the flux of planetary nitrogen remaining constant, although there is no indication for such a drastic change in the noble gas record (e.g., Wieler, 1998). Alternatively, the planetary nitrogen contribution could have varied drastically. Soils that have been excavated recently might have been loaded with non-solar, planetary ^{15}N-rich nitrogen, possibly derived from the impactors that allowed excavation of these soils. Soils would then be progressively loaded with a mixture of solar, ^{15}N-depleted N, and of planetary N supplied by the micrometeoritic background flux. This possibility implies that planetary nitrogen supplied during the excavation event should be progressively replaced by a mixture of solar wind and meteoritic N. An increase of the planetary flux could be also due to a recent (<400 My) increase in the flux of planetary material if planetary N is mostly contributed by the continuous flux of meteorites, micrometeorites, and IDPs at the

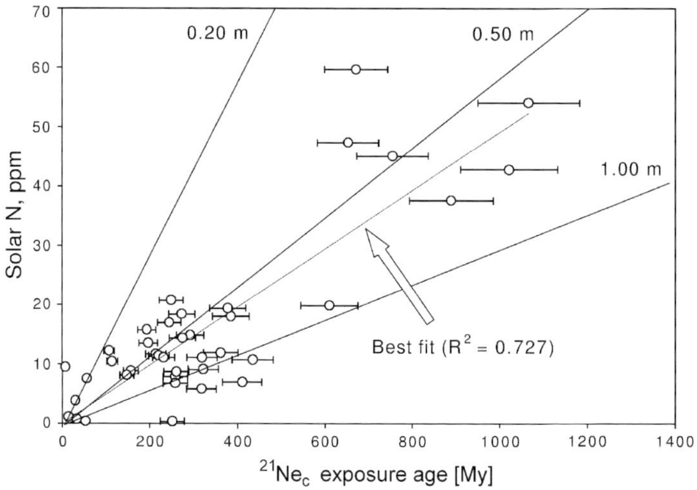

Figure 6. $^{21}Ne_c$ cosmic ray exposure age versus the amount of solar nitrogen for bulk Apollo soils. The latter is computed from the observed N content and $\delta^{15}N$ value assuming that trapped nitrogen is a mixture of solar-derived nitrogen having a $\delta^{15}N$ value of −374‰ (Owen *et al.*, 2001), and non-solar nitrogen with $\delta^{15}N = +100‰$ (this value corresponds to an average of $\delta^{15}N$ values of CM and CR chondrites, the best analogues of micrometeorites (Kurat *et al.*, 1994) on the one hand, and to the upper limit of the $\delta^{15}N$–$^{36}Ar/^{14}N$ correlation of Figure 5 on the other hand). The dotted line represents the best fit of data ($R^2 = 0.727$). The three solid lines represent the concentration of solar N, computed from the measured flux of solar ^4He and the solar wind $^4He/^{14}N$ ratio, which would be loaded in the regolith, for different regolith thickness. The data fit best a thickness of ≈ 0.5 − 1 m, which corresponds roughly to the thickness of regolith in which the production of $^{21}Ne_c$ has been significant. Note that the computed content of solar wind nitrogen is consistent with irradiation at constant flux during ≈ 1.2 Gy.

Moon's surface. In order to test these possibilities, we estimate the strengths of each contribution and evaluate if they can provide the amount and isotope composition of nitrogen trapped in lunar soils.

Assuming that lunar nitrogen is a simple mixture of solar N ($\delta^{15}N \approx -374‰$ from the Jupiter measurement) and planetary N ($\delta^{15}N \approx +100‰$, based on the correlation presented in Figure 4, the end-member of the single grain correlation displayed in Figure 3, and the average of non-solar values from ion probe measurements in Figure 4), we compute for each soil the amounts of solar N and non-solar N from the respective $\delta^{15}N$ values and N contents. The results (Figure 6) show that there is a positive correlation between the exposure age and the solar N content, suggesting a linear increase in solar N with time, i.e., little modulation of the SW N flux. This correlation is in full agreement with models based on noble gas contents of lunar regolith which propose that the flux of the solar radiation has little varied with time during the last 2.6 Gy (Geiss *et al.*, 1973).

The present-day flux at the Moon's surface, estimated from the measured SW proton flux (Hundhausen *et al.*, 1970) the solar abundance N/H ratio, and corrected for differences in shielding between Earth and Moon, is $\approx 4 \times 10^3$ atoms cm^{-2} s^{-1}.

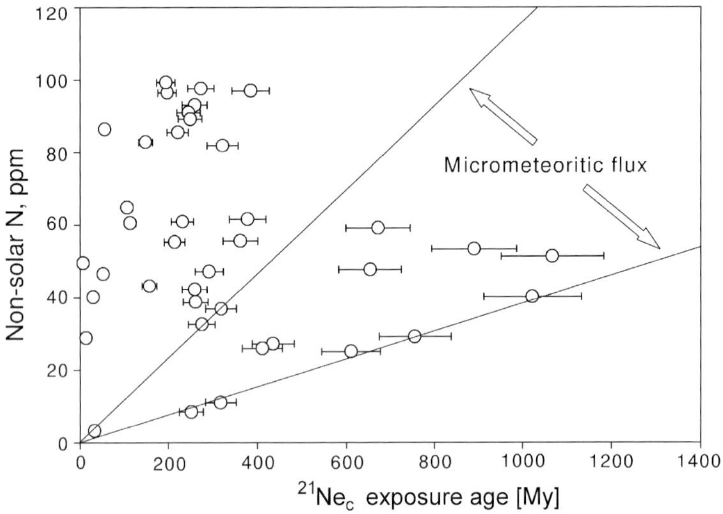

Figure 7. Same exercise as for Figure 6, but for non-solar nitrogen in this case. The two solid lines correspond to the flux of micrometeorites, taking the terrestrial flux of 40000 $\pm$ 20000 tons/y (Love and Brownlee, 1993) corrected for gravitational differences between Earth and Moon. The nitrogen content of micrometeorites is taken as 1000 ppm, which is typical of CM and CR meteorites (the best meteoritic analogue for micrometeorites, Kurat *et al.*, 1994).

The concentration of SW N in a given layer near the regolith surface depends on the average volume of regolith that has been exposed to the solar radiation. Cosmic ray-produced noble gas isotopes accumulate mainly in the first few tens of cm of regolith, so that the cosmic ray exposure ages are relevant for soil grains that has been located in such depth interval. In Figure 6 are drawn lines corresponding to accumulation of solar nitrogen in regolith depths corresponding to 0.2–1 m. The array of soil data is comprised within these lines, suggesting that solar nitrogen has been quantitatively retained in this depth range.

Following the same approach, the amount of non-solar nitrogen can be computed as a function of time (Figure 7), but in this case there is no straightforward correlation. Samples having been exposed more than 400 My may show a rough correlation between non-solar N and exposure age, but many samples exposed during less than 400 My have much higher non-solar N content than that predicted by accumulation at constant rate. This lack of correlation may indicate that the cause of variation of the $\delta^{15}N$ values of lunar soils is due to changes in the amount of non-solar nitrogen at the Moon's surface. For comparison, the flux of micrometeoritic matter falling onto the Earth and scaled to the dimensions and gravitational focusing (comprised between the two solid lines in Figure 7), fits well the accumulation rate computed for the lunar soils for samples having been exposed longer than 400 My, whereas samples showing shorter exposure duration require fluxes. This simple approach suggests that soils exposed during the last 400 My might have seen an increase in the flux of planetary material infalling on to the Moon. This possibility is in agreement with the recent finding of Culler *et al.* (2000) who, from ^{39}Ar–^{40}Ar ages of glassy spherules presumably produced by meteoritic impacts, reported an increase in the frequency of impact during the last 400 My, and with other studies advocating an increase of the flux of extraterrestrial material on Earth in the last few hundreds of My (Grieve and Shoemaker, 1994).

8. Implications for the N Isotope Variability in the Solar System

A strong depletion of the solar wind in ^{15}N relative to planetary N as proposed by Hashizume *et al.* (2000) is in conflict with the recent solar wind N measurement made by the time-of-flight mass spectrometer of the SOHO spacecraft ($\delta^{15}N = +360^{+520}_{-290}$‰, Kallenbach *et al.*, 1998). As discussed by Kallenbach in this volume however, a re-evaluation of the corrections involved in the SOHO measurement leads to a somewhat different value with a higher uncertainty range ($\delta^{15}N = +34 \pm 490$‰) which encompasses the $\delta^{15}N \leq -240$‰ limit given by Hashizume *et al.* (2000). A ^{15}N-depleted composition for solar nitrogen is in agreement with a value of $\delta^{15}N = -480^{+240}_{-280}$‰ for ammonia in the Jupiter atmosphere measured by the ISO spacecraft short wavelength spectrometer (Fouchet *et al.*, 2000) and the Galileo mass spectrometer measurement of $\delta^{15}N = -374 \pm 130$‰ during its descent in the Jupiter atmosphere (Owen *et al.*, 2001). It is hoped that the problem of the nitrogen isotopic composition of the Sun will be resolved in the framework of the Genesis project when the targets irradiated in space by the solar wind during 2 years are analyzed.

The inferred $\delta^{15}N$ values for the Sun and the protosolar nebula are lower than the $\delta^{15}N$ values found in any solid planetary object such as the Earth, meteorites, micrometeorites or interplanetary dust particles (IDPs) (Figure 8). This difference points to a large scale inhomogeneity of the Solar nebula which was not well mixed for N isotopes. In agreement with this view, large N isotopic variations are observed for the different classes of meteorites with for instance $\delta^{15}N$ values ranging from −90 to +150‰ in Iron meteorites (Prombo and Clayton, 1993), from −47.4 to +0.6‰ in Enstatite Chondrites (Grady *et al.*, 1986), from −18‰ to +95‰ in Ordinary Chondrites (Hashizume and Sugiura, 1995), from −43‰ to +190‰ in Carbonaceous Chondrites (Kerridge, 1985), and from +14‰ to +200‰ in Antarctic micrometeorites (Marty *et al.*, 2002). The $\delta^{15}N$ variations in IDPs are among the largest with a range from −100‰ to +813‰ (Messenger, 2000; Stadermann and Floss, 2001), and recent observations show that the large excesses in ^{15}N are associated to large D enrichments of presolar origin and are carried by organic macromolecules (Aléon *et al.*, 2002). These large ^{15}N enrichments strongly suggest the occurrence in the early solar system of N isotopic anomalies of two possible origins. (i) Nucleosynthetic anomalies are indicated by the finding in meteorites of ^{15}N-rich silicon nitrides (Si_3N_4 with $\delta^{15}N$ up to +1000‰) thought to be formed in type II supernovae (Nittler *et al.*, 1995). However, mainstream SiC

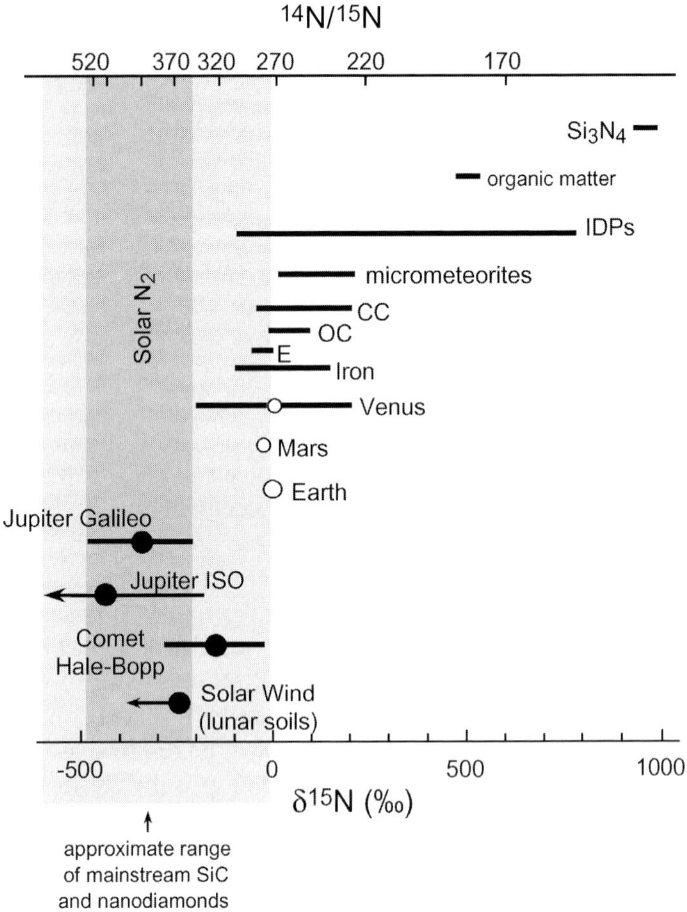

Figure 8. N isotope variability in the solar system. Comparison of N isotopic compositions given either as $\delta^{15}N$ values or as $^{14}N/^{15}N$ ratios (see text for references) determined for the solar wind, the atmosphere of Jupiter from ISO data and Galileo observations, HCN in comet Hale-Bopp, the terrestrial planets (bulk Earth, the mantle of Mars and Venus), meteorites (E for Enstatite Chondrites, OC for Ordinary Chondrites, CC for Carbonaceous Chondrites and Iron for Iron meteorites), micrometeorites, IDPs, presolar organic macromolecules from IDPs and presolar Si_3N_4. Most of the $\delta^{15}N$ variations might result from incomplete homogenisation during the formation of the solar system of N isotopes between ^{15}N depleted N_2 gas and ^{15}N-rich presolar phases, although contributions of ^{15}N-depleted presolar N cannot be ruled out.

grains and nanodiamonds, among the most abundant presolar grains in primitive meteorites, tend to have negative $\delta^{15}N$ values between 0 and -600‰ (e.g., Ott, 2003, and references therein). (ii) Ion-molecule reactions taking place in interstellar molecular clouds could also fractionate intensively nitrogen isotopes (Adams and Smith, 1981), although the present knowledge on these processes predicts a maximum $\delta^{15}N$ variation of (only) 300‰ (Terzieva and Herbst, 2000) (see also Kallenbach *et al.*, 2003, for alternative possibilities). Although the incorporation of such presolar phases carrying ^{15}N excesses could explain part of the range of $\delta^{15}N$ values found either in bulk analyses or very locally at the nm to μm scale in meteoritic matter, other processes such as for instance isotope fractionation of early planetary atmospheres and/or during planetesimal evolution might have played a significant role. Despite these fractionation processes are not well known at present, the observation of a major isotopic heterogeneity between (i) presolar phases enriched in ^{15}N such as silicon nitrides and/or organic macromolecules and (ii) the gas of the nebula depleted in ^{15}N as evidenced by the solar wind, is a strong hint for the occurrence during the formation of early planetary objects of gas-grain exchanges which could be responsible of an incomplete isotopic homogenisation of N isotopes. In this respect the bulk Earth ($\delta^{15}N = 0 \pm 3‰$, Dauphas and Marty, 1999) and the mantle of Mars ($\delta^{15}N \geq -30‰$, Mathew and Marti, 2001b; Murty and Mohapatra, 1997) have $\delta^{15}N$ values in the middle of the range observed within the solar system.

Acknowledgements

We thank the International Space Science Institute for hosting the workshop on the Solar System History and Isotopic Volatile Signatures. Richard Becker kindly supplied his data base for nitrogen and noble gases in Apollo samples. Discussions with R. Becker, D. Gautier, R. Kallenbach, F. Robert, and T. Owen were highly appreciated. Financial support by Centre National d'Etudes Spatiales, Programme National de Planétologie (INSU-CNRS) and Région Lorraine is gratefully acknowledged.

References

Adams, N.G., and Smith, D.: 1981, '$^{14}N/^{15}N$ Isotope Fractionation in the Reaction $N_2H^+ + N_2$: Interstellar Significance', *Astrophys. J.* **247**, L123–L125.

Aléon, J., Robert, F., Chaussidon, M., and Engrand, C.: 2002, '^{15}N Excesses in Deuterated Organics from two Interplanetary Dust Particles', *Lunar Planet Sci. XXXIII*, abstract #1397 (CD ROM).

Anders, E., and Grevesse, N.: 1989, 'Abundances of the Elements: Meteoritic and Solar', *Geochim. Cosmochim. Acta* **53**, 197–214.

Assonov, S.S., Marty, B., Shukolyukov, Y.A., and Semenova, A.S.: 2001, 'Nitrogen and Argon Isotopes in Single Particles from Luna-24 Regolith: Search for the Modern Solar Wind Component', *Lunar Planet Sci.* **XXXIII**, 1798.

Assonov, S.S., Franchi, I.A., Pillinger, C.T., Semenova, A.S., Shukolyukov, Y.A., Verchovsky, A.B., and Iassevitch, A.N.: 2002, 'Nitrogen and Argon Release Profiles in Luna-16 and Luna-24 Regolith Samples: The Effects of Regolith Re-working', *Met. Planet. Sci.* **37**, 27–48.

Becker, R.H., and Clayton, R.N.: 1975, 'Nitrogen Abundances and Isotopic Compositions in Lunar Samples', *Proc. 6th Lunar Sci. Conf.*, 2131–2149.

Becker, R.H., and Clayton, R.N.: 1978, 'Nitrogen Isotope Systematics of two Apollo 12 Soils', *Proc. 9th Lunar Sci. Conf.*, 1619–1627.

Becker, R.H., Clayton, R.N., and Mayeda, T.K.: 1976, 'Characterization of Lunar Nitrogen Components', *Proc. 7th Lunar Planet. Sci. Conf.*, 441–458.
Brenemann, H.H., and Stone, E.C.: 1985, 'Solar Coronal and Photospheric Abundances from Solar Energetic Particle Measurements', *Astrophys. J.* **299**, L57–L61.
Brilliant, D.R., Franchi, I.A., and Pillinger, C.T.: 1994, 'Nitrogen Components in Lunar Soil 12023 — Complex Grains are not the Carrier of Isotopically Light Nitrogen', *Meteoritics* **29**, 718–723.
Chaussidon, M., and Robert, F.: 1999, 'Lithium Nucleosynthesis in the Sun Inferred from Solar-wind $^7Li/^6Li$ Ratio', *Nature* **402**, 270–273.
Clayton, R.N., and Thiemens, M.H.: 1980, 'Lunar Nitrogen: Evidence for Secular Change in the Solar Wind', *in*, R.O. Pepin, J.A. Eddy, and R.B. Merrill (eds.), *The Ancient Sun: Fossil Record in the Earth, Moon and Meteorites*, Pergamon Press, New York and Oxford, pp. 463–473.
Culler, T.S., Becker, T.A., Muller, R.A., and Renne, P.R.: 2000, 'Lunar Impact History from Ar-40/Ar-39 Dating of Glass Spherules', *Science* **287**, 1785–1788.
Dauphas, N., and Marty, B.: 1999, 'Heavy Nitrogen in Carbonatites of the Kola Peninsula: A Possible Signature of the Deep Mantle', *Science* **286**, 2488–2490.
Eberhardt, P., Geiss, J., Graf, H., Grögler, N., Krähenbühl, U., Schwaller, H., Schwarzmüller, J., and Stettler, A.: 1970, 'Trapped Solar Wind Noble Gases, Exposure Age and K/Ar-age in Apollo 11 Lunar Fine Material', *Proc. Apollo 11 Lunar Sci. Conf.*, Vol. 2, pp. 1037–1070.
Engrand, C., and Maurette, M.: 1998, 'Carbonaceous Micrometeorites from Antartica', *Met. Planet. Sci.* **33**, 565–580.
Eugster, O., Terribilini, D., Polnau, E., and Kramers, J.: 2001, 'The Antiquity Indicator Argon-40/Argon-36 for Lunar Surface Samples Calibrated by Uranium-235-Xenon-136 Dating', *Met. Planet. Sci.* **36**, 1097–1115.
Fisk, L. A.: 1978, 'He-3-rich Flares – A Possible Explanation', *Astrophys. J.* **190**, L35–L38.
Fouchet, T., Lellouch, E., Bézard, B., Encrenaz, T., Drossart, P., Feuchtgruber, H., and deGraauw, T.: 2000, 'ISO-SWS Observation of Jupiter: Measurement of the Ammonia Tropospheric Profile and of the $^{15}N/^{14}N$ Ratio', *Icarus* **143**, 223–243.
Franchi, I.A., Wright, I.P., and Pillinger, C.T.: 1986, 'Heavy Nitrogen in Bencubbin – A Light-element Isotopic Anomaly in a Stony-iron Meteorite', *Nature* **323**, 138–140.
Frick, U., Becker, R.H., and Pepin, R.O.: 1988, 'Solar Wind Record in the Lunar Regoltih: Nitrogen and Noble Gases, *Proc. 18th Lunar Planet. Sci. Conf.*, 87–120.
Futagami, T., Ozima, M., and Nakamura, Y.: 1990, 'Helium Ion Implantation Into Minerals', *Earth Planet. Sci. Lett.* **101**, 63–67.
Geiss, J., and Bochsler, P.: 1982, 'Nitrogen Isotopes in the Solar System', *Geochim. Cosmochim. Acta* **46**, 529–548.
Geiss, J., and Gloeckler, G.: 2003, 'Isotopic Composition of H, He and Ne in the Protosolar Cloud', *Space Sci. Rev.*, this volume.
Geiss, J., Gloecker, G., and von Steiger, R.: 1973, 'Solar Wind Composition and Implications About the History of the Solar System', *Proc. 13th Int. Cosmic Ray Conf.*, 3375–3398.
Grady, M.M., and Pillinger, C.T.: 1990, 'ALH 85085: Nitrogen Isotope Analysis of a Highly Unusual Primitive Meteorite', *Earth Planet. Sci. Lett.* **97**, 29–40.
Grady, M.M., Wright, I.P., Carr, L.P., and Pillinger, C.T.: 1986, 'Compositional Differences in Enstatite Chondrites Based on Carbon and Nitrogen Stable Isotope Measurements', *Geochim. Cosmochim. Acta* **50**, 2799–2813.
Grieve, R.A.F., and Shoemaker, E.M.: 1994, 'The Record of Past Impact on Earth', *in* T. Gehrels, M.S. Matthews, and A. Schumann (eds.), *Hazards due to Comets and Asteroids*, Univ. Arizona Press, pp. 417–462.
Haggerty, S.: 1978, 'The Redox State of Planetary Basalts', *Geophys. Res. Lett.* **5**, 443–446.
Hashizume, K., and Sugiura, N.: 1995, 'Nitrogen Isotopes in Bulk Ordinary Chondrites', *Geochim. Cosmochim. Acta* **59**, 4057–4069.

Hashizume, K., Chaussidon, M., Marty, B., and Robert, F.: 2000, 'Solar Wind Record on the Moon: Deciphering Presolar From Planetary Nitrogen', *Science* **290**, 1142–1145.

Hashizume, K., Chaussidon, M., Marty, B., and Robert, F.: 2001, 'Isotopic Variability of Nitrogen in Lunar Regolith – Response', *Science* **293**, 1947a.

Hashizume, K., Marty, B., and Wieler, R.: 2002, 'Analyses of Nitrogen and Argon in Single Lunar Grains: Towards a Quantification of the Asteroidal Contribution to Planetary Surfaces', *Earth Planet. Sci. Lett.* **202**, 201–216.

Humbert, F., Marty, B., and Libourel, G.: 1999, 'Enhanced Solubility of Nitrogen in Basaltic Melt Under Reducing Conditions: A Way to Enrich Nitrogen Relative to Noble Gases during Planetary Formation', *Lunar Planet. Sci. XXX*, abstract #1955 (CD ROM).

Hundhausen, A.J., Bame, S.J., Ashbridge, J.R., and Sydoriak, S.J.: 1970, 'Solar Wind Proton Properties: Vela 3 Observations From July 1965 to July 1967', *J. Geophys. Res.* **75**, 4643–4651.

Jewitt, D.C., Matthews, H.E., Owen, T., and Meier, R.: 1997, 'Measurements of C-12/C-13, N-14/N-15, and S-32/S-34 Ratios in Comet Hale-Bopp (C/1995 O1)', *Science* **278**, 90–93.

Kallenbach, R.: 2003, 'Isotopic Fractionation by Plasma Processes', *Space Sci. Rev.*, this volume.

Kallenbach, R., Geiss, J., Ipavich, F.M., Gloeckler, G., Bochsler, P., Gliem, F., Hefti, S., Hilchenbach, M., and Hovestadt, D.: 1998, 'Isotopic Composition of Solar Wind Nitrogen: First in Situ Determination with the CELIAS/MTOF Spectrometer on Board SOHO', *Astrophys. J.* **507**, L185–L188.

Kallenbach, R., Robert, F., Geiss, J., Herbst, E., Lammer, H., Marty, B., Millar, T., Ott, U., and Pepin, R.O.: 2003, 'Sun and Protosolar Nebula', *Space Sci. Rev.*, this volume.

Keller, L.P., and McKay, D.S.: 1997, 'The Nature and Origin of Rims on Lunar Soil Grains', *Geochim. Cosmochim. Acta* **61**, 2331–2341.

Kerridge, J.F.: 1975, 'Solar Nitrogen: Evidence for a Secular Change in the Ratio of Nitrogen-15 to Nitrogen-14', *Science* **188**, 162–164.

Kerridge, J.F.: 1985, 'Carbon, Hydrogen, and Nitrogen in Carbonaceous Chondrites: Abundances and Isotopic Compositions in Bulk Samples', *Geochim. Cosmochim. Acta* **49**, 1707–1714.

Kerridge, J.F.: 1993, 'Long-term Compositional Variation in Solar Corpuscular Radiation: Evidence From Nitrogen Isotopes in the Lunar Regolith', *Rev. Geophys.* **31**, 423–437.

Kerridge, J.F., Kaplan, I.R., Petrowski, C., and Chang, S.: 1975, 'Light Element Geochemistry of the Apollo 16 Site', *Geochim. Cosmochim. Acta* **39**, 137–162.

Kerridge, J.F., Eugster, O., Kim, J.S., and Marti, K.: 1991, 'Nitrogen Isotopes in the 74001/74002 Double-drive Tube From Shorty Crater', *Proc. 21st Lunar Planet. Sci. Conf.*, 291–299.

Kim, J.S., Kim, Y., Marti, K., and Kerridge, J.F.: 1995, 'Nitrogen Isotope Abundances in the Recent Solar Wind', *Nature* **375**, 383–385.

Kurat, G., Koeberl, C., Presper, T., Branstätter, F., and Maurette, M.: 1994, 'Petrology and Geochemistry of Antarctic Micrometeorites', *Geochim. Cosmochim. Acta* **58**, 3879–3904.

Love, S., and Brownlee, D.E.: 1993, 'A Direct Measurement of the Terrestrial Mass Accretion Rate of Cosmic Dust', *Science* **262**, 550–553.

Manka, R.H., and Michel, F.C.: 1971, 'Lunar Atmosphere as a Source of Lunar Surface Elements', *Proc. 22nd Lunar Sci. Conf.*, Vol. 2, 1717–1728.

Marty, B., and Zimmermann, L.: 1999, 'Volatiles (He, C, N, Ar) in Mid-ocean Ridge Basalts: Assesment of Shallow-level Fractionation and Characterization of Source Composition', *Geochim. Cosmochim. Acta* **63**, 3619–3633.

Marty, B., Matrajt, G., Zimmermann, L., Engrand, C., and Duprat, J.: 2002, 'Nitrogen and Noble Gas Isotopes in Antarctic Micrometeorites', *Lunar Planet. Sci. Conf.* **XXXIII**, abstract #1578 (CD ROM).

Mathew, K.J., and Marti, K.: 2001a, 'Lunar Nitrogen: Indigenous Signature and Cosmic-ray Production Rate', *Earth Planet. Sci. Lett.* **184**, 659–669.

Mathew, K.J., and Marti, K.: 2001b, 'Early Evolution of Martian Volatiles: Nitrogen and Noble Gas Components in ALH84001 and Chassigny.', *J. Geophys. Res.* **106**, 1401–1422.

Mathew, K.J., Kerridge, J.F., and Marti, K.: 1998, 'Nitrogen in Solar Energetic Particles: Isotopically Distinct From Solar Wind', *Geophys. Res. Lett.* **25**, 4293–4296.

Maurette, M.: 1998. 'Carbonaceous Micrometeorites and the Origin of Life', *Orig. Life Evol. Biosphere* **28**, 385–412.

Messenger, S.: 2000, 'Identification of Molecular-cloud Material in Interplanetary Dust Particles', *Nature* **404**, 968–971.

Mewaldt, R.A., Ogliore, R.C., Gloeckler, G., and Mason, G.M.: 2001, 'A New Look at Neon-C and SEP-neon', *AIP Conf. Proc.* **598**, 393–398.

Morgan, J.W., Krähenbühl, U., Ganapathy, R., and Anders, E.: 1972, 'Trace Elements in Apollo 15 Samples: Implications for Meteoritic Influx and Volatile Depletion on the Moon', *Proc. 3rd Lunar Sci. Conf.*, 1361–1376.

Murty, S.V.S., and Goswami, J.N.: 1992, 'Nitrogen, Noble Gases and Nuclear Tracks in Lunar Meteorites MMAC88104/105', *Proc. 22nd Lunar Planet. Sci. Conf.*, 225–237.

Murty, S.V.S., and Mohapatra, R.K.: 1997, 'Nitrogen and Heavy Noble Gas Components in the ALH84001: Signatures of Ancient Martian Atmosphere', *Geochim. Cosmochim. Acta* **61**, 5417–5428.

Nier, A.O., and McElroy, M.B.: 1977, 'Composition and Structure of Mars' Upper Atmosphere: Results From the Neutral Mass Spectrometers on Viking 1 and 2', *J. Geophys. Res.* **82**, 4341–4349.

Nittler, L.R., Hoppe, P., Alexander, C.M., Amari, S., Eberhardt, P., Gao, X., Lewis, R.S., Strebel, R., Walker, R.M., and Zinner, E.: 1995, 'Silicon Nitride From Supernovae', *Astrophys. J.* **453**, L25–L28.

Ott, U.: 2003, 'Isotopes of Volatiles in Pre-solar Grains', *Space Sci. Rev.*, this volume.

Owen, T., Mahaffy, P.R., Niemann, H.B., Atreya, S., and Wong, M.: 2001, 'Protosolar Nitrogen', *Astrophys. J.* **553**, L77–L79.

Paesold, G., Kallenbach, R., and Benz, A.O.: 2002, 'Acceleration and Enrichment of ^{3}He in Impulsive Flares by Electron Firehose Waves', *Astrophys. J.*, in press.

Prombo, C.A., and Clayton, R.N.: 1985, 'A Striking Nitrogen Anomaly in the Bencubbin and Weatherford Meteorites', *Science* **230**, 935–937.

Prombo, C.A., and Clayton, R.N.: 1993, 'Nitrogen Isotopic Compositions of Iron Meteorites', *Geochim. Cosmochim. Acta* **57**, 3749–3761.

Reedy, R.C.: 1981, 'Cosmic-ray-produced Stable Nuclides: Various Production Rates and their Implications', *Proc. 12th Lunar Planet. Sci. B*, Lunar Planet. Inst., Houston, 1809–1823.

Stadermann, F.J., and Floss, C.: 2001, 'Heating Experiments of Individual Particles From the Renazzo Matrix: SIMS Measurements of Nitrogen Isotopes', *Met. Planet. Sci.* **36**, A196–A197.

Terzieva, R., and Herbst, E.: 2000, 'The Possibility of Nitrogen Isotopic Fractionation in Interstellar Clouds', *Month. Not. Roy. Astron. Soc.* **317**, 563–568.

Thiemens, M.H., and Clayton, R.N.: 1980, 'Solar and Cosmogenic Nitrogen in the Apollo 17 Deep Drill Core', *Proc. 11th Lunar Planet. Sci. Conf.*, 1435–1451.

Wieler, R.: 1998, 'The Solar Noble Gas Record in Lunar Samples and Meteorites', *Space Sci. Rev.* **85**, 303–314.

Wieler, R., and Baur, H.: 1994, 'Krypton and Xenon From the Solar Wind and Solar Energetic Particles in two Lunar Ilmenites of Different Antiquity', *Meteoritics* **29**, 570–580.

Wieler, R., and Heber, V.S.: 2003, 'Noble Gas Isotopes on the Moon', *Space Sci. Rev.*, this volume.

Wieler, R., Kehm, K., Meshik, A.P., and Hohenberg, C.: 1996, 'Secular Changes in the Xenon and Krypton Abundances in the Solar Wind Recorded in Single Lunar Grains', *Nature* **384**, 46–49.

Wieler, R., Humbert, F., and Marty, B.: 1999, 'Evidence for a Predominantly Non-solar Origin of Nitrogen in the Lunar Regolith Revealed by Single Grain Analyses', *Earth Planet. Sci. Lett.* **167**, 47–60.

Address for Offprints: Bernard Marty, CRPG-CNRS, BP 20, 54501 Vandoeuvre-lès-Nancy Cedex, France; bmarty@crpg.cnrs-nancy.fr

NOBLE GAS ISOTOPES ON THE MOON

R. WIELER and V.S. HEBER
ETH Zürich, Isotope Geology, NO C61, CH-8092 Zürich, Switzerland (wieler@erdw.ethz.ch)

Received: 11 April 2002; Accepted in final form: 30 May 2002

Abstract. We discuss some of the major noble gas components on the Moon and their implications on the history of Moon and Sun. He, Ar, and Rn have been detected in the tenuous lunar atmosphere. The Ar and Rn abundances suggest that a sizeable fraction of the lunar interior presently loses all its freshly produced radiogenic noble gases. Part of the radiogenic Ar and Xe (the latter from now extinct radioactive isotopes of I and Pu) outgassed from the lunar interior later became retrapped in the dust grains on the lunar surface. These "parentless" gases may also be used to constrain the degassing history of the Moon, although a quantitative understanding is lacking. The dust layer on the lunar surface contains large amounts of noble gases implanted by the solar wind. The lunar regolith therefore constitutes the best available archive of the solar history of the past four billion years.

1. Introduction

The major noble gas components in the Moon are: i) gases in the tenuous lunar atmosphere, potentially constraining the current degassing of the Moon (^{4}He, ^{36,40}Ar, ^{222}Rn), ii) "parentless" radiogenic and fissiogenic isotopes residing near grain surfaces and being useful to constrain when a sample was exposed to the solar wind, the lunar degassing history and possibly also the time of lunar formation (^{40}Ar, ^{129}Xe, $^{131-136}$Xe), iii) solar wind gases implanted into soils, providing a precious archive of solar history (all isotopes), iv) in-situ produced radiogenic isotopes, used for dating (notably ^{40}Ar), and v) cosmic-ray produced isotopes, used to study the exposure history of the lunar regolith (e.g., ^{21}Ne, ^{38}Ar).

Here we will discuss the first three of these components. Dating of lunar samples with radiogenic ^{40}Ar has recently been reviewed by Bogard (1995), Stöffler and Ryder (2001), and Swindle (2002). The lunar cosmogenic noble gas record has been summarized by Wieler (2002), whereas Polnau and Eugster (1998) compiled cosmic ray exposure ages of lunar meteorites.

Prominently absent in the list above are "indigenous" lunar noble gases, those that would have been present in the Moon since its formation. Despite some early reports on the possible detection of indigenous lunar xenon with similar composition as terrestrial atmospheric xenon (Lightner and Marti, 1974; Leich and Niemeyer, 1975), indigenous lunar noble gases have never unambiguously been found. One reason for this is the later finding that terrestrial atmospheric Xe may become tightly adsorbed during lunar sample preparation (Niemeyer and Leich, 1976; Niedermann and Eugster, 1992). It appears timely, however, to resume the

Space Science Reviews **106:** 197–210, 2003.

TABLE I

Noble Gases in the Lunar Exosphere and in the Lunar Regolith (Isotopic and elemental ratios and further details compiled in, e.g., Wieler, 2002)

Exosphere			*Regolith*	
He	Density at surface	2000 cm^{-3} (day) †)	Mean concentrations ($\pm 1\sigma$) at Apollo sites ◇):	
		40000 cm^{-3} (night) †)		
	Loss rate to space	9×10^{23} s^{-1} ♯)		
Ar	Density at surface	<100 cm^{-3} (night)	^{4}He	$7.8 \pm 6.3 \times 10^{-2}$ cc/g
		>35000 cm^{-3} (sunrise) ♭)	^{20}Ne	$1370 \pm 900 \times 10^{-6}$ cc/g
	Loss rate to space	2×10^{21} s^{-1} ‡)	^{36}Ar	$310 \pm 120 \times 10^{-6}$ cc/g
	^{40}Ar/^{36}Ar	≈ 10 ♮)	^{84}Kr	$14.5 \pm 5.8 \times 10^{-8}$ cc/g
^{222}Rn	Effusion rate	$(7 \pm 4) \times 10^{15}$ s^{-1} △)	^{132}Xe	$2.4 \pm 1.0 \times 10^{-8}$ cc/g

†) Stern (1999); ‡) Hodges (1975); ♮) Hoffman *et al.* (1973); ◇) Fegley and Swindle (1993);
♯) Further $\approx 5 \times 10^{23}$ atoms/s probably lost from nonatmospheric component (Hodges, 1975);
♭) Lower limit corresponds to detector saturation (Stern, 1999; Hoffman *et al.*, 1973);
△) Effusion rate from lunar interior (atoms/s), based on ^{210}Po decay rate (Hodges, 1975).

search for lunar indigenous noble gases with present-day analytical tools, also in view of reports of indigenous lunar nitrogen in several rocks and volcanic glasses (Kerridge *et al.*, 1991; Mathew and Marti, 2001; Marty *et al.*, 2003). Clearly, lunar indigenous noble gas signatures would offer crucial constraints on the formation and evolution of the Moon.

2. Noble Gases in the Lunar Exosphere

Atoms and molecules in the tenuous lunar atmosphere interact predominantly with the lunar surface rather than with each other. This atmosphere is therefore called an exosphere. Its characteristics have recently been reviewed by Stern (1999), and its noble gas inventory is discussed by Wieler (2002). The stable noble gas isotopes ^{4}He, ^{36}Ar, and ^{40}Ar have unambiguously been detected during the Apollo missions, whereas early reports of Ne in the atmosphere have not been confirmed (Stern, 1999). Furthermore, radioactive ^{222}Rn (half-life 3.8 d) from the decay of ^{238}U in the lunar interior has been detected by alpha-spectrometry both directly and by its decay product ^{210}Po (half-life 138 d, but production held up by the nuclide ^{210}Pb with a half-life of 21 y). The first such measurements were made during the Apollo missions (Bjorkholm *et al.*, 1973), recent analyses have been carried out with Lunar Prospector (Lawson *et al.*, 2002). Table I gives some basic data.

2.1. Sources and Sinks

The exospheric ^{36}Ar represents solar wind particles once implanted into dust grains at the lunar surface and later released, e.g. by sputtering (Killen and Ip, 1999; Stern, 1999). In contrast, the exospheric ^{40}Ar is from radioactive decay of ^{40}K (half-life 1.27 Gy) degassed from the lunar interior. Part of this ^{40}Ar may be freshly outgassed, part may have been recycled between soil and atmosphere by the Manka-Michel mechanism (see Section 3). The ^{40}Ar/^{36}Ar ratio has been measured at the Apollo 17 site shortly before sunrise, during release of Ar frozen at night (Hoffman *et al.*, 1973). The value of $\approx$ 10 is close to the high end of the range of values observed for Ar trapped on grain surfaces (Section 3) and considerably larger than the mean ratio of $\approx$ 1 in the regolith. This indicates that most of the ^{40}Ar in the exosphere is freshly degassed and only a minor part is recycled. Also Hodges (1975) argued against a substantial recycling of this isotope, based on the observed variation of exospheric ^{40}Ar at the Apollo 17 site during a period of several months.

It seems clear that, like ^{36}Ar, most of the ^{4}He in the lunar atmosphere is also solar wind gas released from the soil, with a possible minor contribution of radiogenic ^{4}He from the lunar interior (Hodges, 1975).

Exospheric Ar is primarily lost by photoionisation, followed by acceleration in the electromagnetic field of the solar wind and escape to space or reimplantation into the regolith. Helium is lost primarily by Jeans escape (Stern, 1999). Hodges (1975) modeled the escape of lunar atmospheric He and Ar (see Table I).

2.2. Exospheric ^{40}Ar and Lunar Degassing

The ^{40}Ar concentration in the present-day and past exosphere constrains the degassing history of the Moon. This has been discussed in most detail by Hodges (1975, 1977). Here we therefore mainly discuss these two papers. Hodges (1977) estimated that the average atmospheric loss rate of ^{40}Ar of 2×10^{21} atoms/s is balanced by degassing of $\approx$ 6% of the present production of radiogenic ^{40}Ar in the entire Moon. He concluded that the most probable source of the exospheric ^{40}Ar is a semi-molten center region (corresponding to some 6% of the total lunar volume, see also Hodges, 1975). The inferred loss fraction of 6% seems to be astonishingly large, as has also been noted by Hodges (1977). Therefore, one may ask whether his basic assumption is justified, which is that the present-day ^{40}Ar loss rate is proportional to its present-day production rate rather than its present-day inventory. To illustrate, let us assume that almost all of the $\approx 1.5 \times 10^{40}$ atoms of radiogenic ^{40}Ar ever produced in the moon (assuming $\approx$ 100 ppm K) would still be retained. Then, the 2×10^{21} atoms/s escaping now would correspond to a relative loss rate of $\approx 1.3 \times 10^{-19}$ s^{-1}. If this rate had remained constant, only $\approx$ 1.3% of the total ^{40}Ar would have been lost throughout lunar history. Even if the Moon would have lost, say, 99% of all its radiogenic ^{40}Ar ever produced, the relative loss rate would still only be $\approx 1.3 \times 10^{-17}$ s^{-1} (99% loss is a very conservative estimate, since the remaining 1% would correspond to what has been accumulated in <1.5 km of lunar highland crust since 3.9 Gy, a typical ^{40}Ar-^{39}Ar age of highland samples; Bogard, 1995). One therefore does not need to postulate from the exospheric ^{40}Ar alone that a sizeable part of the interior completely degasses at present.

However, the presence of ^{222}Rn in the atmosphere provides support for Hodges' interpretation. In contrast to stable nuclides, it is clear that the emission rate of radioactive ^{222}Rn is related to its current production rate. Based on the ^{210}Po activity reported by Bjorkholm *et al.* (1973), Hodges (1977) estimated a ^{222}Rn production rate for the Moon of 5.4×10^{22} atoms/s averaged over the past few decades. Neglecting outgassing, this yields a current ^{222}Rn inventory of 2.6×10^{28} atoms. Hodges also estimated a ^{222}Rn emission rate of $(7 \pm 4) \times 10^{15}$ atoms/s. If no Rn atoms decayed while on the way to the surface, the present-day relative Rn degassing rate would roughly be 3×10^{-13} s^{-1}, orders of magnitude higher than that of $\approx 10^{-19} - 10^{-17}$ s^{-1} inferred above for ^{40}Ar. This relative ^{222}Rn loss rate is probably a very conservative lower limit, since the overwhelming part will have decayed between source and lunar surface. For example, Hodges calculates a transit time of only 71 days ($\approx$ 19 half-lives of ^{222}Rn) by assuming that the Rn in the atmosphere originates from complete degassing of the same innermost 6% of the Moon which he infers to be the source of the ^{40}Ar. As it is highly unlikely that Rn degasses many orders of magnitude faster than Ar, the above comparison of inferred loss rates for the two isotopes thus indicates that indeed a sizeable fraction of the freshly produced radiogenic Ar in the Moon is lost essentially instantaneously.

The $^{40}Ar/^{36}Ar$ ratio of Ar trapped in lunar soils has often been used to support this interpretation. As discussed in the next section, this ratio has decreased throughout lunar history, roughly in parallel with the decreasing ^{40}K abundance. In contrast, if the ^{40}Ar degassing were governed by the total ^{40}Ar present in the Moon at any time, $^{40}Ar/^{36}Ar$ would be expected to increase. On the other hand, other factors may also influence this ratio. First, secular cooling of the Moon should have lead to a decrease of the ^{40}Ar degassing efficiency and therefore to lower $^{40}Ar/^{36}Ar$ ratios in samples irradiated later. Second, the flux of solar wind ^{36}Ar may have been higher in the past when the Sun was more active, such that $^{40}Ar/^{36}Ar$ may have become higher in more recently irradiated samples, due to decreasing ^{36}Ar trapped per unit time. However, this latter effect might have been compensated or overcompensated, because the flux of ionizing photons from the Sun in the past was also higher than today (Walter and Barry, 1991), or because a stronger solar magnetic field in the past might have led to a better efficiency of the Manka-Michel process, e.g. due to a lower gyroradius of the ions (R. F. Wimmer-Schweingruber, personal communication, 2002). We are unaware of attempts to assess all these effects and conclude that the $^{40}Ar/^{36}Ar$ ratio in lunar samples is not understood well enough to quantitatively constrain the lunar degassing history.

The exospheric ^{40}Ar concentration is quite variable on a timescale of months (Hodges, 1975; Hodges, 1977). The author notes that this implies that the supply of ^{40}Ar from the Moon is episodic, also supporting a localized deep reservoir, from which gas release may be coupled to seismic processes. Gorenstein *et al.*

(1973) note that "hotspots" of Rn and Po occur in the vicinity of craters such as Aristarchus. The authors even suggest a spatial correlation between such hot spots and reported transient optical events. Recently, Lawson *et al.* (2002) confirmed the present-day emanation of ^{222}Rn near craters Aristarchus and Kepler.

In summary, it appears that the Moon at present indeed loses a sizeable fraction of perhaps some 6% of the radiogenic ^{40}Ar produced today. Probably, this Ar is supplied from the deep interior. Such a loss fraction seems astonishingly high, but possibly even more astonishing is the very short transit time of some 70 days inferred from ^{222}Rn. Hodges (1977) proposes that this requires the existence of fractures extending deep into the lunar interior.

3. Parentless Noble Gases on Grain Surfaces

Lunar dust grains contain near their surfaces "parentless" ^{40}Ar in amounts much too high to be explained by in-situ decay of ^{40}K. The surface-sited ^{40}Ar doubtlessly is radiogenic Ar degassed from the lunar interior and retrapped onto the grains (Section 2). Most probably, exospheric ^{40}Ar was ionized by solar UV and then accelerated in the electromagnetic field of the solar wind and reimplanted into the soil (Manka and Michel, 1971). A few regolith samples also contain, near their grain surfaces, ^{129}Xe from the decay of the "short-lived" nuclide ^{129}I (half-life 15.7 My) as well as $^{131-136}$Xe from fission of ^{244}Pu, a now extinct nuclide with a half-life against α-decay of 80 My (Swindle *et al.*, 1986). The radiogenic and fissiogenic portions of these nuclides (labelled ^{129}Xe* and $^{131-136}$Xe*) are also parentless, as they cannot have been derived from in-situ decay of their parent nuclides. They probably were trapped very early in lunar history. Some soils contain other possibly parentless noble gas isotopes from nuclear processes, which are not further discussed here. An example is a ^{126}Xe anomaly in many soils, presumably due to solar proton irradiation of iodine once deposited on grain surfaces but lost later (Pepin *et al.*, 1995).

Parentless ^{40}Ar is used as semiquantitative measure of the "antiquity" of a soil, which is the time in the past when a sample was exposed to the solar wind at the top of the regolith. The ^{40}Ar concentration needs to be normalized to the length of time a sample spent on the surface, for which ^{36}Ar from the solar wind is used as proxy. The ^{40}Ar/^{36}Ar ratio of surface-sited Ar in samples irradiated several billion years ago is higher than in samples which received their solar wind later. A recent calibration is by Eugster *et al.* (2001). Samples with an independently estimated antiquity of $\approx$ 3.7 Gy have ^{40}Ar/^{36}Ar ratios of $\approx$ 13, samples with an antiquity on the order of 10 My have values around 0.5. However, there are only very few samples with a reliable independent antiquity estimate. Furthermore, the antiquity of a soil may not be very well defined, i.e. samples or their individual grains may have acquired solar gases over one or several extended periods. The ^{40}Ar/^{36}Ar ratio is therefore a semiquantitative antiquity estimate only.

It is usually assumed that the decrease of $^{40}Ar/^{36}Ar$ with time reflects a decrease of the ^{40}Ar abundance in the atmosphere rather than an increase of the solar wind flux of ^{36}Ar, because the solar activity is expected to have decreased with time. If the ^{36}Ar concentration is a good proxy for the duration a sample was exposed to the solar wind, i.e. if the flux of solar ^{36}Ar has not varied much during the time for which we have antiquity-calibrated lunar samples (or if a higher ^{36}Ar flux in the past has been exactly compensated by a higher ^{40}Ar ionisation rate in the exosphere; Section 2.2.), then the ^{40}Ar concentration in the atmosphere had decreased by a factor of $\approx 25 - 30$ during the past ≈ 3.7 Gy or the past ≈ 3 half-lives of ^{40}K. The decrease of $^{40}Ar/^{36}Ar$ has thus been about three times faster than that of the lunar ^{40}K abundance. Eugster *et al.* (2001) suggested that this difference is due to a long-term decrease of the Moon's degassing efficiency. However, in view of the uncertainties discussed above, we reiterate the conclusion in Section 2.2 that the temporal trend of $^{40}Ar/^{36}Ar$ in lunar soils is not well understood quantitatively.

Parentless Xe is less widespread than parentless ^{40}Ar. It has been observed mostly in regolith breccias (compacted soil) from the highlands at Apollo 14 and 16 (Drozd *et al.*, 1972; Bernatowicz *et al.*, 1979). Although most work has concentrated on just two or three samples, parentless Xe may be a global component, since it has been found in all highland breccias rich in solar wind noble gases (Swindle *et al.*, 1986). These samples have among the highest observed $^{40}Ar/^{36}Ar$ ratios, independently indicating an exceptional antiquity. The isotopic composition shows that the fission component is from now extinct ^{244}Pu. The ratio of $^{129}Xe^*$ to fissiogenic Xe varies between samples, with some not containing any $^{129}Xe^*$. This ratio is also variable in different extraction steps of single samples, which forms the basis of attempts to date them by the I-Pu-Xe system. In the two best studied samples (14301 and 14318), low-temperature steps have lower $^{129}Xe^*/^{136}Xe^*$ ratios than high-temperature steps (Bernatowicz *et al.*, 1979). Swindle *et al.* (1986) point out that this would be expected if the excess Xe was acquired over tens of millions of years and if the gas trapped first would have become more tightly bound with time than the gas trapped last. If the $^{129}Xe^*/^{136}Xe^*$ ratio of the gas trapped at any time were to correspond to the production ratio at that time, gas trapped later would have a lower $^{129}Xe^*/^{136}Xe^*$, because of the different half-lives of the parent nuclides. Most importantly, this would imply that the trapping occurred in the first few 10 My after the birth of the solar system, while both parents were still extant. Assuming that parentless Xe degassed from a substantial fraction of the Moon, i.e. that the appropriate initial $^{129}I/^{244}Pu$ ratio is the bulk Moon value, Swindle *et al.* concluded that retention of parentless Xe in 14301 started 63 ± 42 My after the formation of primitive meteorites. If real, this age would be a lower limit to the age of the Moon, in agreement with the age of ≈ 50 My inferred from the Hf-W system (Halliday, 2000). This old I-Pu-Xe age would also imply that very old grain surfaces have been preserved to the extent that stepwise heating still reflects the successive trapping of Xe with decreasing $^{129}Xe^*/^{136}Xe^*$. So the crucial question is whether dating with parentless Xe really yields quantitative age information.

Swindle *et al.* (1986) argue that it does, although they also critically discuss the assumptions involved. The ages would imply that enough highland crust had solidified from a magma ocean within a few 10 My after lunar formation such that this very old material constitutes a sizeable fraction ($<15\%$ according to Swindle *et al.*) of what is now breccia 14301. Also, very old grain surfaces would have to have been preserved in a way that the degassing order of the parentless Xe in the laboratory still reflects the reverse of the trapping sequence. Note that the ^{40}Ar-^{39}Ar age of 14301 is only 3.7 ± 0.1 Gy (Megrue, 1973), an age which Eugster *et al.* (2001) adopt as its solar wind antiquity. It may be possible that the parentless Xe partly survived the event which reset the K-Ar clock, but this last heating would then have to have been so mild as not to completely remix the two parentless Xe components and at the same time severe enough to release all radiogenic ^{40}Ar.

The key question is thus whether the different release temperatures of excess ^{129}Xe* and $^{131-136}$Xe* indeed indicate trapping of these gases while their parents were extant. With stepwise etching in-vacuum rather than stepwise heating, Wieler *et al.* (1993) found that fission Xe from bulk 14301 was released prior to solar wind Xe. This apparently confirmed that the parentless fission Xe was less deeply implanted than the solar Xe, just as would be expected from the analogous situation with solar-wind-derived ^{36}Ar and retrapped ^{40}Ar. The fission Xe was also released earlier than the ^{129}Xe*, in agreement with the results from stepwise heating. However, in-vacuo etching of mineral separates from the same sample showed a reverse picture, since solar Xe was released prior to fission Xe (Wieler and Baur, 1994b). The solar gas appeared to reside mostly in mineral grains proper, whereas fission Xe presumably mostly came from interior surfaces of small-grain aggregates contaminating the mineral grains. This decoupling shows that solar gases need not have been trapped while ^{244}Pu was extant. Unfortunately, no clear-cut conclusion could be derived on the relative release sequence of ^{129}Xe* and Pu-fission Xe from the mineral separates. However, these etch experiments at least suggest that the different peak release temperatures of ^{129}Xe* and fission Xe in stepwise heating do not necessarily indicate a temporal sequence of very early trapping, but may also be due to different carriers for the two parentless components. If so, trapping of the two components might have occurred well after production.

Another possibility might be that the gas release sequence in fact indicates a sequential trapping of I- and Pu-derived Xe, respectively, but that the two components nevertheless were stored in the lunar interior for a considerable time, before being trapped at their present locations. If so, the reservoirs formed earlier with higher ^{129}Xe*/^{136}Xe* would have been degassed prior to reservoirs that sampled Xe with lower ^{129}Xe*/^{136}Xe*. It is unclear whether this might have happened, perhaps by early reservoirs having formed at shallower depth than reservoirs created later.

In summary, it seems that further work is required before the very old I-Pu-Xe ages of Apollo 14 regolith breccias can be accepted. It remains uncontested, though, that these samples contain very ancient solar wind of at least some 3.7 Gy.

4. The Moon as Archive of Solar History

The dust layer on the lunar surface has been stirred so intensively by meteorite impacts that almost every grain once trapped ions from the solar wind at the immediate surface. This solar contribution can be detected for the few elements intrinsically very rare in the samples. By far most important are the noble gases, where the solar wind component constitutes for most isotopes the overwhelming fraction in most soils. Some of them trapped their share of solar wind in the relatively recent past, whereas others contain solar wind that is several billion years old. As already noted, the $^{40}Ar/^{36}Ar$ ratio forms a tool to date this "antiquity" semiquantitatively (Eugster *et al.*, 2001). The lunar regolith therefore constitutes a precious archive of solar history. Here we summarize some earlier findings and then concentrate on a few recent results. More comprehensive reviews are given by Wieler (1998, 2002). Lunar nitrogen is discussed by Marty *et al.* (2003).

4.1. Evidence for a Higher Solar Wind Flux in the Past?

Comparing the amounts of solar wind noble gases in the regolith column with those expected during four billion years of regolith lifetime assuming present-day solar wind flux may allow us to rule on the average long-term flux. This comparison is necessarily quite uncertain because only the topmost ≈ 2.5 m of the roughly 10 m thick regolith have been sampled. It is not clear whether gas concentrations in the unsampled portions are similar to those near the top, although one might expect them to be lower (Fegley and Swindle, 1993). Geiss (1973) estimated that the Xe in the Apollo 15 deep drill core corresponds to a solar wind irradiation of 2.6 Gy assuming present-day flux and thus concluded that the average solar wind intensity in the past several billion years was at least as large and probably larger than today. Wieler (1997) came to a similar conclusion, but emphasized that a constant flux is also consistent with the data. This also suggests that essentially all solar Xe ions that ever hit the lunar surface may still be retained in the regolith.

4.2. Two Different Solar Noble Gas Components?

The first lunar sample analyses already hinted at the existence of two separate components of implanted solar Ne, differing in isotopic composition and implantation depth (Pepin *et al.*, 1970a; Reynolds *et al.*, 1970; Black, 1972). Although early workers cautioned that the data may also indicate isotopic fractionation of a single solar wind component (e.g., Pepin *et al.*, 1970b), the existence of the second component now appears to be well established for all noble gases, mainly by in-vacuo etching experiments (Benkert *et al.*, 1993; Wieler and Baur, 1994a) and recently also by stepwise heating analyses of single grains (Palma *et al.*, 2002). This component is known by the acronym "SEP", for solar energetic particles. It is poorly understood, however, since its abundance appears to be too high by orders of magnitude to be reconciled with fluxes of SEPs measured in space (Wieler,

1998; Wieler, 2002). Wimmer-Schweingruber and Bochsler (2001) therefore proposed that "SEPs" may represent interstellar pick-up ions. These would have a galactic rather than solar origin, and hence the lunar regolith would also represent an archive of the galactic environment of the Sun. A problem with this view is that the isotopic ratios of "SEPs" and solar wind noble gases show a simple mass-dependent relationship (Wieler and Baur, 1994a). Mewaldt *et al.* (2001) proposed that "SEPs" in fact represent solar particles, but that the inferred difference in isotopic composition to the solar wind is an artifact caused by different penetration depths of different isotopes. Mewaldt *et al.* also concluded, though, that the present-day flux of particles with energies higher than those of the solar wind falls short by two orders of magnitude to explain the apparent abundance of the "SEP" component. Hopefully the Genesis mission currently collecting solar particles in space will shed light on this enigmatic component.

4.3. Elemental Composition, FIP Effect

In order to deduce elemental abundances of noble gases in the solar wind from regolithic samples, one needs minerals that have retained their solar noble gases without elemental fractionation. There is hardly a mineral in the lunar regolith that retains unfractionated solar He and Ne, but metallic Fe-Ni in regolithic meteorites seems to conserve the true He/Ne/Ar abundance ratios in the solar wind (Murer *et al.*, 1997; Pedroni and Begemann, 1994). These long term average values agree with those derived from the Apollo Solar Wind Composition (SWC) experiment (Geiss, 1973). Comparing He/Ar in the solar wind with estimates of this ratio in the Sun suggests a depletion of He in the solar wind by a factor of ≈ 0.7 (Wieler, 2002), within the large uncertainties consistent with the lower-than-solar He/H ratios in the present-day solar wind observed by spacecraft (Aellig *et al.*, 2001).

On the other hand, solar Ar, Kr, and Xe appear to have been retained unfractionated in mineral separates and even bulk samples (Wieler and Baur, 1995; Wieler *et al.*, 1996). These data can thus be used to compare the relative abundances of the heavy noble gases in the solar wind with those inferred for the Sun. Values vary with antiquity. Xenon is enhanced in the solar wind relative to Ar by a factor of ≈ 4, similar to elements with a first ionisation potential (FIP) of <10 eV in the interstream (slow) solar wind (von Steiger *et al.*, 1997). Krypton is also enhanced, but less than Xe and with more variation between samples. These enhancements appear to be surprising, because FIPs of Kr and Xe are both above the 10 eV threshold. However, models with the ionisation time in the chromosphere as the actual parameter governing this "FIP" fractionation actually predict Kr and Xe enrichments similar to observed values (Geiss *et al.*, 1994; Wieler, 1998). New data from meteorites indicate that the Xe and Kr enrichments in the solar wind are even more variable than inferred from lunar samples, and that the FIP effect may have existed since at least 4 Gy ago (Heber *et al.*, 2001).

4.4. Isotopic Compositions

Lunar soils are a major source of information on the isotopic composition of solar noble gases. They essentially provide the only data so far on the composition of Kr and Xe in the Sun. It should not be taken for granted that solar noble gases in dust grains represent the unfractionated solar wind composition, or nearly so. However, this appears to be indeed the case if samples are well chosen and experiments well designed. This is concluded by comparing He, Ne, and Ar isotopic compositions deduced from lunar samples with values measured in the Apollo SWC aluminium foils (Geiss, 1973) or by instruments on space missions such as SOHO (Kallenbach *et al.*, 1997). The best comparison is possible for Ne (cf. Wieler, 2002), indicating that initial gas fractions from stepwise in-vacuo etch runs (Benkert *et al.*, 1993) and the low temperature steps of single grain analyses (Palma *et al.*, 2002) both represent the true ^{20}Ne/^{22}Ne ratio in the solar wind. He, Ne, and Ar isotopic ratios in the solar wind are compiled in Wieler (2002). This good agreement gives us confidence that lunar soils also provide reliable isotopic compositions of solar wind Kr and Xe, for which no other information is available. It is widely accepted that the first steps of in-vacuo etch runs currently yield the best values (Wieler and Baur, 1994a; Pepin *et al.*, 1995; Wieler, 2002). Note that the isotopic ratios of any element in the solar wind are not necessarily identical to bulk Sun values or those of the outer convective zone. However, possible fractionations should be $<$1–2%/amu (Bochsler, 2000; Kallenbach, 2001). An exception is helium, since ^{3}He is thought to be enriched by about 9% in the fast solar wind (Bodmer and Bochsler, 1998). Mg and Si data from SOHO confirm that the fractionation between Sun and solar wind is small but probably not negligible. Most likely is an enrichment of the light isotopes by $\approx$ 0.5%/amu (R. Kallenbach, personal communication, 2002).

4.5. A Secular Variation of the ^{3}He/^{4}He Ratio in the Solar Wind?

The He isotopic composition in the solar wind and the question whether it varied with time are of particular interest and have been extensively studied with lunar samples. Geiss and Reeves (1972) used the ^{3}He/^{4}He ratio in the solar wind to determine the protosolar D/H ratio, a value of fundamental importance in cosmology and cosmochemistry. To be able to do so, one needs confidence that the present-day ^{3}He/^{4}He ratio in the solar wind (and hence the outer convective zone) is essentially identical to the protosolar (D+^{3}He)/^{4}He ratio. This is not necessarily so, because during the Sun's main sequence phase ^{3}He becomes enriched below the convective zone by incomplete hydrogen burning. The maximum enrichment after 4.6 Gy is a factor of $\approx$ 30 at $\approx$ 0.3 solar radii (Bochsler *et al.*, 1990). If such ^{3}He were admixed to the convective zone, the solar wind would have a higher ^{3}He/^{4}He ratio than that of the early Sun immediately after conversion of D to ^{3}He. Geiss (1973) studied this by compiling ^{3}He/^{4}He ratios of lunar samples with different antiquities. These data are reproduced in Figure 1. Geiss concluded that the He isotopic composition in the solar wind had not changed much in the past few billion years, indicating

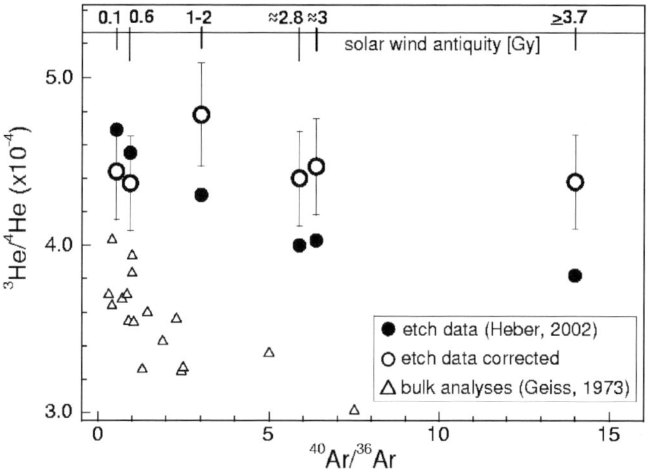

Figure 1. Helium isotopic composition of the solar wind component derived from lunar samples, versus the antiquity indicator $^{40}Ar/^{36}Ar$. The approximate antiquity scale in Gy according to Eugster *et al.* (2001) is shown on the upper abscissa. Bulk analyses compiled by Geiss (1973) indicated a secular increase of $^{3}He/^{4}He$ in the solar wind by about 10% per Gy. New noble gas analyses by in-vacuo etching at face value appear to confirm such an increase, though of only ≈ 5% per Gy. However, these data corrected by assuming $^{20}Ne/^{22}Ne$ in the solar wind to have remained constant in time (*open circles*) suggest that the apparent increase of $^{3}He/^{4}He$ is an artifact, and hence that mixing of fresh ^{3}He into the convective zone has been negligible throughout the past ≈ 3.7 Gy.

that ^{3}He in the convective zone had not dramatically increased due to mixing from below and hence that He in the solar wind is indeed useful to determine protosolar D/H. This important result also excluded non-standard solar models invoking strong internal mixing or high mass loss after onset of H-burning (Bochsler *et al.*, 1990). However, the data suggested a small increase of the $^{3}He/^{4}He$ ratio of about 30% during the last ≈ 3 billion years (Figure 1), hence a minor amount of deep mixing. Later studies indicated that such an increase must have been <5% per Gy (e.g. Benkert *et al.*, 1993; Wieler, 1998). These studies also hinted at a small secular change in the isotopic composition of Ne and Ar in the solar wind. This would, however, not be an expected result of mixing. Therefore, the question arose whether the apparent secular changes might be artifacts, e.g. because samples with high antiquity had lost relatively much of the solar wind noble gas component residing very close to the surface, such that initial etch fractions might already contain discernible amounts of the more deeply sited, isotopically heavier "SEP" component (Section 4.2.). In-vacuo etch analyses of mineral separates with very high depth resolution probably confirm this suspicion (Heber, 2002; Figure 1). At face value, they also suggest a secular increase of $^{3}He/^{4}He$ of ≈ 5%/Gy. However, in a diagram $^{3}He/^{4}He$ vs. $^{20}Ne/^{22}Ne$ (not shown), inferred He and Ne compositions of the solar wind component for each sample fall onto the same array as the data points of individual etch steps. The latter indicate a progressive admixing of SEP gases with increasing depth of etching. The coincidence of the two trends thus strongly suggests that the apparent secular change of the $^3He/^4He$ ratio is an artifact. Helium isotopic compositions corrected by means of Ne compositions (assuming $^{20}Ne/^{22}Ne$ in the solar wind to have remained constant) are shown in Figure 1 by open circles. The corrected data indicate that the He isotopic composition in the solar wind has remained constant within a few percent during the past $\approx$ 4 Gy. Mixing from below into the convective zone can thus essentially be neglected and the present-day solar wind indeed faithfully reflects the protosolar $(D + ^3He)/^4He$ abundance.

Acknowledgements

The first author would like to thank ISSI and its staff for the hospitality during the workshop. We thank Heinrich Baur for his constant expert support and advise. Reviews of the manuscript by Uli Ott and Tim Swindle are greatly appreciated. This work has been supported by the Swiss National Science Foundation.

References

Aellig, M.R., Lazarus, A.J., and Steinberg, J.T.: 2001, 'The Solar Wind Helium Abundance: Variation with Wind Speed and the Solar Cycle', *Geophys. Res. Lett.* **28**, 2767–2770.

Benkert, J.-P., Baur, H., Signer, P., and Wieler, R.: 1993, 'He, Ne, and Ar from the Solar Wind and Solar Energetic Particles in Lunar Ilmenites and Pyroxenes', *J. Geophys. Res.* **98**, 13147–13162.

Bernatowicz, T. J., Hohenberg, C.M., and Podosek, F.A.: 1979, 'Xenon Component Organization in 14301', *Proc. Lunar Planet. Sci. Conf. 10th*, 1587–1616.

Black, D. C.: 1972, 'On the Origins of Trapped He, Ne and Ar Isotopic Variations in Meteorites - I. Gas-rich Meteorites, Lunar Soil and Breccia', *Geochim. Cosmochim. Acta* **36**, 347–375.

Bjorkholm, P.J., Golub, L., and Gorenstein, P.: 1973, 'Distribution of ^{222}Rn and ^{210}Po on the Lunar Surface as Observed by the Alpha Particle Spectrometer', *Proc. Lunar Sci. Conf. 4th*, 2793–2802.

Bochsler, P.: 2000, 'Abundances and Charge States of Particles in the Solar Wind', *Rev. Geophys.* **38**, 247–266.

Bochsler, P., Geiss, J., and Maeder, A.: 1990, 'The Abundance of ^3He in the Solar Wind – a Constraint for Models of Solar Evolution', *Solar Phys.* **128**, 203–215.

Bodmer, R., and Bochsler, P.: 1998, 'The Helium Isotopic Ratio in the Solar Wind and Ion Fractionation in the Corona by Inefficient Coulomb Drag', *Astron. Astrophys.* **337**, 921–927.

Bogard, D.D.: 1995, 'Impact Ages of Meteorites: A Synthesis', *Meteoritics* **30**, 244–268.

Drozd, R., Hohenberg, C.M., and Ragan, D.: 1972, 'Fission Xenon from Extinct ^{244}Pu in 14301', *Earth Planet. Sci. Lett.* **15**, 338–346.

Eugster, O., Terribilini, D., Polnau, E., and Kramers, J.: 2001, 'The Antiquity Indicator $^{40}Ar/^{36}Ar$ for Lunar Surface Samples Calibrated by ^{235}U-^{136}Xe Dating', *Met. Planet. Sci.* **36**, 1097–1115.

Fegley, B., and Swindle, T.D.: 1993, 'Lunar Volatiles: Implications for Lunar Resource Utilization', *in* J. Lewis, M. S. Matthews, and M.L. Guerrieri (eds.), *Resources of Near-Earth Space*, Univ. Arizona Press, Tucson AZ, pp. 367–426.

Geiss, J.: 1973, 'Solar Wind Composition and Implications About the History of the Solar System', *Conf. Pap. Int. Cosmic Ray Conf. 13th*, 3375–3398.

Geiss, J., and Reeves, H.: 1972, 'Cosmic and Solar System Abundances of Deuterium and Helium-3', *Astron. Astrophys.* **18**, 126–132.

Geiss, J., Gloeckler, G., and von Steiger, R.: 1994, 'Solar and Heliospheric Processes from Solar Wind Composition Measurements', *Phil. Trans. Royal Soc. London* **A 349**, 213–226.

Gorenstein, P., Golub, L., and Bjorkholm, P.J.: 1973, 'Spatial Features and Temporal Variability in the Emission of Radon from the Moon: An Interpretation of Results from the Alpha Particle Spectrometer', *Proc. Lunar. Sci. Conf. 4th*, 2803–2809.

Halliday, A.N.: 2000, 'Terrestrial Accretion Rates and the Origin of the Moon', *Earth Planet. Sci. Lett.* **176**, 17–30.

Heber, V.S.: 2002, 'Ancient Solar Wind Noble Gases in Lunar and Meteoritic Archives and Tests for Modern Solar Wind Collection with the GENESIS Mission', *PhD Thesis ETH Zürich Nr. 14579*, 154 pp.

Heber, V.S., Baur, H., and Wieler, R.: 2001, 'Heavy Solar Noble Gases in Meteorites – New Insights into a Unique Archive of Solar Wind', *in* R.F. Wimmer-Schweingruber (ed.), *Solar and Galactic Composition*, AIP Conf. Proc. **598**, pp. 387–392.

Hodges, R.R.: 1975, 'Formation of the Lunar Atmosphere', *Moon* **14**, 139–157.

Hodges, R.R.: 1977, 'Release of Radiogenic Gases from the Moon', *Phys. Earth Planet. Int.* **14**, 282–288.

Hoffman, J.H., Hodges, R.R., Johnson, F.S., and Evans, D.E.: 1973, 'Lunar Atmospheric Composition Results from Apollo 17', *Proc. Lunar Sci. Conf. 4th*, 2865–2875.

Kallenbach, R.: 2001, 'Isotopic Composition Measured *In-situ* in Different Solar Wind Regimes by CELIAS/MTOF on board SOHO', *in* R.F. Wimmer-Schweingruber (ed.), *Solar and Galactic Composition*, AIP Conf. Proc. **598**, pp. 113–119.

Kallenbach, R., Ipavich, F.M., Bochsler, P., *et al.*: 1997, 'Isotopic Composition of Solar Wind Neon Measured by CELIAS/MTOF on Board SOHO', *J. Geophys. Res.* **102**, 26895–26904.

Kerridge, J.F., Eugster, O., Kim, J.S., and Marti, K.: 1991, 'Nitrogen Isotopes in the 74001/74002 Double-drive Tube from Shorty Crater, Apollo 17', *Proc. Lun. Planet. Sci.* **21**, 291–299.

Killen, R.M., and Ip, W.-H.: 1999, 'The Surface-bounded Atmospheres of Mercury and the Moon', *Rev. Geophys.* **37**, 361–406.

Lawson, S.L., Feldman, W.C., Lawrence, D.J., Moore, K.R., Maurice, S., Belian, R.D., and Binder, A.B.: 2002, 'Maps of Lunar Radon-222 and Polonium-210', *Lunar Planet. Sci.*, XXXIII, abstract #1835, Lunar Planet. Institute, Houston (CD-ROM).

Leich, D.A., and Niemeyer, S.: 1975, 'Trapped Xenon in Lunar Anorthositic Breccia 60015', *Proc. Lunar Sci. Conf. 6th*, 1953–1965.

Lightner, B.D., and Marti, K.: 1974, 'Lunar Trapped Xenon', *Proc. Lunar Sci. Conf. 5th*, 2023–2031.

Manka, R.H., and Michel, F.C.: 1971, 'Lunar Atmosphere as a Source of Lunar Surface Elements', *Proc. Lunar Sci. Conf. 2nd*, 1717–1728.

Marty, B., Hashizume, K., Chaussidon, M., and Wieler, R.: 2003, 'Nitrogen Isotopes on the Moon: Archives of the Solar and Planetary Contributions to the Inner Solar System', *Space Sci. Rev.*, this volume.

Mathew, K.J., and Marti, K.: 2001, 'Lunar Nitrogen: Indigenous Signature and Cosmic-ray Production Rate', *Earth Planet. Sci. Lett.* **184**, 659–669.

Megrue, G.H.: 1973, 'Spatial Distribution of $^{40}Ar/^{39}Ar$ Ages in Lunar Breccia 14301', *J. Geophys. Res.* **78**, 3216–3221.

Mewaldt, R.A., Ogliore, R.C., Gloeckler, G., and Mason, G.M.: 2001, 'A new Look at Neon-C and SEP-Ne', *in* R.F. Wimmer-Schweingruber (ed.), *Solar and Galactic Composition*, AIP Conf. Proc. **598**, pp. 393–398.

Murer, C.A., Baur, H., Signer, P., and Wieler, R.: 1997, 'He, Ne, and Ar Abundances in the Solar Wind: in Vacuo Etching of Meteoritic Iron-nickel', *Geochim. Cosmochim. Acta* **61**, 1303–1314.

Niedermann, S., and Eugster, O.: 1992, 'Noble Gases in Lunar Anorthositic Rocks 60018 and 65315: Acquisition of Terrestrial Krypton and Xenon Indicating an Irreversible Adsorption Process', *Geochim. Cosmochim. Acta* **56**, 493–509.

Niemeyer, S., and Leich, D.A.: 1976, 'Atmospheric Rare Gases in Lunar Rock 60015', *Proc. Lunar Sci. Conf. 7th*, 587–597.

Palma, R.L., Becker, R.H., Pepin, R.O., and Schlutter, D.J.: 2002, 'Irradiation Records in Regolith Materials, II: Solar-wind and Solar-energetic-particle Components in Helium, Neon, and Argon Extracted from Single Lunar Mineral Grains and from the Kapoeta Howardite by Stepwise Pulse-heating', *Geochim. Cosmochim. Acta* **66**, in press.

Pedroni, A., and Begemann, F.: 1994, 'On Unfractionated Solar Noble Gases in the H3-6 meteorite Acfer111', *Meteoritics* **29**, 632–642.

Pepin, R.O., Nyquist, L.E., Phinney, D., and Black, D.C.: 1970a, 'Isotopic Composition of Rare Gases in Lunar Samples', *Science* **167**, 550–553.

Pepin, R.O., Nyquist, L. E., Phinney, D., and Black, D.C.: 1970b, 'Rare Gases in Apollo 11 Lunar Material', *Proc. Apollo 11 Lunar Sci. Conf.*, 1435–1454.

Pepin, R.O., Becker, R.H., and Rider, P.E.: 1995, 'Xenon and Krypton Isotopes in Extraterrestrial Regolith Soils and in the Solar Wind', *Geochim. Cosmochim. Acta* **59**, 4997–5022.

Polnau, E., and Eugster, O.: 1998, 'Cosmic-ray Produced, Radiogenic, and Solar Noble Gases in Lunar Meteorites Queen Alexandra Range 94269 and 94281', *Met. Planet. Sci.* **33**, 313–319.

Reynolds, J.H., Hohenberg, C.M., Lewis, R.S., Davis, P.K., and Kaiser, W.A.: 1970, 'Isotopic Analysis of Rare Gases from Stepwise Heating of Lunar Fines and Rocks', *Science* **167**, 545–548.

Stern, S.A.: 1999, 'The Lunar Atmosphere: History, Status, Current Problems, and Context', *Rev. Geophys.* **37**, 453–491.

Stöffler, D., and Ryder, G.: 2001, 'Stratigraphy and Isotope Ages of Lunar Geologic Units: Chronological Standard for the Inner Solar System', *Space Sci. Rev.* **96**, 9–54.

Swindle, T.D.: 2002, 'Noble Gases in the Moon and Meteorites – Radiogenic Components and Early Volatile Chronologies', *Rev. Min. Geochem.* **47**, in press.

Swindle, T.D., Caffee, M.W., Hohenberg, C.M., and Taylor, S.R.: 1986, 'I-Pu-Xe Dating and the Relative Ages of the Earth and Moon', *in* W.K. Hartmann, R.J. Phillips, and G.J. Taylor (eds.), *Origin of the Moon*, Lunar Planet. Institute, Houston TX, pp. 331–357.

von Steiger, R., Geiss, J., and Gloeckler, G.: 1997, 'Composition of the Solar Wind', *in* J.R. Jokipii, C.P. Sonett, and M.S. Giampapa (eds.), *Cosmic Winds and the Heliosphere*, Univ. Arizona Press, Tucson AZ, pp. 581–616.

Walter, F.M., and Barry, D.C.: 1991, 'Pre- and Main-sequence Evolution of the Solar Activity', *in* C.P. Sonett, M.S. Giampapa, and M.S. Matthews (eds.), *The Sun in Time*, Univ. Arizona Press, Tucson AZ, pp. 633–657.

Wieler, R.: 1997, 'Why are SEP Noble Gases so Abundant in Extraterrestrial Samples?', *Lunar Planet. Sci. XXVIII*, Lunar Planet. Inst., Houston, TX, pp. 1551–1552.

Wieler, R.: 1998, 'The Solar Noble Gas Record in Lunar Samples and Meteorites', *Space Sci. Rev.* **85**, 303–314.

Wieler, R.: 2002, 'Noble Gases in the Solar System', *Rev. Mineral. Geochem.* **47**, in press.

Wieler, R., and Baur, H.: 1994a, 'Krypton and Xenon from the Solar Wind and Solar Energetic Particles in two Lunar Ilmenites of Different Antiquity', *Meteoritics* **29**, 570–580.

Wieler, R., and Baur, H.: 1994b, 'Parentless Fission Xenon in Mineral Separates of Lunar Breccia 14301 Studied by Closed System Etching', *Meteoritics* **29**, 550.

Wieler, R., and Baur, H.: 1995, 'Fractionation of Xe, Kr, and Ar in the Solar Corpuscular Radiation Deduced by Closed System Etching of Lunar Soils', *Astrophys. J.* **453**, 987–997.

Wieler, R., Baur, H., and Signer, P.: 1993, 'Parentless Fission and Radiogenic Xe in Lunar Breccia 14301 Studied by Closed-system Etching', *Meteoritics* **28**, 459–460.

Wieler, R., Kehm, K., Meshik, A.P., and Hohenberg, C.M.: 1996, 'Secular Changes in the Xenon and Krypton Abundances in the Solar Wind Recorded in Single Lunar Grains', *Nature***384**, 46–49.

Wimmer-Schweingruber, R.F., and Bochsler, P.: 2001, 'Lunar Soils: a Long-term Archive for the Galactic Environment of the Heliosphere?', *in* R.F. Wimmer-Schweingruber (ed.), *Solar and Galactic Composition*, AIP Conf. Proc. **598**, pp. 399–404.

ON NOBLE GAS PROCESSING IN THE SOLAR ACCRETION DISK

R.O. PEPIN

School of Physics and Astronomy, University of Minnesota, Minneapolis, Minnesota 55455

Received: 20 July 2002; Accepted in final form: 20 September 2002

Abstract. Two fractionation models are applied to the problem of generating the widely distributed "Q-component" noble gases in meteorites from the solar-like isotopic and elemental compositions that presumably characterized the early solar accretion disk. Noble gas fractionation by mass-dependent dissipation of the solar nebula, as suggested by Ozima *et al.* (1998), is examined in the context of a model developed by Johnstone *et al.* (1998) for accretion disk photoevaporation driven by intense UV radiation from a neighboring giant star. Hydrodynamic escape of heavier species entrained in hydrogen outflow from the UV-heated outer regions of the disk can generate substantial noble gas fractionations, but they do not match the observed Q-component isotopic pattern and moreover require the physically unrealistic assumption that the fractionated gases are confined to the heated disk boundary zone, without mixing with the interior nebula, for long periods of time. It seems more likely that hydrodynamic outflow is actually established below this zone, in the body of the disk. In this case fractionations are governed by Rayleigh distillation of the entire remaining nebula, and are negligible at the time when disk erosion is halted by the gravitational potential of the young sun embedded in the disk.

A "local" model of noble gas fractionation by hydrodynamic blowoff of transient, methane-rich atmospheres outgassed from the interiors of large primitive planetesimals (Pepin, 1991) is updated and assessed against current data. Degassed atmospheres are assumed to contain isotopically solar noble gases except for an additional nucleogenic Xe component that contributes primarily to the two heaviest isotopes; there is evidence that this same component is present at varying levels in other solar-system volatile reservoirs, possibly reflecting a compositional change with time in the solar nebula. Single fixed values for the two free parameters in the blowoff modeling equations can generate fractionated Xe, Kr, Ar and Ne compositions in the residual atmosphere that closely match observed meteoritic isotopic distributions, and Q-gas elemental ratios are approximated by adsorption of fractionated gases on planetesimal surface grains using plausible values of relative Henry Law constants. Additional requirements for adsorption of sufficient absolute amounts of Q-gases on carrier grains, and their subsequent ejection to space, mixing in the nebula, and dispersal into meteorite bodies, are examined in the context of current models for body sizes and dynamical evolution in an early mass-rich asteroid belt (Chambers and Wetherill, 2001). Despite its ability to replicate isotopic compositions, uncertainties about the environments in which the blowoff model can successfully operate suggest that there is, as yet, no entirely satisfactory understanding of how the Q-component noble gases might have evolved from solar-like precursor compositions.

1. Introduction

The astrophysical environments and mechanisms responsible for generating trapped meteoritic noble gas distributions from more primitive compositions are not well understood. These "primordial" gases are dominated by a widely distributed and,

Space Science Reviews **106:** 211–230, 2003.

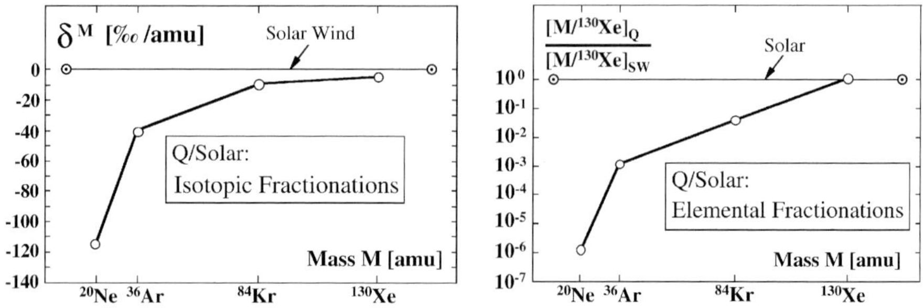

Figure 1. Isotopic and elemental fractionations of Q-gases with respect to solar wind (SW) or solar compositions. Isotopic fractionations (*left panel*) in permil (‰) per amu calculated from the Q-gas $^{124-128}Xe/^{130}Xe$, $^{78-86}Kr/^{84}Kr$, $^{36}Ar/^{38}Ar$, and $^{20}Ne/^{22}Ne$ ratios, all divided by the corresponding SW ratios. Xe and Kr data from Table I, Ar and Ne data from Busemann *et al.* (2000), Pepin *et al.* (1999), and Palma *et al.* (2002). The δ-representation used here and in subsequent figures is defined as $\delta^M = 1000[(R_M/R_{\text{ref}}) - 1]$ ‰ where R_M is the sample composition and R_{ref} is the corresponding isotope ratio in the reference gas. Data for Q-gas elemental fractionations with respect to solar noble gases (*right panel*) from Busemann *et al.* (2000).

for the most part, an isotopically invariant component called Q-gases. Although they could be associated with a separate and so far unidentified minor carrier phase, current evidence suggests that they occupy adsorption sites on the surfaces of carbonaceous materials. Relative to solar compositions (Figure 1), these Q-gases display isotopically heavier isotope ratios and increasingly severe elemental depletions with decreasing atomic mass (e.g. Wieler, 1994; Busemann *et al.*, 2000; Ott, 2002), strongly suggesting that processes favoring loss of lighter species – for example, Rayleigh fractionation – have played a role in generating them from solar-like parental compositions (Pepin, 1991; Ozima *et al.*, 1998).

A "local" fractionation model based on hydrodynamic blowoff of compositionally solar-like degassed planetesimal atmospheres (Pepin, 1991) can generate isotopic distributions in the residual atmospheres that closely resemble meteoritic isotope ratios for Xe, Kr, Ar and Ne. "Local" models of this kind, however, face the intrinsic difficulty that carrier grains loaded with Q-gases on individual planetesimals must later be ejected, mixed, and widely distributed throughout meteorite-forming regions of the accretion disk. Ozima *et al.* (1998) propose a quite different scenario – that Q-gas compositions were established by a "global" Rayleigh distillation of an initially solar-composition gaseous nebula, presumably during its separation and loss from condensed solar-system matter. In this view there is no dispersal and mixing problem; ambient Q-gases pervade the dissipating nebula and are adsorbed or otherwise occluded into carbonaceous phases later incorporated into meteorites. But there are difficulties here as well. There is no known astrophysical mechanism that leads naturally to the required Rayleigh fractionation of the entire gas-phase nebula. Moreover the particular fractionation model utilized by Ozima *et al.* fails to generate the characteristic Q pattern for all noble gas isotopes and elements (Pepin and Porcelli, 2002).

In the following sections I summarize preliminary results of work in progress to model the extent of noble gas fractionations that might be generated during dissipation of the solar accretion disk by a powerful external source of ultraviolet radiation, and update the 1991 planetesimal blowoff model in the light of more recent data on Q-component isotopic compositions.

2. Isotopic Fractionation during Accretion Disk Photoevaporation

Thermal evaporation from the outer edge of an accretion disk, if sufficiently intense, could generate hydrogen escape fluxes large enough to exert upward drag forces on heavier gas-phase nebular constituents and lift them out of the disk boundary, even though they are too massive to escape on their own by thermal evaporation (Jeans escape). Lighter species are entrained and lost with the outflowing hydrogen more readily than heavier ones in this "hydrodynamic escape" mechanism, leading to mass fractionation of the residual gas phase (Hunten *et al.*, 1987). The process has been applied in some detail to evolution of noble gas distributions during escape of hydrogen-rich primordial atmospheres from the terrestrial planets (e.g., Pepin, 1991, 1997, 2000; Pepin and Porcelli, 2002), but not, in the absence of a plausible and sufficiently powerful energy source and a general lack of consensus on dispersal mechanisms, to possible fractionation of residual gases, as suggested by Ozima *et al.* (1998), during dissipation of the solar nebula. However a recent model developed by Johnstone *et al.* (1998) of accretion disk photoevaporation driven by intense UV fluxes radiated from a nearby massive star provides a context for exploring this possibility. There is no independent evidence that the solar accretion disk was dissipated in this way. The approach taken in this paper is to ask what maximum fractionations might have resulted if photoevaporative erosion actually were responsible, without considering the question of how well the process could be accommodated in other models for evolution of the protosolar nebula (e.g., Morfill *et al.*, 1985; Cameron, 1995).

The astrophysical framework of the Johnstone *et al.* (1998) modeling relevant to this discussion is illustrated in the Figure 2 cartoon. Intense UV radiation from a neighboring luminous star – in their work Θ^1 Ori C in the Trapezium region of Orion – impinges on an accretion disk separated from the star by 0.1–1 light-y. Extreme ultraviolet (EUV) ionizes the outer disk boundary (the ionization front IF) and heats it to $\approx$10,000 K. Longer wavelength far-ultraviolet (FUV) radiation penetrates the IF into the photodissociation region (PDR) where it dissociates and heats gases to $\approx$1000 K. Under these conditions the hydrogen outflow through the IF of a newly irradiated disk is enormous, on the order of $\approx 10^{-6}$ solar mass/y from a disk of initial radius $r_d = 300$ AU and nebular mass $M_d \approx 1/7$ solar mass, at a distance d of 0.1 light-y from the UV source and with a 1 solar mass star embedded in the disk. The mass-loss rate is proportional to $r_d^{3/2}$ in the case considered here of loss dominated by the EUV, and so decreases as the disk erodes inward from its

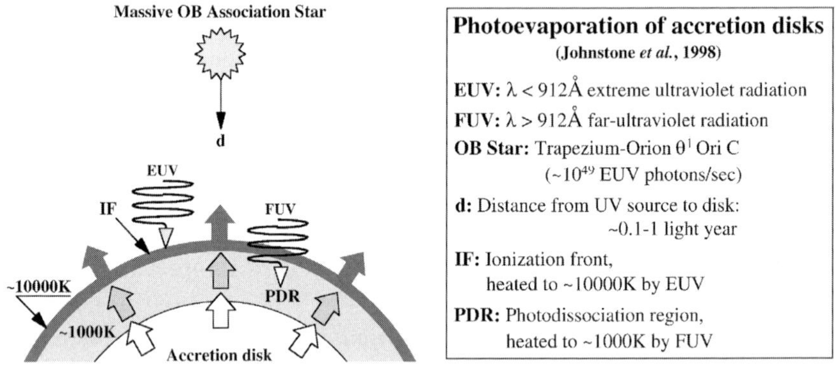

Figure 2. Cartoon of accretion disk photoevaporation driven by a powerful external UV source (after Johnstone *et al.*, 1998). Arrows represent the supply and outflow of hydrogen and entrained heavier gases at the heated disk boundary. Definitions and parameter values used in the analytic formalism developed by Johnstone *et al.* are given in the legend. See Section 2 of the text for application to possible noble gas fractionation in the evaporative loss process.

outer boundary. It falls off more rapidly than this due to gravity effects as the outer disk boundary approaches the embedded star, and terminates when the thermal energy at the IF becomes insufficient to remove hydrogen from the star's potential well. For a 1 solar mass embedded star this gravitational cutoff radius is close to the present orbit of Jupiter. Disk evolution time to the cutoff radius is rapid, $\approx$13 My for EUV-driven erosion of the 300 AU disk specified above, and substantially less than this if the approximation used here of a thin PDR, which yields minimum mass-loss rates (Johnstone *et al.*, 1998), is dropped.

Fractionations of noble gases during hydrogen outflow from the disk boundary are governed by two central equations describing the hydrodynamic escape process (Hunten *et al.*, 1987; Pepin and Porcelli, 2002):

$$F_2 = F_1 \frac{N_2}{N_1} \left[\frac{m_c - m_2}{m_c - m_1} \right] \tag{1}$$

for $m_c \geq m_2$, where F_1, N_1, m_1 and F_2, N_2, m_2 are respectively the outflow fluxes, inventories, and masses of hydrogen and a noble gas isotope. The critical or "crossover" mass m_c represents an escape constraint: if m_c is greater than the mass m_2 of a noble gas isotope entrained in the F_1 outflow, F_2 is > 0 and the isotope escapes; if $m_c \leq m_2$, $F_2 = 0$ and it is retained. The crossover mass is defined (Hunten *et al.*, 1987) as

$$m_c = m_1 + \frac{kTF_1}{bgX_1} \tag{2}$$

where k = Boltzmann's constant, T = gas temperature, g = gravitational acceleration – assumed to be that of the embedded star, X_1 is the mole fraction of hydrogen (which remains near unity throughout the escape episode since the hydrogen inventory dominates in a solar-composition gas mixture), and b is the diffusion parameter (the product of diffusion coefficient and total number density) of a particular noble gas in hydrogen. Note from Equation 2 that a large F_1 yields a large crossover mass m_c; for F_1 high enough that $m_c \gg m_2 > m_1$, losses described by Equation 1 are maximum and $F_2/F_1 \approx N_2/N_1$ – i.e., noble gas species escape close to the ratios of their nebular inventories, essentially without fractionation. Maximum fractionation occurs when F_1 drops to the level where m_c approaches m_2. As the mass-loss rate and therefore F_1 declines during disk evolution, m_c eventually drops below m_2 and the inventory of that noble gas isotope is frozen. But losses of lighter species continue until m_c crosses their masses. This consequence of a declining F_1, together with the increasing escape flux F_2 given by Equation 1 for species with increasingly smaller masses m_2, is responsible for the preferential loss of the lighter noble gas isotopes and elements in hydrodynamic escape.

For the 300 AU disk assumed here for illustration of fractionation effects, the initial mass-loss rate translates to a hydrogen escape flux through the IF of $F_1 \approx 10^{11}$ H atoms $\mathrm{cm^{-2}\,s^{-1}}$. With typical values of the diffusion parameter b for the noble gases (on the order of 10^{18} to 10^{19} $\mathrm{cm^{-1}\,s^{-1}}$), Equation 2 yields crossover masses m_c of approximately 10^7 amu. It is clear from Equation 1, where m_2 for Xe to Ne ranges from 136 to 20 amu and $m_1 = 1$ or 2 amu, that outflow fractionation is negligible – in Ne, for example, it follows directly from Equation 1, written separately for $F_2 = F(^{20}\mathrm{Ne})$, $m_2 = 20$ and $F_2 = F(^{22}\mathrm{Ne})$, $m_2 = 22$, that $F(^{20}\mathrm{Ne})/F(^{22}\mathrm{Ne}) = (10^7 - 20)/(10^7 - 22)$ or essentially 1. Substantial fractionations of residual noble gases, if they occur at all, are thus restricted to the terminal stage of disk evolution near the gravitational cutoff radius, where m_c calculated from Equation 2 falls precipitously as the mass-loss rate – and therefore F_1 – declines sharply and solar gravity g is maximum.

2.1. Fractionation in the PDR

One may calculate fractionation effects in two different disk regimes: in the PDR alone, or in the disk as a whole. The first, which yields maximum effects, assumes that fractionating hydrodynamic outflow is established in the PDR, that there is no back-mixing across the PDR-disk boundary into the much larger volume of the interior disk, and that the only communication across the boundary is a continuous inflow of solar-composition gases – represented by the white arrows in Figure 2 – to replenish those lost through the IF. The mathematical formalism describing this kind of loss-replenishment system and its compositional evolution with time, together with other details of the modeling summarized here, will be reported separately. Because parameters in Equations 1 and 2 change with time as the disk erodes, isotopic evolution is calculated through short sequential time steps within which F_1, g, and m_c are essentially constant.

Results for the initially 300 AU disk are shown in Figure 3 for evolution of Xe and Ne crossover masses and the $^{124}\mathrm{Xe}/^{130}\mathrm{Xe}$ and $^{20}\mathrm{Ne}/^{22}\mathrm{Ne}$ isotope ratios. Significant isotopic fractionation occurs only in the last $\approx$1 My of erosion to the

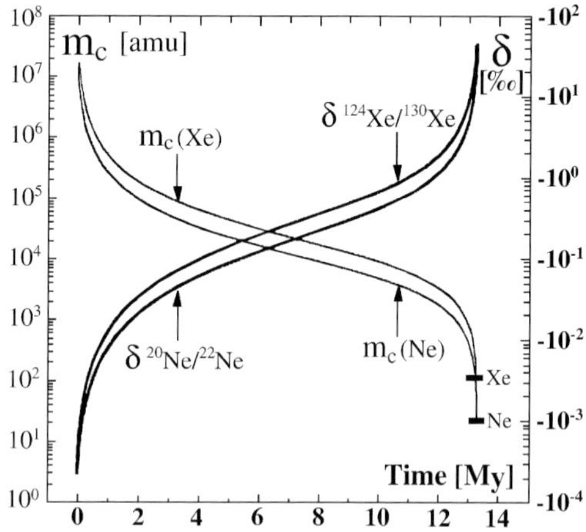

Figure 3. Calculated Xe and Ne crossover masses m_c (*light curves, left scale*) and isotopic fractionations in permil (°/₀₀) of initially solar $^{124}Xe/^{130}Xe$ and $^{20}Ne/^{22}Ne$ ratios (*heavy curves, right scale*) in the photodissociation region (PDR) of an eroding accretion disk, plotted against disk evolution time from an initial radius of 300 AU and mass ≈1/7 solar mass to gravitational cutoff of hydrogen loss at ≈4.5 AU. The labeled "Xe" and "Ne" bars at the lower right mark the times at which the crossover masses for these species cross their isotopic masses, freezing their inventories against further fractionations.

gravitational cutoff at ≈4.5 AU. Xe fractionation exceeds that of Ne for most of the disk evolution time, but the Xe inventory is frozen first and the final fractionations are almost identical for both ratios at approximately −40 °/₀₀ with respect to solar values. For these two Xe isotopes, −40 °/₀₀ compares well to the ≈ −30 °/₀₀ to −45 °/₀₀ observed for meteoritic Xe relative to solar wind Xe (see Table I), but for Ne it is far short of the −230 °/₀₀ difference between the Ne-Q and solar wind $^{20}Ne/^{22}Ne$ ratios shown below in Figure 4 (see also Figure 1, where the difference is plotted as °/₀₀ per amu = −115 °/₀₀). Moreover these PDR fractionations were maximized by choosing m_1 in the PDR to be 2 amu (molecular hydrogen). But the "photodissociation region" is so named because much of the hydrogen is expected to be in atomic form (Johnstone *et al.*, 1998). If so, F_1 is twice as large and noble gas diffusion parameters b are almost an order of magnitude lower than they are for diffusion in H_2. Both changes elevate m_c via Equation 2 and reduce final fractionations correspondingly, to < 10 °/₀₀ for both ratios.

These results change only modestly when the calculations are applied to disks with initial radii ranging from 100 to 1000 AU and masses from ≈0.03 to 0.9 solar masses, and in all cases the permil Xe and Ne isotopic fractionations are approximately equal. Tenfold larger separations d between the UV source and the disk, to ≈1 light-y, increase the final fractionations by roughly a factor of two, but these are unlikely to be realized because the disk evolution timescales to the cutoff radius also increase tenfold, probably beyond the expected lifetime of the massive source star.

It is evident that the Q-gas isotopic fractionation pattern shown in Figure 1 is not replicated, even in this extreme scenario of accumulating fractionations in an isolated PDR. Moreover acquisition of fractionated noble gases by Q-gas carrier grains would have to occur prior to dilution by mixing with the bulk of the remaining nebula, in the hot low-pressure environment of the PDR – conditions not conducive to efficient adsorption on grain surfaces. A more fundamental problem is that this PDR fractionation model is unrealistic in a number of ways. A strictly one-way transport of material into an otherwise isolated PDR, with no mixing of fractionated species back into the body of the disk during its entire erosional timescale, is physically improbable; turbulent back-transport of at least some of the fractionated PDR material and dilution by mixing with unfractionated disk gases would be expected. Most importantly, the assumption that the disk supplies unfractionated gases to the PDR – equivalent to assuming that hydrodynamic entrainment of heavier species is established in the PDR and not in the underlying disk – is almost certainly wrong. One sees from Figure 2 that mass conservation requires F_1 (in #/cm^2-sec) through the PDR-disk boundary to be comparable to the average outward flux in the PDR, which means that the fractionating outflow would actually be established in the disk unless crossover masses given by Equation 2 in the cooler disk environment are too low for noble gas entrainment. Diffusion parameters b, however, are temperature dependent, increasing as T^x with x not greatly below 1; the ratios T/b in Equation 2 consequently vary by factors of ≤ 3 between $T = 1000$ K and $T = 10$ K. Crossover masses in the disk, and outflow fractionations given by Equation 1, would therefore be similar to those in the PDR. In this situation we are dealing with the second of the two regimes noted above – that of fractionations produced in the disk as a whole.

2.2. Whole-Nebula Fractionation

Residual gas fractionations in the disk may be calculated from the Rayleigh distillation equation if the disk volume is turbulent and well-mixed. For the case of fractionation generated by hydrodynamic loss, the Rayleigh equation is (Hunten *et al.*, 1987)

$$\frac{N_2}{N_2^0} = \left[\frac{N_1}{N_1^0}\right]^{\frac{m_c - m_2}{m_c - m_1}} \tag{3}$$

where zero superscripts designate initial inventories. Written separately for the two Ne isotopes and $m_1 = 2$ amu, Equation 3 yields

$$f_{20/22} = \left[\frac{N_1}{N_1^0}\right]^{\frac{22-20}{m_c - 2}} \tag{4}$$

for the ^{20}Ne/^{22}Ne fractionation factor relative to the initial ratio; the corresponding equation for the Xe isotopes may be written by inspection. Total isotopic fractionations are given by the product of fractionation factors generated in short time increments spanning the erosion history of a disk. They are very small indeed – < 1 ‰ for the $^{20}Ne/^{22}Ne$ and $^{124}Xe/^{130}Xe$ ratios – in all the disks. The reason is clear. Mass-loss rates are maximum in the initial stages of disk evaporation, so the change in the hydrogen inventory N_1 is relatively large over a short time period. But crossover masses m_c are also maximum here, and the exponent in Equation 4 is essentially zero. The opposite situation applies near the end of erosion: crossover masses are falling rapidly (Figure 3) and exponents approach their maximum values, but solar gravity throttles mass loss near the cutoff radius and N_1/N_1^0 differs only negligibly from 1.

About 20 Jupiter masses remain after erosion of this example disk from 300 AU to the assumed gravitational truncation of mass loss near ≈4.5 AU. However, Johnstone *et al.* (1998) point out that evaporative loss of hydrogen could perhaps continue at distances as small as ≈40% of this cutoff radius. In the case of disk erosion to $r_d \approx$ 1.8 AU, maximum fractionations in an isolated PDR can exceed 100 ‰, but again are approximately equal for the Ne and Xe isotope ratios and so do not replicate the Q-gas pattern. The interior nebular mass is now roughly 10 Jupiter masses, and Rayleigh fractionations of its noble gases are still negligible. These residual gases must eventually have disappeared, but their dissipation and possible isotopic fractionation by a thermal loss process appears unlikely unless they were transported outward past the gravitational cutoff for photoevaporation by mechanisms such as a stellar wind or tidal interactions – a possibility noted by Johnstone *et al.* Even in this case, matching the Q-gas isotopic distributions requires hydrogen escape fluxes, and therefore crossover masses for the various noble gas species, to decrease with time in a very specific way. Neon loss, for example, would have to continue for a lengthy period after the residual Xe inventory was frozen in order to reach the ≈25-fold greater fractionation, in ‰ per amu, of Q-$^{20}Ne/^{22}Ne$ compared to Q-$^{124}Xe/^{130}Xe$ (Figure 1). This kind of time dependence, which has to be stipulated ad hoc in the nebular photoevaporation model, occurs naturally in the planetesimal blowoff model discussed in the following section.

3. Isotopic Fractionation in Blowoff of Transient Planetesimal Atmospheres

This approach to generating Q-gas compositions from initially solar-like elemental and isotopic distributions was developed in detail in an earlier report (Pepin, 1991). An early generation of asteroid-size objects is assumed to accrete at low ambient temperature from materials rich in icy-mantled extrasolar grains carrying isotopically solar Ne, Ar and Kr, and solar Xe with an added component, resembling H-Xe or the H component of Xe-HL (Pepin *et al.*, 1995), at the heaviest isotopes. Interior warming by extinct radionuclide decay drives hydrothermal circulation beneath a surface ice layer. Interior volatiles are vented to the surface by projectile impact large enough to penetrate the ice cap. A rapid "source-limited" hydrodynamic blowoff of this transient atmosphere then occurs, in which the noble gases are entrained in the outflow of an abundant, light, and gravitationally unbound constituent and are lost at mass-dependent rates ("source-limited" escape fluxes are controlled by the amount of available gas, in contrast to the case for evaporative "energy-limited" nebular loss discussed above where the energy supply controls). The light constituent in primordial planetary atmospheres is generally assumed to be H_2 (e.g., Pepin, 1991), but the free hydrogen abundance supplied to these planetesimals by icy-mantled grains would be small. A thermodynamically attractive alternative to H_2 as an abundant and relatively non-condensable carrier gas in planetesimal interiors is methane. Methane contents of grain mantle ices appear to be low to nil (Tielens and Allamandola, 1987). But at $T \approx 400 - 600°$ C in the planetesimal interior – temperatures that do not seem too high for small bodies at early evolutionary stages, given the existence of basaltic meteorites and spectroscopic evidence for differentiated asteroids such as Vesta – the reaction C[graphite] + $2H_2O$[gas] + 2Fe[metal] $\rightarrow$ CH_4[gas] + 2FeO[in olivine] is potentially important (Holloway, 1988a, b). The first two reactants are abundant in mantled interstellar grains. While there is no direct evidence for metallic iron, the small amount needed (probably $\leq$ 1 wt.%) could have been present in other accretional components. The blowoff model therefore assumes that an outgassed methane-rich atmosphere containing noble gases from grain mantle ices represents the initial state of volatile evolution on a primitive planetesimal.

As discussed in the previous section, fractionations in hydrodynamic escape are governed by crossover masses m_c. Escape fluxes, residual inventories, and crossover masses decay exponentially during a source-limited blowoff episode. The crossover masses therefore sweep sequentially across first the Xe isotopes and later those of lighter species, freezing their atmospheric inventories as they cross. Rayleigh distillations of this type are described by simple equations containing only two free parameters once the mass of the light atmospheric constituent is chosen – the beginning and terminal crossover masses for one of the noble gases (Pepin, 1991). One sees from Equation 2 that values of m_c for each of the noble gases at a given escape flux F_1 of the light carrier gas differ only by their differing diffusion parameters b, and so the choice of a particular m_c for one species, Xe in this case, fixes the crossover masses for all other species S through the known ratios b(Xe)/b(S). The central test of the model therefore lies in whether it can replicate the meteoritic/solar fractionation factors of the Xe, Kr, Ar and Ne isotope ratios – 15 of them – with single values of the two free parameters. The best-fit fractionations yielded by the blowoff model are shown in Figure 4 and compared to observed meteorite isotope ratios.

3.1. Xenon

A component called U-Xe appears to have dominated the primordial terrestrial atmosphere, and is suspected to have been widespread in the solar nebula at some stage of its evolution (Pepin, 1991, 2000). It differs from SW (solar wind)-Xe,

TABLE I

Isotopic compositions of solar-system xenon and krypton components. Uncertainties ($\pm 1\sigma$) in the last two digits of the listed isotope ratios are indicated in parentheses.

Component	^{124}Xe	^{126}Xe	^{128}Xe	^{129}Xe	^{130}Xe	^{131}Xe	^{132}Xe	^{134}Xe	^{136}Xe
Air-Xe [a)]	2.337(07)	2.180(11)	47.15(05)	649.6(06)	≡100	521.3(06)	660.7(05)	256.3(04)	217.6(02)
U-Xe [b)]	2.928(10)	2.534(13)	50.83(06)	628.6(06)	≡100	499.6(06)	604.7(06)	212.6(04)	165.7(03)
SW-Xe [c)]	2.939(70)	2.549(82)	51.02(54)	627.3(42)	≡100	498.0(17)	602.0(33)	220.7(09)	179.7(06)
AVCC-Xe [d)]	2.851(51)	2.512(40)	50.73(38)	653.3 [d)]	≡100	504.3(28)	615.0(27)	235.9(13)	198.8(12)
Xe-Q [e)]	2.810(13)	2.506(12)	50.77(16)	643.6(17)	≡100	505.6(11)	617.7(11)	233.5(08)	195.4(06)
SiC-N-Xe [f)]	≡2.928	2.562(79)	48.6(15)	614(26)	≡100	490(21)	614(21)	258(11)	242(14)
Xe-HL [f)]	1.013(27)	0.443(22)	3.08(16)	11.2(11)	≡0	11.35(41)	7.40(57)	67.70(41)	≡100
Xe-L	≡0	≡0	≡0	≡0	≡0	11.35(41)	7.40(57)	67.70(41)	≡100

Component	^{78}Kr	^{80}Kr	^{82}Kr	^{83}Kr	^{84}Kr	^{86}Kr
Air-Kr [a)]	0.6087(20)	3.960(02)	20.217(21)	20.136(21)	≡100	30.524(25)
SW-Kr [f)]	0.6365(17)	4.088(07)	20.482(27)	20.291(13)	≡100	30.240(50)
Chondritic Kr [g)]	0.599(11)	3.930(33)	20.06(11)	20.08(12)	≡100	30.88(17)
Kr-Q [e)]	0.6030(30)	3.937(07)	20.180(20)	20.180(20)	≡100	30.950(50)

[a)] Basford *et al.* (1973); [b)] Pepin (2000); [c)] Measured solar wind (SW) ratios in acid-etching experiments on lunar ilmenites (Wieler and Baur, 1994; Pepin *et al.*, 1995); [d)] Average Carbonaceous Chondrite Xe (Pepin, 1991, 2000), the listed value for ^{129}Xe is an average for CI and CM carbonaceous chondrites, the approximate range is from 636 to 671 (Mazor *et al.*, 1970); [e)] Busemann *et al.* (2000), Xe-Q composition renormalized to ^{130}Xe = 100; [f)] Pepin *et al.* (1995) and references therein, SiC N-Xe composition slightly adjusted to reflect revised U-Xe composition reported by Pepin (2000); [g)] Average of Kr compositions reported by Krummenacher *et al.* (1962), Eugster *et al.* (1967a, b), Marti (1967), Wieler *et al.* (1992), Wieler (1994), and Busemann *et al.* (2000), uncertainties are 1 standard deviation of the individual ratios ($\pm 1\sigma_i$).

which presumably represents the Sun's Xe inventory, only in significantly lower relative abundances of the two heaviest isotopes, by 37 ‰ and 78 ‰ for the ^{134}Xe/^{130}Xe and ^{136}Xe/^{130}Xe ratios respectively as seen in the Table I data. However the Xe composition supplied by extra-solar grains incorporated into the planetesimal cannot be pure U-Xe. The upper panel of Figure 4 shows average ^{M}Xe/^{130}Xe isotope ratios in carbonaceous meteorites (AVCC-Xe) in units of their permil deviations from the corresponding U-Xe ratios. With a choice for the initial Xe crossover mass of m_c(Xe) = 250 amu, hydrodynamic escape fractionation ("HEF") of U-Xe in atmospheric blowoff reproduces the light-isotope AVCC ratios reasonably well, but a continuation of this fractionation line falls below the heavy isotope ratios, strikingly so for the two heaviest isotopes. An additional component must be added to U-Xe at these masses to match the observed composition. Among

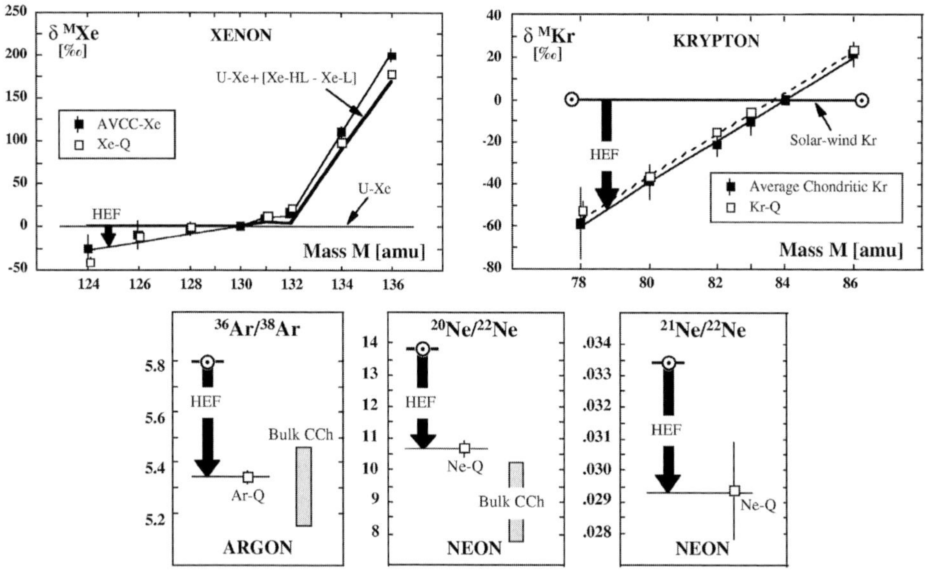

Figure 4. Comparisons of meteoritic Xe, Kr, Ar and Ne isotope ratios with those generated by hydrodynamic escape fractionation (HEF) in the planetesimal blowoff model described in the text. All meteoritic and calculated Xe and Kr ratios are plotted in units of permil deviations from the corresponding U-Xe ratios (*upper left panel*) or SW-Kr ratios (*upper right panel*). Meteoritic data from Table I for Xe and Kr, and from references cited in the text for Ar and Ne. The "Bulk CCh" rectangles in two lower panels represent the spread in Ar and Ne ratios measured in unseparated carbonaceous chondrites by Mazor *et al.* (1970).

other possibilities, the heavy end of the Xe-HL spectrum (identified in Table I and Figures 4 and 5 as Xe-HL - Xe-L) – an extrasolar "exotic" component found in interstellar nanodiamonds (Huss and Lewis, 1994; Ott, 2002) – serves very well (Pepin *et al.*, 1995).

It was originally thought (Pepin, 1991) that the Xe-Q composition reflected a superposition of escape-fractionated U-Xe adsorbed on grain surfaces with an extrasolar heavy-isotope component sited in the grain interiors; in this view, the adsorbed Xe by itself would have the composition of lightly fractionated U-Xe. But it now seems clear that Xe carefully extracted by chemical etching techniques from just the surfaces of carrier grain, unmixed with grain interior gases, already displays the heavy isotope enhancements (e.g., Wieler, 1994; Busemann *et al.*, 2000) – i.e., these enrichments are intrinsic to surface-sited Xe-Q, and not merely a contaminant released by over-etching into carrier grains or otherwise an artifact of the experimental protocol. Therefore, in the context of the planetesimal blowoff model, the Xe-Q parental composition carried into these planetesimals is now best interpreted as a gas-phase mixture of the U-Xe and heavy-isotope components that was subsequently trapped in icy grain mantles. There is evidence, discussed in the following section, that other solar-system volatile reservoirs can be characterized as mixtures, with variable mixing ratios, of these same two Xe components.

The heavy curve in the upper panel of Figure 4 represents U-Xe enriched at the heavy isotopes (by ≈17% at ^{136}Xe) by mixing with Xe-HL - Xe-L. The isotope ratios generated by HEF of this initial composition (*light curve*), with m_c(Xe) chosen to be 250 amu, is seen to be in good accord with measured AVCC-Xe ratios (*filled symbols*), excluding ^{129}Xe which is "polluted" by variable radiogenic contributions. The open symbols represent Busemann *et al.*'s (2000) Xe-Q ratios derived from etching measurements on de-mineralized residues of 8 chondritic meteorites. They fall within the error bars of AVCC-Xe except at the two heaviest isotopes, where their lower values suggest that the actual Xe-Q progenitor contains somewhat less of the heavy-isotope component – this would be understandable since the AVCC-Xe results are from bulk meteorites, and thus would contain Xe components sited in carrier grain interiors. Note, however, that uncertainties in the isotopic data reported by Busemann *et al.* are very small (Table I), and that their light-isotope ratios scatter around the fractionation curve but do not lie on it within the stated errors.

3.2. Krypton

Assuming HEF of an initial composition identical to SW-Kr (Table I), the m_c(Xe) = 250 amu crossover mass – the equivalent m_c(Kr) crossover mass is ≈213 amu, taking into account the differing b values for the two species – generates the very close match seen in the middle panel of Figure 4 to the average chondritic Kr ratios listed and referenced in Table I. There is, however, an unexplained oddity in the isotopic pattern when Busemann *et al.*'s (2000) Kr-Q ratios are plotted by themselves (*open symbols*). All five ratios lie above the HEF fractionation line and are fit almost exactly by the dashed parallel line offset upward from it by 4 ‰, suggesting a lower base isotope (^{84}Kr) abundance by this amount. A choice for m_c(Xe) somewhat below 250 amu rotates the Kr fractionation line clockwise around ^{84}Kr, improving the fit to the Busemann *et al.* $^{78-83}$Kr/^{84}Kr ratios (but not within their errors) while creating a discrepancy at ^{86}Kr/^{84}Kr. The linear array of the open symbols in Figure 4 points strongly to generation of Kr-Q from SW-Kr by mass fractionation, but there appears to be a devil in the details. Perhaps the uncertainties, all < 3 ‰, in the SW-Kr composition estimated by Pepin *et al.* (1995) are too small. New data from direct collection of SW species by the Genesis mission and further experimental studies of the Kr-Q composition should clarify the issue.

3.3. Argon

As shown in a lower panel of Figure 4, escape fractionation – again with m_c(Xe) = 250 amu (m_c(Ar) ≈168 amu) – of a pre-blowoff planetesimal atmosphere containing Ar with ^{36}Ar/^{38}Ar = 5.80 replicates almost exactly the Q-^{36}Ar/^{38}Ar ratio of 5.34 ± 0.02 reported by Busemann *et al.* (2000). This initial ratio is assumed to be equal to the solar wind value obtained from experimental studies of lunar regolith grains by Pepin *et al.* (1999) and Palma *et al.* (2002). The isotopic fractionation factor with m_c(Xe) = 250 amu is fixed at 0.921, so this simple escape model generates the Q-ratio within its present uncertainty only for an initial ratio in the narrow range 5.78 – 5.82.

3.4. NEON

The need for a second free parameter comes into play for the Ne isotopes. If the escape process continues to very low values of the methane escape flux F_1, then from Equation 2 the crossover masses decline correspondingly, and m_c(Ne) eventually sweeps across the masses of the Ne isotopes. This generates a residual ^{20}Ne/^{22}Ne ratio of $\approx$9.4, overfractionated by $\approx 7-12\%$ relative to the observed Ne-Q range of $\approx$10.1 to 10.7 (Busemann *et al.*, 2000). However if F_1 drops abruptly while m_c(Ne) is still above the Ne masses, the value of m_c(Ne) at this terminal stage of evolution may be chosen such that HEF of an initial solar ratio of 13.84 (Palma *et al.*, 2002) exactly matches either end of the Q-^{20}Ne/^{22}Ne range. The fit to Q-^{20}Ne/^{22}Ne $=$ 10.66 shown in a lower panel of Figure 4 requires a terminal m_c(Ne) $\approx$ 23 amu, just above the heaviest Ne isotope (the corresponding final crossover masses for Ar, Kr and Xe all lie well below their isotopic masses, so the residual isotopic distributions of these heavier gases plotted in Figure 4 have already been fixed in the planetesimal atmosphere and are unaffected). Note that while a final Ne crossover mass may be chosen to exactly match Q-^{20}Ne/^{22}Ne, it does not necessarily generate a compatible value of ^{21}Ne/^{22}Ne. The fact that the corresponding fractionation of this ratio from a solar-wind value of 0.0334 (Palma *et al.*, 2002) replicates the current estimate for Ne-Q of 0.0294 (Busemann *et al.*, 2000), as shown in Figure 4, points strongly to derivation of Ne-Q from solar Ne by escape fractionation.

This apparent need to throttle escape at a particular value of m_c(Ne) is an unsatisfying aspect of the present model, although one can imagine circumstances that might have constrained the F_1 outflow flux in the terminal stages of atmospheric loss – one such possibility is a decreasing relative abundance of methane (the atmospheric methane/argon ratio, for example, declines by a factor of $\approx$5 through the escape episode) to the point where it must diffuse through heavier species to escape, resulting in a limited F_1 compared to the free escape flux. Alternatively, it could be that the blowoff episode was not truncated in this way, and that fully escape-fractionated Ne with ^{20}Ne/^{22}Ne $\approx$ 9.4 (and ^{21}Ne/^{22}Ne $\approx$ 0.0281) was adsorbed on Q-carrier grains but was variably polluted by solar Ne before their incorporation into meteorites. Given the strong elemental fractionation of the Q-gases (Figure 1), addition of a solar gas component in the amount needed to elevate these Ne ratios into the Ne-Q range would not contribute significantly to Ar, Kr or Xe. A two-component Ne system of this kind would be one way to account for the observed variability of Ne-Q ratios compared to the isotopic uniformity displayed by the heavier Q-gases (Busemann *et al.*, 2000).

3.5. Pros and Cons of the Model

Planetesimal blowoff does a generally good job in generating Q-gas isotopic distributions from a solar-like initial composition augmented by additions to the heavy isotopes of Xe; as noted in the next section, there is isotopic evidence for the presence of this latter component, in variable mixing ratios with U-Xe, in other volatile systems. A fixed value of a single modeling parameter fractionates the parental Xe, Kr and Ar isotope ratios into reasonably close agreement with the corresponding meteorite ratios; this value was chosen above to be $m_c(\text{Xe}) = 250$ amu, but a range from $\approx$200 to 350 amu is allowed within the data uncertainties. A second parameter appears to be required to match observed Ne-Q isotopes; without it, Ne is overfractionated. It is, however, possible that overfractionated Ne is later contaminated with solar Ne to generate the isotopically lighter and variable Q-^{20}Ne/^{22}Ne ratios in meteorites, and so the need for a second modeling parameter is uncertain. Whatever the details for the evolution of Ne-Q may have been, a central characteristic of the blowoff model, compared to the possible fractionation mechanisms in nebular dissipation discussed in Section 2, is the exponential decline of fluxes and crossover masses during the escape episode that accounts in a natural way for the much larger fractionations of the meteoritic Ne isotopes relative to those displayed by Xe (Figure 1).

Noble gas elemental ratios are only modestly fractionated relative to solar ratios in the blowoff. About 80% of the original atmospheric Xe inventory survives, and Ne/Xe is depleted by a factor of only $\approx$7. This means that the lower panel of Figure 1 is, to a good approximation, a plot of relative elemental fractionations required for the presumably adsorptive process by which Q-gas carrier grains on planetesimal surfaces were loaded from the residual planetesimal atmosphere. (Note that the Ne/Ar fractionation implied by Figure 1 may well be too large; it assumes an initially solar Ne/Ar ratio, but gases carried by icy grain mantles would probably have been deficient in Ne unless trapping and subsequent ambient temperatures were very low (e.g., Owen and Bar-Nun, 1995). The relative Henry constants needed to replicate these elemental distributions fall in the range of experimentally measured values (Pepin, 1991).

To this point the model appears quantitatively promising. However it must also yield absolute as well as relative Q-gas abundances in the proper (i.e., observed) range. Adequate population of Q-gas carrier grains by adsorption requires substantial noble gas partial pressures in the residual planetesimal atmosphere. It can be shown, utilizing a polytrope approximation due to Gross (1974) and Hunten (1979), that the necessary post-blowoff pressures could probably be attained only on bodies at least several time the mass of Ceres (see Pepin, 1991) – but not so large (e.g., approximately lunar size) that timescales for atmospheric loss become unacceptably long. Escape fractionation of degassed atmospheres can certainly occur on smaller objects in the $\approx$ 100 – 500 km diameter range of the larger contemporary asteroids, but here blowoff times are extremely short and essentially nothing is retained. Therefore an additional and central stipulation of the model is that dust grains carrying fractionated Q-type gases must be ejected from relatively large objects that have since vanished from the asteroid belt. Wide dispersal of these ejected carriers throughout the regions of meteorite parent body formation, on timescales consistent with extinct radionuclide constraints on formation ages, is needed to account for the striking uniformity of Q-component isotope and elemental ratios observed in various meteorite classes. Since blowoff occurred in all likelihood on many such primitive bodies, and escape episodes on individual planetesimals would be expected to generate isotopic distributions both over- and under-fractionated with respect to Q-gas compositions, the carrier dust from separate sources must also be well mixed during dispersal.

It is not at all clear that these non-isotopic modeling requirements can readily be met. However, the need for regolith ejection from the postulated super-Ceres planetesimals and the subsequent disappearance of such bodies from the current asteroid population may not necessarily be serious problems. Simulations of early growth in the asteroid belt point to runaway accretion of planet-size bodies that were later removed together with smaller objects by scattering into giant planet resonances, with concurrent perturbation of all asteroidal materials into orbits of increasing eccentricity and inclination (Chambers and Wetherill, 2001, and references therein). Collisions in an early high-mass asteroid belt with these dynamical characteristics would presumably be frequent and energetic, and efficient impact ejection of regolith to space, even from relatively massive (>Ceres) bodies prior to their removal from the belt, seems plausible. The principal problem here, however, is mixing of the Q-carrier grains in the nebula with large amounts of chemically similar dust released from planetesimals that were too small to have generated Q-gas isotopic compositions and elemental abundances by the blowoff-adsorption mechanisms. The fact that characteristic Q-gas signatures are still prominent in the carbonaceous phases of meteorites despite such likely dilution of their carriers is a serious challenge to the blowoff model.

All things considered, fractionation of a dissipating solar nebula would in principle be preferable to planetesimal blowoff as a means for generating Q-component distributions. The process is conceptually and operationally straightforward. All carrier grains would be loaded from ambient nebular gases, so the dilution problem disappears. On the other hand, one notes the isotopic success of the blowoff model and the fact that no plausible way has been found to implement whole-nebula fractionation to the extents observed in the Q-gases. Moreover the need for substantial noble gas partial pressures is common to both models if adsorption is governed by experimentally measured Henry constants – a requirement that is difficult to meet in the open nebula (Pepin, 1991) unless transient high pressures are created, for example by shock (C. KenKnight, personal communication). Given the difficulties encountered by both of the modeling approaches discussed above, there is presently no fully satisfactory model for derivation of the Q-gases from solar-like progenitors.

4. The Xenon Heavy-Isotope Component

The initial Xe composition in the planetesimal atmospheric blowoff model was taken to be a mixture of U-Xe and the heavy-isotope end of the Xe-HL spectrum, there represented as Xe-HL - Xe-L and hereafter called Xe-H. The extrasolar "exotic" Xe-HL component is itself a superposition of products generated by different nuclear processes: p-process for the two lightest isotopes, r-process for the two heaviest, and a mixture of nucleosynthetic products for the intermediate nuclei. Given their origin in distinct nucleosynthetic environments, the individual p- and r-process contributions to Xe-HL could in principle be sited in different carrier grains, but much laboratory effort has so far failed to provide convincing evidence for their separability, at least in the extrasolar diamond carriers found in meteorites. Mixing of U-Xe with Xe-H assumes that at least the heavy end of Xe-HL was present, without its lighter constituents, in volatile reservoirs supplying Xe to the early solar system.

Volatile systems other than the meteoritic Q-gases discussed above may also be viewed as mixtures of U-Xe and Xe-H; here Xe-H is taken to be that part of the Xe-HL spectrum above ^{130}Xe, although just the two heaviest r-process isotopes (H*-Xe in Pepin *et al.*, 1995) would serve about as well. Mixing fits to an inferred Xe composition of the "normal" (N-Xe) component in presolar silicon carbide grains, and to measured solar-wind Xe isotope ratios, are illustrated in the upper and lower panels of Figure 5; the middle panel, showing the match of AVCC-Xe to a slightly fractionated parental U-H mixture as discussed in the previous section, is taken from Figure 4.

Xenon measurements on SiC grains define a series of isotopic correlation lines, anchored on one end by s-process Xe ("Xe-G") synthesized in the He-burning shells of AGB stars and, somewhere along the correlations on the other side of the data field, by a "normal" component thought to characterize the AGB stellar envelopes (Lewis *et al.*, 1994; Ott and Begemann, 2000). Pepin *et al.* (1995) found that a particular location on the correlation lines yielded the N-Xe composition shown in the upper panel of Figure 5. Within the substantial uncertainties associated with the correlations at this remove from the data (about twice the width of the data field), the isotopic pattern is reasonably consistent with a representation of N-Xe as a U-Xe - Xe-H mixture, with ^{136}Xe/^{130}Xe enriched by ≈46% relative to the U-Xe ratio. (Note that the added component cannot contain all of the Xe-HL isotopes: inclusion of the full Xe-HL spectrum (Table I) seriously perturbs the N-Xe light-isotope ratios as indicated by the light "U-Xe + Xe-HL" curve in Figure 5). A closely similar pattern is evident in the bottom panel for Xe in the solar wind, but here the elevation of ^{136}Xe is much less, ≈8.5%. The meteoritic parental Xe composition (*middle panel*) has an intermediate ^{136}Xe enrichment of ≈17%. Primordial terrestrial Xe, in contrast, has no Xe-H component at all (Pepin, 2000).

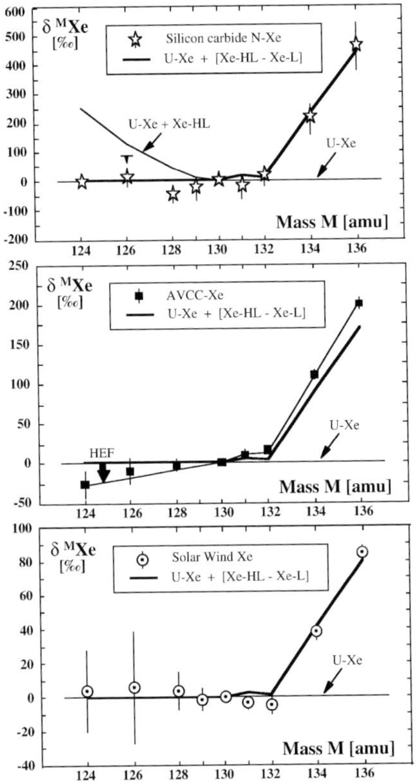

Figure 5. Fits of Xe isotope ratios, inferred for SiC N-Xe (*upper panel*) and measured for AVCC-Xe and SW-Xe (*middle and lower panels*), to the U-Xe + [Xe-HL - Xe-L] = U-Xe + Xe-H mixtures indicated by the solid curves, or, in the case of AVCC-Xe (*middle panel*), to a slightly fractionated parental U + H mixture as discussed in Section 3 of the text. All isotope ratios ^{M}Xe/^{130}Xe are plotted in units of per-mil deviations from the corresponding U-Xe ratios; data from Table I. The light curve labeled U-Xe + Xe-HL in the upper panel shows the light-isotope N-component distribution produced by adding the full Xe-HL isotopic distribution to U-Xe (*see text*). The bar with downward arrow above ^{126}Xe represents the ^{126}Xe/^{130}Xe ratio in N-Xe that would result from maximum spallation correction of the SiC ^{126}Xe abundances; the position of the plotted data point corresponds to a uniform spallogenic enrichment of the SiC samples (see Ott and Begemann, 2000).

Excess ^{136}Xe abundances associated with Xe-H contributions to these reservoirs, in percentages of the amounts of U-^{136}Xe present in the mixtures, are plotted in Figure 6. An uncertain estimate for the Jovian atmosphere, derived from Galileo probe data reported by Mahaffy *et al.* (2000), is also shown. The five cases are ordered along the abscissa only by decreasing H/U mixing ratio. An interesting question, posed along the abscissa, is whether this ordering could imply a temporal sequence of volatile incorporation from a solar nebula in which the H/U ratio decreased with time. In this view the Xe-H-rich N-component in SiC would perhaps represent early infalling material from the parent molecular cloud. Presolar grains later incorporated into first-generation planetesimals and then, after relatively gentle processing, into primitive carbonaceous chondrites carry N-Xe diluted by additional U-Xe (since we are dealing here with a grain carrier rather than the ambient gas phase, the position of the carbonaceous chondrites in Figure 6 does not imply that the meteorites themselves formed before the Sun, only that these grains sampled an earlier nebular composition). The early nebula would have to be rapidly diluted by infall of U-Xe-rich material in order for the sun to acquire a significantly lower H/U ratio. Thereafter, during the planet-building stage of disk evolution, the nebular gas supply becomes even more dominated by U-Xe, to the point where a Xe-H component is not visible in Earth's initial noble gas inventory.

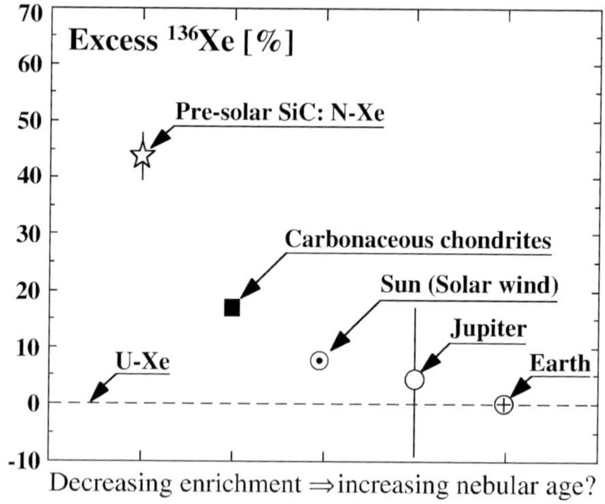

Figure 6. Calculated mixing ratios, in %, of excess Xe-H to U-Xe at ^{136}Xe in five volatile reservoirs. Approximate value for Jupiter derived from Galileo probe data (Mahaffy *et al.*, 2000). The five cases are ordered along the abscissa in descending mixing ratio. See text for discussion.

The Venusian Xe composition is unknown, but in such a sequence one would predict compositionally similar primordial Xe on Venus and Earth since two planets as alike in mass and location would be expected to acquire volatiles from similar sources at approximately the same time. The Jovian Xe-H/U-Xe mixing ratio is important in this context, but, as seen in Figure 6, estimates from Galileo probe data are too imprecise to place Jupiter accurately enough in a possible temporal sequence – if that is, in fact, what Figure 6 implies. It should be noted that the error bars for the Jovian mixing ratio are probably overly conservative; examination of the Galileo probe results for the most abundant isotopes suggests that Xe on Jupiter is most likely to lie somewhere in the SW-Xe to U-Xe compositional range (Pepin and Porcelli, 2002).

An evaluation of Figure 6 as a temporal compositional sequence may be made in the context of models developed by Boss and coworkers for presolar molecular cloud collapse induced by shock from a neighboring supernova or AGB star, with concurrent injection of shock-wave material into the outer regions of the cloud (Boss and Vanhala, 2000, 2001, and references therein). Collapse of the "contaminated" outer cloud to the solar accretion disk lags the arrival of matter from the inner part of the cloud, and so one would expect the nebula to become progressively more enriched over time with shock-wave debris carrying compositional signatures of the triggering star (Boss and Vanhala, 2000). Applied to Figure 6, this argument implies injection of dust grains loaded with U-Xe – somewhat surprising if collapse was triggered by a supernova, where shock front matter would seem more likely to carry the r-process Xe-H component, but perhaps not if an AGB star was responsible; as noted above, a component resembling U-Xe appears to be present in AGB envelopes. Figure 6 suggests that after most of the collapsing matter had been sequestered in the young sun, Xe composition in the residual nebula during planet-building was dominated by late infall of injected U-Xe-rich material. Accurate Xe isotopic data from Venus and Jupiter are clearly needed to test this hypothesis.

Acknowledgements

This research was supported by Grants NAG5-7094 and NAG5-11723 from the NASA Cosmochemistry Program. I thank Igor Tolstikhin for a helpful review.

References

Basford, J.R., Dragon J.C., Pepin R.O, Coscio M.R., Jr., and Murthy V.R.: 1973, 'Krypton and Xenon in Lunar Fines', *Proc. Lunar Sci. Conf.* **4**, 1915–1955.

Boss, A.P., and Vanhala, H.A.T.: 2000, 'Triggering Protostellar Collapse, Injection, and Disk Formation', *Space Sci. Rev.* **92**, 13–22.

Boss, A.P., and Vanhala, H.A.T.: 2001, 'Injection of Newly Synthesized Elements Into the Protosolar Cloud', *Phil. Trans. R. Soc. Lond.* **A359**, 2005–2017.

Busemann, H., Baur, H., and Wieler, R.: 2000, 'Primordial Noble Gases in "Phase Q" in Carbonaceous and Ordinary Chondrites Studied by Closed-system Stepped Etching', *Met. Planet. Sci.* **35**, 949–973.

Cameron, A.G.W.: 1995, 'The First Ten Million Years in the Solar Nebula', *Meteoritics* **30**, 133–161.

Chambers, J.E., and Wetherill, G.W.: 2001, 'Planets in the Asteroid Belt', *Met. Planet. Sci.* **36**, 381–399.

Eugster, O., Eberhardt, P., and Geiss, J.: 1967a, 'The Isotopic Composition of Krypton in Unequilibrated and Gas-rich Chondrites', *Earth Planet. Sci. Lett.* **2**, 385–393.

Eugster, O., Eberhardt, P., and Geiss, J.: 1967b, 'Krypton and Xenon Isotopic Composition in Three Carbonaceous Chondrites', *Earth Planet. Sci. Lett.* **3**, 249–257.

Gross, S.H.: 1974, 'The Atmospheres of Titan and the Galilean Satellites', *J. Atmos. Sci.* **31**, 1413–1420.

Holloway, J.R.: 1988a, 'Planetary Atmospheres during Accretion: the Effect of C-O-H-S Equilibria', *Lunar Planet. Sci.* **XIX**, 499–500.

Holloway, J.R.: 1988b, 'Distribution of H_2O between Early Atmospheres and Magma Oceans', *EOS: Trans. Am. Geophys. Union* **69**, 338.

Hunten, D.M.: 1979, 'Capture of Phobos and Deimos by Protoatmospheric Drag', *Icarus* **37**, 113–123.

Hunten, D.M., Pepin, R.O., and Walker, J.C.G.: 1987, 'Mass Fractionation in Hydrodynamic Escape', *Icarus* **69**, 532–549.

Huss, G.R., and Lewis, R.S.: 1994, 'Noble Gases in Presolar Diamonds I: Three Distinct Components and their Implications for Diamond Origins', *Meteoritics* **29**, 791–810.

Johnstone, D., Hollenbach, D., and Bally, J.: 1998, 'Photoevaporation of Disks and Clumps by Nearby Massive Stars: Application to Disk Destruction in the Orion Nebula', *Astrophys. J.* **499**, 758–776.

Krummenacher, D., Merrihue, C.M., Pepin, R.O., and Reynolds, J.H.: 1962, 'Meteoritic Krypton and Barium Versus the General Isotopic Anomalies in Xenon', *Geochim. Cosmochim. Acta* **26**, 231–249.

Lewis, R.S., Amari, S., and Anders, E.: 1994, 'Interstellar Grains in Meteorites: II. SiC and its Noble Gases', *Geochim. Cosmochim. Acta* **58**, 471–494.

Mahaffy, P.R., Niemann, H.B., Alpert, A., Atreya, S.K., Demick, J., Donahue, T.M., Harpold, D.N., and Owen, T.C.: 2000, 'Noble Gas Abundance and Isotope Ratios in the Atmosphere of Jupiter from the Galileo Probe Mass Spectrometer', *J. Geophys. Res.* **105**, 15061–15071.

Marti, K.: 1967, 'Isotopic Composition of Trapped Krypton and Xenon in Chondrites', *Earth Planet. Sci. Lett.* **3**, 243–248.

Mazor, E., Heymann, D., and Anders, E.: 1970, 'Noble Gases in Carbonaceous Chondrites', *Geochim. Cosmochim. Acta* **34**, 781–824.

Morfill, G.E., Tscharnuter, W., and Völk, H.J.: 1985, 'Dynamical and Chemical Evolution of the Protoplanetary Nebula', *in* D.C. Black and M.S. Matthews (eds.), *Protostars and Planets II*, University of Arizona Press, Tucson, pp. 493–533.

Ott, U.: 2002, 'Noble Gases in Meteorites – Trapped Components', *in* D. Porcelli, C.J. Ballentine, and R. Wieler (eds.), *Noble Gases in Geochemistry and Cosmochemistry, Rev. Mineral. Geochem.* **47**, 71–100.

Ott, U., and Begemann, F.: 2000, 'Spallation Recoil and Age of Presolar Grains in Meteorites', *Met. Planet. Sci.* **35**, 53–63.

Owen, T., and Bar-Nun, A.: 1995, 'Comets, Impacts and Atmospheres', *Icarus* **116**, 215–226.

Ozima, M., Wieler, R., Marty, B., and Podosek, F.A.: 1998, 'Comparative Studies of Solar, Q-gases and Terrestrial Noble Gases, and Implications on the Evolution of the Solar Nebula', *Geochim. Cosmochim. Acta* **62**, 301–3124.

Palma, R.L., Becker, R.H., Pepin, R.O., and Schlutter, D.J.: 2002, 'Irradiation Records in Regolith Materials, II: Solar-wind and Solar-energetic-particle Components in Helium, Neon, and Argon Extracted from Single Lunar Mineral Grains and From the Kapoeta Howardite by Stepwise Pulse-heating', *Geochim. Cosmochim. Acta* **66**, 2929–2958.

Pepin, R.O.: 1991, 'On the Origin and Early Evolution of Terrestrial Planet Atmospheres and Meteoritic Volatiles', *Icarus* **92**, 2–79.

Pepin, R.O.: 1997, 'Evolution of Earth's Noble Gases: Consequences of Assuming Hydrodynamic Loss Driven by Giant Impact', *Icarus* **126**, 148–156.

Pepin, R.O.: 2000, 'On the Isotopic Composition of Primordial Xenon in Terrestrial Planet Atmospheres', *Space Sci. Rev.* **92**, 371–395.

Pepin, R., and Porcelli, D.: 2002, 'Origin of Noble Gases in the Terrestrial Planets', *in* D. Porcelli, C.J. Ballentine, and R. Wieler (eds.), *Noble Gases in Geochemistry and Cosmochemistry, Rev. Mineral. Geochem.* **47**, 191–246.

Pepin, R.O., Becker, R.H., and Rider, P.E.: 1995, 'Xenon and Krypton Isotopes in Extraterrestrial Regolith Soils and in the Solar Wind', *Geochim. Cosmochim. Acta* **59**, 4997–5022.

Pepin, R.O., Becker, R.H., and Schlutter, D.J.: 1999, 'Irradiation Records in Regolith Materials, I: Isotopic Compositions of Solar-wind Neon and Argon in Single Lunar Mineral Grains', *Geochim. Cosmochim. Acta* **63**, 2145–2162.

Tielens, A.G.G.M., and Allamandola, L.J.: 1987, 'Evolution of Interstellar Dust', *in* G.E. Morfill and M. Scholer (eds.), *Physical Processes in Interstellar Clouds*, D. Reidel, Dordrecht, pp. 333–376.

Wieler, R.: 1994, '"Q-gases" as "Local" Primordial Noble Gas Component in Primitive Meteorites', *in* J. Matsuda (ed.), *Noble Gas Geochemistry and Cosmochemistry*, Terra Scientific Publishing, Tokyo, pp. 31–41.

Wieler, R., and Baur, H.: 1994, 'Krypton and Xenon from the Solar Wind and Solar Energetic Particles in two Lunar Ilmenites of Different Antiquity', *Meteoritics* **29**, 570–580.

Wieler, R., Anders, E., Baur, H., Lewis, R.S., and Signer, P.: 1992, 'Characterization of Q-gases and Other Noble Gas Components in the Murchison Meteorite', *Geochim. Cosmochim. Acta* **56**, 2907–2921.

Address for Offprints: R.O. Pepin, School of Physics and Astronomy, 148 Tate Lab. of Physics, University of Minnesota, 116 Church Street SE, Minneapolis, MN 55455, USA; pepin001@tc.umn.edu

ELEMENTAL AND ISOTOPIC ABUNDANCES OF CARBON AND NITROGEN IN METEORITES

MONICA M.GRADY[1] AND IAN P. WRIGHT[2]

[1] *Dept. of Mineralogy, The Natural History Museum, Cromwell Road, London SW7 5BD, UK*

PSSRI, The Open University, Walton Hall, Milton Keynes, MK7 6AA, UK

Received: 9 April 2002; Accepted in final form: 15 October 2002

Abstract. We have taken an inventory of the elemental and isotopic abundances of major carbon- and nitrogen-bearing components in different groups of meteorites. Primary phases, inherited from the solar nebula, are frequently isotopically heterogeneous, and surprisingly resistant to modification through parent body processing. Even melted and recrystallised meteorites retain primordial carbon and nitrogen isotopic signatures.

1. Introduction

Carbon and nitrogen are the fourth and sixth, respectively, most abundant elements in the Solar System. These elements, either together or separately, form a wide variety of different compounds which have a range of characteristics, in terms of their oxidation states (from highly oxidised to highly reduced). Organic species are ubiquitous in the interstellar medium and interstellar clouds; graphite has also been recognised in interstellar space. Carbon-bearing phases such as these, either independently or when combined into mantles on silicate grains, presumably formed the feedstock of dust from which the Solar System ultimately aggregated. Examination of the distribution of carbon and nitrogen in meteorites allows us to trace the evolution of a number of planet-forming processes (aggregation, melting, differentiation, crystallisation and brecciation). In order to follow this evolutionary pathway, we must be able to distinguish between original, primary species, and the products of secondary and tertiary parent body processing. Species from different sources may be recognised by characteristic isotopic compositions, or specific modes of occurrence.

1.1. Measurement Techniques

Early measurements of carbon and nitrogen in meteorites relied on a single-stage combustion to release species for measurement (e.g. Moore and Lewis, 1965; Moore and Gibson, 1969). Unfortunately, however, for all but the most volatile-rich meteorites, this technique leads to an over-estimate of carbon and nitrogen abundances, because no distinction is made between indigenous and (terrestrial)

Space Science Reviews **106:** 231–248, 2003.

contaminant material. Reliable carbon and nitrogen abundances and isotopic compositions are now regularly determined either by multi-step combustion, which allows deconvolution of contaminants from meteoritic species (e.g. Swart *et al.*, 1983), or by *in situ* measurement by ion microprobe-mass spectrometry (Zinner *et al.*, 1987). Organic species are determined using chromatographic methods (gas chromatography-mass spectrometry, pyrolysis techniques, compound-specific isotope analysis, etc.). The determination of the carbon (and oxygen) isotopic composition of carbonates is generally arrived at using acid dissolution (e.g., McCrea, 1950). Nitrogen, which exists at the Earth's surface mainly as the permanent gas N_2, is often released from samples by stepped pyrolysis, increasingly as an adjunct to analyses of noble gases (e.g., Frick and Pepin, 1981).

1.2. Primary Species

Primary carbon- and nitrogen-bearing species in meteorites are materials inherited directly from the molecular cloud that collapsed to form the solar nebula. Such species include organic molecules, amorphous/graphitic carbon, nanodiamonds, SiC, Si_3N_4, etc.). The isotopic composition of primary carbon and nitrogen reflects the original source reservoirs; if several different components are present in a meteorite, they tend to possess different isotopic compositions mirroring the diversity of their origins. However, hydrous and thermal processing on asteroidal parent bodies may lead to modification of the primary species, e.g., the graphitisation of amorphous carbon during metamorphism. These processes may or may not alter primary isotopic compositions.

1.3. Secondary Species

Secondary carbon- and nitrogen-bearing species are entities formed on meteorite parent bodies, e.g., the formation of carbonates by precipitation from oxidising aqueous fluids containing dissolved carbon. The distinction between a modified primary component and a secondary material may not always be clear, especially in processed meteorites. Whilst the abundance and isotopic composition of secondary carbon and nitrogen are related to the initial inventory of species available for modification, the details of the relationship depend on whether the secondary processing took place in open or closed systems. Equilibrium isotope exchange followed by partial or total removal of one particular constituent may have resulted in isotopic effects that are not simple to unravel.

1.4. Tertiary species

Tertiary carbon and nitrogen species are introduced into meteorites during exposure to cosmic radiation, either during residence at the parent body surface, or during transit in interplanetary space. Cosmogenic (or spallation) carbon and nitrogen are produced in stony meteorites by the interaction of cosmic rays with silicate minerals. The abundance of cosmogenic carbon and nitrogen is related to the cosmic ray exposure age of the meteorite, but is generally very low, at the ppb level. However, the production of carbon and nitrogen with $^{12}C/^{13}C$ or $^{14}N/^{15}N$ ratios of approximately unity impart distinctive isotopic signatures that, if care is not taken, can overwhelm primary or secondary species during analysis by combustion or pyrolysis. It is fortunate that cosmogenic species are not released until experimental temperatures exceed $\approx$1200°C, temperatures above those at which most other species combust. Thus, especially if the length of time for which the specimen has been exposed to cosmic rays (the cosmic ray exposure [CRE] age) is known, it is relatively straightforward to correct for contributions from cosmogenic sources.

The non-asteroidal, or 'planetary' meteorites from the Moon and Mars also exhibit tertiary carbon and nitrogen species. Lunar meteorites contain material implanted from the solar wind, whilst impact-generated glassy clasts in Martian meteorites are hosts of gases trapped from the martian atmosphere.

1.5. Contamination and Terrestrial Weathering

In addition to the above, carbon and nitrogen studies of meteorites and meteoritic components are dogged by the possibility of contamination on Earth. For carbon, this is manifested by the presence of extraneous organic materials, mainly biogenic components, including dust and pollen grains and microbes. These may be imparted to the samples during atmospheric entry flight, terrestrial residence and handling by loving curators in museum collections. Nitrogen contamination arises from adsorbed atmospheric gases and is readily removed by heating to 200°C in vacuum. In addition to contamination, meteorites also suffer the effects of weathering or alteration in the terrestrial environment, effects that include the addition of salts such as carbonates or nitrates. Fortunately, contaminant and weathering species can usually be distinguished by stepped combustion techniques, and do not interfere with ion microprobe analyses.

2. Carbon and Nitrogen in Chondritic Meteorites

2.1. Carbonaceous Chondrites

The very name 'carbonaceous' chondrite would seem to imply that these meteorites are rich in carbon. Unfortunately, however, it is a misnomer, since only three of the seven groups of carbonaceous chondrites (CCs) contain carbon at the weight percent level. The two sets of CCs, the carbon-rich (CI, CM and CR chondrites) and the carbon-poor (CO, CV and CK), are therefore considered separately. The seventh group, the CH chondrites, is considered, along with the unusual bencubbinites, as a minor group. One assemblage of carbon- and nitrogen-bearing components, minor in abundance (ppb to ppm levels) but significant in isotopic composition (both $\delta^{13}C$ and $\delta^{15}N$ varying from $\approx -980‰$ to $+12,000‰$; Zinner, 1998), that

TABLE I

Major Components of Carbon and Nitrogen in Meteorites

Group	Main Species	[C] (ppm)	$\delta^{13}C$ (‰)	[N] (ppm)	$\delta^{15}N$ (‰)	Ref.*
Chondrite:						
CI, CM, CR	Total	2–5 wt.%	−12 to −5	500–2000	+20 to +50	1
	Organics:					
	Soluble	40–100	> −5	1–10	> +20	2
	Insoluble	1–4 wt.%	< −10	500–2000	> +20	2
	Carbonate	500-5000	> +20			3
CO, CV, CK	Total	0.1–1.1wt.%	−25 to −8	5–200	−50 to +100	1
	Elemental	10–3000				4, 5
Ordinary	Total	300–6000	−24 to −12	1–25	−20 to +100	6, 7
	Organics:					
	Insoluble	700–4000	−20 to −10	5–50	−10 to +200	8
	Elemental		−65 to +25		−30 to +60	9
	Chondrules	100–3000				4
Enstatite	Graphite	1500–7000	−9 to +4	10–100	−30 to −10	10
	Carbides	20–60	< −15			10
	Nitrides			10–50	−60 to −30	10
CH, Bencubbin	Total	0.35–1.21	+0.3 to +6	20–280	+150 to +900	11–15
Kakangari-type	Total	200–500	−20	10–20	−20 to +10	16, 17
	Graphite	100	−9	6	−80	16, 17
	Carbides/metal	200	−20	3–15	−100 to +40	16, 17
Non-Chondrite:						
Ureilites (main group)	Graphite/diamond	2–7 wt.%	−10 to 0	3–30	−120	18, 19
Ureilites (polymict)		2–6 wt.%	−7 to −2	10	+600	18, 20
HED		10–30	−30 to −20	0.05–1.3	−50 to +20	21, 22
Primitive achondrites:						
Acapulcoites	Graphite	400	−20			16
	Metal or carbides	100	−24	13	< −150	16, 23
	Silicates			3	+15	23
Lodranites	Metal and silicate	40–100	−23	1–17	+2 to +14	24, 25
Winonaites	Graphite	2000–7000	−11 to −4			16
Irons	Graphite		−30 to +4	100–200	−50 to −20	26-28
	Carbides		< −15	< 10	−50 to −40	27
	Nitrides	40–70	−30 to +30		−100 to −50	27
	Taenite	10–50		1000–10000	−200 to +200	27,29,30
	Kamacite	1–10		100–1000		27, 30
Non-Asteroidal						
Martian	Magmatic	15–40	−30 to −20	1–4	−12 to +2	31, 32
	Carbonates	30–60	+10 to +55			33, 34
	Nitrates			0.02–1.25	−11 to −8	35
	Atmosphere		> +30		> +200	36 − 40

*References: 1. Kerridge (1985); 2. Cronin and Chang (1993); 3. Grady *et al.* (1988); 4. Hanon *et al.* (1998); 5. Varela *et al.* (2000); 6. Grady *et al.* (1989); 7. Hashizume and Sugiura (1995); 8. Alexander *et al.* (1998); 9. Mostefaoui *et al.* (2000); 10. Grady *et al.* (1986); 11. Grady and Pillinger (1990a); 12. Grady and Pillinger (1993); 13. Prombo and Clayton (1985); 14. Franchi *et al.* (1986); 15. Sugiura *et al.* (2000); 16. Grady and Pillinger (1986); 17. Weisberg *et al.* (1996); 18. Grady *et al.* (1985); Smith *et al.* (2001); 19. Russell *et al.* (1993); Yamamoto *et al.* (1998); 20. Grady and Pillinger (1988); 21. Grady *et al.* (1997a); 22. Miura and Sugiura (1993); 23. Kim and Marti (1993); 24. Grady *et al.* (1993); 25. Kim and Marti (1994); 26. Zipfel *et al.* (1997); 27. Sugiura (1998); 28. Deines and Wickman (1973, 1975); 29. Franchi *et al.* (1993); 30. Prombo and Clayton (1993); 31. Wright *et al.* (1990); 32. Grady *et al.* (1997b); 33. Wright *et al.* (1992); 34. Jull *et al.* (1995); 35. Grady *et al.* (1995); 36. Carr *et al.* (1985); 37. Becker and Pepin (1984); 38. Murty and Mohapatra (1997); 39. Grady *et al.* (1998); 40. Mathew *et al.* (1998).

might be considered as primary to the chondritic parents, is not considered here. Interstellar grains, including nanometre-sized diamonds, graphite, silicon carbide and silicon nitride, are present in both carbon-rich and carbon-poor CCs, in varying abundances. Discussion of their properties is contained in other papers in this volume, and so is not duplicated here.

2.1.1. *CI, CM and CR Chondrites*

The carbon-rich CI and CM chondrites contain 2–5 wt.% carbon and 500–2000 ppm nitrogen, with overall $\delta^{13}C$ generally around $\approx -5‰$ to $+12‰$ and $\delta^{15}N \approx +20‰$ to $+50‰$ (Kerridge, 1985). Over half of this material occurs as an insoluble organic macromolecular material with $\delta^{13}C < -10‰$ and $\delta^{15}N < +20‰$; the remaining carbon occurs mainly as soluble organics and carbonates. The insoluble macromolecular material can be approximated as a central core of aromatic and heterocyclic molecules bridged together by aliphatic chains. Associated with the macromolecule are less complex species that form side chains off the central core, along with free organic compounds. These soluble components ($\delta^{13}C > -5‰$; $\delta^{15}N > +20‰$) comprise a variety of species, including amino and carboxylic acids, alkanes, ketones, nitrogen heterocyclics etc. (e.g., summarised by Cronin and Chang, 1993). The origin of organic material in CI and CM chondrites is a subject of considerable interest — the traditional view was that it represented a late-stage, low temperature condensate, fully compatible with its formation in the solar nebula (e.g., Hayatsu and Anders, 1981). However, isotopic compositions (particularly its D/H ratio) mitigate against this theory. The current view is that the organic material is partly, if not wholly, primary to the chondritic parent bodies, having been derived directly from interstellar organic molecules (e.g., Alexander *et al.*, 1998); parent body processes (mainly aqueous) have altered the remainder.

The overall nitrogen chemistry of CR chondrites distinguishes them from CI and CM meteorites. All the nitrogen in CRs has elevated $\delta^{15}N$, reaching up to $+200‰$. The precise components in which this nitrogen is located are unknown: it evidently occurs in organic species, and it has been suggested that CR chondrites contain a higher proportion of interstellar organics than other CCs (Alexander *et al.*, 1998).

The asteroidal parents of CCs suffered aqueous alteration to various extents. One of the results of this alteration was the precipitation of carbonates, sometimes in veins, but more often as individual grains scattered throughout the chondritic matrix. Carbonates in CIs are mainly the magnesian and iron carbonates, dolomite, breunnerite and siderite, whilst CM chondrites are characterised by the calcium variety, calcite. Carbonate abundance is variable, between 500–5000 ppm; carbon isotopic composition is unusual, in that it is moderately enriched in ^{13}C, with $\delta^{13}C$ between $+20‰$ and $+80‰$ (Grady *et al.*, 1988). There does not seem to be any systematic relationship of either carbonate abundance or carbon isotopic composition with any measure of the extent to which the chondrites are altered. If the carbonates are produced by hydrous alteration processes on asteroidal parents, then the source of the carbon is presumably from material indigenous to the parent. Oxidation of fragments from the organic component could produce the feedstock for dissolved CO_2. However, the highest $\delta^{13}C$ value recorded for any organic species is $\approx +45‰$ for amino acids (Pizzarello *et al.*, 1991), a value insufficiently enriched in ^{13}C to be a viable source material. However, organics cannot be ruled out totally. If a significant fraction of the macromolecular material is interstellar, then there might be an isotopically heavier component within it that has not yet been identified. One other reservoir of ^{13}C in CI and CM chondrites is that of ^{13}C-enriched presolar graphite and silicon carbide grains. Although these grains are generally resistant to acid and solvent attack, they might be etched by the oxidising, alkaline fluids that are predicted to be in operation on CI/CM parents. If carbonate production is a local process, then it would only take dissolution of a few ppb of a highly ^{13}C-enriched graphite or silicon carbide grain ($\delta^{13}C \geq +1000‰$) to produce a fluid from which carbonates ($\delta^{13}C \leq +100‰$) would be precipitated.

There are several ungrouped carbon-rich CCs that have been identified as having an unusual carbon or nitrogen chemistry. The CO-CM transitional chondrite, Acfer 094 is one such meteorite (Newton *et al.*, 1993), as is the newly-fallen CI2, Tagish Lake (Grady *et al.*, 2002). By and large, these meteorites have the same general components as other carbon-rich CCs, but in higher abundance, or with slightly different isotopic compositions.

2.1.2. *CO, CV, and CK Chondrites*

In contrast to the volatile-rich CI, CM and CR chondrites, the CO, CV and CK meteorites are relatively carbon- and nitrogen-poor, with only 0.1–1.1 wt.% carbon and $\approx$5–200 ppm nitrogen (Kerridge, 1985). Volatile content decreases with increasing petrologic type, and carbon becomes more graphitic in nature: organic species are absent from these meteorites.

There have been few systematic studies of carbon or nitrogen in carbon-poor CCs: most efforts have focussed on study of the rich harvest of interstellar nanodiamonds present in Allende. However, there have also been reports of sub-ppm levels of fullerenes in Allende (Becker *et al.*, 1994, 1999; Harris *et al.*, 2000). Isotopically heterogeneous graphite associated with sulphide grains has been identified in Kainsaz, with $\delta^{13}C$ between $+16‰$ and $+22‰$, and $\delta^{15}N$ at $-13‰$ to $0‰$. As discussed below, this graphite is thought to result from shock-modification of interstellar organic material (Mostefaoui *et al.*, 2000).

There have been several studies of carbon in the silicate fraction of CV chondrites. Abreu and Brearley (2002) used high resolution TEM to image amorphous and poorly-graphitised carbon in voids within olivine grains. Using nuclear reaction analysis, Varela *et al.* (2000) found 10–2000 ppm carbon in glass inclusions in a suite of CV3s whilst Hanon *et al.* (1998) identified 400–3000 ppm carbon in chondrules from Allende by ion microprobe. None of these studies reported isotopic data, or nitrogen abundances for the component, which is assumed to be remnant precursor carbon.

2.2. Ordinary Chondrites

The largest class of meteorites, the ordinary chondrites (OCs), is a federation of three chemical groups, the H, L and LL chondrites, that formed in closely neighbouring localities within the solar nebula. It is, therefore, little surprise to find that the carbon and nitrogen chemistry of these meteorites is similar. Along with variations in composition, the OCs exhibit a textural sequence (petrologic type), resulting from gradual changes brought about by increasing levels of thermal metamorphism. Again, it is no surprise to find that volatile abundances decrease with increasing petrologic type. Thus, the most unaltered OCs, those of type 3 (unequilibrated ordinary chondrites, UOCs) have total carbon abundances $\approx$ 1000–6000 ppm, whilst type 6 are $<$ 300 ppm (Grady *et al.*, 1989). There is no regular variation in overall carbon isotopic composition with increasing metamorphic grade: $\delta^{13}C$ ranges between $\approx$ -24‰ and -12‰ across all grades, implying no systematic isotopic fractionation as carbon is removed by oxidation. Unlike the CCs, carbon in OCs is not particularly well-defined. Primary carbon seems to occur as organic species, or as poorly-graphitised elemental carbon. Carbon is also present as carbides (particularly in assemblages with magnetite), dissolved in metal, and might possibly occur as coatings on silicate grains. Carbon- and nitrogen-bearing interstellar grains are undoubtedly present in the least metamorphosed OCs, similar in composition to but lower in abundance than in CCs.

Nitrogen in OCs has not been studied as closely as carbon, but where present has been shown to occur dissolved in metal, as well as associated with the organic and elemental material. Nitrogen occurs in lower abundance than carbon, and is isotopically much more heterogeneous: total nitrogen content decreases from $\approx$ 25 ppm in UOCs to $<$ 1 ppm in type 6, whilst $\delta^{15}N$ ranges from -20‰ to $+100$‰, a range that has no apparent relationship with chemical group or petrologic type Hashizume and Sugiura, 1995.

Insoluble organic species are present in UOCs, equivalent to 700–4000 ppm carbon and 5–50 ppm nitrogen, with $\delta^{13}C$ of -20‰ to -10‰ and $\delta^{15}N$ of -10‰ to $+200$‰. The most primitive of the UOCs have the heaviest compositions, and might have the highest complement of interstellar organic molecules (Alexander *et al.*, 1998).

Elemental carbon, with a range of crystallinities from almost amorphous to crystalline graphite, has been described in OCs. Results from stepped combustion analysis of OCs reveal that elemental carbon has $\delta^{13}C$ between $\approx$ -17‰ and -15‰ (Grady *et al.*, 1989) and $\delta^{15}N$ between 0‰ and $+50$‰ (Sugiura *et al.*, 1998). High resolution ion microprobe analysis of graphite in OCs shows that within different graphite morphologies, there is an enormous range in isotopic values: in $\delta^{13}C$ from -65‰ to $+25$‰, and in $\delta^{15}N$ from -30‰ to $+60$‰. These variations have been ascribed to heterogeneities in the precursor material, whilst the more extreme ^{15}N-enrichment in carbon veins in the Khohar OC ($\delta^{15}N$ up to $\approx$ $+1300$‰) possibly result from remnant interstellar organic material, decomposed then trapped during a shock event (Mostefaoui *et al.*, 1998, 2000). Ion microprobe analysis has also been undertaken on chondrules from two UOCs; results from this study showed that between 100–3000 ppm carbon were present in chondrules, with reduced chondrules having approximately 3 times as much carbon as oxidised chondrules. It was inferred that the carbon grains were precursor materials present at the time of chondrule formation (Hanon *et al.*, 1998).

2.3. Enstatite Chondrites

Enstatite chondrites are the most reduced of the chondrite classes, a condition manifested by the occurrence of elements combined into sulphides rather than oxides. In keeping with this state of affairs, the distribution of carbon and nitrogen is somewhat different from that in either carbonaceous or ordinary chondrites. Carbon is relatively abundant, at 0.2–0.7 wt.%, occurring as graphite in association with both silicate minerals and within metal grains. The iron carbide, cohenite, $(Fe,Ni)_3C$, is also present. Nitrogen-bearing minerals have been identified, again as reduced species, including osbornite (TiN), nierite (Si_3N_4) and sinoite (Si_2N_2O); isostructural substitution of the N^{3-} ion for O^{2-} within silicate lattices is thought to be an additional site for nitrogen residence.

Most of the carbon in enstatite chondrites occurs as graphite, $\delta^{13}C$ between −9‰ and +4‰, with the heavier values and higher abundances in enstatite chondrites of higher petrologic type (Grady *et al.*, 1986). Diamonds are present in at least one enstatite chondrite, Abee (Russell *et al.*, 1992), with $\delta^{13}C$ similar to that of the co-existing graphite. There is an isotopic fractionation of carbon between graphite and carbides, with the less abundant carbides (20–60 ppm) being isotopically lighter ($\delta^{13}C < -15‰$). This pattern is repeated in the primitive achondrites and in iron meteorites (q.v.).

Stepped combustion and pyrolysis for nitrogen shows that it is distributed within several components (Grady *et al.*, 1986). Total nitrogen abundances are around 100–500 ppm. Much of this nitrogen is associated with graphite ($\delta^{15} \approx N \approx -30‰$ to $-10‰$), but a significant proportion is locked up in isotopically light nitrides ($\delta^{15}N \approx -60‰$ to $-30‰$).

2.4. Minor Chondrite Groups

2.4.1. *CH Chondrites and Bencubbinites*

The CH group of chondrites was recognised by Bischoff *et al.* (1993) to encompass a small set of highly-reduced, carbon-rich CCs. They are closely related to the bencubbinites, a group of metal-rich and brecciated primitive chondrites (Weisberg *et al.*, 2001), and have many characteristics in common with CR chondrites. The carbon abundance of CHs and bencubbinites is 0.35–1.21 wt.%, whilst $\delta^{13}C$ varies from +0.3‰ to +6‰. Not surprisingly, given the reduced nature of these meteorites, some of the carbon is present as graphite, but poorly-graphitised carbon, diamonds and carbides have also been identified (Grady and Pillinger, 1990a,

1993; Sugiura *et al.*, 2000). The outstanding feature of these meteorites is their most unusual nitrogen isotopic composition. Nitrogen abundance is 20–280 ppm, with $\delta^{15}N \approx +15‰$ to $+900‰$, although a minor part of the nitrogen-bearing component must have $\delta^{15}N$ as high as $\approx +1500‰$ (Prombo and Clayton, 1985; Franchi *et al.*, 1986; Grady and Pillinger, 1990a, 1993; Sugiura *et al.*, 2000). Heavy nitrogen is not confined to any one specific component, but is present in graphite, diamond and nitrides, as well as in the metal. Such a pervasive presence of ^{15}N-enriched nitrogen, at such high abundance, has been ascribed to re-mobilisation of primordial interstellar material, implanted during shock brecciation processes.

2.4.2. *K Chondrites*

The K grouplet of chondrites, named for the type specimen Kakangari (Weisberg *et al.*, 1996), is characterised by high metal abundances, and an oxidation state intermediate between H-group OCs and enstatite chondrites. Carbon abundance is 200–500 ppm, with overall $\delta^{13}C \approx -20‰$. Inferences from stepped combustion results are that about 25% of the carbon occurs as graphite, with $^{13}C \approx -9‰$, the remainder being distributed between metal and carbides, with $\delta^{13}C \approx -20‰$ (Grady and Pillinger, 1986). Nitrogen abundance is $\approx$ 10 ppm, with $\delta^{15}N$ of $-20‰$ to $+10‰$. In Kakangari, nitrogen is present in graphite, metal or carbides ($\delta^{15}N \approx -100‰$), whereas in the only other K chondrite analysed, LEW 87232, nitrogen is apparently more ^{15}N-enriched, with $\delta^{15}N \approx +40‰$ (Weisberg *et al.*, 1996). On the basis of data from two meteorites, then, it seems that carbon and nitrogen in K chondrites demonstrate the same sort of heterogeneities exhibited by other reduced, graphite-bearing meteorites.

3. Carbon and Nitrogen in Non-Chondritic Meteorites

Non-chondritic meteorites have been melted to various extents, and subsequently recrystallised. It is likely, then, that most of these will have lost a large proportion of any volatiles which they initially contained from the aggregation process. Despite this, two groups of stony non-chondritic asteroidal meteorites (primitive achondrites and ureilites) have been identified as containing fairly significant quantities of carbon and nitrogen. In the former case, the species are probably primary to the achondrite parents, whereas in the latter, graphite appears to have been introduced after differentiation. The basaltic achondrites, an assemblage of three groups of stony meteorites thought to be petrogenetically related, are meteoritic analogues of terrestrial basalts, and amongst the oldest known Solar System products of melting and differentiation. Although they contain only minor amounts of carbon and nitrogen, the species are present in measurable quantities. Of the two sub-sets of iron meteorites, magmatic and non- magmatic (or impact-generated), the former represent the ultimate stage in fractionation following melting, and because carbon is more soluble in metal than in silicates, iron meteorites are found to contain significant carbon abundances, mainly occurring as graphite and carbides.

3.1. Primitive Achondrites

The term 'primitive achondrite' is used to describe one of several groups of meteorites that exhibit textural evidence of parent body heating, partial melting and re-crystallisation, but compositionally are relatively unfractionated. Three groups of primitive achondrites, the acapulcoites, lodranites and winonaites, contain significant quantities of carbon and nitrogen. The first two of these groups are most probably genetically related, in accordance with their similar oxygen isotope compositions, and presumably formed in similar locations in the solar nebula.

3.1.1. *Acapulcoites and Lodranites*

Stepped combustion analyses (Grady and Pillinger, 1986) show that Acapulco had ≈ 400 ppm carbon, mainly as graphite (overall $\delta^{13}C \approx -20‰$), but with a lesser amount (< 100 ppm) present either dissolved in Fe, Ni metal or as carbides (total $\delta^{13}C \approx -24‰$). Nitrogen is much less abundant, concentrated mainly in graphite or in metal, ≈ 13 ppm with $\delta^{15}N < -150‰$. Silicates are nitrogen-poor (≈ 3 ppm), but what nitrogen there is is slightly enriched in ^{15}N, with $\delta^{15}N \approx +15‰$ (Kim and Marti, 1993).

Lodranites are more thermally-processed than acapulcoites, and have, on average, been heated to higher temperatures and cooled more slowly. It follows, then, that lodranites have lower volatile abundances than acapulcoites ($\approx$ 40–100 ppm carbon; ≈ 1 ppm nitrogen; Grady *et al.*, 1993; Kim and Marti, 1994). Graphite does not appear to be present, and the carbon is associated with metal ($\delta^{13}C \approx -23‰$). Fractionation of nitrogen between metal and silicate is no longer apparent: $\delta^{15}N$ is relatively homogeneous at $+2‰$ to $+14‰$ in the two phases (Kim and Marti, 1994).

More detailed measurements of carbon and nitrogen in Acapulco, made by ion microprobe, have shown that the isotopic composition of the graphite is not homogeneous, with $\delta^{13}C$ varying from around $-35‰$ to $-10‰$, and $\delta^{15}N$ from $-150‰$ to $-70‰$ (El Goresy *et al.*, 1995). Given that the Acapulco parent asteroid melted and recrystallised, it seems that carbon and nitrogen in the graphite did not equilibrate, but that Acapulco retains the isotopic heterogeneity of its precursor components. Similarly detailed studies of graphite in other, even less-processed, meteorites might therefore reveal a similar diversity of isotopic signatures.

3.1.2. *Winonaites*

Meteorites in this group of primitive achondrites come from a different parent body from acapulcoites and lodranites, and have much higher carbon abundances (0.17–0.67 wt.‰; Grady and Pillinger, 1986). Like acapulcoites, carbon in winonaites is dominantly in the form of graphite, but has a heavier isotopic composition: overall $\delta^{13}C \approx -11$ to $-4‰$, with the graphite having $\delta^{13}C$ up to $\approx 0‰$. These values are close to those for graphite from silicates in the main non-magmatic group of iron meteorites (type IAB), inclusions closely related to the winonaites.

3.2. UREILITES

Ureilites are the most carbon-rich of all meteorite groups, with abundances up to 6–7 wt.‰ (Grady *et al.*, 1985; Smith *et al.*, 2001). The origin of the carbon is as uncertain as the genesis of its host material: arguments have been made for the carbon to be a primary magmatic phase, or a tertiary component introduced during shock processes affecting ureilite parent objects (e.g., Goodrich, 1992, and references therein). The occurrence of graphite in veins, and its conversion to diamond in many highly shocked ureilites are observations which are consistent with the carbon being introduced into ureilite parents at a late stage in the formation history of these meteorites.

Oxygen isotope variations (Clayton and Mayeda, 1988; Franchi *et al.*, 1997) indicate that there are several ureilite sub-groups, each of which formed either on a separate parent body or in separate regions on a single parent. If there is any relationship of either carbon abundance or $\delta^{13}C$ value with silicate composition (either mg# or oxygen isotope composition), then it is a complex one that is not repeated consistently between the different ureilite sub-groups. This could be taken as a further indication that the carbon is not a primary phase, although it is clear from reduced rims on the silicates that carbon has played an important role in ureilite petrogenesis. Graphite in ureilites has $\delta^{13}C \approx -10‰$ to 0‰, similar values to enstatite chondrites and winonaites.

Most of the nitrogen present in ureilites occurs associated with the graphite/diamond veins, which have up to 30 ppm N with $\delta^{15}N$ down to $-120‰$ (Russell *et al.*, 1993; Yamamoto *et al.*, 1998). There is no difference in $\delta^{15}N$ between the nitrogen that occurs in the graphite and its shocked diamond counterpart. There are several other nitrogen components present (Yamamoto *et al.*, 1998), of uncertain origin, at least one of which might be related to nitrogen in polymict ureilites.

Polymict ureilites are a sub-set of ureilites characterised by their brecciated nature, containing clasts of material representative of other meteorite types. It is assumed that they were produced within a regolith on a/the ureilite parent body. Carbon in polymict ureilites is similar to that in main group ureilites, graphitic in nature and exhibiting the same range of abundance and isotopic composition (Grady *et al.*, 1985). Nitrogen, however, is very different. Although the ^{15}N-depleted nitrogen associated with graphite veining is present, isotopically heavy nitrogen predominates, with $\delta^{15}N$ values reaching $+600‰$ (Grady and Pillinger, 1988). No specific host for the heavy nitrogen has been identified; rather, it has been suggested that nitrogen of an interstellar nature was introduced into the ureilite parent regolith, mobilised and re-implanted through subsequent impact brecciation. It is likely that the heavy nitrogen in polymict ureilites is related to the heavy nitrogen in other breccias, including Bencubbin and ALH 85085.

3.3. HEDs

Howardites, eucrites (cumulate and non-cumulate) and diogenites make up a suite of rock types (HED) known to be related both through oxygen isotope systematics and by complex processes of partial melting, fractional crystallisation, thermal metamorphism, brecciation and mechanical mixing. They are anhydrous materials, which crystallised at low pressures under moderately reducing conditions. Since the HED meteorites crystallised in vacuo they are well degassed of volatiles. Notwithstanding this, carbon is present and has presumably played an active role in influencing the redox conditions prevailing during formation. The actual form in which carbon occurs in this suite of meteorites is unknown, but, by analogy with terrestrial basalts (e.g., Mattey *et al.*, 1989, may be as a coating of elemental carbon along grain boundaries, or as dissolved $CO/CO_2/CO_3^{2-}$ in silicate minerals or phosphates. The only recent systematic study of carbon in HEDs used stepped combustion to distinguish between (ubiquitous) terrestrial contamination and indigenous material (Grady *et al.*, 1997a). Results from this study showed that eucrites and diogenites contain ca. 10–30 ppm of indigenous carbon with $\delta^{13}C$ values ranging from −29‰ to −19‰. Diogenites have slightly lower carbon abundances than the eucrites, with heavier $\delta^{13}C$, consistent with loss of CO or CO_2 from the mafic silicates during degassing, as the minerals soften at high temperatures. However, the diogenites also appear to contain a discrete carbonaceous component which combusts across the temperature range from 700°C to 800°C, and has an isotopic composition ≈ −27‰ to −23‰. This material is possibly primitive elemental carbon coating the silicate grains, which is not oxidised during subsequent annealing, and which might also host primitive trapped xenon. The carbon chemistry of howardites, mechanical mixtures of eucrites and diogenites produced in a regolith setting, seems to be dominated by components from clasts of CM or CR-related material.

Nitrogen is even less abundant in HEDs than carbon. Again, the only systematic recent study used stepped combustion to distinguish between terrestrial contaminants, spallogenic components and indigenous nitrogen (Miura and Sugiura, 1993). These authors analysed 5 eucrites, and found them to contain only 0.05–1.3 ppm nitrogen, with variable $\delta^{15}N$, between ≈ −54‰ and +20‰. They concluded that some of the nitrogen was trapped, and some present in the (very minor) metal component.

3.4. Iron Meteorites

As in other reduced or processed meteorites, carbon occurs in iron meteorites as graphite, which is also host to nitrogen. Graphite in iron meteorites has a similar carbon isotopic composition to that in other reduced meteorites, with $\delta^{13}C \approx -5$‰ (Deines and Wickman, 1973, 1975; Sugiura, 1998). This seems to be fairly constant amongst the different magmatic iron meteorite groups. Graphite associated with silicates within some of the non-magmatic irons show the same type of variations in $\delta^{13}C$ as does graphite from the related winonaites (Zipfel *et al.*, 1997). The nitrogen isotopic composition of graphite in iron meteorites is light: $\delta^{15}N$ ranges from −50‰ to −20‰ (Zipfel *et al.*, 1997; Sugiura, 1998).

Associated with the graphite are carbides, predominantly the iron, nickel carbide cohenite, $(Fe,Ni)_3C$, although other carbon-bearing minerals such as haxonite, $(Fe,Ni)_{23}C_6$ have been identified. Carbide carbon is isotopically lighter than graphitic carbon, with $\delta^{13}C < -15$‰ (Deines and Wickman, 1973, 1975; Sugiura, 1998). Carbides are depleted in nitrogen relative to graphite, only ≈ 10 ppm nitrogen, with $\delta^{15}N \approx -50$‰ to −40‰. The main discrete nitrogen-bearing species are nitrides, including carlsbergite, CrN and roaldite, $(Fe,Ni)_4N$. These phases appear to be even more depleted in ^{15}N, with $\delta^{15}N$ down to ≈ −100‰ (Sugiura, 1998).

Carbon and nitrogen also occur dissolved in the Fe-Ni metal, although to different extents in the two crystallographic forms (Sugiura, 1998): kamacite is carbon- and nitrogen-poor (1–10 ppm C; 100–1000 ppm N), compared to taenite (10–500 ppm C; 1000–10,000 ppm N). The isotopic composition of nitrogen associated with the metal has an incredibly wide range, $\delta^{15}N$ from −200‰ to +200‰ (Franchi *et al.*, 1993; Prombo and Clayton, 1993; Sugiura, 1998). Some of the magmatic iron meteorites, i.e., those that display the textural and compositional characteristics of fractionated residues from a melt, demonstrate a regular increase in $\delta^{15}N$ with increasing crystallisation, a trend explicable if it is assumed that nitrogen is being lost from a cooling body (Franchi *et al.*, 1993). Mathew *et al.* (2000) found that nitrogen trapped within one sub-group of IIE iron meteorites was close to that of metal from H-group chondrites, whilst nitrogen in a sub-group of IVA meteorites was akin to L-group chondrites. The exact causes of variation in nitrogen isotopic composition are unknown. Nitrogen behaves as a trace element in iron meteorites, in a similar way to Ga and Ge, and thus variations in N abundance and isotopic composition are presumably related to parent body cooling rates on different parent bodies (Franchi *et al.*, 1993; Prombo and Clayton, 1993; Sugiura, 1998).

4. Carbon and Nitrogen in Non-Asteroidal Meteorites

4.1. Martian Meteorites

Martian meteorites exhibit primary (i.e., Martian), secondary (alteration) and tertiary (trapped/implanted) carbon and nitrogen. However, the amounts of these materials are low, and frequently swamped by ubiquitous terrestrial contamination, thus unravelling the light element chemistry of Mars through study of Martian meteorites is something of a challenge.

The occurrence of primary Martian magmatic phases has been inferred from data derived during combustion experiments (e.g. Wright *et al.*, 1990; Grady *et al.*, 1997b), from which it has been deduced that ≈ 15–40 ppm carbon ($\delta^{13}C \approx -30$‰ to −20‰) and 1–4 ppm nitrogen ($\delta^{15}N \approx -12$‰ to +2‰) is present in Martian rocks. These values contrast with magmatic components observed in basalts from Earth and asteroids (as represented by HED meteorites).

Carbon- and nitrogen-bearing Martian weathering products — carbonates and nitrates — occur in Martian meteorites, produced by the action of fluid flow at Mars' surface. The variety of carbonates is similar in major element composition to those in CI, CM and CR chondrites, with Ca-, Mg- and Fe-rich species all present. Zoned carbonate 'rosettes' have been observed in at least one Martian meteorite, ALH 84001. Carbonate abundances range from 30–60 ppm, with $\delta^{13}C \approx +10‰$ to $+55‰$ (e.g. Wright *et al.*, 1992; Jull *et al.*, 1995). Although the $\delta^{13}C$ of Martian carbonates is similar to that of carbonates in CCs, it is presumed that the ^{13}C-enrichments emanate from different CO_2 source reservoirs. In the case of the Martian carbonates, the CO_2 source is the Martian atmosphere, which has $\delta^{13}C > +30‰$ (Carr *et al.*, 1985). The presence of nitrates on Mars has been predicted, but their identification in Martian meteorites is problematic. If they do occur, they are present in very low quantities, calculated as < 1 ppm with $\delta^{15}N \approx -10‰$ (Grady *et al.*, 1995).

Recognition that the SNC meteorites came from Mars was based on identification of Martian atmospheric gases trapped inside impact-produced pockets of melt glass (e.g., Bogard and Johnson, 1983). Martian atmospheric CO_2 and N_2 are distinguished by elevated $\delta^{13}C$ and $\delta^{15}N$ values, respectively. The different crystallisation and shock histories of Martian meteorites have led to suggestions that variations in the isotopic compositions of the trapped species might track changes in the Martian atmosphere through time (e.g. Murty and Mohapatra, 1997; Grady *et al.*, 1998; Mathew and Marti, 2001).

4.2. Lunar Meteorites

The chemistry of carbon and nitrogen in lunar regolith samples returned to Earth by the Apollo and Luna missions is complex, and related to the antiquity and maturity of the samples. Corresponding study of lunar meteorites (which are mainly regolith breccias) has added little to the knowledge base. As for Apollo samples, carbon and nitrogen in lunar meteorites is a mixture of solar wind gases implanted at various stages of the soils' histories, combined with spallogenic components. A report of an unusual carbon component in lunar meteorites, postulated to be remnant from the impactor (Grady and Pillinger, 1990b), was subsequently identified as a laboratory artefact. There appears to be little, if any, indigenous lunar carbon and nitrogen in lunar meteorites.

5. Summary

We have taken an inventory of the elemental and isotopic abundances of major carbon- and nitrogen-bearing components in different groups of meteorites. Primary phases, inherited from the solar nebula, are isotopically heterogeneous, and apparently quite resistant to modification through parent body processing. Even melted and recrystallised meteorites retain primordial carbon and nitrogen isotopic signatures. Graphite is particularly widespread, occurring in many different meteorite groups of varying reduction state, in a variety of morphologies and crystallinities. Broadly speaking, Solar System graphite has $\delta^{13}C \approx -10‰$. This simple statement masks the fact that hidden within graphite grains are regions in which $\delta^{13}C$ can vary from $-65‰$ to $+20‰$. Isotopically light nitrogen, centred around $-80‰$, but ranging from $-200‰$ to $+200‰$ is associated with the graphite. In less reduced meteorites, organic species are the predominant form in which carbon and nitrogen occur; their isotopic signatures also indicate that they have survived from the solar nebula without modification and homogenisation.

Acknowledgements

MMG thanks the Natural History Museum (for letting her go) and the Open University (for letting her stay) for sabbatical leave, during which time the groundwork for this paper was undertaken. Meteorite research at the Natural History Museum and at the Open University is supported by financial assistance from the Particle Physics and Astronomy Research Council.

References

Abreu, N.M., and Brearley, A.J.: 2002, 'Mineralogical Characterization of Vigarano Matrix: HRTEM and EFTEM Observations', *Lunar Planet. Sci.* **33**, #1542.

Alexander, C.M.O'D., Russell, S.S., Arden, J.W., Ash, R.D., Grady, M.M., and Pillinger, C.T.: 1998, 'The Origin of Chondritic Macromolecular Matter; a Carbon and Nitrogen Isotope Study', *Met. Planet. Sci.* **33**, 603–622.

Becker, L., Bada, J.L., Winans, R.E., and Bunch, T.E.: 1994, 'Fullerenes in Allende Meteorite', *Nature* **372**, 507.

Becker, L., Bunch, T.E., and Allamandola, L.J.: 1999, 'Higher Fullerenes in the Allende Meteorite', *Nature* **400**, 227–228.

Becker, R.H., and Pepin, R.O.: 1984, 'The Case for a Martian Origin of the Shergottites — Nitrogen and Noble Gases in EETA 79001', *Earth Planet. Sci. Lett.* **69**, 225–242.

Bischoff, A., Palme, H., Schultz, L., Weber, D., Weber, H.W., and Spettel, B.: 1993, 'Acfer 182 and Paired Samples, an Iron-rich Carbonaceous Chondrite — Similarities with ALH85085 and Relationship to CR Chondrites', *Geochim. Cosmochim. Acta* **57**, 2631–2648.

Bogard, D.D., and Johnson, P.: 1983, 'Martian Gases in an Antarctic Meteorite?', *Science* **221**, 651–654.

Carr, R.H., Grady, M.M., Wright, I.P., and Pillinger, C.T.: 1985, 'Martian Atmospheric Carbon Dioxide and Weathering Products in SNC Meteorites', *Nature* **314**, 248–250.
Clayton, R.N., and Mayeda, T.K.: 1988, 'Formation of Ureilites by Nebula Processes', *Geochim. Cosmochim. Acta* **52**, 1313–1318.
Cronin, J.R., and Chang, S.: 1993, 'Organic Matter in Meteorites: Molecular and Isotopic Analyses of the Murchison Meteorite', *in* J.M. Greenberg *et al.* (eds.), *The Chemistry of Life's Origins*, Kluwer Academic Publishers, Dordrecht, pp. 209–258.
Deines, P., and Wickman, F.E.: 1973, 'The Isotopic Composition of 'Graphitic' Carbon from Iron Meteorites and some Remarks on the Troilitic Sulfur of Iron Meteorites', *Geochim. Cosmochim. Acta* **37**, 1295–1319.
Deines, P., and Wickman, F.E.: 1975, 'A Contribution to the Stable Carbon Isotope Geochemistry of Iron Meteorites', *Geochim. Cosmochim. Acta* **39**, 547–557.
El Goresy, E., Zinner, E., and Marti, K.: 1995, 'Survival of Isotopically Heterogeneous Graphite in a Differentiated Meteorite', *Nature* **373**, 496–499.
Franchi, I.A., Wright, I.P., and Pillinger, C.T.: 1986, 'Heavy Nitrogen in Bencubbin — a Light Element Isotopic Anomaly in a Stony-iron Meteorite', *Nature* **323**, 138–140.
Franchi, I.A., Wright, I.P., and Pillinger, C.T.: 1993, 'Constraints on the Formation Conditions of Iron Meteorites Based on Concentrations and Isotopic Compositions of Iron Meteorites', *Geochim. Cosmochim. Acta* **57**, 3105–3121.
Franchi, I.A., Sexton, A.S., Wright, I.P., and Pillinger C.T.: 1997, 'Resolved Sub-groups within the Ureilite Population', *Meteoritics* **32**, A44.
Frick, U., and Pepin, R.O.: 1981, 'Microanalysis of Nitrogen Isotope Abundances — Association of Nitrogen with Noble Gas Carriers in Allende', *Earth Planet. Sci. Lett.* **56**, 64–81.
Goodrich, C.A.: 1992, 'Ureilites: A Critical Review', *Meteoritics* **27**, 327–352.
Grady, M.M., and Pillinger, C.T.: 1986, 'Carbon Isotope Relationships in Winonaites and Forsterite Chondrites', *Geochim. Cosmochim. Acta* **50**, 255–263.
Grady, M.M., and Pillinger, C.T.: 1988, '^{15}N-enriched Nitrogen in Polymict Ureilites and its Bearing on their Formation', *Nature* **331**, 321–323.
Grady, M.M., and Pillinger, C.T.: 1990a, 'ALH 85085: Nitrogen Isotope Analysis of a Highly Unusual Primitive Chondrite', *Earth Planet. Sci. Lett.* **97**, 29–40.
Grady, M.M., and Pillinger, C.T.: 1990b, 'The Carbon and Nitrogen Stable Isotope Geochemistry of two Lunar Meteorites: ALHA 81005 and Y 86032', *Proc. NIPR Symp. Antarctic Meteorites* **3**, 27–39.
Grady, M.M., and Pillinger, C.T.: 1993, 'Acfer 182: Search for the Location of ^{15}N-enriched Nitrogen in an Unusual Chondrite', *Earth Planet Sci. Lett.* **116**, 165–180.
Grady, M.M., Wright, I.P., Swart, P.K., and Pillinger, C.T.: 1985, 'The Carbon and Nitrogen Isotopic Composition of the Ureilites: Implications for their Genesis', *Geochim. Cosmochim. Acta* **49**, 903–915.
Grady, M.M., Wright, I.P., Carr, L.P., and Pillinger, C.T.: 1986, 'Compositional Differences in Enstatite Chondrites Based on Carbon and Nitrogen Isotope Measurements', *Geochim. Cosmochim. Acta* **50**, 2799–2813.
Grady, M.M., Wright, I.P., Swart, P.K., and Pillinger, C.T.: 1988, 'The Carbon and Oxygen Isotopic Composition of Meteoritic Carbonates', *Geochim. Cosmochim. Acta* **52**, 2855–2866.
Grady, M.M., Wright, I.P., and Pillinger, C.T.: 1989, 'A Preliminary Investigation into the Nature of Carbonaceous Material in Ordinary Chondrites', *Meteoritics* **24**, 147–154.
Grady, M.M., Franchi, I.A., and Pillinger, C.T.: 1993, 'Carbon and Nitrogen Chemistry of Lodranites: Relationship to Acapulco?', *Meteoritics* **28**, 355–356.
Grady, M.M., Wright, I.P., and Pillinger, C.T.: 1995, 'A Search for Nitrates in Martian Meteorites', *J. Geophys. Res.* **100**, 5449–5455.
Grady, M.M., Wright, I.P., and Pillinger, C.T.: 1997a, 'Carbon in Howardite, Eucrite and Diogenite Basaltic Achondrites', *Met. Planet. Sci.* **32**, 863–868.

Grady, M.M., Wright, I.P., and Pillinger, C.T.: 1997b, 'A Carbon and Nitrogen Isotope Study of Zagami', *J. Geophys. Res.* **102**, 9165–9173.

Grady, M.M., Wright, I.P., and Pillinger, C.T.: 1998, 'A Nitrogen and Argon Stable Isotope Study of Allan Hills 84001: Implications for the Evolution of the Martian Atmosphere', *Met. Planet. Sci.* **33**, 795–802.

Grady, M.M., Verchovsky, A.B., Franchi, I.A., Wright, I.P., and Pillinger, C.T.: 2002, 'Light Element Geochemistry of the Tagish Lake CI2 chondrite: Comparison with CI1 and CM2 Meteorites', *Met. Planet. Sci.* **37**, 713–735.

Hanon, P., Robert, F., and Chaussidon, M.: 1998, 'High Carbon Concentration in Meteoritic Chondrules: A Record of Metal-silicate Differentiation', *Geochim. Cosmochim. Acta* **62**, 903–913.

Harris, P.J.F., Vis, R.D., and Heymann, D.: 2000, 'Fullerene-like Carbon Nanostructures in the Allende Meteorite', *Earth Planet. Sci. Lett.* **183**, 355–359.

Hashizume, K., and Sugiura, N.: 1995, 'Nitrogen Isotopes in Bulk Ordinary Chondrites', *Geochim. Cosmochim. Acta* **59**, 4057–4069.

Hayatsu, R., and Anders, E.: 1981, 'Organic Compounds in Meteorites and their Origins', *in 'Topics in Current Chemistry'*, Vol. 99, Springer-Verlag, Berlin, pp. 1–37.

Jull, A.J.T., Eastoe, C.J., Xue, S., and Herzog, G.F.: 1995, 'Isotopic Composition of Carbonates in the SNC Meteorites ALH 84001 and Nakhla', *Meteoritics* **30**, 311–318.

Kerridge, J.F.: 1985, 'Carbon, Hydrogen and Nitrogen in Carbonaceous Chondrites: Abundance and Isotopic Compositions in Bulk Samples', *Geochim. Cosmochim. Acta* **49**, 1707–1714.

Kim, Y., and Marti, K.: 1993, 'Isotopic Signatures and Distribution of Nitrogen and Trapped and Radiogenic Xenon in the Acapulco and FRO 90011 Meteorites', *Lunar Planet. Sci.* **24**, 801–802.

Kim, Y., and Marti, K.: 1994, 'Genetic Relationships of Acapulcoites and Lodranites? A Study of Nitrogen and Xenon Isotopic Signatures', *Lunar Planet. Sci.* **25**, 703–704.

Mathew, K.J., and Marti, K.: 2001, 'Early Evolution of Martian Volatiles: Nitrogen and Noble Gas Components in ALH 84001 and Chassigny', *J. Geophys. Res.* **106**, 1401–1422.

Mathew, K.J., Kim, Y., and Marti, K.: 1998, 'Martian Atmospheric and Indigenous Components of Xenon and Nitrogen in the Shergotty, Nakhla and Chassigny Group Meteorites', *Met. Planet. Sci.* **33**, 655–664.

Mathew, K.J., Palma, R.L., Marti, K., and Lavielle, B.: 2000, 'Isotopic Signatures and Origin of Nitrogen in IIE and IVA Iron Meteorites', *Geochim. Cosmochim. Acta* **64**, 545–557.

Mattey, D.P., Exley, R.A., and Pillinger, C.T.: 1989, 'Isotopic Composition of CO_2 and Dissolved Carbon Species in Basalt Glass', *Geochim. Cosmochim. Acta* **53**, 2377–2386.

McCrea, J.M.: 1950, 'On the Isotopic Chemistry of Carbonates and a Palaeotemperature Scale', *J. Chem. Phys.* **18**, 849–857.

Miura, Y., and Sugiura, N.: 1993, 'Nitrogen Isotopic Compositions in three Antarctic and two Non-Antarctic Eucrites', *Proc. NIPR Symp. Ant. Met.* **6**, 338–356.

Moore, C.B., and Lewis, C.F.: 1965, 'Carbon Abundances in Chondritic Meteorites', *Science* **149**, 317–318.

Moore, C.B., and Gibson, E.K., Jr.: 1969, 'Nitrogen Abundances in Chondritic Meteorites', *Science* **163**, 174–176.

Mostefaoui, S., Hoppe, P., and El Goresy, A.: 1998, 'In situ Discovery of Graphite with Interstellar Isotopic Signatures in a Chondrule-free Clast in an L3 Chondrite', *Science* **280**, 1418–1420.

Mostefaoui, S., Perron, C., Zinner, E., and Sagon, G.: 2000, 'Metal-associated Carbon in Primitive Chondrites: Structure, Isotopic Composition and Origin', *Geochim. Cosmochim. Acta* **64**, 1945–1964.

Murty, S.V.S., and Mohapatra, R.K.: 1997, 'Nitrogen and Heavy Noble Gases in ALH 84001: Signature of Ancient Martian Atmosphere', *Geochim. Cosmochim. Acta* **61**, 5417–5428.

Newton, J., Bischoff, A., Arden, J.W., Franchi, I.A., Geiger, T., Greshake, A., and Pillinger, C.T.: 1993, 'Acfer 094, a Uniquely Primitive Carbonaceous Chondrite from the Sahara', *Meteoritics* **30**, 47–56.

Pizzarello, S., Krishnamurthy, R.V., Epstein, S., and Cronin, J.R.: 1991, 'Isotopic Analyses of Amino Acids from the Murchison Meteorite', *Geochim. Cosmochim. Acta* **55**, 905–910.

Prombo, C.A., and Clayton, R.N.: 1985, 'A Striking Nitrogen Isotope Anomaly in the Bencubbin and Weatherford Meteorites', *Science* **230**, 935–937.

Prombo, C.A., and Clayton, R.N.: 1993, 'Nitrogen Isotopic Compositions of Iron Meteorites', *Geochim. Cosmochim. Acta* **57**, 3749–3761.

Russell, S.S., Pillinger, C.T., Arden, J.W., Lee, M.R., and Ott, U.: 1992, 'A New Type of Diamond in the Enstatite Chondrite Abee', *Science* **256**, 206–209.

Russell, S.S., Arden, J.W., Franchi, I.A., and Pillinger, C.T.: 1993, 'A Carbon and Nitrogen Isotope Study of Carbonaceous Vein Material in Ureilite Meteorites', *Lunar Planet. Sci.* **24**, 1221–1222.

Smith, C.L., Franchi, I.A., Wright, I.P., Grady, M.M., and Pillinger, C.T.: 2001, 'New Data on Carbon Isotopic Compositions of some Ureilites', *Lunar Planet. Sci. Conf.* **32**, #1878.

Sugiura, N.: 1998, 'Ion Probe Measurements of Carbon and Nitrogen in Iron Meteorites', *Met. Planet. Sci.* **33**, 393–409.

Sugiura, N., Kiyota, K., and Hashizume, K.: 1998, 'Nitrogen Components in Primitive Ordinary Chondrites', *Met. Planet. Sci.* **33**, 463–482.

Sugiura, N., Zashu, S., Weisberg, M.K., and Prinz, M.: 2000, 'A Nitrogen Isotope Study of Bencubbinites', *Met. Planet. Sci.* **35**, 987–996.

Swart, P.K., Grady, M.M., and Pillinger, C.T.: 1983, 'A Method for the Identification and Elimination of Terrestrial Contamination During Carbon Isotope Analysis of Extraterrestrial Samples', *Meteoritics* **18**, 137–154.

Varela, M.E., Metrich, N., Bonnin-Mosbah, M., and Kurat, G.: 2000, 'Carbon in Glass Inclusions of Allende, Vigarano, Bali and Kaba (CV3) Olivines', *Geochim. Cosmochim. Acta* **64**, 3923–3930.

Weisberg, M.K., Prinz, M., Clayton, R.N., Mayeda, T.K., Grady, M.M., Franchi, I.A., Pillinger, C.T., and Kallemeyn, G.W.: 1996, 'The K (Kakangari) Chondrite Grouplet', *Geochim. Cosmochim. Acta* **60**, 4253–4263.

Weisberg, M.K., Prinz, M., Clayton, R.N., Mayeda, T.K., Sugiura, N., Zashu, S., and Ebihara, M.: 2001, 'A New Metal-rich Chondrite Grouplet', *Met. Planet. Sci.* **36**, 401–418.

Wright, I.P., Grady, M.M., and Pillinger, C.T.: 1990, 'The Evolution of Atmospheric CO_2 on Mars: The Perspective from Carbon Isotope Measurements', *J. Geophys. Res.* **95**, 14789–14794..

Wright, I.P., Grady, M.M., and Pillinger, C.T.: 1992, 'Chassigny and the Nakhlites: Carbon-bearing Components and their Relationship to Martian Environmental Conditions', *Geochim. Cosmochim. Acta* **56**, 817–826.

Yamamoto, T., Hashizume, K., Matsuda, J.-I., and Kase, T.: 1998, 'Multiple Nitrogen Isotopic Components Coexisting in Ureilites', *Met. Planet. Sci.* **33**, 857–870.

Zinner, E.K.: 1998, 'Stellar Nucleosynthesis and the Isotopic Composition of Presolar Grains from Primitive Meteorites', *Ann. Rev. Earth Planet. Sci.* **26**, 147–188.

Zinner, E., Tang, M., and Anders, E.: 1987, 'Large Isotopic Anomalies of Si, C, N and Noble Gases in Interstellar Silicon Carbide from the Murray Meteorite', *Nature* **330**, 730–732.

Zipfel, J., Hutcheon, I.D., and Marti, K.: 1997, 'Carbon Isotopic Compositions of Graphite Grains in the El Taco IAB Iron Meteorite', *Lunar Planet. Sci.* **28**, #1707.

Address for Offprints: Monica M. Grady, Dept. of Mineralogy, The Natural History Museum, Cromwell Road, London SW7 5BD, UK; mmg@nhm.ac.uk

ISOTOPIC CRITERIA FOR IDENTIFICATION OF ORGANIC CARBON ON EARTH AND METEORITES

E.M. GALIMOV

V.I. Vernadsky Institute of Geochemistry and Analytical Chemistry of the Russian Academy of Sciences, 119991, Moscow, Kosygin 19, Russia

Received: 16 April 2002; Accepted in final form: 15 October 2002

Abstract. Contrary to the often stated view, an enrichment of organic material in the light isotope is not a conclusive evidence of its life-related origin. The $\beta^{13}C$ - $\delta^{13}C$ correlation is a special feature of biological systems. Therefore it can be used as a criterion for identification of organic carbon. A survey of the available isotopic data for organic compounds in meteorites shows that they do not comply with the $\beta^{13}C$ - $\delta^{13}C$ correlation. The prevalence of amino and hydroxy acids in purines and sugars found in carbonaceous meteorites indicates that condensation of HCN and HCHO passed through cyanohydrin reaction, while biological evolution proceeds through formation of adenosine triphosphate (ATP). This, in addition to the isotope criterion, indicates that the organic compounds in carbonaceous chondrites are not life-related substances.

1. Introduction

To date a great variety of organic compounds has been identified in interstellar space, in meteorites and comets (e.g. Irwine, 1998; Zinner, 1997; Cronin and Chang, 1993; Crovisier, 1998). Microstructures resembling microorganisms were reported to be present in the Martian meteorite ALH 840011 (McKay *et al.*, 1996).

The development of a reliable identification of biogenic products is of crucial importance for the evaluation of the origin of life. One of the promising tools is isotopic study. Traditionally carbon isotope analysis is used to search for traces of life in ancient sedimentary rock and extraterrestrial material (e.g. Schidlowski, 1993; Mojzsis *et al.*, 1996; Wright *et al.*, 1989). The belief exists that "isotopically light organic carbon provides a distinct biomark when compared to the inorganic pool" (Report, 1997-1998). However, enrichment of organic carbon in the light isotope is not a special feature of biological material. It occurs both in biogenic and non-biogenic natural substances.

To be able to apply isotope chemistry to distinguish a life-related origin of organic matter one needs to find properties which are unique for the chemistry of life. It is the aim of this paper to discuss aspects of this problem.

Space Science Reviews **106:** 249–262, 2003.

2. Carbon Isotopes

Carbon isotopes are frequently used for the identification of organic carbon because it is known that organisms discriminate against heavy carbon isotope.

There are two main types of isotope effects: kinetic and thermodynamic. For the equilibrium state the relationship is given by

$$\Delta_{p-s} \approx \left(\frac{\beta_p}{\beta_s} - 1\right) \times 10^3 ‰ , \tag{1}$$

Where Δ_{p-s} is the isotope fractionation between product (p) and substrate (s); the β-value is the reduced partition function ratio.

Reactions with isotopically light species are usually faster than those that are isotopically heavy. This is the cause of a kinetic isotope effect.

The magnitude of the isotope fractionation depends on the magnitude of kinetic isotope effects (α_k) and on the coefficient of utilization (f) of the original substance:

$$\Delta_{p-s} = \left(\frac{1}{f}\left[1 - (1 - f)^{\alpha_k}\right] - 1\right) \times 10^3 ‰ . \tag{2}$$

It follows from Equation (2), that the fractionation has a maximum if the source of the substrate is unlimited and the rate of supply of the substrate significantly exceeds the rate of consumption.

As an example, Figure 1 demonstrates carbon isotope fractionation between photosynthesized organisms and their inorganic sources. Atmospheric CO_2 is the common link between organic and inorganic cycles of carbon. It has a $\delta^{13}C$ value of about $-7‰$, while the $\delta^{13}C$ range of -22 to $-32‰$ characterizes the terrestrial plants of C-3 type. The plants of the C-4 type have a $\delta^{13}C$ range with variations from -10 to $-18‰$. The differences between the isotope composition of these two types are primarily due to a more efficient consumption (higher f) of CO_2 in the Hatch-Slack cycle (C-4 plants) compared to the Calvin cycle (C-3 plants). Marine plankton is enriched in the light isotope to a lesser degree than terrestrial plants. This is, at least in part, due to the lower diffusion rate (hence the higher f) of CO_2 in water than in air. Plankton in marine water at higher latitudes is lighter than in waters at lower latitudes. This is because of better solubility of CO_2 in colder water and consequently higher concentration and hence availability (lower f) of CO_2 for the marine organisms.

Therefore many geochemical features of the isotope distribution in the biological world have no biological specificity but present different manifestation of the isotope effect of carbon fixation. The observed $\delta^{13}C$ variations are mostly due to variations of the extent of utilization of the source carbon.

If we now turn to the mantle and the lower crust environment, that is the domain of non-biological processes, we observe again the same type of the isotope

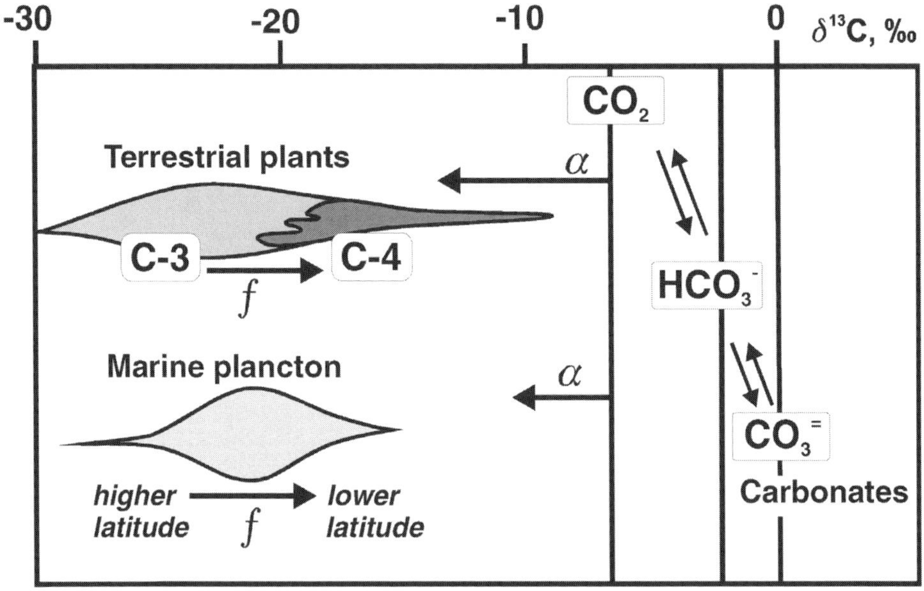

Figure 1. Carbon isotope fractionation between photosynthesizing organisms and the inorganic source of their carbon. The α-arrow is the isotope effect (enrichment in ^{12}C) during CO_2 consumption. Directions of the f-arrows indicate incresase of degree of utilization of ambient CO_2 during its consumption by organisms.

effect. The bitumen, present in igneous rock, is a product of abiogeneous synthesis. It is believed to be synthesized from simple precursors (CO, CH_4, etc.) under high temperature and high-pressure conditions. Endogenous bitumen found as an inclusion in minerals of magmatic rock is enriched in the light carbon isotope almost to the same extent as biogeneous bitumen and kerogen in sedimentary rock (Galimov and Petersilie, 1967; Galimov, 1991). In particular, bitumen inclusions studied in the alkaline ultramafic rock of the Kola Peninsula showed $\delta^{13}C$-values from $-32.3‰$ to $-26.7‰$ while endogenous methane sealed in the same minerals was found to have $\delta^{13}C$ from $-3.2‰$ to $-7.9‰$. Some olivine and garnet crystals from kimberlite rock had gas-liquid inclusions, which contained organic compounds. The measurement showed that the organic carbon was isotopically light ($\delta^{13}C \approx -29‰$).

The kinetic isotope effect of breaking of C-C bond controls the production of natural gaseous hydrocarbons. The isotope distribution pattern of biogenic and nonbiogenic low molecular hydrocarbon gases (C_1-C_3) is different because of different mechanism of their formation. Methane (C_1) as well as ethane (C_2) and propane (C_3) in sedimentary rock are produced via degradation of kerogen, i.e. by breaking of C-C bond. The same compounds in magmatic rock are products of consccutive polymerization through the creation of C-C bond. Therefore their enrichment in the light isotope, in contrast to their sedimentary counterparts, grows

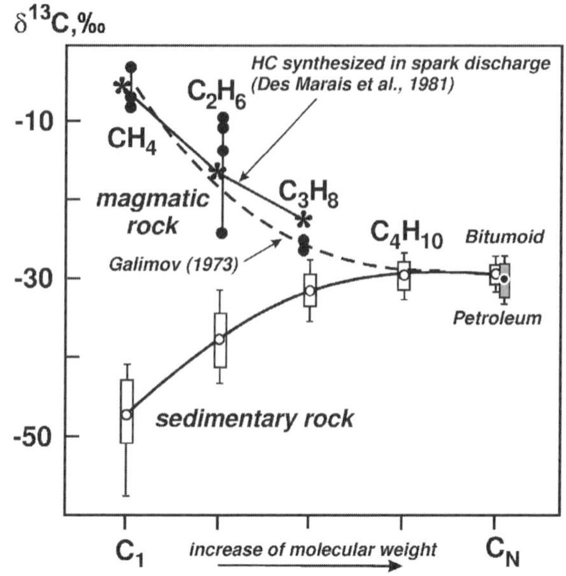

Figure 2. Change of Carbon isotopic composition of hydrocarbons with increase in molecular weight. The open symbols stand for the averaged $\delta^{13}C$ values of hydrocarbons from sedimentary rocks; the filled symbols are $\delta^{13}C$ values found for C_1-C_3 hydrocarbons and averaged value for bitumoid extracted from gas-liquid inclusions in minerals of magmatic rock (Galimov and Petersilie, 1967; Galimov, 1973). Asterisks give trends found for C_1-C_3 hydrocarbons synthesized in spark-discharge experiment (Des Marais *et al.*, 1981).

with the carbon number. Figure 2 is taken from earlier work (Galimov and Shirinski, 1975) and illustrates this phenomenon for natural magmatic hydrocarbons. The original picture is modified here by the data obtained later by Des Marais *et al.* (1981) during spark discharge experiment. The C_1-C_3 hydrocarbons synthesized in that experiment (asterisk) showed the same isotope distribution pattern as their counterparts in igneous rock. The examples presented above show that enrichment in the light carbon isotope is not conclusive evidence for the origin of organic matter.

3. $\beta^{13}C$ - $\delta^{13}C$ Correlation in Biological Material

Some years ago a correlation was found between the reduced partition function ratio ($\beta^{13}C$-factor) and isotopic composition of the corresponding compound in biological systems (Galimov, 1985). Recently it was shown that biological evolution in general and biochemical reactions in particular proceed in a linear field of thermodynamics of irreversible process (Galimov, 2001).

One should keep in mind that the observable essentially non-equilibrium highly organized state of biological system is a result of the specific structure of interactions controlled by a system of enzymes. Each living organism receives the instruction about how this system must work at the moment of its conception through the DNA and the relevant set of enzymes. This instruction has been formed over billions of years of evolution. The elementary chemical reaction proceeding in an organism under control of enzymes represents a non-equilibrium steady state system, which, however, does not need to be far from equilibrium. The presence of reversibility in enzymatic reactions is difficult to evaluate through chemical composition because of the chemical selectivity of enzymes. However, since the isotope composition is not controlled by enzymes (i.e. isotope composition is not encoded, there are no isotope-specific enzymes) the natural distribution of carbon isotopes ($^{13}C/^{12}C$-ratio) in and between biological molecules can provide a unique test of the thermodynamic state of biochemical processes. If any proximity to an equilibrium is a feature of biological reactions, it should be manifested in a tendency towards an equilibrium distribution of isotopes.

In an equilibrium state, the isotope composition of a reaction component is in accordance with the respective thermodynamic isotopic factors (Equation 1). The $\beta^{13}C$ values can be calculated through vibration frequencies, if they are known for the isotopic species, or using a technique developed by Galimov (1985).

It should be stressed once again that the thermodynamic isotopic factors describe the distribution of isotopes in the state of isotope-exchange equilibrium. Complex organic molecules typical for biological systems exchange carbon isotopes very slowly. It is known from the isotope geochemistry that organic compounds can preserve their carbon isotope composition for many million years. Nevertheless, the facts indicate that in many cases, a correlation between $\beta^{13}C$ and $\delta^{13}C$ values is observed.

Some of the results are presented below.

For the first time the $\delta^{13}C$ - $\beta^{13}C$ correlation was shown for different lipid compounds (Galimov and Shirinski, 1975).

The results are shown in Figure 3 (a-d). In all four cases the $\delta^{13}C$ - $\beta^{13}C$ correlation exists. The analytical errors and uncertainty related to the approximate technique of evaluating the β-factor values contribute to the scattering. In spite of these sources of dispersion the relationship between $\beta^{13}C$ and $\delta^{13}C$ values is characterized by relatively high values of the correlation coefficients. Note that the regression equation contains a reducing coefficient κ before the term (in brackets) defining the thermodynamic isotope effect. The coefficient κ varies within 0.31-0.55 (instead of being unity which is characteristic of the equilibrium state). It means that the biochemical reactions contain a component of reversibility. They are not in equilibrium state but they are not far from equilibrium.

Bogacheva and Galimov (1979) studied intramolecular carbon isotope distributions in a chlorophyll molecule. The correlation was found between the measured values of the $\delta^{13}C$ and $\beta^{13}C$ factors evaluated for the corresponding fragments (Figure 4a).

At the same time a group of American scientists headed by Warren Meinschein from the Indiana University studied intramolecular isotope effects in acetoin (Fig-

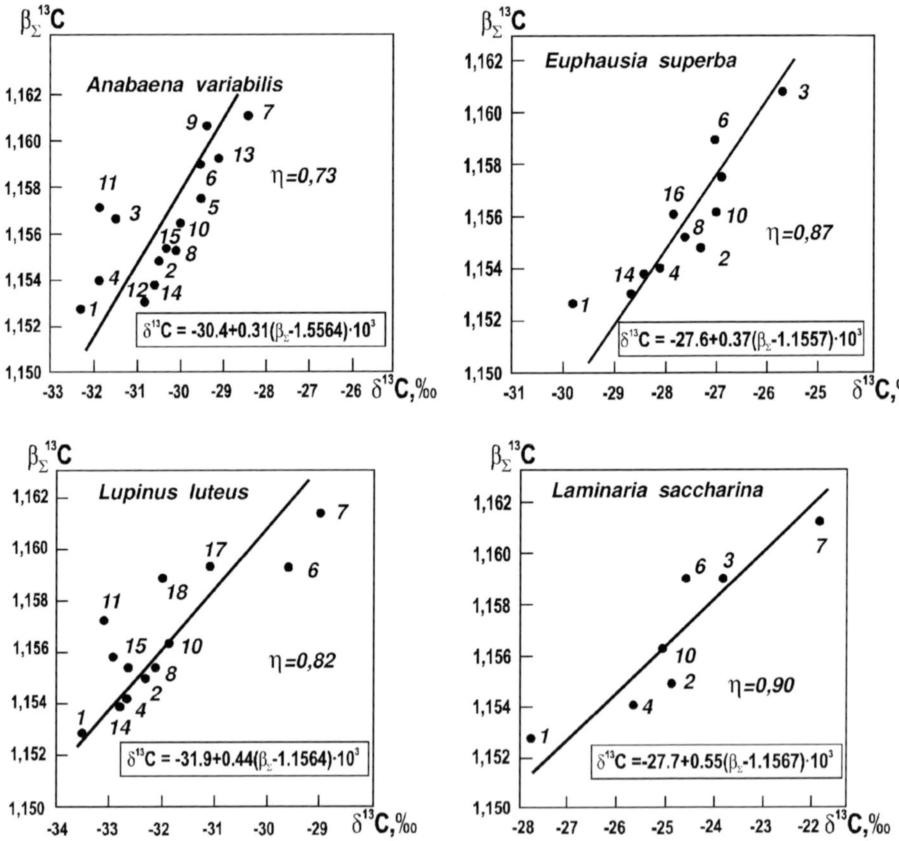

Figure 3. β_{Σ}^{13}C - δ^{13}C correlation of components of the lipid fraction extracted from blue-green alga (Anabaena variabilis), krill (Euphausia superba), lupine (Lupinus luteus) and laminaria (Laminaria saccharina). 1. waxes, hydrocarbons, sterol, alcohols; 2. triglycerides; 3. n-carotene; 4. astatine; 5. carotenoid (fucoxanthine); 6. fatty acids; 7. carotenoid (echinenone); 8. sterol (β-sitosterol); 9. sterol (cholesterol); 10. sterol (β-sitosterol); 11. sterol (fucosterol); 12. chlorophyll; 13. diglycerides; 14. carolenoid (myxoxanthophyll); 15. monoglycerides; 16. phosphatidylserine; 17. sphingomyelin; 18. phosphatidylinositol; 19. lecithin; 20. cephalin; 21. cardiolipin; 22. pigment (lutein); 23. monogalactosylglyceride. The regression equations are in the insertions; η is the coefficient of correlation.

ure 4b) and malic acid (Figure 4c) and concluded that "the ^{13}C contents of the specific carbon atoms in malic acid from apple and sorghum increase in accordance with their β values, as predicted by Galimov" (p. 346, Meinschein *et al.*, 1984).

It does not mean, however, that all biochemical reactions at all times must show such a relationship. Many processes in organism are essentially irreversible and non-linear. It is clear that the observed correlation is not accidental. The β^{13}C - δ^{13}C correlation is inherent to compounds of different structure, related to different biochemical components, observed in organisms of different taxonomy, from different environment, and manifested both on the intermolecular and intramolec-

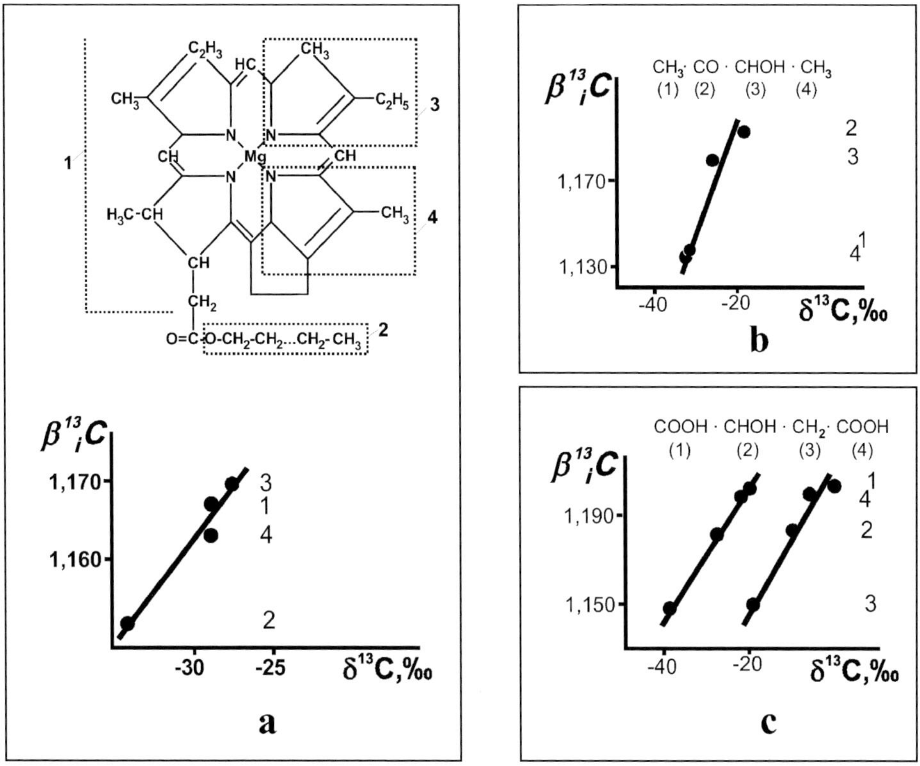

Figure 4. Intramolecular $\beta^{13}C$ - $\delta^{13}C$ correlation: a) intramolecular carbon isotope distribution in chlorophyll molecule, compared to $\beta^{13}C$ values of the corresponding fragments (Bogacheva and Galimov, 1979); b) intramolecular carbon isotope distribution in acetoin molecule (experimental data by Rinaldi *et al.*, 1974) compared to $\beta_i^{13}C$ values related to the carbon atoms in the corresponding positions; c) intramolecular $\delta^{13}C$ - $\beta_i^{13}C$ correlation for malonic acid from C-3 plant (apple) at the left and C-4 plant (sorghum) at the right obtained by Meinschein *et al.* (1984).

ular level. It should be kept in mind that this is not caused by direct isotope exchange between compounds but is a consequence of microscopic reversibility of substrate-product state in enzyme-substrate complex (Galimov, 1985).

During the last 10-15 years the most significant results in the field of study of intramolecular isotope composition of biomolecules has been made by the group of H.-L.Schmidt from the Technical University in Munich. For the first time they measured the isotope composition of glucose in all six carbon positions (Rossman *et al.*, 1991). In that case the $\beta_i^{13}C$ - $\delta^{13}C$ correlation is not sufficient (Figure 5a). It was much better in the case of citric acid (Schmidt and Gleixner, 1998), with a correlation coefficient of $\eta = 0.93$ (Figure 5b).

Analysis of complex caffein molecules (Figure 5c) has been made of preparations from different geographical locations (Weilacher *et al.*, 1996).

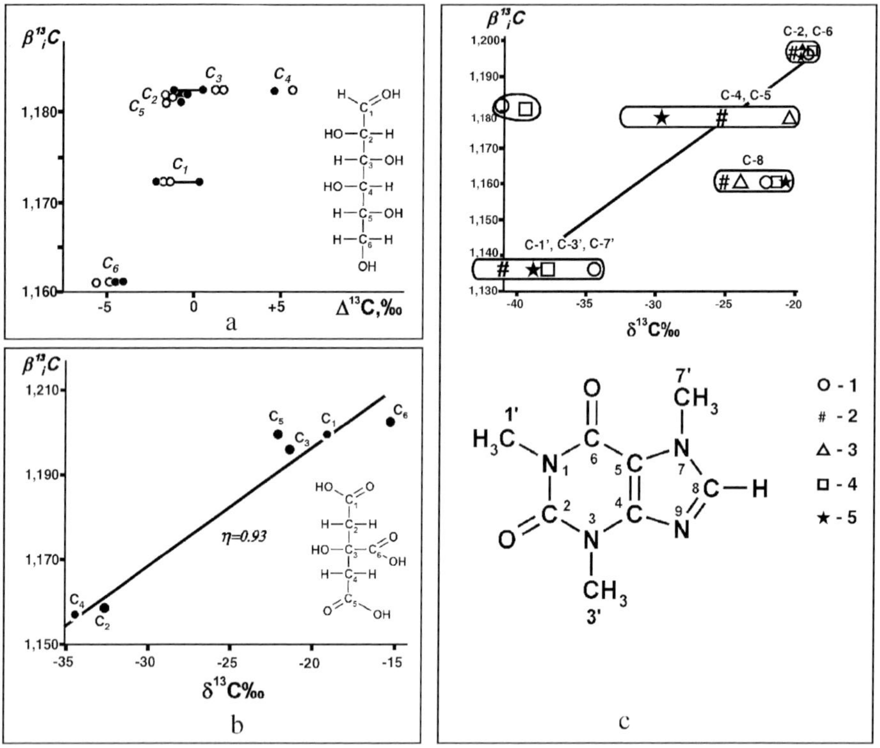

Figure 5. Intramolecular $\delta^{13}C$ - $\beta_i^{13}C$ correlation: a) glucose, extracted from C-4 plant maize (*filled symbols*) and C-3 plant beet (*open symbols*); the measured $\delta^{13}C$-values are adopted from Rossman *et al.* (1991); b) citric acid; the measured $\delta^{13}C$ values are from Schmidt and Gleixner (1998); c) purine alkoloid (caffeine) from different geographical locations: 1 - Sri Lanca; 2 - Darjeeling; 3 - Assam, 4 - China, 5 - USSR; the measured $\delta^{13}C$ values are from Weilacher *et al.* (1996).

In Figure 6a data obtained by Gleixner *et al.* (1998) are presented for different biochemical compounds, extracted from leaves (*filled symbols*) and roots (*open symbols*) of potato. In Figure 6b the data for the amino acids from the same object are shown separately.

More examples are available. The data shown here, however, provide an adequate view on the character of the $\delta^{13}C$ - $\beta^{13}C$ correlation in biosystems.

The most interesting fact, despite many factors affecting isotope composition of biomolecules, the tendency to equilibrium isotope fractionation, is clearly apparent. What is the significance of these examples to the objective of this paper? The $\beta^{13}C$ - $\delta^{13}C$ correlation between complex organic molecules or between different atoms of an organic compound is a **special feature** of biological systems. This phenomenon is related to the mechanism of chemical conversion in a substrate-enzyme complex (Galimov, 1985). It does not occur in abiogeneously synthesized organic compounds. Therefore it can be used as criterion for identification of organic carbon.

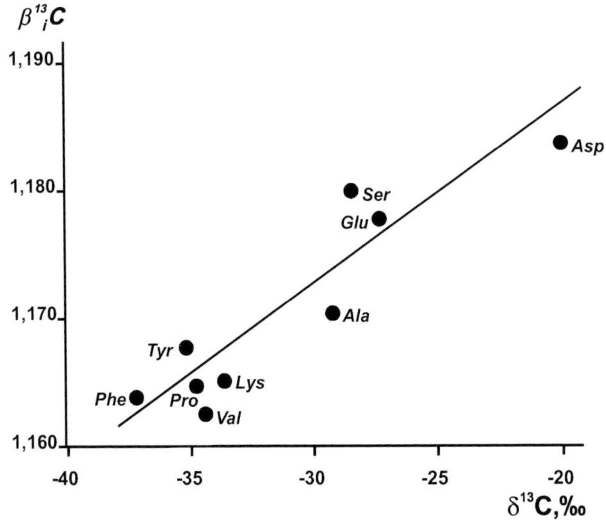

Figure 6. $\beta_i^{13}C$ - $\delta^{13}C$ correlation for amino acids extracted from leaves of potato. The $\delta^{13}C$ values are from Gleixner *et al.* (1998).

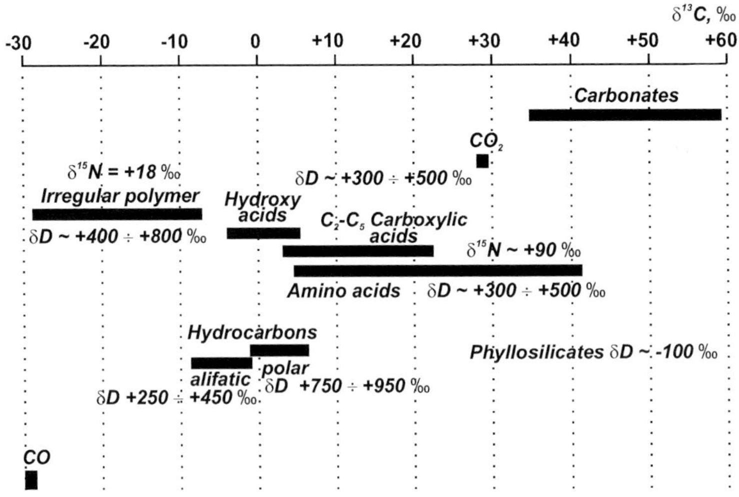

Figure 7. Isotope distribution in organic substances from carbonaceous chondrites. See the text for references.

4. Isotope Composition of Organic Compounds in Carbonaceous Meteorites

Among the available extraterrestrial material carbonaceous chondrites are known as carriers of significant amount of organic compounds. The ranges of isotopic variations in the carbon substances from carbonaceous chondrites are shown in Figure 7. The data are mostly from Cronin and Chang (1993) for the Murchison, with additional data from some other sources cited below.

The carbonaceous matter consists mainly of insoluble structurally heterogeneous macromolecular material.

Its carbon is relatively isotopically light. More than 70 species of amino acids have been identified in carbonaceous meteorites. They show significant enrichment in the ^{13}C isotope both relative to the macromolecular carbon and their terrestrial counterparts. Organic matter from carbonaceous chondrites (amino acids, in particular) shows also significant enrichment in ^{15}N and D (Yang and Epstein, 1983). Hydrogen isotopic composition of amino acids is characterized by δD values from about +700‰ up to +2500‰. The isotope data bear unambiguous evidence of extraterrestrial origin of the amino acids.

Another important individual component, identified in the carbonaceous chondrites, is hydroxy carboxylic acid. It has a structure similar to that of amino acids, but contains an OH-group at the place of the NH_2-group of amino acids. The isotopic composition of hydroxy carboxylic acids is slightly different from that of amino acids. The hydroxy acids from the Murchison are reported to have $\delta^{13}C = +4‰$ and $\delta D = +573‰$.

The substances on Figure 7 are schematized along the vertical scale in approximate accordance with their β-factors: carbonates and CO_2 have the highest β 13C-values. CO is the lowest one, organic carbon is intermediate. Among organic compounds hydrocarbons have $\beta^{13}C$-values lower than amino acids and hydroxy carboxylic acids.

Figure 7 leads to the impression that there is some trend reflecting correspondence between the established range of the $\delta^{13}C$ values for a given type of carbon and its position on the diagram defined by $\beta^{13}C$-characteristic. The polymer carbon however will not comply with this trend, and, when CO and CO_2 (carbonate) are excluded, for which a direct isotope exchange cannot be ruled out, the trend disappears.

Isotope data are available for some individual amino acids (Engel *et al.*, 1990) (Figure 8). There is no $\beta^{13}C$ - $\delta^{13}C$ correlation for meteoritic amino acids. This should be compared with biological amino acids shown in Figure 6b. Thus we can conclude that the criterion stated above being applied to organic material in carbonaceous chondrites does not reveal their biological origin.

5. Adenosine Triphosphate Synthesis as Indicator of Prebiotic Chemical Evolution

One more feature of the model cited above (Galimov, 2001) deserves to be mentioned in the context of the topic discussed, though it is not directly related to the isotopes. It has been argued that adenosine triphosphate (ATP) is a key molecule in prebiotic chemical evolution (Galimov, 2001).

Although ATP is a rather complex molecule both organic moieties of this compound, adenine and ribose, could be synthesized under certain conditions on the

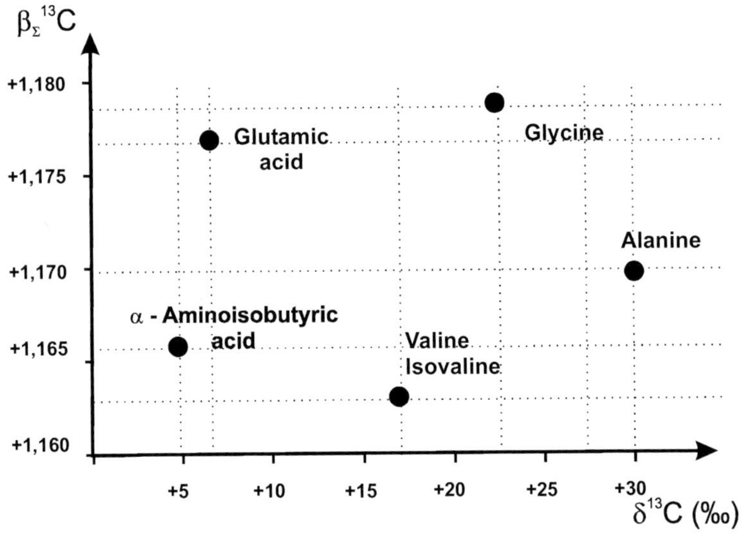

Figure 8. Comparison of $\delta^{13}C$ and $\beta^{13}C$ values for amino acids from Murchison meteorite. See text for references.

early Earth. Adenine is a product of HCN polymerization, while ribose can be synthesized during polymerization of formaldehyde, HCOH.

However, if HCN and formaldehyde are synthesized simultaneously, they tend to interact with the formation of cyanohydrin and this blocks subsequently synthesis of adenosine. Therefore to produce ATP, synthesis of the adenine and the ribose should proceed separately – in different environments. This condition of geochemical uncoupling is fulfilled provided that adenine is synthesized in a reduced atmosphere, whereas formaldehyde formation and its subsequent polymerization to ribose occur in an aqueous medium. Subsequently transfer of adenine from atmosphere to water and its interaction with ribose-phosphate may result in adenosine phosphate formation.

In contrast, synthesis of organic compoundes formed in meteorites should proceed via cianohydrin rather than ATP mechanism. There exists consent that material of carbonaceous meteorites was subjected to aqueous alteration in its early history (e.g., Cronin and Chang, 1993). However, we cannot expect to find atmosphere and water basins on the surface of the meteorite parent bodies. Therefore the cyanohydrin pathway must have been the dominating mechanism of HCN and HCHO interaction:

$$2HCHO + 2HCN \rightarrow \underbrace{2H_2COH - C \equiv N}_{\text{cyanohydrin}}$$

$$2H_2COH - C \equiv N + 3H_2O \rightarrow \underbrace{H_2COHCOOH}_{\text{hydroxyacid}} + \underbrace{H_2CNH_2COOH}_{\text{aminoacid}} + NH_3 \quad (3)$$

Products of the cyanohydrin reaction (Strecker reaction) are amino acids and hydroxy acids (Peltzer *et al.*, 1984), that indeed are common among individual organic compounds in carbonaceous meteorites. In contrast, presence of purines and sugars in carbonaceous chondrites is very low (if any) when compared to amino acids and hydroxy acids (Stoks and Schwartz, 1982). This means that a chemical pathway of forming organic compounds in meteorites was completely different from that, which was characteristic of pre-biological chemical evolution of organic molecules on the early Earth. In addition to isotopic data this indicates a none-life-related origin of organic material in carbonaceous chondrites.

6. Conclusions

1. Contrary to the current belief, relative enrichment in the light isotope of total carbon of organic materials is not a conclusive evidence of its life-related origin. To be able to distinguish a life-related origin of organic matter one needs to find properties which are unique for the chemistry of life.
2. One of them is the $\beta^{13}C$ - $\delta^{13}C$ pattern in carbon isotope distributions in complex organic molecules. The $\beta^{13}C$ - $\delta^{13}C$ correlation is a special feature of biological systems. It has often been observed both on the intermolecular and intramolecular level. In contrast, such reguliarity does not occur in a system of abiogeneously synthesized organic compounds. Therefore it can be used as a criterion for identification of organic carbon.
3. A survey of the available isotope data for organic compounds in meteorites shows that they do not comply with the $\beta^{13}C$ - $\delta^{13}C$ correlation. Thus, one can conclude that the organic material in carbonaceous chondrites is not a life-related substance.
4. The other criterion results from the same concept, which also suggests, that ATP is a key molecule of prebiotic evolution. Conditions for ATP synthesis come into existence on the surface of planets having reduced atmosphere and water basins. In contrast, condensation of HCN and HCHO in meteorites proceeds through cyanohydrin reaction, which leads to amino acids and hydroxy acids, rather than ATP. This process, in addition to the isotope criterion, indicates that the organic compounds in carbonaceous chondrites have an origin not related to life.
5. The techniques considered above are applicable to distinguish a life-related organic carbon from extraterrestrial sources.

References

Bogacheva, M.P., and Galimov, E.M.: 1979, 'Intramolecular Distribution of Carbon Isotopes in Chlorophyll and Hemine', *Geokhimiya* **7**, 1166–1172.

Cronin, J.R., and Chang, S.: 1993, 'Organic Matter in Meteorites: Molecular and Isotopic Analyses of the Murchison Meteorite', *in* J.M. Greenberg *et al.* (eds.), *The Chemistry of Life's Origins*, 209–258.

Crovisier, J.: 1998, 'Solids and Volatiles in Comets', *in* J.M. Greenberg (ed.), *Formation and Evolution of Solids in Space*.

Des Marais, D.J., *et al.*: 1981, 'Molecular Carbon Isotopic Evidence for the Origin of Geothermal Hydrocarbons', *Nature* **292**, 826–828.

Engel, M.H., Macko, S.A., and Sifer, S.A.: 1990, 'Carbon Isotope Composition of Individual Amino Acids From the Murchison Meteorite', *Nature* **348**, 47–49.

Galimov, E.M.: 1973, *Carbon Isotope in Oil and Gas Geology*, M., Nedra Press, 385 p. English translation: NASA TT-682, 1975.

Galimov, E.M.: 1985, 'The Biological Fractionation of Isotopes', Academ. Press, NY Toronto, 262 p.

Galimov, E.M.: 1991, 'Isotope Fractionation Related to Kimberlite Magmatism and Diamond Formation', *Geochim. Cosm. Acta* **55**, 1697–1708.

Galimov, E.M.: 2001, *Phenomenon of Life: Between Equilibrium and Non-linearity; Origins and Principles of Evolution*, URSS Press, M. 254 p.

Galimov, E.M., and Petersilie, I.A.: 1967, 'Carbon Isotope Composition of Methane Isolated in Pores of Some Igneous Minerals', *Doklady AN SSSR* **76**, N 4.

Galimov, E.M., and Shirinsky, V.G.: 1975, 'Ordered Distribution of Carbon Isotopes in Individual Compounds and Componenets of the Lipid Fraction of Organisms', *Geokhimiya* **4**, 503–528.

Gleixner, G., Scrimgeour, C., Schmidt, H.-L., and Viola, R.: 1998, 'Stable Isotope Distribution in the Major Metabolites of Source and Sink Organs of Solanum Tuberosum L.: A powerful Tool in the Study of Metabolic Partitioning in Intact Plants', *Planta* **207**, 241–245.

Irwine, W.M.: 1998, 'Extraterrestrial Organic Matter: A Review', *Origins Life Evol. Biosphere* **28**, 365–383.

McKay, D.S., *et al.*: 1996, 'Search for Past Life on Mars: Possible Relic Biogenic Activity in Martian Meteorite ALH 84001', *Science* **273**, 924–930.

Meinschein, W.G., Hegeman G.D., and Bromley, B.W.: 1984, 'Intramolecular Distribution of Stable Isotopes of Carbon', abstract, 27th Intern. Geolog. Congress, vol. 5, sect. 10,11, Moscow, 4-14 Aug. 1984, pp. 345–346.

Mojzsis, S.J., Arrhenius, G., McKeegan, K.D., Harrison, T.M., Nutman, A.P., and Friend, C.R.L.: 1996, 'Evidence for Life on Earth Before 3,800 Million Years Ago', *Nature* **384**, 55–59.

Peltzer, E.T., Bada, J.L., Schlesinger, G., and Miller, S.L.: 1984, 'The Chemical Conditions on the Parent Body of the Murchison Meteorite: Some Conclusions Based on Amino, Hydroxy, and Dicarboxylic Acids', *Adv. Space Res.* **4**, 69–74.

Rinaldi, G.G., Meinschein, W.G., and Hays, J.M.: 1974, 'Intramolecular Carbon Isotope Distribution in Biologically Produced Acetoin', *Biomed. Mass Spectrometry* **1(6)**, 415–417.

Rossmann, A., Butzenlecher, M., and Schmidt, H.-L.: 1991, 'Evidence for a Nonstatistical Carbon Isotope Distribution in Natural Glucose', *Plan. Physiol.* **96**, 609–614.

Schidlowski, M.: 1993, 'The Beginnings of Life on Earth: Evidence From the Geological Record', *in* J.M. Greenberg *et al.* (eds.), *The Chemistry of Life's Origins*, Kluwer Acad. Publ., 389–413.

Schmidt, H.-L., and Gleixner, G.: 1998, 'Carbon Isotope Effects in Key Reaction in Plant Metabolism and ^{13}C-pattern in Natural Compounds', *in:* H. Griffiths (ed.), *Stable Isotopes*, BIOS Science Publisher Ltd., Oxford, pp. 13–25.

Stoks, P.G., and Schwartz, A.W.: 1982, 'Basic Nitrogen-heterocyclic Compounds in the Murchison Meteorite', *Geochim. Cosmochim. Acta* **46**, 309–315.

Weilacher, T., Gleixner, G., and Schmidt, H.-L.: 1996, 'Carbon Isotope Pattern in Purine Alkoloids: A Key to Isotope Discriminations in C_1 Compounds', *Phytochemistry* **41**, 1073–1077.

Wright, I.P., Grady, M.M., and Pillinger, C.T.: 1989, 'Organic Materials in a Martian Meteorite', *Nature* **340**, 220–222.

Yang, J., and Epstein, S.: 1983, 'Interstellar Organic Matter in Meteorites', *Geochim. Cosmochim. Acta* **47**, 2199–2216.

Zinner, E.: 1997, 'Presolar Material in Meteorites', *in* T. Bernatowicz and E. Zinner (eds.), *Astrophysical Implications of the Laboratory Study of Presolar Materials*, AIP, IUY, 3–26.

Address for Offprints: E.M. Galimov, V.I. Vernadsky Institute of Geochemistry and Analytical Chemistry of the Russian Academy of Sciences, 119991, Moscow, Kosygin 19, Russia; galimov@geokhi.ru

IV. ISOTOPIC FRACTIONATION PROCESSES

OXYGEN ISOTOPE PROCESSES AND TRANSFER REACTIONS

K. MAUERSBERGER, D. KRANKOWSKY AND C. JANSSEN
Max-Planck-Institut für Kernphysik, Division of Atmospheric Physics, D–69029 Heidelberg, Germany

Received: 11 April 2002; Accepted in final form: 1 October 2002

Abstract. A unique kinetic isotope effect has been found in the formation process of ozone molecules. Isotope enrichments of about 10% above statistically expected values were first discovered in atmospheric isotopomers $^{49}O_3$ and $^{50}O_3$ and later in many other molecular combinations. Most recently the source of this effect was identified through measurement of isotope-specific ozone formation rate coefficients which show a large variability of over 50%. Ozone molecule formation is a complex process since different reaction channels contribute to a specific isotopomer. In addition, fast oxygen isotope exchange reactions determine the abundance of atomic oxygen participating in ozone formation. The isotope enrichments observed are both pressure and temperature-dependent and they decrease at pressures above 100 mbar and toward lower temperatures. Ozone possesses not only one of the most unusual isotope anomalies, it also serves as a mediator by transferring heavy oxygen from the O_2 reservoir to other species. Stratospheric isotope composition of CO_2 has been recently measured with high accuracy and a pronounced isotopic signature was found which shows that ^{17}O is preferentially transferred from O_3 into CO_2.

1. Introduction

A large and unusual isotope effect in ozone has been known for about two decades (Mauersberger, 1981; Heidenreich and Thiemens, 1983). Immediately after its discovery the effect was linked to the oxygen isotope anomaly found in meteorites (Clayton *et al.*, 1973; Heidenreich and Thiemens, 1983) although little was known at that time about its origin and the parameters that influence its magnitude. Extensive laboratory work and stratospheric ozone analysis have now resulted in a consistent picture of ozone isotopomer enrichments and their pressure and temperature dependence. Further progress was recently made when the origin of the isotope anomaly was found in the highly variable rate coefficients of isotope specific ozone formation channels (Anderson *et al.*, 1997; Mauersberger *et al.*, 1999; Janssen *et al.*, 1999). Because of competing reactions such as the oxygen isotope exchange processes that are always present in the course of ozone formation, a complex isotope effect in ozone has emerged of which the enrichments measured in atmospheric ozone isotopomers ($^{49}O_3$ and $^{50}O_3$) are just the final result of a substantial number of various reaction channels.

This paper will provide a review of measured enrichments and of parameters that influence their magnitude. After a brief introduction to the ozone molecule, the ozone isotope effect in atmospheric oxygen which is dominated by ^{16}O will be

Space Science Reviews **106:** 265–279, 2003.

presented, followed by a comparison of atmospheric measurements and laboratory-derived predictions. Analysis of ozone that is produced in isotopically enriched oxygen gases will be described next. The discussion on laboratory results will be completed with a presentation of the isotope-specific rate coefficients. Overall, it is a main goal of this paper to clearly separate isotope effects that are caused by the exchange reactions from those that are a result of the intrinsic ozone formation process which is expressed in the values of the different rate coefficients.

2. A Brief Introduction to the Ozone Molecule and the Oxygen Isotope Exchange Reactions

Ozone is the only stable atmospheric molecule containing three atoms of the same element. Chapman proposed in 1930 (Chapman, 1930) the following oxygen reactions that produce and destroy ozone in our atmosphere:

$$O_2 + h\nu \rightarrow O + O \tag{1}$$

$$O + O + M \rightarrow O_2 + M \tag{2}$$

$$O + O_2 + M \rightarrow O_3 + M \tag{3}$$

$$O_3 + h\nu \rightarrow O + O_2 \tag{4}$$

$$O_3 + O \rightarrow 2O_2 \tag{5}$$

Throughout the paper we will study the formation of ozone since this is the process in which the isotope effect is observed. Numerous destruction cycles besides (4) and (5) enhance the role of ozone in our atmosphere, but play no role in the discussion of this paper. While reaction (2) is only found at very high altitudes (> 90 km), reaction (3) leads in a three-body reaction to the formation of ozone throughout the atmosphere.

The source of atomic oxygen is photolysis of O_2 (reaction (1)), of NO_2 or even of O_3 (4). In the atmosphere and in laboratory tests described below ozone is always a minor constituent and its isotope anomaly will not influence the atomic distribution of the gas in which it was formed. Thus, only very small amounts of ozone are produced, collected and analyzed in our laboratory work.

The O_3 molecule forms an open triangle and has a binding energy of 1.1 eV. This low energy compared to O_2 (5.1 eV) or N_2(9.8 eV) makes O_3 a very reactive molecule. Because there are three stable oxygen isotopes – ^{16}O, ^{17}O, and ^{18}O – the ozone molecules can carry a large variety of isotope combinations. In the atmosphere, however, only $^{49}O_3$ and $^{50}O_3$ have any significant abundance since ^{18}O is by a factor of almost 500 and ^{17}O by a factor of 2500 lower than ^{16}O . Because of its triangular geometry the singly-substituted heavy oxygen atom can be found in the apex or at either end of the triangle. In the first case, such an ozone molecule is called symmetric; when the heavy oxygen atom is located at either end, the molecule is asymmetric.

Early theoretical discussions on ozone isotope ratios were focused on the well-known isotope exchange reactions between atomic and molecular oxygen of which two are shown below:

$$^{18}O + {}^{16}O^{16}O \rightleftarrows {}^{18}O^{16}O + {}^{16}O \quad (6)$$

$$^{17}O + {}^{16}O^{16}O \rightleftarrows {}^{17}O^{16}O + {}^{16}O \quad (7)$$

Those fast equilibrium processes determine after photolysis the distribution of atomic oxygen in gas mixtures such as air. Exchange recycles O-atoms numerous times through O_2 before in a three-body collision ozone is formed (Kaye and Strobel, 1983). Because of the different zero point energies of the O_2 molecules that participate in the exchange reactions, the rate coefficients of the different channels are higher for exothermic processes which proceed from left to right in (6) and (7) and lower for endothermic that are right to left. Thus, the distribution of the three oxygen atoms is governed by exchange which will lower ^{18}O and ^{17}O compared to what would be expected from just statistically distributed O-isotopes in molecular oxygen. Their abundance will directly determine the production of ozone molecules and thus the two heavy isotopomers $^{49}O_3$ and $^{50}O_3$ are supposed to be lower than expected. More examples of how the impact of exchange can be observed on the different ozone isotopes will follow later. Here we like to summarize that would an intrinsic ozone isotope effect not exist, the isotope exchange reactions alone would cause a typical mass-dependent depletion in the isotopomers from mass 49 through 54. To demonstrate this once more: According to Anderson *et al.* (1997), the ^{18}O density is about 7% lower in gas mixtures than statistically expected, and since only one reaction channel ($^{18}O + {}^{18}O^{18}O + M$) leads to mass 54, there should be an isotope fractionation of approximately -7% in $^{54}O_3$, assuming that the rate coefficient for the heavy ozone formation process is the same as for $^{48}O_3$. Similarly, Kaye and Strobel (1983) calculated for $^{50}O_3$ a 3% lower than statistically expected value.

3. Measurements of Isotope Enrichments in Atmospheric Oxygen

3.1. Laboratory Investigations

First measurements of isotope ratios in the stratosphere and in laboratory environments were met with surprise since an usually large enrichment and not a depletion in the two heavy isotopomers $^{49}O_3$ and $^{50}O_3$ was observed. Intensive laboratory studies permit now to characterize in sufficient details the pressure and temperature dependence of the enrichments. Shown in Figure 1 are results obtained in two laboratories (Thiemens and Jackson, 1988, 1990; Morton *et al.*, 1990) when ozone was produced over a wide range of gas pressures and at room temperature. The remarkable result is that $^{49}O_3$ is almost as much enriched as $^{50}O_3$ and not only half of that as would be expected in a strictly mass-dependent process. At an ozone

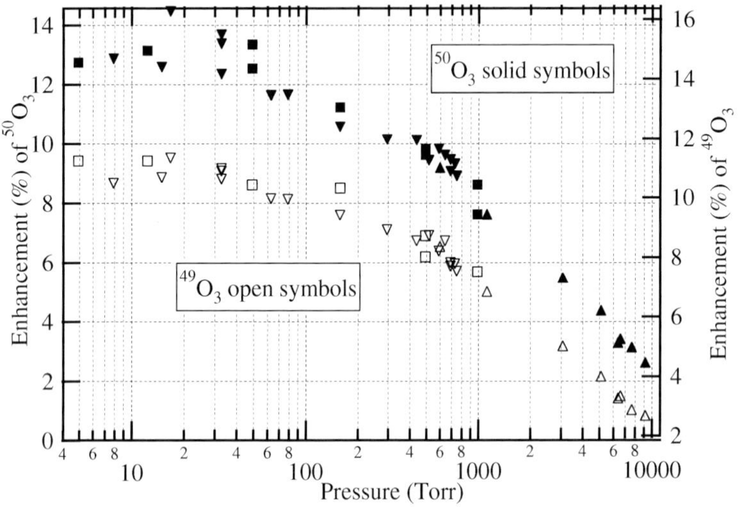

Figure 1. Pressure dependence of ozone isotopomer enrichment in atmospheric oxygen at room temperature. The enrichment is defined as $E(^{M}O_3)(\%) = [(^{M}O_3/^{48}O_3)_{meas} / (^{M}O_3/^{48}O_3)_{calc} - 1] \times 100\%$ with M = 49 or 50. The calculated ratios are based on the atmospheric partitioning of the oxygen isotopes ($[^{16}O] = 99.756\%$, $[^{17}O] = 0.038\%$, and $[^{18}O] = 0.205\%$) and assuming statistical partitioning for the ozone isotopomers. Square symbols are from Morton *et al.* (1990), downside triangles from Thiemens and Jackson (1988) and upside triangles from Thiemens and Jackson (1990).

formation pressure below 100 mbar the enrichment for $^{49}O_3$ is 11.5% and for $^{50}O_3$ about 13.0%.

It was soon recognized that the temperature of the gas in which ozone is formed also plays a crucial role. Shown in Figure 2 is the change of enrichment of the two important atmospheric isotopomers over a wide temperature range and at low ozone formation pressure. Morton *et al.* (1990) recognized that the abundance of isotopomers is also strongly influenced by the temperature dependence of the exchange reactions.

3.2. Atmospheric Measurements

While laboratory measurements are well controlled and, if needed, can be repeated and changed to check for consistent results, atmospheric ozone isotope data are much more difficult to obtain, and thus initial results were sparse and carried large uncertainties. All techniques employed, such as mass spectrometry (Mauersberger, 1981, 1987) or infrared measurements (Abbas *et al.*, 1987; Goldman *et al.*, 1989), provided data that scattered considerably and showed in general large enrichments in $^{50}O_3$. The emerging laboratory data soon permitted a prediction for atmospheric pressure and temperature conditions (Morton *et al.*, 1990). In the stratosphere temperatures and pressures are low and laboratory data would predict enrichments near 10% or lower for the two heavy ozone isotopomers. Until 1990

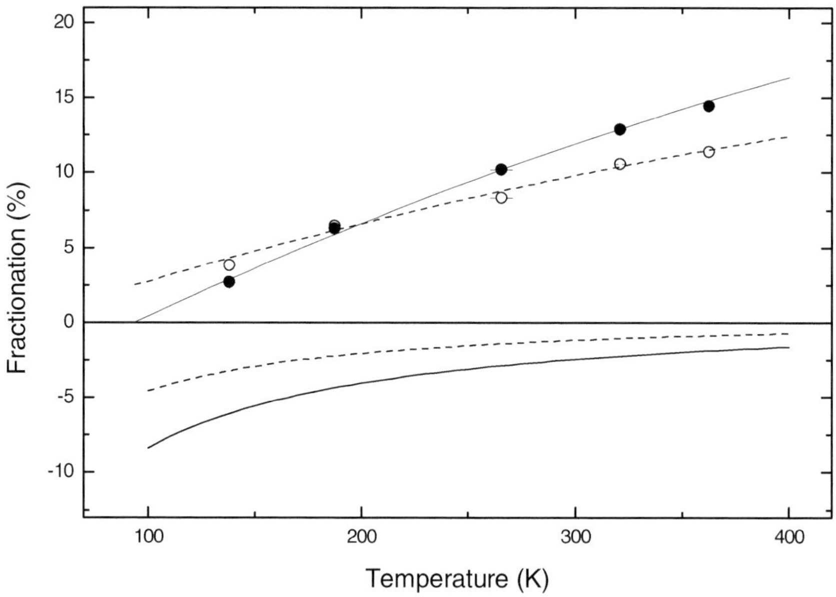

Figure 2. Measured temperature dependence of $^{49}O_3$ (*open circles*) and $^{50}O_3$ (*black dots*) enrichments. The two lower curves are the expected fractionation of $^{49}O_3$ (*solid curve*) and $^{50}O_3$ (*dashed curve*) resulting from atomic oxygen fractionation alone due to the exchange reactions (Morton *et al.*, 1990).

atmospheric measurements were in average considerably higher (Irion *et al.*, 1996). After 1990, a stratospheric ozone collector system was developed that promised to provide the most reliable data for both heavy isotopes (Schueler *et al.*, 1990). During the last five years a similar system was repeatedly flown onboard a balloon gondola and collected separate samples of CO_2 and ozone from Kiruna/Sweden and Aire-sur-l'Adour/France. Krankowsky *et al.* (2000) showed that the isotope analysis from those collector flights clearly linked stratospheric enrichments to atmospheric temperatures and pressures. Figure 3 shows a direct comparison of enrichments measured during laboratory investigations versus temperatures and, using the same scale, stratospheric results. The predictions made from laboratory measurements were proved to be correct. In a most recent paper Mauersberger *et al.* (2001) reviewed all past stratospheric measurements and concluded that indeed the early results were in error and should be rejected. Typical values of enrichments around 30 km are 8.0% for $^{49}O_3$ and 9.0 to 10.0% for $^{50}O_3$ (Figure 3).

Ozone in the troposphere is only present in parts per billion and thus samples are difficult to collect and to analyze. Krankowsky *et al.* (1995) were first to analyze a comprehensive set of nearly 50 tropospheric ozone samples collected over 4 months which showed that the enrichments in both heavy isotopes follow laboratory predictions when pressure and temperatures ($\approx$ 20 °C) are included. Figure 4 shows that no statistically significant departures in the samples were

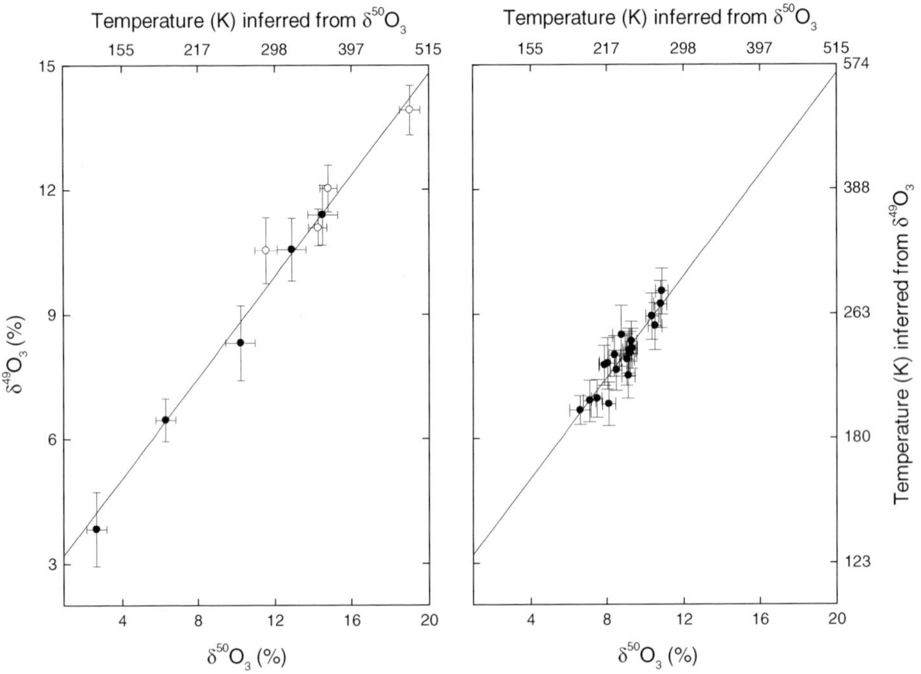

Figure 3. Temperature dependence of heavy ozone isotope ratios measured in the laboratory (*left*) and in the stratosphere (*right*). Solid symbols are data from Morton *et al.* (1990), open symbols from Burghardt (1997). Error bars are 2σ statistical errors. Note: Corresponding temperature scales are above and to the right for $\delta^{50}O_3$ and $\delta^{49}O_3$. The stratospheric ozone isotope data were collected during balloon flights between 20 and 34 km (Krankowsky *et al.*, 2000). The data follow the same linear relation as suggested from the laboratory studies shown on the left.

found, irrespective of the abundance of tropospheric ozone (Krankowsky *et al.*, 1995). Average enrichment values are 7.1% for $^{49}O_3$ and 9.1% for $^{50}O_3$. Later in a paper by Johnston and Thiemens (1997) departures from those toward lower values were measured in three different locations and were linked to secondary effects influencing the ozone isotope distribution. Unfortunately, no other follow-up studies have been made. Those values may have suffered from measurement problems since the xenon gas collected with the ozone samples was not removed before ozone isotope analysis, and the collection efficiency was generally very low which might have induced additional isotope effects.

4. Ozone Produced in Isotopically Enriched Oxygen Mixtures

The combination of the three stable oxygen isotopes permits the formation of ozone molecules far exceeding the two isotopomers found in the atmosphere. A careful selection of the isotope composition within an oxygen gas can lead to an unambiguous determination of ozone molecules from mass 48 through 54 using

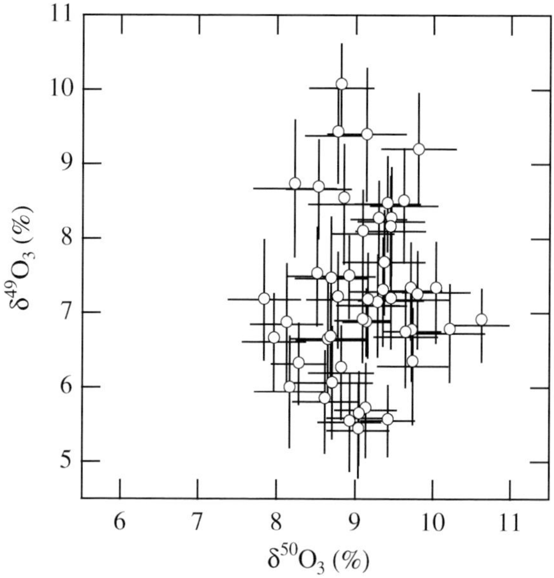

Figure 4. Enrichment in $^{50}O_3$ and $^{49}O_3$ measured in tropospheric air during the period from July to September, 1994 (Krankowsky *et al.*, 1995). The δ-values are calculated with respect to the isotopic composition of molecular oxygen in air. Error bars are 1σ statistical errors. The mean values in enhancement are (9.1 ± 0.2)% and (7.1 ± 0.3)% in $^{50}O_3$ and $^{49}O_3$, respectively.

mass spectrometry. For example, in a ^{17}O-dominated gas mixture, $^{16}O^{17}O^{17}O$ will be much more frequently produced than $^{16}O^{16}O^{18}O$ of the same mass. To perform such investigations, the O_2 gas has to be well scrambled and its oxygen isotope composition must be precisely known. The expected distribution of the O_3 molecules can be calculated using the rules of combinatoric.

The amount of ozone produced will always be much smaller than the amount of O_2 gas present. Thus, the formation of ozone will not significantly change the isotopic distribution of the original gas. O_3 is collected at liquid N_2-temperatures by flowing the gas mixture through a cold trap (Mauersberger *et al.*, 1993). After removing any residual O_2, ozone will be released by warming the trap and a mass spectrometric analysis will be performed. It is important to note that ozone is a very reactive constituent and can only be exposed to the ion source of a mass spectrometer in a 'fly-through' fashion without wall interactions. Such an experiment called a mass spectrometer beam system was developed at the University of Minnesota (Mauersberger, 1977) and has served as a direct precision analysis instrument ever since. Before an ozone isotope analysis is made, the mass-scale of the instrument is calibrated with ozone gas composed of pure $^{48}O_3$ and of pure $^{54}O_3$.

Results of such investigations are summarized in Figure 5 which displays the departure of the measured from the calculated isotopomer distribution. Three special features are immediately visible in the figure: 1) there are two isotopomers $^{17}O^{17}O^{17}O$ and $^{18}O^{18}O^{18}O$ that are lower than predicted, the first one about half as

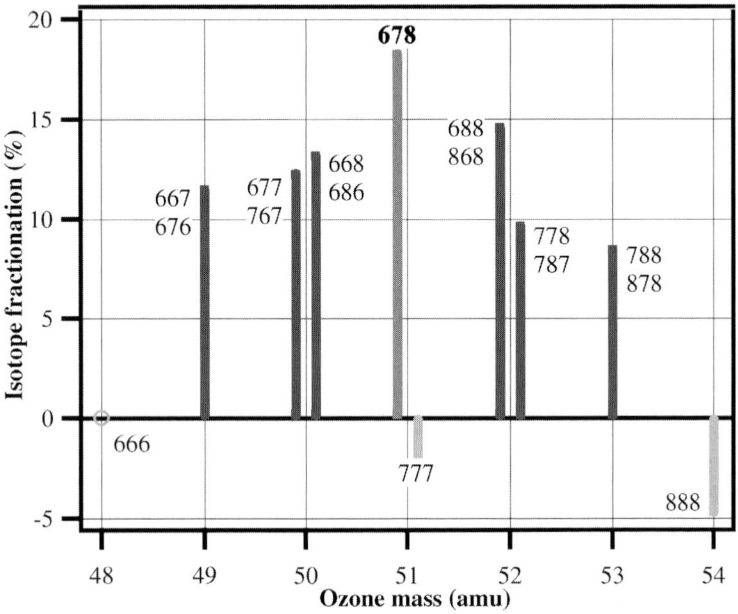

Figure 5. Measured enrichment or depletion of all possible ozone isotopomers. The labels 6, 7, 8 stand for ^{16}O, ^{17}O, and ^{18}O, respectively. Ozone was produced in two isotopically enriched oxygen mixtures of well-known composition at 70 torr and room temperature. The statistical abundance of each isotopomer was calculated and used to derive from the measured values the isotopic fractionation. The values in the graph are normalized so that $^{48}O_3$ has zero enrichment.

much as the latter one; 2) ozone with the atomic combination $^{16}O^{17}O^{18}O$ has the highest enrichment of approximately 18%; 3) all other isotopomers are about two third as much enriched as $^{16}O^{17}O^{18}O$.

Those results led us initially to the conclusion that molecular symmetry effects may play a critical role in the ozone isotope formation (Anderson *et al.*, 1992). Purely symmetric molecules ($^{17}O^{17}O^{17}O$) appear to be formed according to the conventional theory (Kaye and Strobel, 1983), while the molecular combination of mass 51 that contains only asymmetric molecules shows the highest enrichment and thus the largest isotope effect. As the following results in the next section will show, this conclusion was unfortunately wrong. Of importance, however, is also the result that both $^{16}O^{16}O^{17}O$ of mass 49 and $^{16}O^{16}O^{18}O$ of mass 50 which are the most abundant isotopomers in our atmosphere (see Section 3) have the same enhancements of 11 and 13% respectively, even when produced in highly enriched mixtures that are very different from the oxygen composition in the atmosphere. Thus, the ozone isotope effect does not depend on the isotope composition of the O_2 gas. It is an intrinsic molecular effect: whenever an isotopomer is formed its enrichment or depletion is determined by the pressure and temperature of the gas in which ozone is produced.

5. The Isotopically Specific Ozone Rate Coefficients

The fundamental quantity which best describes a chemical reaction is the rate coefficient which may be pressure or temperature-dependent and in either case can be related to a specific isotope formation channel. To investigate symmetry effects or other parameters influencing ozone formation, new experiments were constructed and a substantial number of relative rate coefficients measured (Anderson *et al.*, 1997; Janssen *et al.*, 1999).

During the first round of new experimental investigations, four specific rate coefficients were determined:

$$^{16}O + {}^{16}O^{16}O + M \rightarrow {}^{16}O^{16}O^{16}O + M \tag{8}$$

$$^{16}O + {}^{18}O^{18}O + M \rightarrow {}^{16}O^{18}O^{18}O + M \tag{9}$$

$$^{18}O + {}^{16}O^{16}O + M \rightarrow {}^{18}O^{16}O^{16}O + M \tag{10}$$

$$^{18}O + {}^{18}O^{18}O + M \rightarrow {}^{18}O^{18}O^{18}O + M \tag{11}$$

Of the four molecules formed two are symmetric (8 and 11) and two are asymmetric (9 and 10). Because of our preconceived notion that the asymmetric molecules carry the enhancements, the asymmetric molecules should show high rate coefficients that would contribute to the observed enrichments in the isotopomers of Figure 5. To our great surprise, $^{18}O^{16}O^{16}O$ had a rate coefficient of 0.92, smaller than the one for the standard molecule in reaction (8) which was normalized to 1.00 while the molecule produced in reaction (9) had a rate coefficient of 1.53. Detailed analysis actually permitted the determination of absolute coefficients which resulted in a value for reaction (8) of 6.04×10^{-34} cm^6 s^{-1}, in good agreement with literature data (Anderson *et al.*, 1997). Finally, for the last reaction (11) a relative rate coefficient of 1.03 was determined. The large difference between 0.92 and 1.53 of the two asymmetric molecules eliminated a symmetry-driven explanation for the ozone isotope effect that was suggested by the results shown in Figure 5.

Furthermore, for the two reactions (9) and (10) involving homonuclear oxygen molecules the pressure dependence of the relative rate coefficients (Günther *et al.*, 1999) and its third-body dependence (Günther *et al.*, 2000) has been experimentally studied. While the rate coefficient for reaction (9) compared to reaction (8) decreases as the total pressure increases, the rate coefficient of reaction (10) compared to reaction (11) remains unchanged. Whereas absolute rate coefficients for ozone formation differ by a factor of four between Ar and CO_2 as third-body, the relative rate coefficients – reaction (9) compared to reaction (8) and reaction (10) relative to (11) – remain the same.

More relative rate coefficients have been measured which are summarized in Table I. Using the mass spectrometer experiment, additional reactions with homonuclear oxygen molecules ($^{16}O^{16}O$, $^{17}O^{17}O$, and $^{18}O^{18}O$) were studied, resulting in asymmetric ozone molecules (Mauersberger *et al.*, 1999). A complete accounting of the isotope-specific channels contributing to the isotopomer $^{50}O_3$ requires two

TABLE I

Reaction channels of all measured oxygen isotope combinations leading to ozone molecules. *Right column:* Rate coefficients (Mauersberger *et al.*, 1999; Janssen *et al.*, 1999) relative to the standard reaction of $^{16}O + {}^{16}O^{16}O + M$ are shown. Data were obtained in N_2/O_2 mixtures at 200 Torr and at room temperature. †For these reactions which involve heteronuclear oxygen molecules the relative reaction probability is shown while relative rate coefficients are obtained by dividing the quoted numbers by two. ‡Those rates may contain small contributions from the subsequent symmetric molecules (Janssen *et al.*, 1999).

Mass	Reaction	Relative rate coeff. rate coefficient
48	$^{16}O + {}^{16}O^{16}O \rightarrow {}^{16}O^{16}O^{16}O$	1.00
49	$^{17}O + {}^{16}O^{16}O \rightarrow {}^{17}O^{16}O^{16}O$	1.03‡
50	$^{16}O + {}^{17}O^{17}O \rightarrow {}^{16}O^{17}O^{17}O$	1.23‡
	$^{16}O + {}^{16}O^{18}O \rightarrow {}^{16}O^{16}O^{18}O$	1.45†
	$\rightarrow {}^{16}O^{18}O^{16}O$	1.08†
	$^{18}O + {}^{16}O^{16}O \rightarrow {}^{18}O^{16}O^{16}O$	0.92
	$\rightarrow {}^{16}O^{18}O^{16}O$	0.006
51	$^{17}O + {}^{17}O^{17}O \rightarrow {}^{17}O^{17}O^{17}O$	1.02
52	$^{16}O + {}^{18}O^{18}O \rightarrow {}^{16}O^{18}O^{18}O$	1.50
	$\rightarrow {}^{18}O^{16}O^{18}O$	0.029
	$^{18}O + {}^{16}O^{18}O \rightarrow {}^{18}O^{16}O^{18}O$	1.04†
	$\rightarrow {}^{18}O^{18}O^{16}O$	0.92†
	$^{18}O + {}^{17}O^{17}O \rightarrow {}^{18}O^{17}O^{17}O$	1.03‡
53	$^{17}O + {}^{18}O^{18}O \rightarrow {}^{17}O^{18}O^{18}O$	1.31‡
54	$^{18}O + {}^{18}O^{18}O \rightarrow {}^{18}O^{18}O^{18}O$	1.03

more rate coefficients for the reactions

$$^{16}O + {}^{16}O^{18}O + M \rightarrow {}^{16}O^{16}O^{18}O + M \tag{12}$$

$$^{16}O + {}^{18}O^{16}O + M \rightarrow {}^{16}O^{18}O^{16}O + M\,. \tag{13}$$

In a separate experiment, tunable diode laser technique was employed (Janssen *et al.*, 1999) and the rate coefficients for (12) and (13) as well as for other reactions determined. This experiment also showed that the ozone formation process $O + O_2$ is dominated by end-on reactions and not by insertion. With a set of 15 relative rate coefficients available (Table I), certain isotopomer enrichments shown in Figure 5 can now be calculated. However, besides the rate coefficients for ozone formation the oxygen isotope exchange reactions have also to be considered. The molecule $^{54}O_3$ on the other hand shows in Figure 5 a depletion of about 4.5%. This can now be understood as a result of the relative rate coefficient of 1.03 and a 7% lower ^{18}O-density in the gas mixture explained above. Recently Janssen *et al.* (2001)

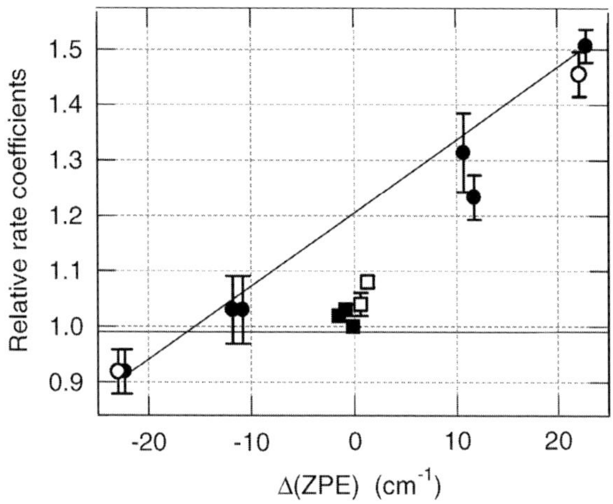

Figure 6. Relative rate coefficients versus zero point energy (ZPE) differences (in cm^{-1}) in oxygen molecules of the corresponding isotope exchange reactions. Error bars represent 1σ statistical uncertainties. Exothermic or endothermic isotope exchange reactions (*circles*) can be distinguished from energetically neutral collisions (*squares*). All of the former reactions ($\Delta(\mathrm{ZPE}) \neq 0$) which are collisions with homonuclear diatoms (*full symbols*) and heteronuclear molecules (*open symbols*) clearly show a linear correlation with the ZPE change, such that endothermic exchange corresponds to fast and exothermic exchange to slow ozone formation. The line is a best fit to the collisions with homonuclear oxygen molecules (*full circles*) with reaction $^{16}O + {}^{17}O^{17}O \rightarrow {}^{16}O^{17}O^{17}O$ being excluded. While all of the neutrals (shifted in ZPE for clarity) are well below this line, there is apparently a small difference between the reactions resulting from collisions with homonuclear O_2 (*full squares*) and heteronuclear molecules (*open squares*).

offered an explanation, based on experimental results that point to a fundamental correlation in the ozone formation process. When isotope exchange processes are considered, two of which are shown in (6) and (7), a relationship emerges that shows a close correlation between the values of the rate coefficients (Table I) and the kind of oxygen molecules participating in an exchange process that may lead to ozone. For processes that are exothermic (10) the rate coefficients are low, for endothermic processes (9) they are high.

Shown in Figure 6 are all known relative rate coefficients, plotted versus differences in ZPE of the oxygen molecules. For eight coefficients, a linear relationship is found. Those are all asymmetric ozone molecules, while for symmetric molecules the rates are below the straight line by approximately 15%.

6. Oxygen Isotope Transfer Into Stratospheric CO_2

Ozone is present with 90% of its atmospheric content in the stratosphere reaching mixing ratios as high as 10 ppm around 25 km. Because of its numerous photochemical reactions, the isotope anomaly can be transferred to other oxygen-bearing minor constituents. While some models have explored this mechanism

(Lyons, 2001), experimental work has focused on analysis of stratospheric CO_2 samples that were collected during balloon flights either as whole-air samples or in CO_2 traps when ozone collectors were flown. Collection, transfer and isotope ratio measurements were even for CO_2 a challenging task and resulted initially in very variable enrichments of ^{17}O and ^{18}O (Thiemens *et al.*, 1991). They always showed, however, that ^{17}O was considerably enhanced in stratospheric CO_2 compared to ^{18}O.

Yung and coworkers first proposed in 1991 (Yung *et al.*, 1991) a chemical transfer mechanism that resulted in ^{18}O enrichments in CO_2 via the following processes:

$$O_3 + h\nu \rightarrow O(^1D) + O_2 \tag{14}$$
$$O(^1D) + CO_2 \rightarrow CO_3^* \tag{15}$$
$$CO_3^* \rightarrow CO_2 + O(^3P)\,. \tag{16}$$

As the ozone sample analysis became more reliable (Krankowsky *et al.*, 2000; Mauersberger *et al.*, 2001), the isotope ratios determined from stratospheric CO_2 samples resulted in an interesting relationship that is shown in Figure 7. Starting at the tropopause (near 12 to 15 km), the isotope ratios rapidly increase resulting in a gradient $\delta^{17}O/\delta^{18}O$ of approximately 1.7, considerably different from older values that were closer to 1.0. It appears from the data collected at midlatitude and high northern latitudes and during different seasons that this gradient is representative of the isotope enrichment found in CO_2 throughout the stratosphere. Clearly, there is a preferred transfer mechanism of ^{17}O over ^{18}O into CO_2 (Lämmerzahl *et al.*, 2002). A detailed kinetical model to describe such an isotope transfer process has not yet been developed.

7. Conclusions

The isotope effect observed in the formation process of ozone is complex and unusual: It does not depend on the mass of the ozone molecules, but in a very subtle way on the oxygen molecules participating in the formation of ozone. A correlation between those molecules and the relative rate coefficients of isotope-specific formation channels has been identified (Janssen *et al.*, 2001). The complexity of the effect is further compounded since a number of formation channels contribute to the ozone molecule of a specific mass. The measured enrichments or depletions are a result of those contributions and of the oxygen isotope exchange reactions. Finally the rate coefficients and, subsequently, the isotope enrichments show a strong dependence on pressure and temperature of the gas in which the ozone molecules are formed. In our opinion the fundamental ozone isotope effect is expressed in the rate coefficients. There exist, however, a large variety of ozone isotope data (such as shown in Figures 1 and 5) that combine different contributions and are often referred to as the 'ozone isotope anomaly'.

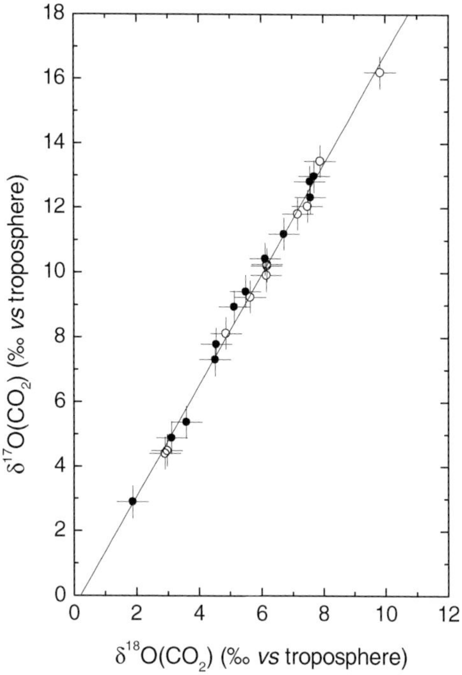

Figure 7. Three-isotope plot of stratospheric CO_2 (Table I). The error bars represent 1σ-values. Data shown were obtained at different dates and altitudes from two locations Kiruna, Sweden (67.9° N, 21.1° E) open circles and Aire-sur-l'Adour, France (43.7° N, 0.3° W) solid circles. A very tight correlation exists with a steep gradient of 1.71 ± 0.03 while passing through the tropospheric values of $\delta^{17}O \approx 21$‰ and $\delta^{18}O \approx 41$‰ which are set equal to zero.

A number of attempts have been made to develop a comprehensive theoretical model for the observed isotopomer's enrichment. Only after the discovery of the rate coefficient variability as the source of the ozone isotope effect theoretical research has become more focused and shown some very interesting results. In a number of papers by Marcus and coworkers rate coefficients have been calculated by applying a modified RRKM-theory of unimolecular reactions to the ozone formation process. Their calculations of relative rate coefficients have been based on normalization to the highest and lowest measured values (Table I, reactions $^{16}O + {}^{18}O^{18}O \rightarrow {}^{16}O^{18}O^{18}O$ and $^{18}O + {}^{16}O^{16}O \rightarrow {}^{18}O^{16}O^{16}O$) and by applying a non-RRKM factor of about 18% to the symmetric molecules. Hathorn and Marcus (1999; 2000) first pointed out the complex dependency of the rate coefficients on the molecular mass of the O_2 molecules that contribute to ozone.

Recently, another successful attempt has been made by Robert and Camy-Peyret (2001) who used a numerical application of differential scattering cross sections to calculate rate coefficients of distinguishable and indistinguishable isotopes. Some features of the enrichments were well reproduced. A "first-principle approach" to the problem is still waiting and may be complicated by the fact that not only the isotopomers show a temperature dependence but the different rate coefficients as well. More experimental work in this area which also will relate to other possible applications such as the early solar nebula may be highly desirable.

References

Abbas, M.M., Guo, J., Carli, B., Mencaraglia, F., Carlotti, M., and Nolt, I.G.: 1987, 'Heavy Ozone Distribution in the Stratosphere From Far-infrared Observations', *J. Geophys. Res.* **92**, 13231–13239.

Anderson, S.M., Mauersberger, K., Morton, J., and Schueler, B.: 1992, 'Heavy Ozone Anomaly', *ACS Symp. Series, Isotope Effects in Gas-Phase Chemistry* **502**, 155.

Anderson, S.M., Hülsebusch, D., and Mauersberger, K.: 1997, 'Surprising Rate Coefficients for Four Isotopic Variants of $O+O_2+M$', *J. Chem. Phys.* **107**, 5385–5392.

Burghardt, Renata: 1997, 'Untersuchungen zur Ozonisotopie: Experimente und Vergleich mit bisherigen Ergebnissen', PhD thesis, Ruprecht-Karls University at Heidelberg, Germany.

Chapman, S.: 1930, 'A Theory of Upper-atmosphere Ozone', *Mem. Roy. Meteorol. Soc.* **3**, 103–125.

Clayton, R.N., Grossman, L., and Mayeda, T.K.: 1973, 'A Component of Primitive Nuclear Composition in Carbonaceous Meteorites', *Science* **182**, 485–488.

Gao, Yi Quin, and Marcus, R.A.: 2002, 'On the Theory of the Strange and Unconventional Isotopic Effects in Ozone Formation', *J. Chem. Phys.* **116**, 137–154.

Goldman, A.F., Murray, J., Murcray, D.G., Kosters, J.J., Rinsland, C.P., Flaud, J.-M., Camy-Peyret, C., and Barbe, A.: 1989, 'Isotopic Abundances of Stratospheric Ozone From Balloon-borne High Resolution Infrared Solar Spectra', *J. Geophys. Res.* **94**, 8467–8473.

Günther, J., Krankowsky, D., and Mauersberger, K.: 2000, 'Third-body Dependence of Rate Coefficients for Ozone Formation in $^{16}O-^{18}O$ Mixtures', *Chem. Phys. Lett.* **324**, 31–36.

Günther, J., Erbacher, B., Krankowsky, D., and Mauersberger, K.: 1999, 'Pressure Dependence of two Relative Ozone Formation Rate Coefficients', *Chem. Phys. Lett.* **306**, 209–213.

Hathorn, B.C., and Marcus, R.A.: 1999, 'An Intramolecular Theory of the Mass-independent Isotope Effect for Ozone.', *J. Chem. Phys.* **111**, 4087–4100.

Hathorn, B.C., and Marcus, R.A.: 2000, 'An Intramolecular Theory of the Mass-independent Isotope Effect for Ozone. II. Numerical Implementation at Lower Pressures Using a Loose Transition State.', *J. Chem. Phys.* **113**, 9497–9509.

Heidenreich, J.E., III, and Thiemens, M.H.: 1983, 'A Non-mass-dependent Isotope Effect in the Production of Ozone From Molecular Oxygen', *J. Chem. Phys.* **78**, 892–895.

Irion, F.W., Gunson, M.R., Rinsland, C.P., Yung, Y.L., Abrams, M.C., Chang, A.Y., and Goldman, A.: 1996, 'Heavy Ozone Enrichments From ATMOS Infrared Solar Spectra', *Geophys. Res. Lett.* **23**, 2377–2380.

Janssen, C., Günther, J., Krankowsky, D., and Mauersberger, K.: 1999, 'Relative Formation Rates of $^{50}O_3$ and $^{52}O_3$ in ^{16}O–^{18}O Mixtures', *J. Chem. Phys.* **111**, 7179–7182.

Janssen, C., Günther, J., Mauersberger, K., and Krankowsky, D.: 2001, 'Kinetic Origin of the Ozone Isotope Effect: A Critical Analysis of Enrichments and Rate Coefficients', *Phys. Chem.* **3**, 4718–4721.

Johnston, J.C., and Thiemens, M.H.: 1997, 'A Non-mass-dependent Isotope Effect in the Production of Ozone From Molecular Oxygen', *J. Geophys. Res.* **102**, 25395–25404.

Kaye, J.A., and Strobel, D.F.: 1983, 'Enhancement of Heavy Ozone in the Earth's Atmosphere?', *J. Geophys. Res.* **88**, 8447–8452.

Krankowsky, D., Bartecki, F., Klees, G.G., Mauersberger, K., Schellenbach, K., and Stehr, J.: 1995, 'Measurement of Heavy Isotope Enrichment in Tropospheric Ozone', *Geophys. Res. Lett.* **22**, 1713–1716.

Krankowsky, D., Lämmerzahl, P., and Mauersberger, K.: 2000, 'Isotopic Measurements of Stratospheric Ozone', *Geophys. Res. Lett.* **271**, 2593–2595.

Lämmerzahl, P., Röckmann, T., Brenninkmeijer, C.A.M., Krankowsky, D., and Mauersberger, K.: 2002, 'Oxygen Isotope Composition of Stratospheric Carbon Dioxide', *Geophys. Res. Lett.*, in press.

Lyons, J.R.: 2001, 'Transfer of Mass-independent Fractionation in Ozone to Other Oxygen-containing Radicals in the Atmosphere', *Geophys. Res. Lett.* **28**, 3231–3234.

Mauersberger, K.: 1977, 'Mass Spectrometer Beam System for Applications in the Stratosphere', *Rev. Sci. Instrum.* **48**, 1169–1173.

Mauersberger, K.: 1981, 'Measurement of Heavy Ozone in the Stratosphere', *Geophys. Res. Lett.* **8**, 935–937.

Mauersberger, K.: 1987, 'Ozone Isotope Measurements in the Stratosphere', *Geophys. Res. Lett.* **14**, 80–83.

Mauersberger, K., Morton, J., Schueler, B., Stehr, J., and Anderson, S.M.: 1993, 'Multi-isotope Study of Ozone: Implications for the Heavy Ozone Anomaly', *Geophys. Res. Lett.* **20**, 1031–1034.

Mauersberger, K., Erbacher, B., Krankowsky, D., Günther, J., and Nickel, R.: 1999, 'Ozone Isotope Enrichment: Isotopomer-specific Rate Coefficients', *Science* **283**, 370–372.

Mauersbrger, K., Lämmerzahl, P., and Krankowsky, D.: 1999, 'Stratospheric Ozone Isotope Enrichments – Revisited', *Geophys. Res. Lett.* **28**, 3155–3158.

Morton, J., Barnes, J., Schueler, B., and Mauersberger K.: 1990, 'Laboratory Studies of Heavy Ozone', *J. Geophys. Res.* **95**, 901–907.

Robert, F., and Camy-Peyret, C.: 2001, 'Ozone Isotope Composition: An Angular Effect in Scattering Processes?', *Annales Geophysicae* **19**, 229–244.

Schueler, B., Morton, J., and Mauersberger, K.: 1990, 'Measurement of Isotopic Abundances in Collected Stratospheric Ozone Samples', *Geophys. Res. Lett.* **17**, 1295–1298.

Thiemens, M.H., and Jackson, T.: 1988, 'New Experimental Evidence for the Mechanism for Production of Isotopically Heavy O_3', *Geophys. Res. Lett.* **15**, 639–642.

Thiemens, M.H., and Jackson, T.: 1990, 'Pressure Dependency for Heavy Isotope Enhancement in Ozone Formation', *Geophys. Res. Lett.* **17**, 717–719.

Thiemens, M.H., Jackson, T., Mauersberger, K., Schueler, B., and Morton, J.: 1991, 'Oxygen Isotope Fractionation in Stratospheric CO_2', *Geophys. Res. Lett.* **18**, 669–672.

Thiemens, M.H., Jackson, T., Zipf, E.C., Erdman, P.W., and van Egmond, C.: 1995, 'Carbon Dioxide and Oxygen Isotope Anomalies in the Mesosphere and Stratosphere', *Science* **270**, 969–972.

Yung, Y., DeMore, W.B., and Pinto, J.P.: 1991, 'Isotopic Exchange Between Carbon Dioxide and Ozone via $O(^1D)$ in the Stratosphere', *Geophys. Res. Lett.* **18**, 13–16.

Address for Offprints: Konrad Mauersberger, Max-Planck-Institut für Kernphysik, Division of Atmospheric Physics, Saupfercheckweg 1, Postfach 103 980, D-69029 Heidelberg, Germany;
Konrad.Mauersberger@mpi-hd.mpg.de

ISOTOPIC FRACTIONATION BY GRAVITATIONAL ESCAPE

H. LAMMER[1] and S.J. BAUER[2]

[1] *Space Research Institute, Department of Extraterrestrial Physics, Austrian Academy of Sciences, Schmeidlstr. 6, A-8042 Graz, Austria , Graz, Austria*

[2] *Institute for Geophysics, Astrophysics and Meteorology, University of Graz, Universitätsplatz 5, A-8010, Graz, Austria*

Received: 9 April 2002; Accepted in final form: 25 October 2002

Abstract. Present natural data bases for abundances of the isotopic compositions of noble gases, carbon and nitrogen inventories can be found in the Sun, the solar wind, meteorites and the planetary atmospheres and crustal reservoirs. Mass distributions in the various volatile reservoirs provide boundary conditions which must be satisfied in modelling the history of the present atmospheres. Such boundary conditions are constraints posed by comparison of isotopic ratios in primordial volatile sources with the isotopic pattern which was found on the planets and their satellites. Observations from space missions and Earth-based spectroscopic telescope observations of Venus, Mars and Saturn's major satellite Titan show that the atmospheric evolution of these planetary bodies to their present states was affected by processes capable of fractionating their elements and isotopes. The isotope ratios of D/H in the atmospheres of Venus and Mars indicate evidence for their planetary water inventories. Venus' H_2O content may have been at least 0.3% of a terrestrial ocean. Analysis of the D/H ratio on Mars imply that a global H_2O ocean with a depth of ≤ 30 m was lost since the end of hydrodynamic escape. Calculations of the time evolution of the $^{15}N/^{14}N$ isotope anomalies in the atmospheres of Mars and Titan show that the Martian atmosphere was at least ≥ 20 times denser than at present and that the mass of Titan's early atmosphere was about 30 times greater than its present value. A detailed study of gravitational fractionation of isotopes in planetary atmospheres furthermore indicates a much higher solar wind mass flux of the early Sun during the first half billion years.

1. Introduction

It is generally thought that planetary atmospheres derive from source reservoirs of primordial solar system volatiles via physical and chemical processes which operate during the pre- and post-accretionary histories of the planets. In that case the chemical and isotopic compositions of present-day atmospheres should provide clues on the characteristics of the primordial source reservoir and the nature of the subsequent processing of its volatiles.

In current models for the formation and evolution of terrestrial planetary atmospheres the primordial volatile sources include the solar nebula, the solar wind, cometary matter and volatile rich meteoroids. The primordial reference for atmospheric deuterium (D) to hydrogen (H) anomalies on Venus and Mars is terrestrial ocean water. Table I shows various observed isotope ratios in the atmospheres of Venus, Mars and Titan, as well as their anomalies compared to the Earth values.

Space Science Reviews **106:** 281–291, 2003.

Interpretations of the mass spectrometry data of Pioneer Venus regarding the D/H ratio suggest that Venus once may have had more H_2O, corresponding to at least 0.3% of Earth-like oceans. The D/H ratio on Venus agrees with two possibilities, where Venus' once had a similar water abundance as Earth and Mars (Dayhoff *et al.*, 1967; Walker *et al.*, 1970), or that Venus was formed from condensates in the solar nebula that contained little or no water (Lewis, 1970; Lewis, 1974a).

If Venus was wet the planet must have lost most of its water during its history. The suggested loss mechanism is the photodissociation of water vapor in the upper atmosphere, followed by the escape of hydrogen to space and the reaction of oxygen with reduced gases in the atmosphere or minerals in the crust (McElroy *et al.*, 1982; Donahue *et al.*, 1982; Kasting and Pollack, 1983).

Kasting and Pollack (1983) and Kasting (1988) argue that Venus may have had once abundant hot water stable on the surface but a runaway moist greenhouse effect shifted the climate to an extremely hot and dry planet. Another alternative would be a humid greenhouse effect where the water in the Venus atmosphere was, because of the high temperature, mainly in the vapor phase. In both cases hydrogen and oxygen could have been lost by hydrodynamic escape driven by an early active Sun.

Further indication that the early solar radiation and particle environment affected the Venus atmosphere is suggested by Bauer (1983) and Chassefière (1997) who argued that the early loss of H and O equivalent as much as up to 2% of a terrestrial ocean would have been possible, if our early Sun went through an active phase where a dense solar wind mass flux swept away atmospheric O^+ from the Venus atmosphere.

Measurements by the Viking spacecraft and ground based observations, as well as studies of volatiles trapped in SNC meteorites (e.g., Mathews and Marti, 2001) which are believed to be from Mars give information about the isotope ratio in the Martian atmosphere and its interior.

The D/H isotope ratio in the present Martian atmosphere H_2O vapor is $8.1 \pm 0.3 \times 10^{-4}$ which is greater than the terrestrial value by a factor of 5.2 ± 0.2 (e.g. Owen *et al.*, 1988; Yung *et al.*, 1988; Krasnopolsky *et al.*, 1997). Modeling the atmospheric D/H ratio by using different methods result in a total H_2O loss of a 3.6 - 50 m global layer of H_2O from Mars during the past 3.5 Gy (e.g. Yung *et al.*, 1988; Lammer *et al.*, 1996; Kass and Yung, 1999; Krasnopolsky and Feldman, 2001; Bertaux and Montmessin, 2001).

For $^{15}N/^{14}N$ isotope anomalies on terrestrial planetary bodies the ratio on Earth is commonly used as the primordial unfractionated value and should therefore reflect the material out of which the planets in our solar system were formed.

However, the variations of the $^{14}N/^{15}N$ ratio in solar system samples are larger than a factor 2 (see, e.g., Figure 8 of Marty *et al.*, 2003). These differences may indicate that there were at least two separate reservoirs of nitrogen in the early solar system. It seems possible that one of these was dominated by molecular nitrogen,

TABLE I
Isotope ratios in terrestrial-like planetary atmospheres.

Isotope ratios	Venus	Earth	Mars	Titan
D/H	$1.6 - 2.2 \times 10^{-2}$	1.5×10^{-4}	$8.1 \pm 0.3 \times 10^{-4}$	1.6×10^{-4}
$^{15}N/^{14}N$	$\approx 3.5 \times 10^{-3}$	3.7×10^{-3}	$6.4 - 5.0 \times 10^{-3}$	≈ 0.017
$^{18}O/^{16}O$	$\approx 2.0 \times 10^{-3}$	2.04×10^{-3}	$1.89 \pm 0.2 \times 10^{-3}$	
$^{13}C/^{12}C$	$1.14 \pm 0.02 \times 10^{-2}$	1.12×10^{-2}	$1.18 \pm 0.12 \times 10^{-2}$	

a highly volatile gas. Generally it is expected that the same source dominates the atmospheres of Mars, Venus and Titan.

Isotope abundances in Venus' atmosphere have been determined by Earth-based near infrared spectroscopy and in situ mass spectrometer measurements. Hoffmann *et al.* (1979) deduced from the $^{29}N_2/^{28}N_2$ peak ratio observed by their mass spectrometer that the isotope ratio $^{15}N/^{14}N$ on Venus is within 20% of the terrestrial $^{15}N/^{14}N$ value. Brinkmann (1982) considered that substantial quantities of nitrogen have escaped from Mars via dissociative recombination in producing nitrogen atoms with energies exceeding the escape energy of about 1.72 eV for the ^{14}N isotope. The Viking mission provided support for his escape hypothesis, since the mass spectrometer recorded an anomalous $^{15}N/^{14}N$ ratio equal to 1.62 ± 0.16 times the Earth value (Nier, 1976; Nier *et al.*, 1976). Processes for producing energetic nitrogen, including photoelectron impact dissociation (Nier, 1976), dissociative recombination of molecular N_2^+ ions (Yung *et al.*, 1977; Fox and Dalgarno, 1983; Lammer and Bauer, 1991; Fox and Hác, 1997) and chemical reactions have also been studied (Fox and Dalgarno, 1983).

The $^{15}N/^{14}N$ anomaly on Mars is an important indicator for the history of the Martian atmosphere. Recently Fox and Hác (1997) investigated the $^{15}N/^{14}N$ isotope fractionation in dissociative recombination of N_2^+ and came to the conclusion that the Martian nitrogen isotope anomaly may indicate that Mars had in its past a much denser atmosphere. They found in their detailed study that dissociative recombination over-fractionates the nitrogen isotopes so that the $^{15}N/^{14}N$ ratio is larger than the presently measured value for an initial isotope value assumed to be that of the Earth. In order to reproduce the currently measured nitrogen isotope ratio they invoke a scenario in which the nitrogen is embedded in a much denser atmosphere of at least 114 mbar from which escape is presumed to occur above the exobase (Fox and Hác, 1997).

In contrast to nitrogen isotopes, the relative abundances of O and C isotopes on Mars appear similar to observed values on Earth and seem therefore buffered by surface reservoirs. Ar and Xe isotopes are mass fractionated and should not be buffered by a surface reservoir (Pepin, 1991). Hutchins and Jakosky (1996) estimated that about $90 \pm 5\%$ of ^{36}Ar and about $80 \pm 10\%$ of ^{40}Ar has been lost by atmospheric sputtering from the Martian atmosphere after the intrinsic magnetic field vanished about 3.5 Gy ago (Acuña *et al.*, 1998).

Hidayat *et al.* (1997) determined the ratio of $^{15}N/^{14}N$ for the first time in Titan's atmosphere, via Earth-based millimetric wavelength spectroscopic observations of $HC^{15}N/HC^{14}N$. If one assumes that there is no fractionation of the nitrogen isotopic ratio between HCN and N_2, their measurements indicate that the bulk nitrogen is enriched in the heavy isotope ^{15}N by about 4.5 times relative to the Venus and Earth value and 2.77 times the Martian value.

Our work investigates and summarizes processes leading to the gravitational fractionation of isotopes if they reach the exobase level of a planetary atmosphere. Below the exobase and above the homopause diffusive separation of the isotopes according to their atomic mass occurs. Therefore, diffusive separation between isotopes is also investigated. We also show that numerical Monte Carlo techniques are well suited to simulate diffusion processes of isotopes in planetary atmospheres.

2. Diffusive Isotope Separation in Planetary Atmospheres

The isotopes in planetary atmospheres which escape via nonthermal processes are assumed to be escaping at the exobase in atomic form. Between the homopause and the exobase diffusive separation of the different atmospheric constituents according to their atomic mass occurs. The diffusion occurs due to a gradient in relative concentration arising from slight deviations from a *Maxwellian* distribution (e.g., Chamberlain, 1963).

In considering the behavior of a minor gas constituent under diffusion we can assume that the total pressure remains partially constant, while the minor constituent diffuses upward or downward depending on local sources and sinks and the divergence of the diffusive flux.

The atmospheric level where the *molecular* diffusion coefficient D_i is equal to the *Eddy* diffusion coefficient K_D corresponds to the homopause. Below the homopause the scale heights of light constituents are approximately that of the main atmospheric constituents due to turbulent mixing, whereas at altitudes above the homopause they approach that of a diffusive equilibrium.

The actual fractionation of isotopes by escape processes which takes place in the exosphere depends on the altitude difference Δh between the homopause and the exobase or critical level. Table II shows homopause and exobase altitudes, as well as Δh for Venus, Earth, Mars and Saturn's large moon Titan (Banks and Kockarts, 1973; Bougher *et al.*, 1998; Bougher *et al.*, 2000; Smith *et al.*, 1982; Strobel *et al.*, 1992). The exobase is defined by:

$$\int_{z_c}^{\infty} \frac{dz}{l(z)} = \int_{z_c}^{\infty} n(z)\sigma dz = n_c H dz \equiv \frac{H}{l} = 1, \tag{1}$$

where $l(z)$ is the particle mean free path at altitude z, $n(z)$ the total number density and σ is the gas-kinetic collision cross section. The index c corresponds to the

TABLE II
Homopause and exobase altitude differences.

Planetary bodies	Homopause [km]	Exobase [km]	Δh [km]
Venus	$\approx$135-150	$\approx$205	$\approx$55-70
Earth	$\approx 100 - 110$	$\approx$500	$\approx$390-400
Mars	$\approx$115-130	$\approx$230	$\approx$100-115
Titan	500, 985 - 1125	$\approx$1400	900, 275 - 415

exobase or critical level. Exospheric conditions are assumed to hold at altitudes where the particle mean free path is equal to or greater than the local scale height H. The density at the exobase is given by:

$$n_c = \frac{1}{\sigma H}. \tag{2}$$

For mass-driven diffusive separation the fractionation factor d_f (Hunten *et al.*, 1989) is:

$$d_f = \exp\left[\frac{\Delta h(m_X - m_Y)g}{kT}\right] - 1, \tag{3}$$

with Δh the distance between the homopause and the exobase, k the *Boltzmann* constant, T the temperature, g the gravitational acceleration and m_X and m_Y the isotope masses, where index X corresponds to the heavier isotope. Table II shows homopause and exobase altitudes, as well as Δh for Venus, Earth, Mars and Titan.

Titan's homopause altitude has been calculated from Voyager ultraviolet spectrometer profiles of CH_4 and other data to lie between 985 kilometer and 1125 kilometer (Smith *et al.*, 1982; Strobel *et al.*, 1992). Steiner and Bauer (1990) had derived the *Eddy* diffusion coefficient from CH_4 and C_2H_x mixing ratios, calculating homopause altitudes between 500 kilometer and 950 kilometer.

The fractionation of an atmospheric species due to loss to space of a fractionated component of that species can be described by a Rayleigh law (Hunten *et al.*, 1989):

$$\frac{N_2}{N_2^0} = \left(\frac{N_1}{N_1^0}\right)^{\frac{1}{1+d_f}}, \tag{4}$$

where N_1 and N_2 are the inventories of the light and heavy gas, N_1^0 and N_2^0 are their initial values and d_f is the fraction factor between the homopause and the exobase. Their loss rates are related by:

$$\frac{dN_2}{dN_1} = \frac{N_2}{N_1(1 + d_f)} \tag{5}$$

The diffusive separation effect enhances the importance of all atmospheric escape processes that act at the exobase level or at least above the homopause.

At atmospheric conditions where the escape parameter $X(r)$ is:

$$X(r) = \frac{v_\infty^2}{v_0^2} = \frac{GM_p m}{rkT_\infty} \leq 1.5, \tag{6}$$

the diffusive fractionation according to atomic or molecular mass vanishes, since the level at which atmospheric constituents escape moves below the exobase into the bulk atmosphere (hydrodynamic escape). v_∞ is the planetary escape velocity, v_0 the most probable *Maxwellian* velocity at an exospheric temperature T_∞, G is the universal gravitational constant, M_p the mass of a planet, m the mass of an atmospheric particle, k the *Boltzmann* constant and r the radial distance from the planetary center. The critical value of $X(r)$=1.5 represents *Öpik's* condition of *blow off* (Öpik, 1963).

If water vapor is a major atmospheric constituent up to atmospheric altitudes where it could rapidly be photodissociated by UV radiation, atomic hydrogen became the dominant constituent of the upper atmosphere and the escape can become hydrodynamic if the exosphere temperature is high enough. The hydrodynamic formulation is appropriate when collisions between the atmospheric constituents occur frequently up to high altitudes so that fluid dynamics rather than kinetic theory may be used to describe the behavior of the escaping gas. For studying hydrodynamic escape one must solve a set of one-dimensional, steady-state equations governing the vertical composition and motion of an atmosphere from a cold trap out to infinity (e.g., Kasting and Pollack, 1983). The cold trap at a terrestrial planet is defined as the altitude at which the saturation water mixing ratio reaches its minimum value.

3. Isotope Separation by Chemical and Sputtering Processes

For investigating in detail how efficient nonthermal escape processes lead to the required fractionation of isotopes in planetary atmospheres one can use a Monte Carlo technique which simulates the particle movement, collisions and energy loss of the particles in an atmosphere (e.g. Ip, 1988; Lammer and Bauer, 1991; Lammer *et al.*, 2000a, 2000b).

At a planetary exobase all isotopes with energies greater than the escape energy are lost to space. After its generation in lower atmospheric levels, a newly released isotope may collide with the neutral background atmospheric gas, may change its direction, lose its energy, or the isotope may travel long distances without collisions. One can express the collisional probability P of the isotope with the ambient atmosphere as:

$$P = 1 - \exp\left(-\int_{z_1}^{z_2} \frac{dz}{l(z)}\right), \tag{7}$$

with $l(z)$ being the mean free path between the isotope and the background gas:

$$l(z) = [n(z)\sigma]^{-1}, \tag{8}$$

where σ is the momentum transfer cross section of about $3\times10^{-15}\text{cm}^2$ (e.g. Ip, 1988; Lammer and Bauer, 1991).

In the Monte Carlo simulation, each collision step is generated by equating a random number R_1 to the following integral:

$$R_1 = exp\left(-\int_{z_1}^{z_2}\frac{dz}{l(z)}\right). \tag{9}$$

The path that a single isotope follows from altitude z_1 to altitude z_2 may be highly erratic. At atmospheric levels where the mean free path is short collisions occur so that we can assume that the particle gets thermalized eventually by a series of elastic hard sphere collisions with the background gases. Inelastic collisions are negligibly small at these low energies (Ip, 1988).

If a collision occurs the direction of the changing velocity vector will be simulated with a second random number R_2. If R_2 is between 0 and 0.5 the new direction of the isotope points downwards. If the random number is between 0.51 and 1 the isotope travels upwards. At the ith collision step, the energy loss from hard sphere collisions between the isotope and the surrounding atmosphere is taken to be (e.g. Ip, 1988; Lammer *et al.*, 2000a, 2000b):

$$E_i = R_3\frac{4m_{X_j}m_Y}{(m_{X_j}+m_Y)^2}E_{i-1}, \tag{10}$$

with m_{X_j} the mass of the isotopes (^{15}N and ^{14}N, etc.) and m_Y the mass of the surrounding gas. Random number R_3 representing the type of collision and stimulating the effectiveness of the energy transfer between the colliding particles. This simulates everything between a full head-on and a glancing collision. The random numbers R_k (k=1,2,3) are from a sample of random numbers uniformly distributed over the interval between 0 and 1.

The simulations are started below the exobase levels at altitudes where the atomic isotopes are produced. After the particles are released by chemical reactions the isotopes are traced in the atmosphere up to the exobase level with a discrete step size Δz. A simulation with $\geq 3\times10^5$ particles matches nature very well.

One of the most important nonthermal escape processes for atoms X (mostly N and O) and their isotopes from planetary atmospheres is dissociative recombination. The newly born atoms may be formed in ^4S, ^2D, or ^2P states. In planetary atmospheres the kinetic energy release in the dissociative recombination process can play an important role in generating the anomalous ^{15}N/^{14}N isotope ratios due to the preferred escape of the lighter ^{14}N isotope. Dissociative recombination of N_2^+ can proceed via three possible channels (e.g. Kella *et al.*, 1996; Fox and Hác, 1997):

$$N_2^+ + e \rightarrow N(^4S) + N(^2D) + 3.44\text{eV} \;; \tag{11}$$

$$N_2^+ + e \rightarrow N(^4S) + N(^2P) + 2.24\text{eV} \; ; \qquad (12)$$

$$N_2^+ + e \rightarrow N(^2D) + N(^2D) + 1.06\text{eV} \; . \qquad (13)$$

The escape energy of a ^{14}N isotope in the Martian exobase is about 1.722 eV. In dissociative recombination of $^{29}N_2^+$ a ^{15}N isotope from Equation (11) is released with 14/29$\times$3.44 eV, which yields an energy of 1.66 eV, significantly less than the Martian escape energy of about 1.844 eV. It is important to note that this is not the case on Titan. The reactions in Equations (12) and (13) release the isotopes with energies of 1.081 eV and 0.511 eV which are greater than Titan's escape energy of about 0.3 eV.

Kella *et al.* (1996) investigated the branching ratios for dissociative recombination of nitrogen isotopes and found from measurements that the reactions in Equations (11) and (13) are more important than (12) since the reaction in Equation (12) has a much smaller branching ratio.

Lammer *et al.* (2000a) showed from such Monte Carlo studies that isotope separation by dissociative recombination, impact dissociation and atmospheric sputtering is not very efficient if the newly released isotopes have energies which are much higher than their escape energies.

If the isotopes have energies much closer to the escape energy, the separation effect is relevant at altitudes below the exobase level. Fewer collisions are needed for the isotopes in the lower energy regime to get energies lower than their escape energy. At altitudes where the mean free path is larger than the scale height, only few or no collisions and therefore no isotope separation occurs anymore.

It is also important to note that fractionation can occur in altitude regions where only a small fraction of isotopes reaches the exosphere with energies greater than the escape energy. This effect compensates the separation effect by the low energy reactions like the one in Equation (13). Therefore, Lammer *et al.* (2000a) showed that dissociative recombination and atmospheric sputtering do not contribute much to nitrogen isotope separation on Titan.

However, the nonthermal escape processes together with the diffusive separation effect contribute to an enhancement in heavier isotopes in planetary atmospheres over time.

4. Discussion and Conclusion

Our study shows that anomalous ^{15}N/^{14}N ratios in the atmospheres of Mars and Titan are powerful indicators of their atmospheric evolution. Diffusive separation between their homopause and exobase levels and nonthermal atmospheric escape processes together indicate that the Martian atmosphere was at least ≥ 20 times denser and that Titan's atmospheric mass was about 30 times greater than its present values.

By assuming that Titan's nitrogen isotope ratio was comparable with the present terrestrial value 4.0 Gy ago we have to expect that Titan was exposed to a much higher solar wind above its exobase during the first 500 million years after the origin of the solar system resulting in much higher solar wind induced sputtering and pick-up loss rates of nitrogen isotopes. Therefore, Titan's enrichment in ^{15}N isotopes may be an indication that our Sun experienced a very active early period.

These result agree well with actual multi-wavelength studies of solar-like G-type stars at several stages of their main sequence evolution. These studies indicate that our early Sun must have undergone a highly active phase in its particle and radiation environment 3.5-4.5 Gy ago. An overview of that problem can be found e.g. in Guinan and Ribas (2002) and in Kallenbach *et al.* (2003). Detailed observations of such stars by the ROSAT and ASCA X-ray satellites show that the X-ray luminosity may have been several hundred times higher than today.

Wood *et al.* (2002) found a correlation between these high X-ray fluxes and mass loss rates. They estimated from these mass loss rates of young solar like stars with an age between 100 to 300 million years, solar wind densities 1000 to 10000 times higher than today. Both isotope anomalies in planetary atmospheres and actual data from solar like young stars support the hypothesis that the radiation and particle environment of the early Sun had a large effect on the paleoatmospheres of terrestrial planets.

D/H isotope anomalies in the atmospheres of Venus and Mars imply that both planets may have lost a large amount of H_2O to space, corresponding to about 2% of a terrestrial ocean on Venus and a global H_2O ocean with a depth of $\leq$ 30 m on Mars, since the end of hydrodynamic escape.

References

Acuña, M.H., *et al.*: 1998, 'Magnetic Field and Plasma Observations at Mars: Initial Results of the Mars Global Surveyor Mission', *Science* **279**, 1676–1680.

Banks, P.M., and Kockarts, G.: 1973, *Aeronomy*, Academic Press, London.

Bauer, S.J.: 1983, 'Water on Venus: Lack or Loss?', *Annales Geophysicae* **1**, 477–480.

Bertaux, J.-L., and Montmessin, F.: 2001, 'Isotopic Fractionation Through Water Vapor Condensation: The Deuteropause, a Cold Trap for Deuterium in the Atmosphere of Mars', *J. Geophys. Res.* **106**, 32,879–32,884.

Bougher, S.W., Hunten, D.M., Phillips, R.J.: 1998, 'Venus II', University of Arizona Press, Tucson, USA.

Bougher, S.W., Engel, S., Roble, R.G., Foster, B.: 2000, 'Comparative Terrestrial Planet Thermospheres 3. Solar Cycle Variation of the Upper Atmosphere Temperature of Mars', *J. Geophys. Res.* **105**, 17,669–17,692.

Brinkmann, R.T.: 1982, 'Mars: Has Nitrogen Escaped?', *Science* **179**, 944–945.

Chamberlain, J.W.: 1963, 'Planetary Coronae and Atmospheric Evaporation', *Planet. Space Sci.* **11**, 901–960.

Chassefière, E.: 1997, 'Loss of Water on the Young Venus: The Effect of a Strong Primitive Solar Wind', *Icarus* **126**, 229–232.

Dayhoff, M.O., Eck, R.V., Lippincott, F.R., and Sagan, C.: 1967, 'Venus: Atmospheric Evolution', *Science* **155**, 556.

Donahue, T.M., Hoffman, J.H., and Watson, A.J.: 1982, 'Venus was Wet: A Measurement of the Ratio of Deuterium to Hydrogen', *Science* **216**, 630–633.

Fox, J.L., and Dalgarno, A.: 1983, 'Nitrogen Escape From Mars', *J. Geophys. Res.* **88**, 9027–9032.

Fox, J.L., and Hác, A.: 1997, 'The $^{15}N/^{14}N$ Isotope Fractionation in Dissociative Recombination of N_2^+', *J. Geophys. Res.* **102**, 9191–9204.

Guinan, E.F., and Ribas, I.: 2002, 'Our Changing Sun: The Role of Solar Nuclear Evolution and Magnetic Activity on Earth's Atmosphere and Climate', *in* B. Montesinos, A. Giménez, and E.F. Guinan (eds.), *The Evolving Sun and its Influence on Planetary Environments*, ASP Conf. Series **269**, pp. 85–106.

Hidayat, T., Marten, A., Bezard, B., Gautier, D., and Owen, T.: 1997, 'Millimeter and Submillimeter Heterodyne Observations of Titan: Retrieval of the Vertical Profile of HCN and the $^{12}C/^{13}C$ Ratio', *Icarus* **126**, 170–182.

Hoffmann, J.H., Oyama, V.I., and von Zahn, U.: 1979, 'Measurements of the Venus Lower Atmosphere Composition: A Comparison of Results', *J. Geophys. Res.* **85**, 7871–7880.

Hunten, D.M., Donahue, T.M., Walker, J.C.G., and Kasting, J.F.: 1989, 'Origin and Evolution of Planetary Atmospheres', *in* S.K. Atreya, J.P. Pollack, and M.S. Matthews (eds.), University of Arizona Press, Tucson, pp. 170–182.

Hutchins, K.S., and Jakosky, B.M.: 1996, 'Evolution of Martian Atmospheric Argon: Implication for Sources of Volatiles', *J. Geophys. Res.* **101**, 14,933–14,949.

Ip, W.-H.: 1988, 'On a Hot Oxygen Corona of Mars', *Icarus* **76**, 135–145.

Jakosky, B.M., Pepin, R.O., Johnson, R.E., and Fox, J.L.: 1994, 'Mars Atmosphere Loss and Isotopic Fractionation by Solar-wind-induced Sputtering and Photochemical Escape', *Icarus* **111**, 271–281.

Johnson, R.E.: 1990, *Energetic Charged Particle Interactions with Atmospheres and Surfaces*, Springer Publishing House, Heidelberg, Germany.

Kallenbach, R., Robert, F., Geiss, J., Herbst, E., Lammer., H., Marty, B., Millar, T., Ott, U., and Pepin, R.O.: 2003, 'Sun and Protosolar Nebula', *Space Sci. Rev.*, this volume.

Kass, D.M., and Yung, Y.L.: 1999, 'Water on Mars: Isotopic Constraints on Exchange Between the Atmosphere and Surface', *Geophys. Res. Lett.* **26**, 3653–3656.

Kasting, F.J., and Pollack, J.B.: 1983, 'Loss of Water from Venus. I. Hydrodynamic Escape of Hydrogen', *Icarus* **53**, 479–508.

Kasting, F.J.: 1988, 'Runaway and Moist Greenhouse Atmospheres and the Evolution of Earth and Venus', *Icarus* **74**, 472–494.

Kella, D., Johnson, P.J., Pedersen, H.B., Vejby-Christensen, L., and Andersen, L.H.: 1996, 'Branching Ratios for Dissociative Recombination of $^{15}N^{14}N^+$', *Phys. Rev. Lett.* **77**, 2432–2435.

Krasnopolsky, V.A., Bjoraker, G.L., Mumma, M.J., and Jennings, D.E.: 1997, 'High Resolution and Detection of HDO', *J. Geophys. Res.* **102**, 6524–6534.

Krasnopolsky, V.A., and Friedman, P.: 2001, 'Early Mars More Water than Earth?', *Science* **294**, 1914–1917.

Lammer, H., and Bauer, S.J.: 1991, 'Nonthermal Atmospheric Escape from Mars and Titan', *J. Geophys. Res.* **96**, 1819–1825.

Lammer, H., and Bauer, S.J.: 1993, 'Atmospheric Mass Loss from Titan by Sputtering', *Planet. Space Sci.* **41**, 657–663.

Lammer, H., Stumptner, W., and Bauer, S.J.: 1996, 'Loss of H and O from Mars: Implications for the Planetary Water Inventory', *Geophys. Res. Lett.* **23**, 3353–3356.

Lammer, H., Stumptner, W., Molina-Cuberos, G.J., Bauer, S.J., and Owen, T.: 2000a, 'Nitrogen Isotope Fractionation and its Consequence for Titan's Atmospheric Evolution', *Planet. Space Sci.* **48**, 529–543.

Lammer, H., Stumptner, W., and Bauer, S.J.: 2000b, 'Upper Limits for the Martian Exospheric Number Density During the Planet B/Nozomi Mission', *Planet. Space Sci.* **48**, 1473–1478.
Lewis, J.S.: 1970, 'Venus: Atmospheric Lithospheric Composition', *Earth Planet. Sci. Lett.* **10**, 73–80.
Lewis, J.S.: 1974a, 'The Temperature Gradient in the Solar Nebula', *Science* **186**, 440–443.
Lewis, J.S.: 1974b, 'Volatile Element Influx on Venus from Cometary Impact', *Earth Planet. Sci. Lett.* **22**, 239–244.
Marty, B., Hashizume, K., Chaussidon, M., and Wieler, R.: 2003, 'Nitrogen Isotopes on the Moon: Archives of the Solar and Planetary Contributions to the Inner Solar System', this volume.
Mathew, K.J., and Marti, K.: 2001, 'Early Evolution of Martian Volatiles: Nitrogen and Noble Gas Components in ALH84001 and Chassigny', *J. Geophys. Res.* **106**, 1401–1422.
McElroy, M., Yung, Y.L., and Nier, A.O.: 1976, 'Isotopic Composition of Nitrogen: Implications for the Past History of Mars' atmosphere', *Science* **194**, 70–72.
McElroy, M.B., Prather, M.J., and Rodriguez, J.M.: 1982, 'Escape of Hydrogen from Venus', *Science* **215**, 1614–1617.
Nier, A.O.: 1976, 'Isotope Composition of Nitrogen: Implications for the Past History of Mars' Atmosphere', Science **194**, 70–72.
Nier, A.O., McElroy, M.B., and Yung, Y.L.: 1976, 'Isotopic Composition of the Martian Atmosphere', *Science* **194**, 68–70.
Öpik, E.J.: 1963, 'Selective Escape of Gases', *Geophys. J.* **7**, 490.
Owen, T., Maillard, J.-P., de Bergh, C., and Lutz, B.L.: 1988, 'Deuterium on Mars – The Abundance of HDO and the Value of D/H', *Science* **240**, 1767–1770.
Pepin, R. O.: 1991, 'On the Origin and Early Evolution of Terrestrial Planetary Atmospheres and Meteoritic Volatiles', *Icarus* **92**, 2–79.
Smith, G.R., Strobel, D.F., Broadfoot, A.L., Sandel, B.R., Shemansky, D.E., and Holberg, J.B.: 1982, 'Titan's Upper Atmosphere: Composition and Temperature from the EUV Solar Occultation Results', *J. Geophys. Res.* **87**, 1351–1359.
Steiner, G., and Bauer, S.J.: 1990, 'Molecular and Eddy Diffusion in the Atmosphere of Titan', *Ann. Geophys.* **8**, 473–476.
Strobel, D.F., Summers, M.E., and Zhu, X.: 1992, 'Titan's Upper Atmosphere: Structure and Ultraviolet Emissions', *Icarus* **100**, 512–526.
Walker, J.C.G., Turekian, K.K., and Hunten, D.M.: 1970, 'An Estimate of the Present-day Deep-mantle Degassing Rate from Data on the Atmosphere of Venus', J. Geophys. Res. **75**, 3558–3561.
Wood, B.E., Müller, H.-R., Zank, G.P., and Linsky, J.L.: 2002, 'Measured Mass Loss Rates of Solar-like Stars as a Function of Age and Activity', *Astrophys. J.*, in press.
Yung, Y.L., Strobel, D.F., Kong, T.Y., and McElroy, M.B.: 1977, 'Photochemistry of Nitrogen in the Martian Atmosphere', *Icarus* **30**, 26–41.
Yung, Y.L., Chen, J.S., Pinto, J.P., Allen, M., Pierce, K.K., and Paulson, S.: 1988, 'HDO in the Martian Atmosphere. Implications for the Abundance of Crustal Water', *Icarus* **76**, 146–159.

Address for Offprints: Helmut Lammer, Space Research Institute, Austrian Academy of Sciences, Schmeidlstr. 6, A-8042 Graz, Austria; helmut.lammer@oeaw.ac.at

ISOTOPIC FRACTIONATION BY ION-MOLECULE REACTIONS

ERIC HERBST
Departments of Physics and Astronomy, The Ohio State University, Columbus, OH 43210 USA

Received: 15 April 2002; Accepted in final form: 29 May 2002

Abstract. Isotopic fractionation in interstellar clouds can occur by ion-molecule reactions at low temperatures. The major effect is not kinetic but thermodynamic in origin in that it arises from the difference in rate coefficients between forward and backward directions in reactions which exchange isotopic atoms. In this article, we concentrate on deuterium fractionation in interstellar clouds; this effect enhances the relative abundances of deuterated isotopomers to their normal counterparts by up to four orders of magnitude as compared with the basic D/H elemental abundance ratio. We also discuss the fractionation of ^{15}N and ^{13}C.

1. Introduction

Dense interstellar clouds of gas and dust, characterized by average gas densities of perhaps 10^3 cm^{-3} and temperatures from 10-30 K, are rather inhomogeneous regions in which astronomers are very interested because they are the sites of star formation. The gas in these regions is overwhelmingly molecular in nature. Although densities are low by terrestrial standards, there is a huge amount of matter contained in dense clouds, up to a million times the mass of the sun. Molecules of up to 13 atoms in size have been detected via high-resolution spectroscopy in the gas phase, but the most abundant species by far is simple molecular hydrogen (Herbst, 2001). Molecules are also detected in the form of ice mantles on dust particles; among the main constituents of the mantles are water and dry ice. The particles themselves are mainly silicates and carbonaceous material (Williams and Herbst, 2002). Molecules trace the varying conditions in dense clouds and are particularly useful in elucidating the various phases of star formation, a process which starts when particularly dense regions of clouds, known as cold cores, begin to collapse. Among the best-studied cores are two known simply as TMC-1 and L134N. These nuggets, each of which possesses a gas density of 10^4 cm^{-3} and a temperature of 10 K, are not yet actively collapsing to form stars. Cores in an active state of collapse are often referred to as "protostars."

The chemistry that occurs in cold cores, indeed in all cool regions of dense clouds, must be efficient at temperatures as low as 10 K (Herbst and Klemperer, 1973). Most chemical reactions are not efficient at low temperatures because they are either endothermic or, even if exothermic, they possess so-called "activation energy" barriers, which are closely related to saddle points in the potential energy surface connecting reactants and products. The rate coefficients of binary chemical

Space Science Reviews **106:** 293–304, 2003.

reactions follow the so-called Arrhenius rate law, where the coefficient k is given by the expression (Herbst, 1996)

$$k = A(T) \times \exp(-E_a/k_B T). \quad (1)$$

In this expression, $A(T)$ is the pre-exponential factor, which possesses at most a weak dependence on temperature, k_B is the Boltzmann constant, T is the temperature, and E_a is the activation energy. A typical value for E_a is a few eV for an exothermic reaction; with an activation energy of this size, the Boltzmann term in Equation (1) essentially makes the rate coefficient zero at low temperatures. The constraint that reactions must have no activation energy to be efficient at low temperatures does not rule out all chemical processes, despite the widespread view to the contrary by astronomers before interstellar molecules were discovered. Although certain classes of reactions involving two neutral species are now known to be barrierless, ion-molecule reactions constitute the major class of such reactions in interstellar clouds (Le Teuff *et al.*, 2000). In these reactions, a charged species collides with a neutral one under the influence of a strong long-range potential. Positive molecular ions are produced in clouds mainly via bombardment by relativistic nuclei known as cosmic rays (Herbst and Klemperer, 1973).

Ion-molecule reactions have been well studied in the laboratory, and, if exothermic, can be rapid at low temperatures (Rowe, 1988). For the case of a non-polar neutral reactant, the rate coefficient k is often well approximated by a simple formula first derived by Langevin (Herbst, 1996):

$$k = 2\pi e(\alpha/\mu)^{1/2}, \quad (2)$$

where e is the electronic charge in esu, α is the polarizability in cm^3, and μ is the reduced mass of the reactants in gm. The formula derives from the cross section for capture by a $1/R^4$ attractive potential, which pertains to the long-range interaction between a charged species and a non-polar neutral with isotropic dipole polarizability α. In the derivation, short-range forces are ignored. In cgs units, k is typically 1-2 $\times 10^{-9}$ cm^3 s^{-1}, a number significantly larger than those for barrierless neutral neutral reactions, and one that possesses no temperature dependence at all! If the neutral reactant possesses a permanent dipole moment, the long-range attractive potential for an ion is stronger, and the "capture" rate coefficient, which is inversely proportional to the square root of the temperature, can be as large as 10^{-7} cm^3 s^{-1} at 10 K (Clary, 1988).

The potential surfaces of many if not most ion-molecule reactions possess a potential minimum between reactants and products, rather than a saddle point; the structure corresponding to the minimum is known as a collision complex, and often lives long enough during a collision and has sufficient energy so that some degree of randomization or scrambling can occur. Based loosely on a statistical model known as the phase-space theory (Herbst, 1996), it is typically assumed in ion-molecule reactions involving the transfer, or exchange, of isotopic atoms that they are incorporated into products in a statistical fashion. As an example, consider the exothermic, rapid reaction

$$H_3^+ + CO \longrightarrow HCO^+ + H_2. \tag{3}$$

If, instead of H_3^+, the isotopomer H_2D^+ is used as the reactant, the statistical assumption implies that the product DCO^+ is produced 1/3 of the time.

Ion-molecule exchange reactions are the primary cause of fractionation in interstellar clouds, as illustrated below for the case of deuterium. Consider the following system of forward (exothermic) and backward (endothermic) reactions in which deuterium and hydrogen atoms are exchanged:

$$AH^+ + BD \rightleftharpoons AD^+ + BH + \Delta E_0. \tag{4}$$

Here A and B stand for one or more atoms, and ΔE_0 is the exothermicity at 0 K (defined to be positive). If the forward reaction is exothermic by an amount $\Delta E_0 \gg k_B T$, the backward reaction is slow at such temperatures. The exothermicity generally derives from differences in zero-point vibrational energies, but the Pauli Principle can also play a role (Sidhu *et al.*, 1992) if the ground rotational state of a reactant or product is forbidden, as is the case for the important ion H_3^+. If the neutral reactant BD is abundant enough to be considered a reservoir for deuterium, and if the backward reaction is sufficiently endothermic, then the forward process can convert a large portion of the "trace" AH^+ ion into the deuterated isotopomer AD^+, and the ratio of AD^+ to AH^+ can greatly exceed that of elemental D to H. Once AD^+ is produced, it can react with other neutrals in the gas to cause "secondary" fractionation, so that high relative abundances of deuterated isotopomers can exist for most trace species.

Now let us be more quantitative. The ratio of the forward to the backward rate coefficient under thermal equilibrium conditions is defined as the equilibrium coefficient $K(T)$, which is given by the equation

$$K(T) = k_f / k_b = \exp(-\Delta G / k_B T), \tag{5}$$

where ΔG is the change in Gibbs free energy between products and reactants. At low temperature, it can often be safely assumed that the change in Gibbs free energy is essentially the same as the negative of the exothermicity at 0 K, an assumption equivalent to ignoring entropy differences. With the additional assumption that the forward rate coefficient occurs at the capture value k_C, the backward rate coefficient can then be written as

$$k_b = k_C \exp(-\Delta E_0 / k_B T). \tag{6}$$

The result at low temperature is a vanishingly small value for the backward rate coefficient if $\Delta E_0 \gg k_B T$. Indeed, the value can be so small that the deuterated ion AD^+ is typically destroyed by other reactions. For some systems, however, the approximation that the entropy term can be ignored is not good enough, and the more detailed statistical formulation of the equilibrium coefficient is preferable (Herbst, 1982). In this case,

$$K(T) = f_{\mathrm{trans}} f_{\mathrm{rot}} f_{\mathrm{vib}} f_{\mathrm{elec}} \exp(\Delta E_0 / k_{\mathrm{B}} T) \tag{7}$$

where the factors f refer to ratios of partition functions of the products to those of the reactants for assorted modes of energy (translational, rotational, vibrational, electronic). At low temperatures, the species all lie in their ground electronic-vibrational states, so that it is only necessary to consider rotational and translational motion. The latter leads just to a ratio of masses raised to the 3/2 power. Care must be taken with rotational partition function for light species at low temperature since the normal integral approximation is inaccurate and since ortho/para effects must be taken account of in detail (Sidhu *et al.*, 1992).

Even with a careful determination of the equilibrium coefficient at low temperatures, problems may arise in the estimation of the forward and backward rate coefficients. First of all, it is not obvious that the forward rate coefficient will equal the Langevin value of $\approx 10^{-9}$ cm^3 s^{-1} (as given in Equation 2) or an analogous capture value for polar reactants. Most measurements of isotopic exchange reactions have been made at temperatures of 80 K and above, where the forward rate coefficient often tends to be smaller than the capture value. Secondly, the determination of the small backward rate coefficient may be strongly influenced, as we shall see, by small non-equilibrium effects. Since these effects are present in the laboratory as well, the correlation between laboratory and interstellar rates must also be carefully examined for endothermic reactions.

2. Deuterium Fractionation

In dense interstellar clouds, the main reservoir of deuterium is the molecule HD, which takes up much of the elemental abundance of D. While H_2 is produced on the surfaces of dust grains via the association of diffusing H atoms, HD is produced both on the surfaces of dust grains and via gas-phase reactions leading from the precursor H_2. Measurements of the HD to H_2 abundance ratio show it to be $\approx 3 \times 10^{-5}$, or double the elemental abundance ratio, confirming the theoretical understanding. We now must consider the reactions of assorted interstellar ions AH^+ with HD, as in Equation (4) (B=H). Although it is generally thought that all exothermic ion-molecule reactions proceed without activation energy, this is hardly the case. It is especially untrue for exchange reactions involving HD, where barriers often occur (Henchman *et al.*, 1988). The result is that only three major molecular ions in dense interstellar clouds are known to be able to exchange deuterons with HD - these are H_3^+, CH_3^+, and $C_2H_2^+$ (Millar *et al.*, 1989). The reaction systems with their exothermicities (K) are

$$H_3^+ + HD \rightleftharpoons H_2D^+ + H_2 + 232\,\mathrm{K}, \tag{8}$$

$$CH_3^+ + HD \rightleftharpoons CH_2D^+ + H_2 + 370\,K, \tag{9}$$

and

$$C_2H_2^+ + HD \rightleftharpoons C_2HD^+ + H_2 + 550\,K. \tag{10}$$

Of the three ions, H_3^+ is the most abundant, and so is most important at low temperatures. As the temperature rises, the fairly low exothermicity of the first reaction lessens its importance in isotopic fractionation compared with the other two.

Until recently, the forward and backward reactions for system (8) had only been studied at 80 K and above. Modellers extrapolated these results to a near Langevin value for the forward rate coefficient at 10 K, and used the equilibrium coefficient ($K \approx 10^{10}$) to determine an ultra-low value for the backward rate coefficient of $\approx 2 \times 10^{-18}$ cm^3 s^{-1} at this temperature (Millar *et al.*, 1989; Sidhu *et al.*, 1992). Under these conditions, the ion H_2D^+ in dense interstellar clouds is depleted mainly by reaction with heavy neutrals such as CO and O (which have a total fractional abundance, or concentration, of 10^{-4}) and secondarily by reaction with electrons (which have a fractional abundance of 10^{-7} - 10^{-8}). Still, the ratio of H_2D^+ to H_3^+ can be enhanced by a factor of over 3000 with respect to the HD/H_2 ratio (Millar *et al.*, 2000). This result can be motivated by first writing an approximate kinetic equation for the rate of change of the concentration, designated by (...), of H_2D^+:

$$\frac{d(H_2D^+)}{dt} = k_f(H_3^+)(HD) - k_b(H_2D^+)(H_2) - k_M(H_2D^+)(M), \tag{11}$$

where the "M" stands for all heavy neutrals, then setting the time derivative equal to zero (the steady-state approximation). One obtains a formula for the concentration ratio of H_2D^+ to H_3^+ as the product of an enhancement factor S, which is much greater than unity, and the ratio of HD to H_2 (Millar *et al.*, 2000). The high abundance of H_2D^+ produces enhanced amounts of other deuterated species through the reactions with heavy species: viz.,

$$H_2D^+ + CO \longrightarrow DCO^+ + H_2 : HCO^+ + HD, \tag{12}$$

although the statistical approximation discussed in the Introduction lessens the fractionation for these secondary species since deuterated isotopomers are produced on fewer than 100% of reactive collisions. Thus, one expects the concentration ratio DCO^+/HCO^+ to be one-third that of H_2D^+/H_3^+. Moreover, dissociative recombination of molecular ions with electrons leads to enhanced abundances of deuterated neutrals. One such enhanced product is the deuterium atom, produced in reactions such as

$$DCO^+ + e \longrightarrow CO + D. \tag{13}$$

The abundance of atomic D can in fact reach a significant fraction of the residual abundance of atomic H, and can cause further deuteration both by gas-phase reactions and, after accretion onto the mantles of dust particles, by diffusive surface reactions. The latter process for deuteration is especially important in star-forming regions where the deuterated surface species are released into the gas-phase via rising temperatures or shock waves (Charnley *et al.*, 1997; Caselli *et al.*, 2002).

Gas-phase models of cold cores within dense clouds show that, for many trace ions and neutrals, there is significant deuterium fractionation, in agreement with observation (Millar *et al.*, 1989; Roberts and Millar, 2000a). Furthermore, in dense clouds where a significant depletion of heavy molecules has occurred from the gas phase onto the grains, the destruction of H_2D^+ is no longer dominated by reaction with heavy neutrals but by the slower process of dissociative recombination with electrons. The result is a strong increase in deuterium fractionation (Roberts and Millar, 2000b) such that the ratio of deuterated to normal trace species can reach near unity and, in addition, doubly deuterated species can be detected! The cold core L134N is an example of a source with enhanced deuteration (Roueff *et al.*, 2000). A calculation of how interstellar fractionation of deuterium is modified during the collapse of an interstellar cloud core into a solar-type nebula with attendant accretion onto dust grains has been reported by Aikawa and Herbst (1999).

2.1. Problems Arising From New Laboratory Work

New laboratory work by Gerlich *et al.* (2002) using an ion trap at 10 K has shown that k_f for reaction (8) is 3.5×10^{-10} cm^3 s^{-1}, or 1/5 of the Langevin value. Even more damaging at first glance to the importance of system (8) for deuterium fractionation, they measured two rather high values for k_b, depending on the type of molecular hydrogen used. Indeed, both measured values of k_b are orders of magnitude greater than the equilibrium value used in interstellar models. With *normal*-H_2, in which there are three parts *ortho* to one part *para*, k_b is 4.9×10^{-11} cm^3 s^{-1}, whereas with "pure" *para*-H_2, the value of the backward rate coefficient drops to 7.3×10^{-13} cm^3 s^{-1}. Remember that *ortho*-H_2 consists of the rotational levels with odd quantum number J, whereas *para*-H_2 consists of the levels with even J. *Normal*-H_2 at 10 K has 75% of its molecules in their first excited (J=1) rotational states, while para-H_2 is pure J=0. The rotational excitation of molecular hydrogen in dense clouds is not very well constrained, but the hydrogenic rotational populations are thought to be near the thermal equilibrium value, which is close to pure *para* at 10 K. The current best estimate is that there is approximately 1 part in 1000 *ortho*-H_2 (Le Bourlot, 1991).

In order to determine what value of k_b to use in interstellar models, Gerlich *et al.* simulated the backward process in their laboratory trap and found that the high rates are caused by "*ortho*" heating, an effect in which the J=1 states of H_2 transfer some of their rotational excitation to similar *ortho* states of H_2D^+; the net result being that when excited H_2 and H_2D^+ react, the reaction is no longer endothermic and proceeds efficiently. They were able to determine, moreover, that the supposedly pure *para*-H_2 source used in the laboratory contains an *ortho* impurity of 1%. Finally, they were able to estimate the appropriate backward rate coefficient for interstellar models at 10 K as 7×10^{-14} cm^3 s^{-1}, a value lower than either laboratory measurement but higher than the equilibrium value by more than four orders of magnitude! Moreover, the uncertainty in the laboratory simulation means that the estimated rate for interstellar purposes may well be an order of magnitude lower, making it only more than three orders of magnitude higher than the equilibrium value!

The effect of the new rate coefficients on interstellar models is currently being evaluated. A preliminary study was done by Gerlich *et al.* (2002), while a more detailed study was undertaken by Roberts *et al.* (2002). One must remember that even if the backward rate is much larger than heretofore assumed, if it is still smaller than the rates for the other routes for the destruction of H_2D^+, the model results will not be greatly affected. In fact, current model runs show that the diminution of a factor of five in the *forward* rate coefficient from previous estimates is by far the more important change and results in a lowering of deuterium fractionation such that agreement between model results and observation is significantly worse than heretofore in sources with the most extreme fractionation!

Much more work is needed to understand both the forward and backward rate coefficients for reaction (8) at a variety of temperatures as well as the non-reactive collisional excitations for H_2D^+ and H_2 in both the laboratory and in space before our current understanding of the reactive processes for interstellar modelling can be confirmed. The potential surface for the reaction is a rather complicated one in which there are potential minima in both the entrance and exit channels plus a saddle point barrier connecting them. The barrier occurs below the reactants' energy, but can still affect the reaction dynamics (Gerlich *et al.*, 2002). The conservation of nuclear spin during the course of the reaction is another constraint (Cordonnier *et al.*, 2000). With these complexities, understanding may come slowly.

3. Fractionation of ^{15}N

The ^{15}N/^{14}N isotopic ratio in the solar system is highly variable (Kallenbach *et al.*, 1998) and is undoubtedly caused by a variety of effects. One possible cause (Owen *et al.*, 2001) is that fractionation from the interstellar period was at least partially preserved in the solar nebula, that this fractionation was different for different molecular species, and that different species were responsible for the bulk of the elemental N in different solar system bodies. Interestingly, nitrogenic isotopic variations can be seen over short distance scales in interstellar sources, where they are ascribed to either fractionation or to actual differences in the total isotopic abundances.

To investigate the amount of ^{15}N fractionation achievable by ion-molecule reactions in a low temperature interstellar cloud, Terzieva and Herbst (2000) calculated the equilibrium coefficients for some possibly important exchange reactions involving nitrogenic isotopes. The reservoirs of ^{15}N were considered to be N^{15}N and

TABLE I

Important nitrogen exchange reactions

Reaction	$\Delta E_0/k_B$	$K(10\text{ K})$	k_f (cm^3 s^{-1})	k_b (cm^3 s^{-1})
$N^{15}N + HN_2^+ \rightleftharpoons N_2 + H^{15}NN^+$	10.7	1.44	4.0(-10)	2.8(-10)
$N^{15}N + HN_2^+ \rightleftharpoons N_2 + HN^{15}N^+$	2.3	0.63	4.0(-10)	6.4(-10)
$^{15}N + HN_2^+ \rightleftharpoons N + H^{15}NN^+$	36.1	35.8	4.0(-10)	1.1(-11)
$^{15}N + HN_2^+ \rightleftharpoons N + HN^{15}N^+$	27.7	15.6	4.0(-10)	2.6(-11)
$^{15}N + HCNH^+ \rightleftharpoons N + HC^{15}NH^+$	35.9	35.1	4.0(-10)	1.1(-11)

atomic ^{15}N. The reaction exothermicities are much smaller than in the case of deuterium; these were calculated by ab initio quantum chemical techniques. Some reactions, their exothermicities, and their calculated equilibrium coefficients at 10 K (see Equation 7) are shown in Table I. Of these reactions, only those involving $N^{15}N$ have been studied in the laboratory, albeit at higher temperatures. It can be seen that the greater equilibrium coefficients are those involving atomic N. Forward and backward rate coefficients at 10 K are also included in Table I. Terzieva and Herbst (2000) assumed that the forward rate coefficients are all approximately equal to one-half the Langevin value; the backward rate coefficients can then be obtained from the equilibrium coefficients. The assumption of rapid forward rate coefficients is highly debatable for two of the atomic N reactions because more than one chemical bond must be broken during the course of the reactions in order to substitute one N isotope for the other, an effect that normally leads to activation energy barriers. Thus, any fractionation calculated with these exchange reactions is actually an upper limit. Given the small differences between forward and backward rate coefficients, no non-equilibrium effects were considered.

The nitrogen exchange reactions in Table I, as well as a few less important ones, were included by Terzieva and Herbst (2000) in a large gas-phase model of dense interstellar clouds at 10 K including many ion-molecule reactions involving both ^{14}N and ^{15}N. The elemental $^{15}N/^{14}N$ abundance ratio was varied between values of 1/200 and 1/600. The results show that atomic N is more important as a reservoir of ^{15}N than molecular N, but that fractionation is still not large despite the greater equilibrium coefficients for exchange reactions involving atomic N, and despite the fact that the abundance of atomic N exceeds that of molecular N at most times. The basic problem is that there is too little atomic N to produce much in the way of ^{15}N isotopomers, because what is produced is depleted rapidly by processes other than the back reaction.

To understand this in more detail, we will consider, as an example, the exchange reaction between ^{15}N and $HCNH^+$. If equilibrium could be reached and the N atom could be viewed as a reservoir, then the ratio between $HC^{15}NH^+$ and $HCNH^+$ could be enhanced by a factor of 35.1 compared with the atomic $^{15}N/^{14}N$ ratio. But since

TABLE II
Fractionation of ^{15}N

Species	Enhancement			
	10^4 y	10^5 y	10^6 y	10^7 y
^{15}N	1.00	1.00	1.00	0.93
$HC^{15}N$	1.05	1.27	1.30	1.10
$^{15}NH_3$	1.02	1.25	1.01	0.99
$H^{15}NN^+$	1.04	1.05	1.05	1.05
$N^{15}N$	1.99	2.00	2.03	2.03
$HC^{15}NH^+$	1.14	1.46	1.41	1.11

$HC^{15}NH^+$ reacts more rapidly with electrons than with atomic N, it is enhanced to a much smaller degree. This can be shown quantitatively if one considers only the forward and backward reactions plus the (dissociative) recombination with electrons. Specifically, the steady-state enhancement factor S, defined as the factor by which the $^{15}N/^{14}N$ ratio in $HCNH^+$ exceeds that in atomic N, is given by the expression

$$S = \frac{k_f}{k_b + k_e(e)/(^{14}N)}, \tag{14}$$

where k_f and k_b are the forward and backward rate coefficients for the N-$HCNH^+$ system, k_e is the rate coefficient for the reaction between $HC^{15}NH^+$ and electrons, and the symbols (...) stand once again for concentration. This expression is derived by setting the derivative of the concentration of $HC^{15}NH^+$ with time equal to zero (see Equation 11). Substituting the rate coefficients from Table I, the rate coefficient for dissociative recombination with electrons (2×10^{-6} cm^3 s^{-1}) and typical concentrations for dense interstellar clouds so that $(e)/(^{14}N) = 3 \times 10^{-4}$ yields an enhancement near if not even below unity. (The actual enhancement is somewhat larger because some of the products of the dissociative recombination reaction are recycled to form $HC^{15}NH^+$ by subsequent ion-molecule reactions.) Exchange reactions with molecular nitrogen as the reservoir are hardly more important since the equilibrium coefficients are close to unity.

Table II shows some values for isotopic fractionation from the complete model of Terzieva and Herbst (2000) at assorted times. Here a slightly different enhancement factor – the abundance ratio between the ^{15}N and normal N isotopomers divided by the elemental abundance ratio – appears. The results do not appear to be sensitive to the elemental abundance ratio chosen in the range $1/200 - 1/600$. As expected, isotopic fractionation as caused by ion-molecule exchange reactions is not a large effect, although it can reach 50% for selected species. Moreover, it must be remembered that the values shown are upper limits to the deviations from unity (or two, for the case of N_2) since the exchange reactions with atomic N may not proceed rapidly. Note that even the so-called reservoirs exhibit a small fractionation at long times. Finally, the statistical assumption made for the secondary reactions following exchange deepens the uncertainties.

4. Fractionation of ^{13}C

The interstellar fractionation of carbon has been studied mainly in diffuse clouds where much of the gas is atomic and the dominant form of the element carbon is the atomic ion C^+. With $^{13}C^+$ as the reservoir, fractionation can occur in such carbon-containing molecules as CO. The exchange reactions

$$^{13}C^+ + CO \rightleftharpoons C^+ + ^{13}CO \tag{15}$$

have a left-to-right exothermicity of 35 K and an equilibrium coefficient at 50 K, a typical temperature for diffuse regions, of 2.0, so that if the reactions can reach equilibrium, the abundance ratio $^{13}CO/^{12}CO$ can be enhanced by a factor of two (Langer *et al.*, 1984). In diffuse clouds, however, the photodissociation rate of species that are not abundant enough to be self-shielded is large, so that ^{13}CO is destroyed more rapidly by photons than by reaction with C^+. If one attempts to calculate an enhancement factor as was done in the previous section, it turns out to be less than unity. A somewhat more detailed method of analysis is to assume that both CO and ^{13}CO are produced via similar ion-molecule pathways from precursors such as C^+ and OH, and that both are mainly depleted via photodissociation (Lambert *et al.*, 1994). With such an analysis, the dominant effect is the differential rate of photodissociation between the two isotopomers of CO – the normal isotopomer is more self-shielded and photodissociated more slowly so that the abundance ratio is actually somewhat less than the elemental abundance ratio! Recent measurements by Federman *et al.* (2002) confirm this view.

In dense clouds, the reservoir of ^{13}C is carbon monoxide. Fractionation in trace species such as the formyl ion (HCO^+) can occur via ion-molecule exchange reactions; viz.,

$$HCO^+ + ^{13}CO \rightleftharpoons H^{13}CO^+ + CO, \tag{16}$$

which possesses an equilibrium coefficient of 2.5 at 10 K (Langer *et al.*, 1984). Assuming $C^{18}O$ to be the repository of oxygen-18 in dense clouds, a similar exchange reaction involving this species can be important. This latter reaction has a calculated equilibrium coefficient of 4.1 at 10 K, and can again lead to some fractionation. Since CO is an abundant interstellar molecule, these exchange systems may come close to equilibrium. Detailed models of carbon and oxygen fractionation in dense interstellar clouds were undertaken nearly twenty years ago (Langer *et al.*, 1984), and new models should be considered.

5. Conclusions

There exists a complex relationship between isotopic fractionation via ion-molecule reactions in the interstellar medium, and isotopic variations in the solar system today. Interstellar fractionation is especially significant for deuterium, where, despite some uncertainties, it is generally accepted that ion-molecule exchange reactions can produce very large enhancements in the D/H abundance ratio involving isotopomers of trace species. The deuterium fractionation achieved in the gas can actually be enhanced by both gas-phase and grain-surface processes as gas-phase material accretes onto grains. It is also generally accepted that material in the solar system with high deuterium content is pristine interstellar material, although few models have been undertaken to probe whether isotopic fractionation from interstellar clouds and protostars can be preserved through the solar-nebula stage. For other elements, such as nitrogen, carbon, and oxygen, interstellar fractionation via ion-molecule exchange reactions is not a large effect and other, competitive sources of fractionation such as differential photodissociation must be considered even in interstellar clouds. We are still a long way from being able to understand the complete role interstellar fractionation played in the original inventory of isotopes in the solar system.

Acknowledgements

We thank the International Space Science Institute for their hospitality during the 2002 workshop on Solar System History from Isotopic Volatile Signatures. This work was supported in part by the National Science Foundation (US).

References

Aikawa, Y., and Herbst, E.:1999, 'Deuterium Fractionation in Protoplanetary Disks', *Astrophys. J.* **526**, 314–326.

Caselli, P., Stantcheva, T., Shalabiea, O., Shematovich, V.I., and Herbst, E.:2002, 'Deuterium Fractionation on Interstellar Grains Studied with Modified Rate Equations and a Monte Carlo Approach', *Plan. Space Sci.*, in press.

Charnley, S.B., Tielens, A.G.G.M., and Rodgers, S.D.: 1997, 'Deuterated Methanol in the Orion Compact Ridge', *Astrophys. J.* **482**, L203–L206.

Clary, D.C.: 1988, 'Theory of Reactive Collisions at Low Temperatures', *in* T.J. Millar and D.A. Williams (eds.), *Rate Coefficients in Astrochemistry*, Kluwer, Dordrecht, pp. 1–16.

Cordonnier, M., Uly, D., Dickman, R.M., Kerr, K.E., Zhang, Y., and Oka, T.: 2000, 'Selection Rule fo Nuclear Spin Modifications in Ion-Neutral Reactions Involving H_3^+', *J. Chem. Phys.* **113**, 3181–3193.

Federman, S.R., Lambert, D.L., Sheffer, Y., Cardelli, J.A., Andersson, B.-G., van Dishoeck, E.F., and Zsargo, J.: 2002, 'Further Evidence for Chemical Fractionation from Ultraviolet Observations of Carbon Monoxide', *Astrophys. J.*, in press.

Gerlich, D., Herbst, E., and Roueff, E.:2002, '$H_3^+ + HD \rightleftharpoons H_2D^+ + H_2$: Low-temperature Laboratory Measurements and Interstellar Implications', *Plan. Space Sci.*, in press.

Henchman, M.J., Paulson, J.F., Smith, D., Adams, N.G., and Lindinger, W.: 1988, 'Chemical Pathways for Deuterium Fractionation in Interstellar Molecules', *in* T.J. Millar and D.A. Williams (eds.), *Rate Coefficients in Astrochemistry*, Kluwer, Dordrecht, pp. 201–208.

Herbst, E.: 1982, 'The Temperature Dependence of the HCO^+/DCO^+ Abundance Ratio in Dense Interstellar Clouds', *Astron. Astrophys.* **111**, 76–80S.

Herbst, E.: 1996, 'Gas Phase Reactions', *in* G.W.F. Drake (ed.), *Atomic, Molecular, and Optical Physics Handbook*, Amer. Inst. of Physics, Woodbury, NY, pp. 429–440.

Herbst, E.: 2001, 'The Chemistry of Interstellar Space', *Chem. Soc. Rev.* **30**, 168–176.

Herbst, E., and Klemperer, W.: 1973, 'The Formation and Depletion of Molecules in Dense Interstellar Clouds', *Astrophys. J.* **185**, 505–533.

Kallenbach, R. *et al.*: 1998, 'Isotopic Composition of Solar Wind Nitrogen: First In Situ Determination With The CELIAS/MTOF Spectrometer Aboard SOHO', *Astrophys. J.* **507**, L185-L188.

Lambert, D.L., Sheffer, Y., Gilliland, R.L., and Federman, S.R.: 1994, 'Interstellar Carbon Monoxide Toward Zeta Ophiuchi', *Astrophys. J.* **420**, 756–771.

Langer, W.D., Graedel, T.E., Frerking, M.A., and Armentrout, P.B.: 1984, 'Carbon and Oxygen Isotope Fractionation in Dense Interstellar Clouds', *Astrophys. J.* **277**, 581–604.

Le Bourlot, J.: 1991, 'Ammonia Formation and the Ortho-to-Para Ratio of H_2 in Dark Clouds', *Astron. Astrophys.* **242**, 235–240.

Le Teuff, Y.H., Millar, T.J., and Markwick, A.J.: 2000, 'The UMIST Database for Astrochemistry 1999', *Astron. Astrophys. Suppl. Ser.* **146**, 157–168.

Millar, T.J., Bennett, A., and Herbst, E.: 1989, 'Deuterium Fractionation in Dense Interstellar Clouds', *Astrophys. J.* **340**, 906–920.

Millar, T.J., Roberts, H., Markwick, A.J., and Charnley, S.B.: 2000, 'The Role of H_2D^+ in the Deuteration of Interstellar Molecules', *Phil. Trans. Roy. Soc. Lond. A* **358**, 2535–2547.

Owen, T., Mahaffy, P.R., Niemann, H.B., Atreya, S., and Wong, M.: 2001, 'Protosolar Nitrogen', *Astrophys. J.* **553**, L77–L79.

Roberts, H., and Millar, T.J.: 2000a, 'Modelling of Deuterium Chemistry and its Application to Molecular Clouds', *Astron. Astrophys.* **361**, 388–398.

Roberts, H., and Millar, T.J.: 2000b, 'Gas-phase Formation of Doubly Deuterated Species', *Astron. Astrophys.* **364**, 780–784.

Roberts, H., Herbst, E., and Millar, T.J.: 2002, 'The Importance of new Rate Coefficients for Deuterium Fractionation Reactions on Interstellar Chemistry', *Month. Not. Roy. Astron. Soc.* **336**, 283–290.

Roueff, E., Tiné, S., Pineau des Forêts, G., Falgarone, E., and Gerin, M.: 2000, 'Detection of Doubly Deuterated Ammonia in L134N', *Astron. Astrophys.* **354**, L63–L66.

Rowe, B.R.: 1988, 'Studies of Ion-Molecule Reactions at $T < 80$ K', *in* T.J. Millar and D.A. Williams (eds.), *Rate Coefficients in Astrochemistry*, Kluwer, Dordrecht, pp. 135–152.

Sidhu, K.S., Miller, S., and Tennyson, J.: 1992, 'Partition Functions and Equilibrium Constants for H_3^+ and H_2D^+', *Astron. Astrophys.* **255**, 453–456.

Terzieva, R., and Herbst, E.: 2000, 'The Possibility of Nitrogen Isotopic Fractionation in Interstellar Clouds', *Month. Not. Roy. Astron. Soc.* **317**, 563–568.

Williams, D.A., and Herbst, E.: 2002, 'It's a Dusty Universe: Surface Science in Space', *Surface Science* **500**, 823–837.

Address for Offprints: E. Herbst, Departments of Physics and Astronomy, The Ohio State University, Columbus, OH 43210, USA; herbst@mps.ohio-state.edu

ISOTOPIC FRACTIONATION BY PLASMA PROCESSES

REINALD KALLENBACH (kallenbach@issi.unibe.ch)
International Space Science Institute, Hallerstr. 6, CH-3012 Bern, Switzerland

Received: 15 April 2002; Accepted in final form: 24 July 2002

Abstract. The Sun is the largest reservoir of matter in the solar system, which formed 4.6 Gy ago from the protosolar nebula. The solar wind carries a nearly unfractionated sample of heavy isotopes at energies of about 1 keV/amu from the Sun into interplanetary space. Data from space missions and theoretical models indicate that the isotopes of the volatile elements N, O, Ne, and Ar are fractionated by at most a few percent per atomic mass unit in different solar wind regimes. In contrast, isotopic abundances of solar and heliospheric energetic particles at energies larger than about 100 keV/amu are observed to strongly vary relative to solar abundances. Processes such as resonant acceleration or pre-acceleration by plasma waves, first-order Fermi acceleration, or propagation in the interplanetary plasma are discussed as causes for charge-to-mass dependent fractionation.

1. Introduction

Solar system samples are usually compared to terrestrial material as a reference. However, as the Sun contains more than 99% of solar system matter, any non-nuclear fractionation process during the formation of the inner solar system from the early solar nebula could have changed the terrestrial isotopic composition, but practically not the solar isotopic composition. Consequently, solar matter is a better witness of the composition of the protosolar nebula than terrestrial material.

For instance, collisions between planetesimals and planetary 'embryos', and subsequent hydrodynamic escape (Pepin, 2000) have depleted the terrestrial atmosphere in volatile elements, in particular in their light isotopes. The ratio $^{20}Ne/^{22}Ne$ in the Earth's atmosphere is 9.78 ± 0.03, whereas the solar wind $^{20}Ne/^{22}Ne$ ratio amounts to 13.7 ± 0.3, as derived from the Apollo Solar Wind foil Collection (SWC) experiment (Geiss *et al.*, 1972) and confirmed by SOHO data (Kallenbach *et al.*, 1997). Isotopic ratios of chemically active volatiles such as N vary even more in different solar system samples (e.g., Marty *et al.*, 2003).

This review summarizes solar wind isotopic abundance data from sensors on board SOHO, WIND, ACE, and ULYSSES. Isotopic abundance variations in the solar wind are compared to theoretical considerations on plasma processes. This constrains the strength of isotopic fractionation in the inner corona, and the solar isotopic composition, most notably that of volatile elements, can be inferred.

The possible origin of the "SEP" noble gas component implanted into lunar soil is discussed in the context of models of isotopic fractionation in solar energetic particles (SEPs) and of related spacecraft observations.

Space Science Reviews **106:** 305–316, 2003.

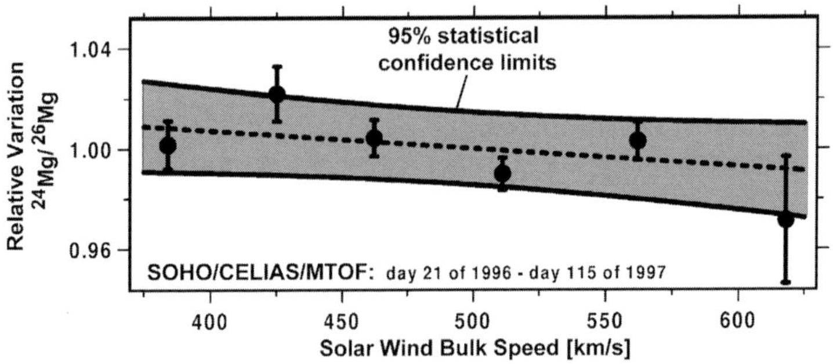

Figure 1. Relative variation of the solar wind $^{24}Mg/^{26}Mg$ ratio versus the solar wind bulk speed. The plotted experimental uncertainty is of statistical nature only. Instrumental systematic background and calibration offsets are at most 3% (Table I), but cancel out to large extent when comparing the $^{24}Mg/^{26}Mg$ ratio in different solar wind regimes. Data from Kallenbach *et al.* (1998a).

2. Variability of Isotopic Abundances in the Solar Wind

Since the launch of the WIND, SOHO, and ACE spacecraft, the measurements of the solar wind isotope abundances for the refractory element Mg serve as a calibration benchmark for the data from a new type of high-resolution isochronous time-of-flight (TOF) mass spectrometer (e.g., Hovestadt *et al.*, 1995), because 1) the minor isotopes ^{25}Mg and ^{26}Mg are fairly abundant, 2) there is little instrumental interference of Mg isotopes with other species, and 3) the mean terrestrial and solar Mg isotopic compositions are expected to agree.

Indeed, the mean solar wind $^{25}Mg/^{24}Mg$ and $^{26}Mg/^{24}Mg$ ratios agree with their terrestrial ratios within a few percent (Bochsler *et al.*, 1996; Kucharek *et al.*, 1998). In search of solar wind isotopic fractionation processes, the ratios $^{25}Mg/^{24}Mg$ and $^{26}Mg/^{24}Mg$ have then been analyzed in different types of streams such as coronal-hole type ('fast') and streamer-belt ('slow') solar wind, and in the ejecta of coronal mass ejections (CMEs). The data shown in Figure 1 suggest a variation of the $^{24}Mg/^{26}Mg$ ratio of (1.4 ± 1.0)%/amu (statistical 1σ-error) in the range of solar wind bulk speeds between 350 km/s and 620 km/s. The heavier isotopes are slightly depleted in the 'slow' solar wind compared to the 'fast' wind, although the error bars of the data do not exclude the assumption of no isotopic fractionation.

If the background and the uncertainties in the instrument efficiencies can be reduced, it may be possible to find significant correlations between variations in the $^{24}Mg/^{26}Mg$ ratio and parameters to distinguish 'slow' and 'fast' solar wind. Better indicators than solar wind speed are abundance ratios of low-FIP (First Ionization Potential) and high-FIP elements such as Mg/O (von Steiger and Geiss, 1989) and Fe/O (Aellig *et al.*, 1999), which are enhanced by up to a factor four in the 'slow' wind compared to the 'fast' wind. Relative to the isotopic fractionation as a function of solar wind speed, similar (Kallenbach, 2001) or stronger (Kucharek *et al.*, 2001) isotopic fractionation has been observed in correlation with the Mg/O and the Fe/O variation, respectively, but the study is still ongoing.

The CME ejecta belong to a special class of solar wind plasma. Wimmer-Schweingruber *et al.* (1999) have constrained the fractionation of Mg and Ne isotopes in the bulk solar wind during the November 1997 CME to less than 5%/amu. However, as the isotopic composition in CME ejecta has not been studied rigorously yet, any identified CME-related solar wind has been excluded from the SOHO data used to determine the solar isotopic composition.

3. Models of Isotopic Fractionation Processes in the Solar Wind

The processes of variable Coulomb friction and wave heating in the inner corona, and of gravitational settling in closed coronal magnetic loops are relevant in the context of SOHO data on solar wind isotopes. The FIP effect, depending on the chemical properties of elements, and gravitational settling in the Sun's outer convective zone fractionate heavy isotopes by only about 0.1%/amu (e.g., Bochsler, 2000) and are not measurable with SOHO/CELIAS/MTOF.

3.1. Variable Coulomb Friction

In the source regions of the 'slow' solar wind, the protons dominate the plasma dynamics and drag the heavier minor ions from the solar surface by Coulomb friction. In the "high-flux approximation", and neglecting thermal diffusion, the parameter $H = (2A - Q - 1)/Q^2$ orders the depletion of minor elements and isotopes with atomic mass A and charge Q (Geiss *et al.*, 1970). This depletion mainly occurs in regions of superradial solar wind expansion, where the proton flux density decreases rapidly. More refined models include the alpha-particles as dominant ions besides the protons (Bürgi and Geiss, 1986).

The model by Bodmer and Bochsler (2000) treats only the protons as dominant ions, but includes the process of thermal diffusion. Assuming that all ions are isothermal in the source regions of the 'slow' solar wind (e.g., Hefti *et al.*, 1998), they derive $H = \sqrt{(A+1)/A}\,(2A - Q - 1)/Q^2$ as parameter roughly ordering the depletion of minor heavy ions. Due to variable Coulomb friction the He/H ratio varies in the 'slow' solar wind (von Steiger *et al.*, 1995) between 0.01 and the photospheric value 0.05 which is observed in the 'fast' solar wind. Thus, the isotopic ratios of an element in the 'slow' solar wind should vary with the depletion factor $f_{\mathrm{HeH}} = (\mathrm{He/H})/0.05$. The model predictions in Figure 2 are consistent with the observed variations of ^{24}Mg/^{26}Mg in the solar wind (Figure 1). They also roughly agree with the ^{3}He/^{4}He variations (Table I), comparing 'slow' and 'fast' solar wind, although the $^{4}He^{2+}$ ions cannot always be treated as minor ions.

In the source regions of the 'fast' solar wind, isotopic temperatures are rather proportional to mass because of Alfvén wave heating (Bochsler *et al.*, 1985). If the 'fast' solar wind is powered to a large extent by waves, the isotopes of heavier elements are sufficiently heated to leave the solar surface without Coulomb drag, so that the mean isotopic fluxes remain unfractionated.

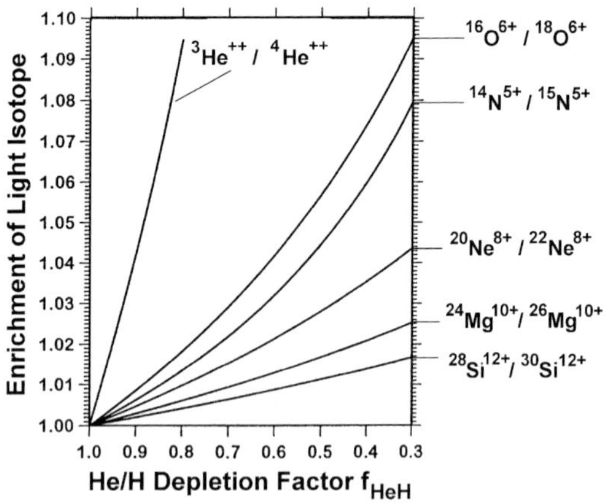

Figure 2. Isotopic fractionation by variable Coulomb friction. From Bochsler (2000).

3.2. Wave Heating

With SOHO/UVCS strong wave heating, presumably ion cyclotron damping of Alfvén waves, has been observed in the source regions of the 'fast' solar wind (Cranmer, 2000). As lower frequencies dominate the Alfvén wave spectrum, ions with low Q/A ratio such as O^{5+} are heated more than ions with higher Q/A such as protons. In general, not much fractionation in the total flux of ion species seems possible in the steady flow of the 'fast' solar wind; for instance, there is only little FIP fractionation (von Steiger *et al.*, 2000). On the other hand, wave heating rises the ion species' temperatures far above the typical coronal values of a few MK.

3.3. Gravitational Settling

Above a certain height r_0, turbulent mixing inside large closed coronal magnetic loops is usually neglectable. When such a loop opens up, it may feed the 'slow' solar wind with statically stratified material (Fisk *et al.*, 1998). This may occur in form of a CME, ejected after reconnection at the loop's footpoints, or a 'blob', released near the heliospheric current sheet at the loop's tip and dragged away by the 'slow' wind as dense elongated structure (Sheeley *et al.*, 1997). If the temperature T [K] (or T_6 [MK]) is constant along the loop's magnetic field between heights $r_1 > r_0$ and $r_2 > r_1$, the pressure of a species with atomic mass A varies as

$$p_A(r_2) = p_A(r_1)\exp\left(-\int_{r_1}^{r_2}\frac{dr\,r_\odot^2}{\Lambda_{0,A}r^2}\right) = p_A(\rho_1)\exp\left(-\frac{14A}{T_6}\frac{\rho_2-\rho_1}{\rho_2\rho_1}\right) \quad (1)$$

(Priest, 1982), where $r_\odot = 6.96\times10^8$ m denotes the solar radius, $\rho = r/r_\odot$, and $\Lambda_{0,A} = 50T/A$ [m] is the scale height, which may also depend on the Q/A ratio of the ion species due to wave heating (Zurbuchen *et al.*, 1998).

Depletions of O relative to H by a factor 10 below photospheric O/H ratios are observed at $\rho_2 \approx 1.7$ in the cores of large coronal loops with $T_6 \approx 1.3$ (Ray-

TABLE I

Mean isotopic abundance ratios in the solar wind compared to standard solar system values.

Volatile Isotopes			Refractory Isotopes		
Ratio	Solar wind	Solar system	Ratio	Solar wind	Solar system
$^{3}He/^{4}He^{(s)}$	0.00041±0.000025 [1]	0.000142 [2]	$^{25}Mg/^{24}Mg$	0.1260±0.0014 [3]	0.12658 [2]
$^{3}He/^{4}He^{(f)}$	0.00033±0.000027 [1]	0.000142 [2]	$^{26}Mg/^{24}Mg$	0.1380±0.0031 [3]	0.13947 [2]
$^{15}N/^{14}N$	0.0038±0.0018 [4]	0.0023(3) [5]	$^{29}Si/^{28}Si$	0.03344±0.00024 [3]	0.033612 [2]
$^{18}O/^{16}O$	0.0022±0.00055 [6]	0.0020 [2]	$^{30}Si/^{28}Si$	0.05012±0.00072 [3]	0.050634 [2]
$^{21}Ne/^{20}Ne$	0.0023±0.0006 [3]	0.0024(3) [7]	$^{42}Ca/^{40}Ca$	0.00657±0.00017 [3]	0.006621 [2]
$^{22}Ne/^{20}Ne$	0.0728±0.0013 [3]	0.0726(3) [7]	$^{44}Ca/^{40}Ca$	0.0209±0.0011 [3]	0.021208 [2]
$^{38}Ar/^{36}Ar$	0.183±0.018 [8]	0.1880 [2]	$^{54}Fe/^{56}Fe$	0.068±0.004 [9]	0.06327 [2]
			$^{57}Fe/^{56}Fe$	0.025±0.005 [9]	0.02339 [2]

1) Gloeckler and Geiss (1998); 2) Anders and Grevesse (1989); 3) Kallenbach (2000; 2001); 4) this work; 5) Owen *et al.* (2001); 6) Collier *et al.* (1998), Wimmer-Schweingruber *et al.* (2001); 7) mean of solar wind ratios determined by other methods (Geiss *et al.*, 1972; Palma *et al.*, 2002); 8) Weygand *et al.* (2001); 9) Ipavich *et al.* (2001); (f): 'fast' solar wind; (s): 'slow' solar wind

mond *et al.*, 1997). This would yield $\rho_2 - \rho_1 \approx 0.04$ in Equation (1) and, hence, a fractionation of e.g. isothermal Mg isotopes of up to 20%/amu. However, the element Mg must also be depleted in the loop at $\rho_2 \approx 1.7$, and 'blobs' from large loops feed only a small fraction of the 'slow' solar wind, so that mean solar wind abundance ratios of heavy isotopes are affected by less than 1%/amu. This estimate is confirmed by long-time averages of solar wind isotopic ratios (Figure 1). The composition of individual 'blobs' remains to be studied in more detail, though.

The ejecta of CMEs may either be depleted, enriched, or not fractionated in the heavier isotopes of an element, depending on the altitude of the material in the pre-CME loop, and on the temperature profile and other properties of the loop.

4. The Solar Isotopic Composition Inferred from Spacecraft Observations

The SOHO data on Mg isotopes and theoretical models suggest that the isotopic abundance ratios of heavy elements in the solar wind are the same as the solar ratios within a few percent. This is confirmed by the data in Table I, which includes solar system (i.e. solar) isotopic ratios inferred from meteorites and planetary samples. The ratio $^{3}He/^{4}He$ in the unfractionated 'fast' solar wind equals $(D + {}^{3}He)/^{4}He$ in the protosolar gas, when corrected for intersolar diffusion and gravitational settling (Geiss and Gloeckler, 2003). In the Sun's core all D has been burnt to ^{3}He.

The $^{15}N/^{14}N$ ratio from SOHO is re-evaluated in this work. The value $^{15}N/^{14}N = 0.005 \pm 0.0014$ of Kallenbach *et al.* (1998b), representing solar wind data of the year 1996 and the first 115 days of 1997, agrees with $^{15}N/^{14}N$ in young lunar soils

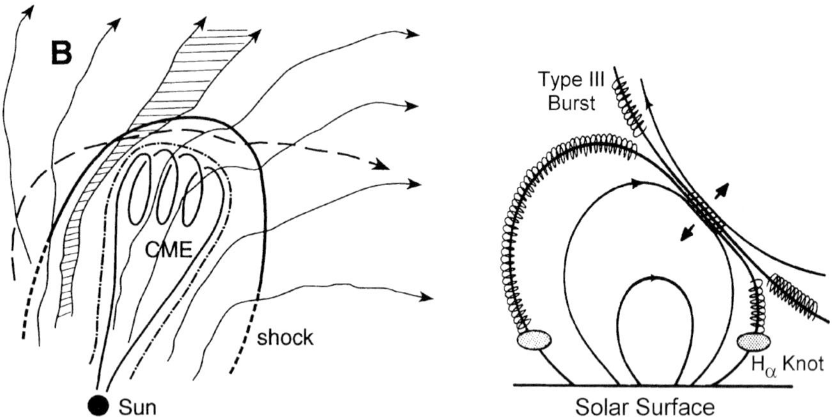

Figure 3. Left: Interplanetary configuration of a coronal mass ejection (Lee, 1997). *Right:* Configuration of a single-loop flare during its impulsive phase (Priest, 1982).

(Kim *et al.*, 1995). It also agrees with the protosolar $^{15}N/^{14}N$ derived by Geiss and Bochsler (1982) and with $^{15}N/^{14}N$ in pristine solar system material (Kung and Clayton, 1978; Sugiura *et al.*, 1998). However, from ratios measured in lunar ilmenites (Hashizume *et al.*, 2000; Marty *et al.*, 2003), in HCN of comet Hale-Bopp (Jewitt *et al.*, 1997), and based on ISO-SWS data for the Jovian troposphere (Fouchet *et al.*, 2000), and Galileo Probe measurements in the Jovian atmosphere, Owen *et al.* (2001) derive a protosolar ratio $^{15}N/^{14}N = (2.3 \pm 0.3) \times 10^{-3}$.

A tentative study with SOHO data of the years 1996 – 1998 yields $^{15}N/^{14}N_{SW} = (3.8 \pm 1.8) \times 10^{-3}$. The large uncertainty is due to electronic noise in the analog-to-digital converter (ADC) determining the time of flight of a single ion event. This noise yields an uncertainty in the background below the peaks of $^{29}Si^{2+}$ and $^{15}N^{+}+^{30}Si^{2+}$ ($^{30}Si^{2+}$ interferes with $^{15}N^{+}$ and is determined by the $^{29}Si^{2+}$ counts) that is about four times larger than the uncertainty from counting statistics. Originally, this "ADC effect" had been neglected. The new tentative SOHO value for the solar wind $^{15}N/^{14}N$ ratio, $(3.8 \pm 1.8) \times 10^{-3}$, roughly agrees with the protosolar $^{15}N/^{14}N$ ratio of Owen *et al.* (2001).

5. Charge-to-mass Fractionation in "Solar" Energetic Particles

In datasets of the IMP, ISEE, ULYSSES, WIND, SOHO, and ACE space missions, isotopic abundances in solar energetic particles (SEPs) vary greatly from event to event, usually classified as gradual (Figure 3, *left*), and impulsive (Figure 3, *right*) or ^{3}He-rich SEP events. Isotopic fractionation apparently occurs in course of the acceleration: In the bulk ejecta, i.e. in the bulk solar wind, of the November 1997 CME, the ratios $^{26}Mg/^{24}Mg$ and $^{22}Ne/^{20}Ne$ were consistent with the solar ratios within 5%/amu (Wimmer-Schweingruber *et al.*, 1999), but they were enhanced by a factor 2 – 3 in the SEPs of the same event (Leske *et al.*, 1999).

Among the ions that are accelerated out of the bulk solar wind energy range while propagating from the Sun into interplanetary space, the terms SEP and suprathermal ions may distinguish ions with energies larger than 100 keV/amu, and ions in the energy gap between solar wind energies and 100 keV/amu, respectively.

Solar wind ions are pre-accelerated by plasma waves such as turbulent waves downstream of an interplanetary shock (in Figure 3, *left*, the sunward region between the shock and the spiral structure of a magnetic cloud). Other suprathermal ion populations are also present in interplanetary space (e.g., Mason, 2001) and further accelerated near shocks: 1) Remnant particles from impulsive flares (Mason *et al.*, 1999), 2) interstellar (Möbius *et al.*, 1985) and inner-source (Geiss *et al.*, 1995) pick-up ions, and even 3) planetary (Grünwaldt *et al.*, 1997), lunar (Hilchenbach *et al.*, 1991), and cometary (Gloeckler *et al.*, 2000) pick-up ions.

The efficiency for stochastic pre-acceleration in Alfvénic turbulence depends on the ion's charge-to-mass ratio Q/A and on the spectral distribution of the waves (e.g., Möbius *et al.*, 1982), whereas transit-time damping in magnetosonic waves (Schwadron *et al.*, 1996) does not discriminate ions by their Q/A. The latter process is effective for ions faster than the Alfvén speed in the solar wind plasma frame, in particular for the pick-up ions (Gloeckler *et al.*, 1994).

The larger the ion speed v, the larger is its mean free path, which is also a function of Q/A. At threshold energies of $10 - 100$ keV/amu, the ion's mean free path becomes larger than the shock width between the interplanetary plasma upstream and downstream with speeds V_1 and V_2 normal to the shock, respectively. Due to any scattering process, the ion then bounces back and forth between upstream and downstream plasma, gaining a speed increment $\Delta V \approx |V_1 - V_2|$ each time. In a stationary situation, this so-called first-order Fermi acceleration leads to power law spectra $\propto v^{-\gamma}$ with spectral index $\gamma = 3V_1/\Delta V$ (Forman and Webb, 1985).

In practice, the mean free path depending on the Q/A of the ion and its speed v, the finite spatial structure of the upstream and downstream plasma, e.g. the spatial distribution of turbulence, and the Q/A-dependent propagation in interplanetary space (e.g., Reames *et al.*, 2000) after escape from the first-order Fermi process, modify the spectra such that they vary spatially and even with time. The Q/A ratio of the ion enters as an important parameter (Lee, 2000).

The SEP flux during the solar system's history is probably dominated by gradual events. It may be sampled in the "SEP" noble gas component implanted into lunar soil (Benkert *et al.*, 1993). From Voyager 1 and 2 data, Breneman and Stone (1985) derived mean SEP abundances that scale with $(Q/A)^{0.66\pm0.17}$ relative to photospheric abundances, whereas Reames (1999) concludes that there is no mean Q/A-fractionation. The enrichment of heavy isotopes in "SEP" Ar and Ne (^{36}Ar/^{38}Ar $\approx$ 4.8, ^{20}Ne/^{22}Ne $\approx$ 11.4) may be an implantation effect (Mewaldt *et al.*, 2001). If it roughly scales as A^2, then the solar wind ^{3}He/^{4}He ratios (Table I) suggest ^{3}He/^{4}He$_{\text{“SEP”}} < 2.7 \times 10^{-4}$, as observed in lunar soils (Heber *et al.*, 2001).

However, the "SEP" noble gases are more abundant than expected from the present-day SEP flux (Wieler, 1997), and there is still ambiguity, how much the

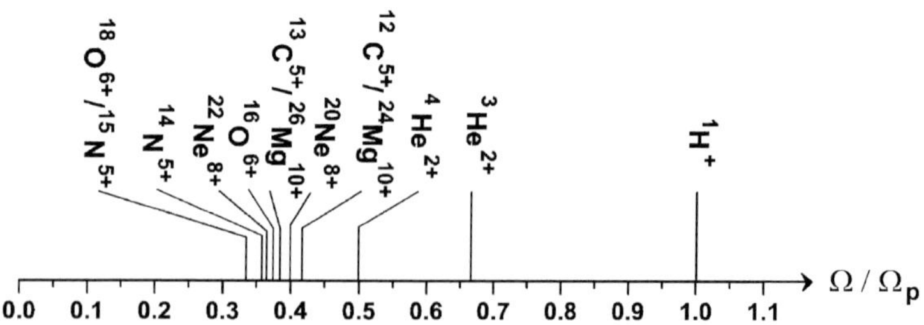

Figure 4. Gyro-frequencies $\Omega_p Q/A$ of ions in the solar corona in units of the proton gyro-frequency Ω_p. The acceleration efficiency for different Q/A-species depends on the spectral distribution of plasma waves. Heavy isotopes may interact with other types of waves than $^3He^{2+}$, but waves that gyro-resonate with $^3He^{2+}$ at $\Omega \approx 2\Omega_p/3$ also may drive ions at $\Omega \approx \Omega_p/3$ via the "$n = 2$" resonance. Charge states correspond to a coronal temperature of 2.5 MK. The quiet corona may be somewhat cooler and an impulsive flare source plasma may be hotter.

$^3He/^4He$ ratio in SEPs changed during the past 4.6 Gy due to 1) processes in the Sun (Geiss and Gloeckler, 2003), 2) varying amounts of interplanetary suprathermal seed ions admixed to the solar wind, and 3) a variable impulsive flare activity.

In impulsive flares, gyro-resonant plasma waves (Fisk, 1978; Temerin and Roth, 1992; Paesold *et al.*, 2002) seem to selectively accelerate ions (Figure 4) to MeV-energies in a single stage, enhancing $^3He/^4He$ by orders of magnitude (Hsieh and Simpson, 1970) and $^{22}Ne/^{20}Ne$ by up to a factor 10 (Dwyer *et al.*, 2001). These enhancements cannot be described by a power law in Q/A (Mason *et al.*, 2002). They are observed in both impulsive and gradual events and supported by models of single-stage acceleration by plasma waves at the impulsive flare site (Bochsler and Kallenbach, 1994) and of a two-stage process of resonant pre-acceleration in impulsive flares and injection into first-order Fermi acceleration at interplanetary CME shocks (Kallenbach, 2002). Although impulsive events contribute up to 5% of SEPs in gradual events (Tylka *et al.*, 2001), the resulting enrichment of ^3He in the "SEP" noble gases apparently remained small, assuming little change in the photospheric $^3He/^4He$ ratio over the Sun's lifetime (Geiss and Gloeckler, 2003). Enrichments of "SEP" ^{22}Ne due to impulsive flares should be very minor.

Wimmer-Schweingruber and Bochsler (2001) deal with another peculiarity of the SEP seed ions in interplanetary space. They argue, that the abundance of volatile isotopes in the observed "SEPs" is close to their interstellar abundance. Some time in the past, the interstellar medium may have been up to a hundred times denser than now and may have supplied more neutrals to interplanetary space, which were ionized, picked up by the solar wind, accelerated, and implanted as "HEPs" (heliospheric energetic particles) into lunar soil. This view is related to the theory that pick-up ions derived from the interstellar gas, pre-accelerated in the heliosphere and re-accelerated at the termination shock, are the dominant source of the anomalous Cosmic rays (Fisk *et al.*, 1974). Indeed, in suprathermal particles from co-rotating interaction regions and CMEs (Bamert *et al.*, 2002), the interstellar $^4He^+$ pick-up ions are almost as abundant as the $^4He^{2+}$ solar wind ions.

6. Conclusions

The solar wind is an authentic witness of the isotopic composition of heavy elements in the protosolar nebula. It is wise, though, to exclude the CME-related solar wind from the analysis of samples to be returned by the Genesis mission.

Solar (heliospheric) energetic particles (SEPs) from the dominant gradual events are fractionated with their charge-to-mass ratio as ordering parameter. They are "polluted" by admixtures of interstellar and planetary material or remnant particles of ^{3}He-rich impulsive flares. Nevertheless, solar coronal isotopic abundances derived from gradual ^{3}He-poor SEP events (Leske *et al.*, 2001) agree with solar wind isotopic abundances (Kallenbach, 2000), when corrected for Q/A-fractionation.

At present, there is no unambiguous identification of the "SEP" component in lunar soils. Although the isotopic fractionation in SEPs strongly varies from event to event, it neither seems to account for the large differences between the solar and the mean "SEP" isotopic ratios, nor does it account for the large variations in the ratio $^{15}N/^{14}N$ in lunar and other solar system samples.

Acknowledgements

This work was supported by INTAS grant 2001-0270.

References

Aellig, M.R., Hefti, S., Grünwaldt, H., Bochsler, P., Wurz, P., Ipavich, F.M., and Hovestadt, D.: 1999, 'The Fe/O Elemental Abundance Ratio in the Solar Wind as Observed with SOHO/CELIAS/-CTOF', *J. Geophys. Res.* **104**, 24769–24780.

Anders, E., and Grevesse, N.: 1989, 'Abundances of the Elements – Meteoritic and Solar', *Geochim. Cosmochim. Acta* **53**, 197–214.

Bamert, K., Wimmer-Schweingruber, R.F., Kallenbach, R., Hilchenbach, M., Klecker, B., Bogdanov, A., and Wurz, P.: 2002, 'Origin of the May 1998 Suprathermal Particles: SOHO/CELIAS/-(H)STOF Results, *J. Geophys. Res.* **107**, 10129–10144.

Benkert, J.-P., Baur, H., Signer, P., and Wieler, R.: 1993, 'He, Ne, and Ar from the Solar Wind and Solar Energetic Particles in Lunar Ilmenites and Pyroxenes', *J. Geophys. Res.* **98**, 13147–13162.

Bochsler, P., and Kallenbach, R.: 1994, 'Fractionation of N Isotopes in SEPs', *Meteorit.* **29**, 653–658.

Bochsler, P.: 2000, 'Abundances and Charge States of Particles in the Solar Wind', *Rev. Geophys.* **38**, 247–266.

Bochsler, P., Geiss, J., and Joos, R.: 1985, 'Kinetic Temperatures of Heavy Ions in the Solar Wind', *J. Geophys. Res.* **90**, 10779–10789.

Bochsler, P., Gonin, M., Sheldon, R.B., Zurbuchen, T., Gloeckler, G., Hamilton, D.C., Collier, M.R., and Hovestadt, D.: 1996, 'Abundance of Solar Wind Magnesium Isotopes Determined with WIND/MASS', *in:* D. Winterhalter *et al.* (eds.), *Proc. 8th Solar Wind Conf., AIP* **382**, 199–202.

Bodmer, R., and Bochsler, P.: 2000, 'Influence of Coulomb Collisions on Isotopic and Elemental Fractionation in the Solar Wind Acceleration Process', *J. Geophys. Res.* **105**, 47–60.

Breneman, H.H., and Stone, E.C.: 1985, 'Solar Coronal and Photospheric Abundances from Solar Energetic Particle Measurements', *Astrophys. J.* **299**, L57–L61.

Bürgi, A., and Geiss, J.: 1986, 'Helium and Minor Ions in the Corona and Solar Wind – Dynamics and Charge States', *Sol. Phys.* **103**, 347–383.

Collier, M.R., Hamilton, D.C., Gloeckler, G., Ho, G., Bochsler, P., Bodmer, R., and Sheldon, R.: 1998, 'Oxygen 16 to Oxygen 18 Abundance Ratio in the Solar Wind Observed by Wind/MASS', *J. Geophys. Res.* **103**, 7–14.

Cranmer, S.R.: 2000, 'Ion Cyclotron Wave Dissipation in the Solar Corona: The Summed Effect of more than 2000 Ion Species', *Astrophys. J.* **532**, 1197–1208, and references therein.

Dwyer, J.R., Mason, G.M., Mazur, J.E., Gold, R.E., Krimigis, S.M., Möbius, E., and Popecki, M.: 2001, 'Isotopic Composition of Solar Energetic Particle Events Measured by Advanced Composition Explorer/ULEIS', *Astrophys. J.* **563**, 403–409.

Fisk, L.A.: 1978, 'He-3-rich Flares – A Possible Explanation', *Astrophys. J.* **224**, 1048–1055.

Fisk, L.A., Kozlovsky, B., and Ramaty, R.: 1974, 'An Interpretation of the Observed Oxygen and Nitrogen Enhancements in Low-Energy Cosmic Rays', *Astrophys. J.* **190**, L35–L38.

Fisk, L.A., Schwadron, N.A., and Zurbuchen, T.H.: 1998, 'On the Slow Solar Wind', *Space Sci. Rev.* **86**, 51–60.

Forman, M.A., and Webb, G.M.: 1985, 'Acceleration of Energetic Particles', *in:* R.G. Stone and B.T. Tsurutani (eds.), *Collisionless Shocks in the Heliosphere: A Tutorial Review*, AGU Press, Washington, pp. 91–114.

Fouchet, T., Lellouch, E., Bézard, B., Encrenaz, T., Drossart, P., Feuchtgruber, H., and de Graauw, T.: 2000, 'ISO-SWS Observations of Jupiter: Measurement of the Ammonia Tropospheric Profile and of the $^{15}N/^{14}N$ Isotopic Ratio', *Icarus* **143**, 223–243.

Geiss, J., and Bochsler, P.: 1982, 'N Isotopes in the Solar System, *Geoch. Cosm. Acta* **46**, 529–548.

Geiss, J., and Gloeckler, G.: 2003, 'Protosolar Isotope Abundances of H, He, and Ne', this volume.

Geiss, J., Hirt, P., and Leutwyler, F.: 1970, 'On Acceleration and Motion of Ions in Corona and Solar Wind', *Sol. Phys.* **12**, 458–483.

Geiss, J., Bühler, F., Cerutti, H., Eberhardt, P., and Filleux, C.: 1972, 'Solar Wind Composition Experiment', *NASA SP* **315**, 14.1.

Geiss, J., Gloeckler, G., Fisk, L.A., and von Steiger, R.: 1995, 'C^+ Pick-up Ions in the Heliosphere and their Origin', *J. Geophys. Res.* **100**, 23373–23377.

Gloeckler, G., and Geiss, J.: 1998, 'Measurement of the Abundance of Helium-3 in the Sun and in the Local Interstellar Cloud with SWICS on ULYSSES', *Space Sci. Rev.* **84**, 275–284.

Gloeckler, G., Geiss, J., Roelof, E.C., Fisk, L.A., Ipavich, F.M., Ogilvie, K.W., Lanzerotti, L.J., von Steiger, R., and Wilken, B.: 1994, 'Acceleration of Interstellar Pickup Ions in the Disturbed Solar Wind Observed on Ulysses', *J. Geophys. Res.* **99**, 17637–17643.

Gloeckler, G., Geiss, J., Schwadron, N.A., Fisk, L.A., Zurbuchen, T.H., Ipavich, F.M., von Steiger, R., Balsiger, H., and Wilken, B.: 2000, 'Interception of Comet Hyakutake's Ion Tail at a Distance of 500 Million Kilometers', *Nature* **404**, 576–578.

Grünwaldt, H., *et al.*: 1997, 'Venus Tail Ray Observation Near Earth', *Geophys. Res. Lett.* **24**, 1163.

Hashizume, K., Chaussidon, M., and Marty, B.: 2000, 'N Isotope Analyses of Lunar Regolith Using an Ion Microprobe; in Search of the Solar Wind Component', *31st Lun. Planet. Sci. Conf.*, #1565.

Heber, V.S., Baur, H., and Wieler, R.: 2001, 'Is There Evidence for a Secular Variation of Helium Isotopic Composition in the Solar Wind?', *Met. Planet. Sci.* **36**, A76.

Hefti, *et al.*: 1998, 'Kinetic Properties of Solar Wind Minor Ions and Protons Measured with SOHO/CELIAS', *J. Geophys. Res.* **103**, 29697–29704.

Hilchenbach, M., Hovestadt, D., Klecker, B., and Möbius, E.: 1991, 'Detection of Singly Ionized Energetic Lunar Pick-up Ions Upstream of Earth's Bow Shock', *Solar Wind* **VII**, 349–355.

Hovestadt, D., *et al.*: 1995, 'Charge, Element, and Isotope Analysis System onboard SOHO', *Sol. Phys.* **162**, 441–481.

Hsieh, K.C., and Simpson, J.A.: 1970, 'The Relative Abundances and Energy Spectra of ^3He and ^4He from Solar Flares', *Astrophys. J.* **162**, L191–L194.

Ipavich, F.M., Paquette, J.A., Bochsler, P., Lasley, S.E., and Wurz, P.: 2001, 'Solar Wind Iron Isotopic Abundances: Results from SOHO/CELIAS/MTOF', *AIP Conf. Proc.* **598**, 121–126.

Jewitt, D.C., Matthews, H.E., Owen, T., and Meier, R.: 1997, 'Measurements of $^{12}C/^{13}C$, $^{14}N/^{15}N$, and $^{32}S/^{34}S$ Ratios in Comet Hale-Bopp (C/1995 O1)', *Science* **278**, 90.

Kallenbach, R., *et al.*: 1997, 'Isotopic Composition of Solar Wind Ne Measured by CELIAS/MTOF on Board SOHO', *J. Geophys. Res.* **102**, 26895–26904.

Kallenbach, R., *et al.*: 1998a, 'Limits on the Fractionation of Isotopes in the Solar Wind as Observed with SOHO/CELIAS/MTOF, *Proc. 31st ESLAB Symp., 22-25 Sept. 1997, ESA SP* **415**, 33–38.

Kallenbach, R., Geiss, J., Ipavich, F.M., Gloeckler, G., Bochsler, P., Gliem, F., Hefti, S., Hilchenbach, M., and Hovestadt, D.: 1998b, 'Isotopic Composition of Solar Wind N: First In Situ Determination with the CELIAS/MTOF Spectrometer on board SOHO', *Astrophys. J.* **507**, L185–L188.

Kallenbach, R.: 2000, *Isotopic Composition of the Solar Wind*, Habilitationsschrift, Univ. of Bern.

Kallenbach, R.: 2001, 'Isotopic Composition Measured In-Situ in Different Solar Wind Regimes by CELIAS/MTOF on board SOHO', *in* R.F. Wimmer-Schweingruber (ed.), *Solar and Galactic Composition*, *AIP Conf. Proc.* **598**, 113–119.

Kallenbach, R.: 2002, 'Charge-to-mass Fractionation in SEPs', *Adv. Space Res.* **30**, 91–96.

Kim, J.S., Kim, Y., Marti, K., and Kerridge, J.F.: 1995, 'Nitrogen Isotope Abundances in the Recent Solar Wind', *Nature* **375**, 383.

Kucharek, *et al.*: 1998, 'Magnesium Isotopic Composition as Observed with the CELIAS/MTOF Experiment on the SOHO Spacecraft', *J. Geophys. Res.* **103**, 26805–26812.

Kucharek, H., Ipavich, F.M., Kallenbach, R., Klecker, B., Grünwaldt, H., Aellig, M.R., and Bochsler, P.: 2001, 'Isotopic Fractionation in Slow and Coronal Hole Associated Solar Wind', *Proc. 203rd IAUS Symp.*, 562.

Kung, C.C., and Clayton, R.N.: 1978, 'Nitrogen Abundances and Isotopic Compositions in Stony Meteorites', *Earth Planet. Sci. Lett.* **38**, 421–435.

Lee, M.A.: 1997, 'Particle Acceleration and Transport at CME-driven Shocks', *in* N.U. Crooker, J.A. Joselyn, and J. Feynman (eds.), *Coronal Mass Ejections*, AGU Press, pp. 227–234.

Lee, M.A.: 2000, 'Acceleration of Energetic Particles on the Sun, in the Heliosphere, and in the Galaxy', *in* R.A. Mewaldt *et al.* (eds.), *Acceleration and Transport of Energetic Particles Observed in the Heliosphere: ACE 2000 Symposium*, AIP press, pp. 3–18.

Leske, R.A., Cohen, C.M.S., Cummings, A.C., Mewaldt, R.A., Stone, E.C., Dougherty, B.L., Wiedenbeck, M.E., Christian, E.R., and von Rosenvinge, T.T.: 1999, 'Unusual Isotopic Composition of SEPs Observed in the November 6, 1997, Event', *Geophys. Res. Lett.* **26**, 153–156.

Leske, R.A., Mewaldt, R.A., Cohen, C.M.S., Christian, E.R., Cummings, A.C., Slocum, P.L., Stone, E.C., von Rosenvinge, T.T., and Wiedenbeck, M.E.: 2001, 'Isotopic Abundances in the Solar Corona as Inferred from ACE Measurements of SEPs', *AIP Conf. Proc.* **598**, 127–132.

Marty, B., Hashizume, K., Chaussidon, M., and Wieler, R.: 2003, 'Nitrogen Isotopes on the Moon: Archives of the Solar and Planetary Contributions to the Inner Solar System', this volume.

Mason, G.M.: 2001, 'Heliospheric Lessons for Galactic Cosmic-ray Acceleration', *Space Sci. Rev.* **99**, 119–133.

Mason, G.M., Mazur, J.E., and Dwyer, J.R.: 1999, '^{3}He Enhancements in Large Solar Energetic Particle Events', *Astrophys. J.* **525**, L133–L136.

Mason, G.M., Mazur, J.E., and Dwyer, J.R.: 2002, 'A New Heavy Ion Abundance Enrichment Pattern in ^{3}He-rich Solar Particle Events', *Astrophys. J.* **565**, L51–L54.

Mewaldt, R.A., Ogliore, R.C., Gloeckler, G., and Mason, G.M.: 2001, 'A New Look at Neon-C and SEP-Neon', *AIP Conf. Proc.* **598**, 393–398.

Möbius, E., Scholer, M., Hovestadt, D., Klecker, B., and Gloeckler, G.: 1982, 'Comparison of Helium and Heavy Ion Spectra in He-3-rich Solar Flares with Model Calculations Based on Stochastic Fermi Acceleration in Alfvén Turbulence', *Astrophys. J.* **259**, 397–410.

Möbius, E., Hovestadt, D., Klecker, B., Gloeckler, G., and Ipavich, F.M.: 1985, 'Direct Observation of He(+) Pick-up Ions of Interstellar Origin in the Solar Wind', *Nature* **318**, 426–429.

Owen, T., Mahaffy, P.R., Niemann, H.B., Atreya, S., and Wong, M.: 2001, 'Protosolar Nitrogen', *Astrophys. J.* **553**, L77–L79.

Paesold, G., Kallenbach, R., and Benz, A.O.: 2002, 'Acceleration and Enrichment of ^{3}He in Impulsive Flares by Electron Firehose Waves', *Astrophys. J.*, in press.

Palma, R.L., Becker, R.H., Pepin, R.O., and Schlutter, D.J.: 2002, 'Irradiation Records in Regolith Materials, II: Solar-wind and Solar-energetic-particle Components in He, Ne, and Ar Extracted from Single Lunar Mineral Grains and from the Kapoeta Howardite by Stepwise Pulse-heating', *Geochim. Cosmochim. Acta*, in press.

Pepin, R.O.: 2000, 'On the Isotopic Composition of Primordial Xenon in Terrestrial Planet Atmospheres', *Space Sci. Rev.* **92**, 371–395.

Priest, E.R.: 1982, *Solar Magneto-Hydromagnetics, Geophys. Astrophys. Monogr.* **21**, 118.

Raymond, J.C., *et al.*: 1997, 'Composition of Coronal Streamers from the SOHO Ultraviolet Coronagraph Spectrometer', *Sol. Phys.* **175**, 645–665.

Reames, D.V.: 1999, 'Particle Acceleration at the Sun and in the Heliosphere', *Space Sci. Rev.* **90**, 413–491.

Reames, D.V., Ng, C.K., and Tylka, A.J.: 2000, 'Initial Time Dependence of Abundances in Solar Energetic Particle Events', *Astrophys. J.* **531**, L83–L86.

Schwadron, N.A., Fisk, L.A., and Gloeckler, G.: 1996, 'Statistical Acceleration of Interstellar Pick-up Ions in Corotating Interaction Regions', *Geophys. Res. Lett.* **23**, 2871–2874.

Sheeley, N.R., Jr., *et al.*: 1997, 'Measurements of Flow Speeds in the Corona between 2 and 30 $R_\odot$, *Astrophys. J.* **484**, 472–478, 1997.

Sugiura, N., Kiyota, K., and Hashizume, K.: 1998, 'Nitrogen Components in Primitive Ordinary Chondrites', *Met. Planet. Sci.* **33**, 463–482, and references therein.

Temerin, M., and Roth, I.: 1992, 'The Production of He-3 and Heavy Ion Enrichment in He-3-rich Flares by Electromagnetic Hydrogen Cyclotron Waves', *Astrophys. J.* **391**, L105–L108.

Tylka, A.J., Cohen, C.M.S., Dietrich, W.F., Maclennan, C.G., McGuire, R.E., Ng, C.K., and Reames, D.V.: 2001, 'Evidence for Remnant Flare Suprathermals in the Source Population of Solar Energetic Particles in the 2000 Bastille Day Event', *Astrophys. J.* **558**, L59–L63.

von Steiger, R., and Geiss, J.: 1989, 'Supply of Fractionated Gases to the Corona', *Astron. Astrophys.* **225**, 222–238.

von Steiger, R., Wimmer-Schweingruber, R.F., Geiss, J., and Gloeckler, G.: 1995, 'Abundance Variations in the Solar Wind', *Adv. Space Res.* **15**, (7)3–(7)12.

von Steiger, R., Schwadron, N.A., Fisk, L.A., Geiss, J., Gloeckler, G., Hefti, S., Wilken, B., Wimmer-Schweingruber, R.F., and Zurbuchen, T.H.: 2000, 'Composition of Quasi-stationary Solar Wind Flows from SWICS/ULYSSES', *J. Geophys. Res.* **105**, 27217–27238.

Weygand, J.M., Ipavich, F.M., Wurz, P., Paquette, J.A., and Bochsler, P.: 2001, 'Determination of the ^{36}Ar/^{38}Ar Isotopic Abundance Ratio of the Solar Wind Using SOHO/CELIAS/MTOF', *Geochim. Cosmochim. Acta* **65**, 4589–4596.

Wieler, R.: 1997, 'Why are SEP Noble Gases So Abundant in Extraterrestrial Samples?', *Proc. 28th Lunar Planet. Sci. Conf., March 17–21, 1997, Houston, TX*, p. 1551.

Wimmer-Schweingruber, R.F., and Bochsler, P.: 2001, 'Lunar Soils: A Long-Term Archive for the Galactic Environment of the Heliosphere?', *AIP Conf. Proc.* **598**, 399–404.

Wimmer-Schweingruber, R.F., Bochsler, P., Gloeckler, G., Ipavich, F.M., Geiss, J., Kallenbach, R., Fisk, L.A., Hefti, S., and Zurbuchen, T.H.: 1999, 'On the Bulk Isotopic Composition of Mg and Si during the May 1998 CME: ACE/SWIMS, *Geophys. Res. Lett.* **26**, 165–168.

Wimmer-Schweingruber, R.F., Bochsler, P., and Gloeckler, G.: 2001, 'The Isotopic Composition of Oxygen in the Fast Solar Wind: ACE/SWIMS', *Geophys. Res. Lett.* **28**, 2763–2766.

Zurbuchen, T.H., Fisk, L.A., Gloeckler, G., and Schwadron, N.A.: 1998, 'Element and Isotopic Fractionation in Closed Magnetic Structures', *Space Sci. Rev.* **85**, 397–406.

WORKING GROUP REPORTS

SUN AND PROTOSOLAR NEBULA

Working Group Report

R. KALLENBACH[1], F. ROBERT[2], J. GEISS[1], E. HERBST[3], H. LAMMER[4], B. MARTY[5], T.J. MILLAR[6], U. OTT[7] and R.O. PEPIN[8]

[1] *International Space Science Institute, Hallerstrasse 6, CH-3012 Bern*

[2] *Lab. de Minéralogie, Muséum National d'Histoire Naturelle, 61, rue Buffon, F-75005 Paris*

[3] *The Ohio State Physics Dept. Smith Lab., College of Mathematical and Physical Sciences, 174 W. 18th Ave., Columbus, OH 43210, USA*

[4] *Space Research Institute, Austrian Academy of Sciences, A-8042 Graz, Austria*

[5] *Centre de Recherches Pétrographiques et Géochimiques, CNRS UPR, 9046, rue Notre-Dame des Pauvres BP 20, F-54501 Vandoevre-les Nancy-Cedex*

[6] *Department of Physics, UMIST, P.O. Box 88, Manchester M60 1QD, UK*

[7] *Max-Planck-Institut für Chemie, Postfach 3060, D-55020 Mainz, Germany*

[8] *School of Physics and Astronomy, University of Minnesota, Minneapolis, Minnesota 55455*

Received: 9 October 2002; Accepted in final form: 31 October 2002

Abstract. Volatile isotope abundances are tracers for the evolutionary processes of the solar system. At the same time they carry information on the galactic nucleosynthetic sources, from which solar matter originates. This working group report summarizes the present knowledge and addresses unresolved issues regarding fractionation of isotopes of volatile elements in the solar system.

1. Introduction

Volatile isotopes in the solar system originate from many galactic nucleosynthetic sources. They have been incorporated into the interstellar medium by stellar winds and explosive events. The Sun formed from a cold and dense core of an interstellar molecular cloud rich in gas and dust. The core of this cloud collapsed when the magnetic support against its self-gravity was gradually lost through the process of ambipolar diffusion. The collapse led to a disk-shaped nebula which may eventually correspond to the theoretical disk surrounding the early Sun and referred to as the "Protosolar Nebula" (Figure 1). The total "solar mix" of volatile isotopes in this nebula may have been fractionated with respect to the original interstellar material, depending on the physical and chemical conditions in the core of the molecular cloud before and during its collapse (Evans, 1999; van Dishoeck and Blake, 1998; Aikawa *et al.*, 2001). Remainders of fractionation between different components of the interstellar medium are preserved, for instance, in some interplanetary dust particles (IDPs), representing specific compounds condensed in interstellar dust grains (Messenger *et al.*, 2003). The subsequent evolution of the protosolar nebula (PSN), i.e., the formation of the Sun and of the planetary system (Benz *et al.*, 2000), caused further fractionation within solar system mat-

Space Science Reviews **106:** 319–376, 2003.

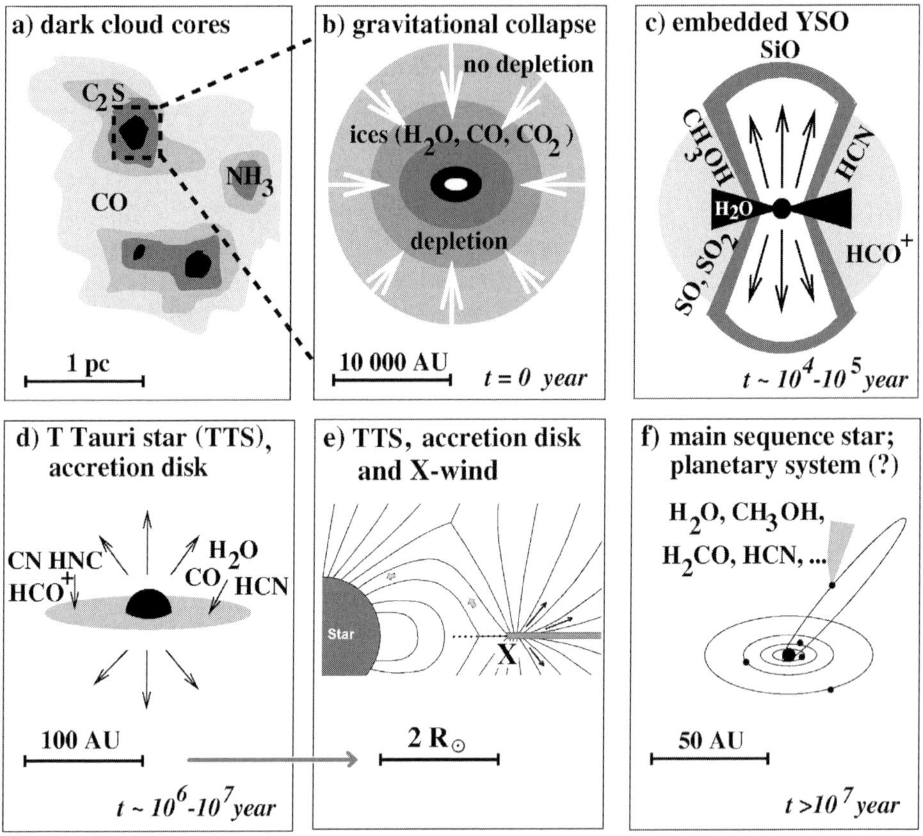

Figure 1. Scenario for the formation of low-mass, isolated stars (adapted from van Dishoeck and Blake, 1998, and Shu *et al.*, 1997). A cold interstellar molecular cloud core collapses (panels a and b) to form a young stellar object dissipating angular momentum into its bipolar outflow (panel c) and into its surrounding accretion disk (panel d). While the young star reaches its classical T Tauri stage, its magnetosphere interacts with the surrounding protoplanetary disk in the hot X-region (panel e). After at most ten million years, the solar system has formed, consisting of the Sun, terrestrial (solid and liquid) planets, giant gas planets, comets, and asteroids (panel f).

ter, influencing mainly the isotopic composition of volatile elements in planetary samples. The strength of isotopic fractionation processes inside the Sun remained below 1%/amu over the lifetime of the Sun, except for the D/H and ^{3}He/^{4}He ratios (Geiss and Gloeckler, 2003).

Based on a summary of the knowledge presented in the introductory articles of this volume and other reviews, this working group article discusses unresolved issues regarding fractionation of volatile isotopes during the formation and the subsequent history of the solar system. These issues include 1) How well is the isotopic "solar mix" related to its Galactic nucleosynthetic sources? 2) Was there any fractionation during the collapse of the pre-solar molecular cloud, which lasted only a few hundred thousand years? 3) What was the structure and evolution of the PSN, what happened to volatile material during the few million years when the young Sun interacted with the surrounding accretion disk? 4) Was there any special fractionation process during the first hundred million years of the lifetime of the solar system due to an active early Sun? 5) To what extent have nuclear processes in the core of the Sun and at the solar surface influenced abundances of volatile isotopes in the solar system? 6) Were there any short-time or continuous sources of "pollution" to the solar system from the outside, e.g. from supernovae or from the variable interstellar medium?

2. Pre-solar Grains and the Origin of the Solar System

The cold pre-solar interstellar cloud incorporated the gaseous and granular debris of many stars, expelled in the form of stellar winds during late stages of stellar evolution or by nova- and supernova-type explosions. In the dense disk-shaped PSN, which presumably had a temperature gradient, with the hottest regions located in the inner parts of the disk, much of the pre-solar grains were evaporated, so that the information on their characteristic isotopic composition was lost. Some grains, however, have been preserved in chondritic meteorites as an archive of the mineralogical, chemical and isotopic features of solid matter that predates the solar system.

2.1. Nucleosynthetic Sources of Pre-solar Grains

The isolation of pre-solar grains in primitive meteorites has been a highlight of modern meteoritic research (Lewis *et al.*, 1987; Anders and Zinner, 1993). The pre-solar grains identified so far are largely made up of elements such as C, N, and O that are usually considered volatile but can be highly refractory in the right chemical environment. The C/O ratio is decisive: if the abundance of C exceeds that of O during the cooling of matter ejected from stars, refractory C phases such as *graphite* and refractory *carbides* are the first minerals to form. If O is more abundant, refractory oxides form instead. Accordingly, pre-solar grains in meteorites consist of refractory C compounds, mainly *graphite*, *diamond*, and *silicon carbide* (SiC), and refractory *oxides* such as *corundum* (Al_2O_3), *spinel* ($MgAl_2O_4$), and *hibonite* ($CaAl_{12}O_{19}$), but also *silicon nitride* (Si_3N_4) (Zinner, 1998; Hoppe and Zinner, 2000).

Interestingly, the likely most abundant pre-solar (interstellar) material – silicates – is missing in the inventory of pre-solar grains found so far in meteorites. On the contrary, IDPs seem to contain abundant pre-solar *silicate* grains (Messenger *et al.*, 2002). This finding may perhaps indicate that more solid material in the solar system is inherited from stages before the formation of the PSN than currently anticipated. Possibly, most silicates were in fact formed in pre-solar nebula stages, and they simply crystallized in the PSN.

TABLE I

Characteristic isotopic signatures and inferred stellar sources of major types of identified pre-solar dust grains in meteorites (Hoppe and Zinner, 2000). Enhancements and depletions are expressed relative to solar system composition.

Mineral	**Diagnostic Isotopic Signatures**	**Stellar Source**	**Percentage**
Diamond	Xe-HL	Supernovae	?
SiC	enhanced ^{13}C, ^{14}N, ^{22}Ne, s-process elements;	AGB stars (1 – 3 $M_\odot$, different metallicities);	> 90%
	low $^{12}C/^{13}C$, often enhanced ^{15}N;	J-type C-stars (?);	< 5%
	enhanced ^{12}C, ^{15}N, ^{28}Si, extinct ^{26}Al and ^{44}Ti;	supernovae;	1%
	low $^{12}C/^{13}C$, low $^{14}N/^{15}N$.	novae.	0.1%
Graphite	enhanced ^{12}C, ^{15}N, ^{28}Si, extinct ^{26}Al, ^{41}Ca, ^{44}Ti;	supernovae, Wolf-Rayet stars (?);	> 80%
	Kr-S;	AGB stars;	< 10%
	low $^{12}C/^{13}C$;	J-type C-stars (?);	< 10%
	low $^{12}C/^{13}C$, Ne-E(L).	novae.	2%
Corundum	enhanced ^{17}O, moderately depleted ^{18}O;	RGB and AGB stars (1–9 $M_\odot$, different metallicities);	> 70%
	enhanced ^{17}O, strongly depleted ^{18}O;	AGB stars (< 1.5 $M_\odot$, cool bottom processing);	20%
	enhanced ^{16}O.	supernovae	1%
Silicon Nitride	enhanced ^{12}C, ^{15}N, ^{28}Si, extinct ^{26}Al.	supernovae	100%

In the context of pre-solar grain analysis, the term isotopic anomalies designates the compositions found in meteoritic or in IDP minerals which are not related to the chondritic (and presumably solar) compositions by mass-dependent relationships. Since no process is known to depart strongly from the mass-dependent fractionation during physico-chemical reactions, these isotopic anomalies are ascribed to stellar nucleosynthesis. In some cases, elements such as C or N with only two isotopes, exhibit variations which are so large (orders of magnitude) that – although the diagnosis of mass-dependence cannot be made on a two-isotope system – they are also ascribed to nuclear processes.

An overview of the most abundant grain types, the inferred stellar sources, and the percentage deriving from a given source is given in Table I (see also Zinner, 1998; Hoppe and Zinner, 2000; Bernatowicz and Zinner, 1997; Ott, 2003a). Isotopic features in volatile elements in particular are discussed by Ott (2003b).

2.1.1. *AGB stars*

More than 90% of all *SiC* grains, the second most abundant of the grains listed in Table I, are derived from carbon-rich Red Giants of low to intermediate mass on the Asymptotic Giant Branch (AGB). These stars burn He and H in shells around a C-O core, with convection bringing He-burning products to the surface ("third dredge-up"). Thus the surface is enriched in ^{12}C, and characteristic isotopic abundance patterns are established in a variety of elements.

These carbon stars are identified as the dominant nucleosynthetic sources of the *SiC* grains by the following indications: 1) Most of them have lower $^{12}C/^{13}C$ (< 80) and higher $^{14}N/^{15}N$ (up to 20000) ratios than the solar system due to the input from H burning via the CNO cycle, which, in steady state, produces $^{12}C/^{13}C \approx 3.5$ and $^{14}N/^{15}N > 10^4$ (Clayton, 1983; Section 7.1). 2) The isotopes of many elements show the features of the s-process (slow neutron capture). The neutrons for this process are released predominantly by ^{13}C when forming ^{16}O with α-particles. 3) The observed Ne-E(H) (identical to Ne-G, i.e., almost monoisotopic ^{22}Ne released at high temperatures) is efficiently formed from ^{14}N and two α-particles (Gallino *et al.*, 1990). 4) The distribution of $^{12}C/^{13}C$ ratios in singly analyzed pre-solar *SiC* grains is similar to that observed in carbon star atmospheres (Alexander, 1993). 5) AGB stars are the main contributors of carbonaceous dust to the interstellar medium (Henning and Salama, 1998) and show the 11.2 μm emission feature of *SiC* grains in their atmospheres (Little-Marenin, 1986).

AGB stars have also probably contributed to those *graphite* grains that show the s-process signatures in Kr (Kr-S, also called Kr-G) and Xe (Amari *et al.*, 1995).

Low to intermediate mass Red Giants in an earlier evolutionary stage – with oxygen-rich surfaces – are the source of most pre-solar *oxide* grains identified so far (e.g., Hoppe and Zinner, 2000). In these stars partially H-burned material is mixed into their envelopes during the "first dredge-up", resulting in lower than solar $^{16}O/^{17}O$ and higher than solar $^{16}O/^{18}O$ in most *oxide* grains (Nittler *et al.*, 1997). Comparison with detailed stellar evolution models allows drawing conclusions regarding the mass and metallicity of the stars providing specific grains. In some cases, evidence for H burning under non-standard conditions – Hot Bottom Burning (HBB) and Cool Bottom Processing (CBP) – is found.

2.1.2. *Novae*

Monoisotopic ^{22}Ne-E(L), presumably the radiogenic (hence also called Ne-R) decay product of ^{22}Na and released at low temperatures, indicates that a minor fraction of *graphites* comes from novae. Novae are expected to produce grains with the signatures of explosive H burning, i.e. low $^{12}C/^{13}C$ and low $^{14}N/^{15}N$ (< 10). Many type A and B *SiC* grains have low $^{14}N/^{15}N$ (Amari *et al.*, 2001a). However, only two of them have been observed with $^{14}N/^{15}N < 10$ (Amari *et al.*, 2001b), and J-type Carbon Stars (Lambert *et al.*, 1986) give a better match to their isotopic features. Therefore, novae are excluded as a dominant source of *SiC* grains.

2.1.3. *Supernovae, Wolf-Rayet Stars*

Massive stars such as supernovae and perhaps also Wolf-Rayet stars are important sources for the grains listed in Table I, in particular for the *graphites*.

Supernovae may also be the source of *nano-diamonds* which are nominally the most abundant pre-solar grains in meteorites. An indicator for the origin of *nano-diamonds* in explosive stellar events are the hardly separable anomalous Xe-H and Xe-L noble gas components (Reynolds and Turner, 1964). Xenon-L is enriched in ^{124}Xe, exclusively made by the p-process (proton capture), compared to the s-process isotopes. Xenon-H is rich in ^{136}Xe, produced solely by the r-process (neutron capture). Both may have been produced by a neutron burst acting on the He shell of an exploding supernova (Clayton, 1989; Clayton *et al.*, 1995) or by the standard r-process, if they are separated from their radioactive pre-cursors within a few hours after the supernova explosion (Ott, 1996; Richter *et al.*, 1998).

One may be skeptical about the pre-solar origin of *nano-diamonds* because their ^{12}C/^{13}C ratio of $\approx$ 92 is solar (note that analysis of individual *nano-diamonds* is not possible, since they consist of on the average $\approx$1000 C atoms only) and because their N isotopic composition (^{14}N/^{15}N $\approx$ 400) may be a typical "solar mix" (Marty *et al.*, 2003). Clayton *et al.* (1995) have developed a model for the growth of *nano-diamonds* during the expansion of supernova interiors. Non-equilibrium chemical vapor deposition accounts for the small sizes of the grains, their C/N, ^{12}C/^{13}C, and ^{14}N/^{15}N ratio, as well as for Xe H and Xe L. It remains puzzling, though, that there is an H part of Kr in *nano-diamonds* but no L part (Ott, 2003b).

Isotopic signatures in type X *SiC* grains, representing $\approx$1% of all pre-solar *SiC* grains, are more clearly indicative for a supernova origin: high ^{12}C/^{13}C, large overabundance of ^{28}Si, and radiogenic ^{26}Mg and ^{44}Ca due to decay in the grains of ^{26}Al ($T_{1/2}$ = 0.7 My) and ^{44}Ti ($T_{1/2}$ = 60 y), with ^{26}Al/^{27}Al and ^{44}Ti/^{48}Ti at the time of grain formation being extremely high. Similar signatures in many *graphite* and virtually all *silicon nitride* grains show that they too originate in supernovae.

2.1.4. *Pollution of Solar-nebula Xe by Late Infall of Supernova or AGB Envelope Debris*

Xenon compositions in several solar-system volatile reservoirs – the SiC "normal" (N-Xe) component, meteoritic Xe-Q, solar/solar wind-Xe, and the Earth's primordial atmospheric Xe – may be interpreted as reflecting variable mixing ratios of a heavy-isotope Xe component (Xe-H) with a component called U-Xe, as noted by Pepin (2003) in this volume. (U-Xe differs compositionally from solar wind-Xe, which presumably represents the inventory of the Sun, only in significantly lower relative abundances of the two heaviest isotopes, Pepin, 2000). An interesting question is whether this variability points to a decrease with time in the H/U mixing ratio of nebular Xe. Pepin (2003) discusses the possibility that the H/U trend may reflect a temporal compositional sequence in the context of models developed by Boss and coworkers for pre-solar molecular cloud collapse induced by shock from a neighboring supernova or AGB star, with concurrent injection of shock-wave material into the outer regions of the cloud (Boss and Vanhala, 2000, 2001, and references therein). If it does, data presently available suggest that after most of the collapsing matter had been sequestered in the young Sun, Xe composition in the residual nebula during planet-building was dominated by late infall of U-Xe-rich material injected into the periphery of the pre-solar cloud. Accurate measurements of Xe compositions in Jupiter and Venus are critical to the question of whether or not this inference of a temporal trend is correct.

2.1.5. *Unresolved Sources or Nucleosynthetic Processes?*

Surprising in the context of supernova grains has primarily been the extent of mixing between products synthesized in different zones as indicated e.g., by the former presence of ^{44}Ti in carbonaceous grains (Hoppe and Zinner, 2000).

Heavy element isotopic patterns indicate additional complexity. For one, they indicate that there may be more processes involved in their synthesis than the traditional s-, r- and p-processes. Molybdenum compositions in *SiC*-X grains (Pellin *et al.*, 1999) seem best to be explained by overabundances from a neutron burst (intermediate between r- and s-process; Meyer *et al.*, 2000). A similar process (or a process modifying isotopic ratios produced by the r-process; Ott, 1996) may also be needed to explain the observed overabundances of heavy isotopes of *diamond*-Xe ("Xe-HL") and Te. The same data indicate that the mix of s-, r-, and p-process material that the parent stars inherited when they formed shows some variations from the solar mix. The Mo data of X-grains indicates that in their parent stars the ratio of p- to s- and r-process nuclei (to which the neutron burst products were admixed) may have been lower than in the solar mix (Figure 1 of Pellin *et al.*, 1999; slightly revised, see Ott, 2002). On the other hand, in the Xe (Lewis *et al.*, 1994; Ott, 2003b) that accompanies the s-process Xe in mainstream *SiC* (the "primary" Xe of the parent star), r-process (or neutron burst?) products seem enhanced relative to s- and p-Xe (not, however in other elements such as Mo and Zr; e.g. Nicolussi *et al.*, 1998).

2.2. Links Between Pre-solar Grain, Solar, Galactic, and Galactic Cosmic Ray Composition

Compositional variations within pre-solar grains are very large and complicated due to the contributions of material of many stellar sources. On the other hand, the mean isotopic composition of volatiles in pre-solar grains may represent at least part of the PSN "mix". This nebula formed at some Galactocentric distance $R_{i,\odot}$ 4.6 Gy ago. The interstellar medium (ISM) presently surrounding the solar system at $R_0 \approx 8.5$ kpc reflects the Galactic chemical evolution (GCE) during these 4.6 Gy, if a clear relation between the GCE and the Galocto-centric distance exists. In fact, Wielen and Wilson (1997) derive $R_0 - R_{i,\odot} \approx 1.9 \pm 0.9$ kpc from observations on C, N, and O isotopic abundances in the ISM.

Some of the pre-history of GCE, the nucleosynthetic and mixing processes in stars before the birth of the Sun, can be studied in the archive of pre-solar grains. The Galactic cosmic rays (GCRs) carry information on the isotopic composition of distant Galactic sources and complement our knowledge on GCE.

2.2.1. *Clues on the Galactic Chemical Evolution in Pre-solar Grains*

The variation of the Si isotopic composition among "mainstream" *SiC* grains (the 90% group in Table I) is clearly related to the initial composition of the stars and hence is governed by GCE. Nucleosynthetic processes in AGB stars cause only minor changes in the abundances of Si isotopes, so that it became feasible to model their GCE (Timmes and Clayton, 1996). Assuming a purely "primary" nucleosynthetic origin of ^{28}Si and a "secondary" origin of ^{29}Si and ^{30}Si, the deviations of isotopic ratios $\delta^{29,30}_{\mathrm{ISM}} = 1000\left[\left({}^{29,30}\mathrm{Si}/{}^{28}\mathrm{Si}\right) / \left({}^{29,30}\mathrm{Si}/{}^{28}\mathrm{Si}\right)_{\mathrm{ISM}} - 1\right]$ evolve temporally according to $\delta^{29}_{\mathrm{ISM}} = m\delta^{30}_{\mathrm{ISM}}$ with a slope $m = 1$. Expressed in deviations $\delta^{29,30}_{\odot}$ to solar isotopic ratios, this yields a slope $m = 4/3$ for the three-isotope plot of abundances in *SiC* grains formed at different times and locations in the Galaxy (Clayton and Timmes, 1997). The best fit for observed Si isotopic ratios in *SiC* grains (Hoppe and Zinner, 2000) in fact corresponds to $m \approx 4/3$.

Similarly, besides the need for "non-standard" H burning in HBB and CBP (see above), oxygen isotopic ratios in *oxide* grains also show such an influence of GCE (Huss *et al.*, 1994; Nittler and Cowsik, 1997). The O isotopes are predominantly synthesized in massive stars and distributed into the interstellar medium by Type II supernovae.

For the C and N isotopes the situation is more complicated. The relative abundances of N and O are of particular interest, since N isotopes are synthesized in different astrophysical sites than O isotopes. A distinction is frequently made between "primary" nucleosynthetic products, which can be synthesized directly from H and He in Population III stars, and "secondary" products, which require a "seed" heavy nucleus to be initially present in the star where its synthesis occurs. By this definition, oxygen is a primary species. Nitrogen is secondary when made as a by-product of CNO-cycle hydrogen burning. According to the simple closed-box model, the abundance of a secondary species is quadratic, so that if N is secondary and O primary, then $(\mathrm{N/H}) \propto (\mathrm{O/H})^2$, or $(\mathrm{N/O}) \propto (\mathrm{O/H})$. The N/O ratio does appear to approach this behavior, for H II regions with moderately high O/H values in M 101 (Torres-Peimbert *et al.*, 1989) and in the Milky Way (Lester *et al.*, 1987). However, below a certain value of O/H, it appears that N/O is constant; these low-metallicity H II regions occur mostly in low-mass galaxies. Thus, it is becoming clear that nitrogen is not purely a secondary nucleosynthetic product. Indeed, N may be produced within intermediate-mass stars by an effectively primary process, if C synthesized within the star by the triple-alpha reaction is later subjected to the CN cycle (Alloin *et al.*, 1979; Renzini and Voli, 1981). Nitrogen made by this process would be primary, but there might be a time-delay in building up its abundance relative to the nuclear products of supernovae, because of the longer lifetimes of the source stars (Edmunds and Pagel, 1978; Serrano and Peimbert, 1983; Diaz and Tosi, 1986).

The $^{14}N/^{15}N \approx 400$ ratio in *nano-diamonds* is close to the newest estimate of the solar ratio (Hashizume *et al.*, 2000; Owen *et al.*, 2001), and is also compatible with the local interstellar ratio $^{14}N/^{15}N = 450 \pm 22$ (Wilson and Rood, 1994). On the other hand, large variations in the $^{14}N/^{15}N$ ratio between 200 and 600 are observed in the ISM, and it is not clear, how much of these variations is due to chemical fractionation within different species and how much is caused by nucleosynthesis in stars.

A large fraction of both stable C isotopes, ^{12}C and ^{13}C are synthesized in AGB stars. The mean $^{12}C/^{13}C$ ratio of pre-solar *SiC* grains, originating in AGB stars, is between 40 and 50 (Swart *et al.*, 1983; Alexander, 1993). There are other sources (see below) that produce much higher ratios $^{12}C/^{13}C$. If all *nano-diamonds* were pre-solar, the pre-solar grains' mean $^{12}C/^{13}C$ would be their value of $\approx$92, which is essentially the presently assumed solar value (Owen and Encrenaz, 2003). At present, there is no clear-cut interpretation of the difference between the solar $^{12}C/^{13}C$ ratio and the $^{12}C/^{13}C$ ratio in the ISM, which is 65.0 ± 11.2 (Bensch *et al.*, 2001). It may represent the mean GCE of the ISM during the 4.6 Gy since the birth of the Sun. Note, that there is no direct determination, yet, of the solar $^{12}C/^{13}C$ ratio, i.e., there is no measurement of $^{12}C/^{13}C$ in the solar wind, and that there are variations in $^{12}C/^{13}C$ in the ISM.

2.2.2. *Nucleosynthetic Sources of the Galactic Cosmic Rays?*

Besides AGB stars, supernovae and Wolf-Rayet (WR) stars are major sources of interstellar dust. Presumably, the dust of supernovae or WR stars inside superbubbles is injected into the Galactic Cosmic Ray (GCR) acceleration at supernova shocks rather than the interstellar medium (Higdon *et al.*, 1998; Parizot, 2001).

Wolf-Rayet stars are believed to be the source of about 25% of the heavy elements in the GCRs (Mewaldt, 1989). Meynet *et al.* (2001) conclude from their stellar models that WR stars emit Ne with $^{22}Ne/^{20}Ne$ enriched by a factor 3 over its solar value. Enhancements of ^{22}Ne at the GCR source are correlated with $^{58}Fe/^{56}Fe$ enhancements by about 70% relative to the solar ratio as expected from WR stars (Wiedenbeck *et al.*, 2001). The $^{12}C/^{13}C$ ratio in GCRs is higher by a factor 2 than the solar ratio in agreement with the models on nucleosynthesis and dusty winds of WR stars (Meynet *et al.*, 2001). Unfortunately, abundances of Kr and Xe isotopes can presently not be studied in GCRs to identify their sources.

Lingenfelter *et al.* (1998) argue that the WR star scenario is not compatible with the Galactic evolution of the Be/Fe ratio. They propose that the refractory elements in GCRs are accelerated out of the erosion products of supernova grains in ejecta-enriched superbubbles. This hypothesis explains many signatures of GCR data such as 1) the enrichment of refractory elements (Meyer *et al.*, 1997) and the delay of $\approx 10^5$ y before the GCR source material is accelerated (Wiedenbeck *et al.*, 2001). The problematic C/O ratio in GCRs is a consequence of refractory *oxide* and *graphite* grain abundances, with all C residing in *graphites*. However, the $^{54}Fe/^{56}Fe$ ratio at the GCR source is predicted to be higher than solar, which is not confirmed by ACE measurements (Wiedenbeck *et al.*, 2001).

Both sources, supernovae and WR stars would supply material with $^{12}C/^{13}C$ rather larger than solar. This matches the mean isotopic composition of pre-solar *graphite* grains and could balance the low $^{12}C/^{13}C$ of pre-solar *SiC* grains. Overall, the connections between the volatile isotopic compositions of GCRs, the Sun, and pre-solar grains are not obvious. As mentioned, there has not been a direct measurement of $^{12}C/^{13}C$ in the solar wind, yet. It may differ from $^{12}C/^{13}C \approx 92$.

Note, that volatile isotopes may undergo fractionation as a consequence of their acceleration to energies of observable GCRs (e.g., Mason, 2001). The noble gases and a fraction of C, N, and O are presumably accelerated as single ionized atoms, so that their atomic charge-to-mass ratio is an ordering parameter for their acceleration efficiencies. However, a large fraction of C, N, and O isotopes is accelerated in condensed form inside grains with high mass-to-charge ratio (Meyer *et al.*, 1997). The refractory elements are all enriched by about a factor 10 with respect to H, when compared to a "solar mix". Fractionation of isotopes residing inside grains, including a fraction of C, N, and O, is discussed in the context of pre-solar grains.

2.2.3. *Isotopic Fractionation in Pre-solar Grains?*

When precisely comparing the mean isotopic composition of pre-solar grains with the "solar mix", it needs to be considered whether the quantities of pre-solar grain types found in meteorites are representative for the dust in the pre-solar molecular cloud – for instance, they simply may have had different resistance to evaporation in the PSN – or whether they are – as debated for the *nano-diamonds* – truly of pre-solar origin. It also needs to be questioned whether pre-solar grains, which are refractory samples, have incorporated unfractionated samples of volatile isotopes of the various nucleosynthetic sources. Isotopic fractionation may have occurred

1. by implantation of energetic particles from exploding stars into grains with efficiencies that depend on isotopic mass A. Implantation depths, for instance, may scale with A or A^2 depending on the ion speed (Mewaldt *et al.*, 2001).
2. during the condensation of pre-solar grains from the gas phase or adsorption on them. Elemental fractionation is known from meteoritic material formed inside the solar system. The abundances of noble gases and the other volatile elements H, Li, Be, C, N, and O in these meteorites differ from solar photospheric abundances (e.g., Busemann *et al.*, 2001). Only for H, D, He, and the Li isotopes the solar nucleosynthetic processes can account for these compositional differences, whereas the other volatiles are fractionated by physical or chemical processes during the formation of the first solids in the PSN. The isotopic fractionation strength of the process of condensation or adsorption in the solar system is considered to be weak (e.g., Wimmer-Schweingruber,

2002). However, in analogy to the pre-solar material in IDPs, pre-solar grains may have preferentially incorporated some fractionated compounds.

3. because of incomplete condensation or metamorphosis of the parent bodies of the pre-solar grains. Huss (1997) has categorized the resistance of pre-solar grains to metamorphism of their host meteorites. In most meteorite classes, *graphite* and the *diamond* fraction that carries P3 noble gases (Ott, 2003b) are least resistant, *SiC* is moderately resistant, and *diamond* and *corundum* are most resistant, whereas in the highly reduced EH chondrites, *SiC* and *diamond* are both more stable, and *SiC* becomes more resistant to metamorphism than *diamond* (Huss and Lewis, 1995). The Orgueil meteorite and CM2 chondrites are the samples with least solar-nebular processing containing original dust from the parent molecular cloud of the solar system.

In fact, the study of all these processes in the particular case of our solar system is just the topic of this volume.

The signatures of nucleosynthetic processes, such as the s-, r-, and p-processes, can be best distinguished in the isotope pattern from other fractionation processes, if the diagnosed element has a large number of stable isotopes. The signatures of mass-dependent fractionation may be separable from those of non-mass dependent fractionation by e.g. the combination of photochemical and kinetic processes. The benchmark for these studies are the O isotopic abundances. Based on their $^{18}O/^{16}O$ and $^{17}O/^{16}O$ ratios, Clayton *et al.* (1976) have grouped the meteorites and planets formed inside the solar system into six categories that cannot be derived by mass fractionation or differentiation from each other. These at least six types indicate a spatial heterogeneity of the PSN that cannot be explained by the spatial variation of physical parameters such as temperature and pressure in the PSN alone. The newest directions in the research on O isotopes are reviewed by Clayton (2003).

3. Fractionation in Molecular Clouds

Although their temperatures are extremely low, cold dense interstellar clouds are sites for active chemistry in both the gas phase and on the surfaces of the tiny dust particles, or grains. In other words, the rates of chemical processes are larger than the inverse lifetimes of the clouds against collapse (typically 10^{6-8} y). Table II lists the basic categories among the thousands of molecular processes that form and destroy molecules in the interstellar gas and on the surfaces of dust particles. For each category, rate coefficients at 10 K are provided. These rate coefficients are for the case of exothermic reactions and no activation energy barrier. Most neutral-neutral reactions do have such barriers, so that the rate coefficient given must be multiplied by a Boltzmann factor of the type $\exp(-E_a/kT)$, where E_a is the barrier. At 10 K, a typical activation energy barrier is so much greater than kT that any reaction that possesses one will be unimportant.

TABLE II

Processes to form and destroy molecules in dense interstellar clouds.

Chemical Process	**Name**	**Rate Coefficient at 10 K**
Gas-phase reactions		units - see text
$A + B \rightarrow AB + h\nu$	radiative association	10^{-10}–10^{-17}
$AB + h\nu \rightarrow A + B$	photodissociation (external)	$\mathit{10^{-9}}\exp(-2A_v)$
$AB + h\nu \rightarrow A + B$	photodissociation (internal)	$\mathit{10^4}\,\zeta$
$AB^+ + e \rightarrow A + B$	dissociative recombination	10^{-6}
$A^+ + BC \rightarrow AB^+ + C$	ion-molecule reaction	10^{-7}–10^{-9}
$A^+ + BC \rightarrow A + BC^+$	charge transfer reaction	10^{-9}
$A + BC \rightarrow AB + C$	neutral-neutral reaction	$\leq 10^{-10}$
Grain-surface reactions		units - see text
A:grain + B:grain → AB:grain	accretion & diffusion	$\leq 10^{-17}$
AB:grain → AB + grain	evaporation	*0*
AB:grain → AB + grain	non-thermal desorption	$\leq \mathit{10^{-12}}$

For basic bimolecular chemical processes, the units of the rate coefficients are cm^3s^{-1} and an overall rate per reactant can be obtained by multiplying the rate coefficient by the density of the other reactant partner. For example, if we consider an ion that reacts with H_2, the dominant neutral species, the collision rate is the ion-molecule rate coefficient times the density of hydrogen, or 10^{-5} s^{-1}, and the collision time is its inverse, or 10^5 s. To obtain this number, we have used a typical density of 10^4 cm^{-3} and a rate coefficient of 10^{-9} cm^3 s^{-1}, appropriate for non-polar neutrals. The time scale for an ion to collide with a less abundant neutral is, of course, much longer.

For processes such as photodissociation, the rate coefficient given is in units of inverse time because the strength of the field has already been taken into account. Such rate coefficients are listed in italics. The rate coefficient for photodissociation by external photons is a function of the extinction in the cloud, A_v, while that for internal photons generated by cosmic ray bombardment is in terms of ζ, the cosmic ray ionization rate, which is typically 1.3×10^{-17} s^{-1}. The rate coefficient for thermal evaporation is listed as zero, although this is strictly true only for heavy species; hydrogen and helium do evaporate at competitive rates at 10 K. The rate coefficient for non-thermal desorption is obtained by analyzing the effects of cosmic ray bombardment (Hasegawa and Herbst, 1993).

Three-body processes can be neglected at typical molecular cloud densities (Dalgarno, 1987; van Dishoeck and Blake, 1998), unless the third body is a dust particle. The maximum rate coefficient given for reactions on dust surfaces pertains simply to the collision frequency between a neutral gas-phase species and a dust particle; if the subsequent efficiency of reaction is less than unity, the overall rate coefficient will be even less. In the units given, one multiplies the rate coefficient by the the overall gas density to get an overall rate for a given species. Thus, for a species in a cloud of standard density, the time scale for sticking to any grain (with 100% efficiency assumed) is approximately 10^{13} s, or 300,000 y, which is typically smaller than the cloud lifetime.

3.1. Fractionation in Cold Interstellar Cloud Cores

During the first stage of star formation (Figure 1a), gas-phase reactions, mainly via ion-molecule and dissociative recombination reactions, create small radicals and unsaturated molecules, while reactions on the surfaces of dust particles lead to the production of molecular hydrogen and a variety of saturated ices such as water, ammonia, and methane. Molecules formed in the gas, such as CO, can also accrete onto dust particles. Ions are produced by collisions of atoms and molecules with cosmic rays, which are high energy nuclei, mainly protons, present throughout the universe. The cosmic rays ionize a small fraction of the He atoms and H_2 molecules in a cloud, and even break up the H_2 to form H and H^+ (Herbst and Klemperer, 1973). Because atomic hydrogen is converted efficiently into molecular hydrogen in dense clouds, the H_3^+ ion, produced via the ion-molecule reaction between H_2 and H_2^+, plays a pivotal role in the subsequent gas-phase ion-molecule chemistry. Interestingly, grains are themselves formed in the atmospheres of older (e.g. AGB) stars and stellar explosions under high-density conditions and wafted into space. Unlike gas-phase molecules, also formed under these conditions, grains are not subsequently destroyed by the strong radiation field present in unshielded interstellar space. Eventually, under the influence of gravity, clouds form from grains and gas-phase matter, which is by now mainly atomic.

Some major products of the gas-phase chemistry are the complex hydrocarbons (e.g., Ehrenfreund and Charnley, 2000). They are formed along three types of ion-molecule and neutral-neutral pathways (Herbst, 1995): 1) C insertion reactions (e.g. $C^+ + CH_4 \rightarrow C_2H_2^+ + H_2$ or $C + C_2H_2 \rightarrow C_3H + H$); 2) condensation reactions (e.g. $CH_3^+ + CH_4 \rightarrow C_2H_5^+ + H_2$ or $C_2H + C_2H_2 \rightarrow C_4H_2 + H$); and 3) radiative association reactions (e.g. $C^+ + C_n \rightarrow C_{n+1}^+ + h\nu$). Insertion reactions dominate and produce unsaturated hydrocarbons because unsaturated hydrocarbon ions react slowly with H_2 if at all. A recent review article on interstellar gas-phase chemistry is Herbst (2001).

The complexity of the chemistry utilized in modelling a source depends on the complexity of a source; i.e., whether it is constant or changing in its density and temperature, and whether it is homogeneous or heterogeneous. The simplest models contain only gas-phase processes in a homogeneous and time-independent setting, whereas the most complex consider both gas-phase and surface chemistry in a heterogeneous setting that may even change with time. Up to the present, most models on chemical fractionation in interstellar molecular clouds have contained only gas-phase processes, although the role of dust particles is currently being re-evaluated. In these simple models, the chemistry proceeds with fixed physical conditions from given initial concentrations at a low temperature until a steady-state is reached. If dust grains are included, their role can either be passive or active; the former adjective applies to how accretion affects gas-phase fractionation while the latter applies to actual grain chemistry involving isotopomers. Two models with both spatial- and temporal-dependence in which grains play a passive role in the fractionation of deuterium are those of Aikawa and collaborators (Aikawa and Herbst, 1999; Aikawa *et al.*, 2001). In one of these, deuterium fractionation is considered during the formation of a protoplanetary disk, while in the other such fractionation is considered in the earliest stages of collapse of a core to form a star. These models are discussed in Section 3.2.

3.1.1. *Gas-phase Fractionation*

Fractionation of elements at low temperatures occurs in trace species given a suitable reservoir of the minor isotope. The best studied example is that of deuterium, in which the reservoir is HD. As is discussed by both Millar and Herbst in this volume, deuterium fractionation results mainly from the reactions of three ions with HD, the most important of which is H_3^+. The reaction between H_3^+ and HD to form H_2D^+ and H_2 is exothermic by 232 K so that the backwards reaction does not occur appreciably at low interstellar temperatures. The H_2D^+ ion is depleted by reactions with heavy species to form a variety of enhanced deuterated isotopomers. Observed ratios of deuterated to normal isotopomers can be enhanced 10^{3-4} times over the HD/H_2 ratio. Even doubly and triply deuterated isotopomers have been detected.

For the case of nitrogen, both ^{15}N and $N^{15}N$ act as reservoirs (Terzieva and Herbst, 2000). Since the exchange reactions with ions such as HN_2^+ are much less exothermic than is the case for deuterium exchange reactions, the fractionation is much smaller. Details are discussed in the article by Herbst. For the case of carbon, the major reservoir of ^{13}C in dense clouds is ^{13}CO, while the reservoir in diffuse clouds is $^{13}C^+$. Fractionation of CO in diffuse clouds by exchange reaction with $^{13}C^+$ is countered by an opposing process in which ^{13}CO is photodestroyed more rapidly than the normal isotopomer.

3.1.2. *Passive and Active Grain-surface Fractionation*

Carbonaceous dust grains form from long carbon-chains in an environment that is rich in elemental C, especially the envelopes of AGB stars, while silicate grains form in similar objects rich in elemental oxygen. During the pre-collapse (Figure 1a) and cold collapse (Figure 1b) phases of a molecular cloud core, many molecules (e.g. CO) accrete onto the grains and, along with the products of surface chemical reactions, form icy mantles. The accretion of heavy species such as CO onto grains actually enhances deuterium fractionation in the gas phase because ions such as H_2D^+ are not depleted as readily by these heavy species. The resulting increase in the H_2D^+/H_3^+ abundance ratio leads to increased fractionation (Millar, 2003). This passive effect is to be distinguished from an active effect in which fractionation actually takes place on grain surfaces. The most important active mechanism involves atomic deuterium, which is produced in the gas phase by dissociative recombination of ions such as H_2D^+. Atomic deuterium can achieve an abundance almost as large as atomic hydrogen. When these two atoms accrete onto dust particle surfaces, they react to deuterate and hydrogenate unsaturated species. For example, a nitrogen atom landing on a grain surface can be converted into a variety of deuterated isotopomers of ammonia, and the ratios of deuterated to normal isotopomers (e.g. NH_2D/NH_3) can reach high levels. Observation does require, however, a mechanism for desorbing the isotopomers back into the gas phase, where they can be detected with more sensitive spectroscopic techniques. Such desorption occurs during star formation, especially for high mass stars (Charnley *et al.*, 1997). Active effects involving nitrogen and carbon fractionation are doubtless much smaller although they have not been studied in great detail.

An initial model that does consider passive grain effects in the fractionation of nitrogen is that of Charnley and Rodgers (2002). This model is considered in the next section because it occurs in a post-molecular cloud stage.

3.2. Post-Molecular Cloud Fractionation

Among the D/H ratios in primitive solar system samples, the ones in IDPs with interstellar provenance are larger and closer to the molecular cloud values (Messenger *et al.*, 2003) than those of the comets and the thermally and aqueously altered meteorites. The pre-solar IDPs appear to have suffered less chemical processing and are likely to be more representative of protosolar volatile isotopic abundance ratios, such as $^{14}N/^{15}N$, than meteorites.

Anticipating the protosolar $^{14}N/^{15}N$ ratio of ≈ 400 by Hashizume *et al.* (2000) and by Owen *et al.* (2001), the $^{14}N/^{15}N$ ratios of less than 200 in some IDPs indicate very strong chemical fractionation or a variety of nucleosynthetic sources. The main argument against nucleosynthesis as the cause of the $^{14}N/^{15}N$ variations in solar system samples is that no such variations are seen for e.g. $^{12}C/^{13}C$.

However, the models on chemical fractionation reviewed by Herbst (this volume) predict a maximum enrichment of ^{15}N by at most 30% in interstellar HCN and by up to 50% in other rare compounds in the ISM. This is not sufficient to explain the $^{14}N/^{15}N$ variations in IDPs, but these models ignore passive grain effects.

Recently, Charnley and Rodgers (2002) developed a chemical fractionation scenario similar to that in cold interstellar cloud cores but occuring in the outer PSN, where temperatures remained below $\approx$ 30 K. The variations in $^{14}N/^{15}N$ in the outer solar system samples depend on the budget of the N components that have contributed to the gas, the first solids, and the planetesimals. Three N components

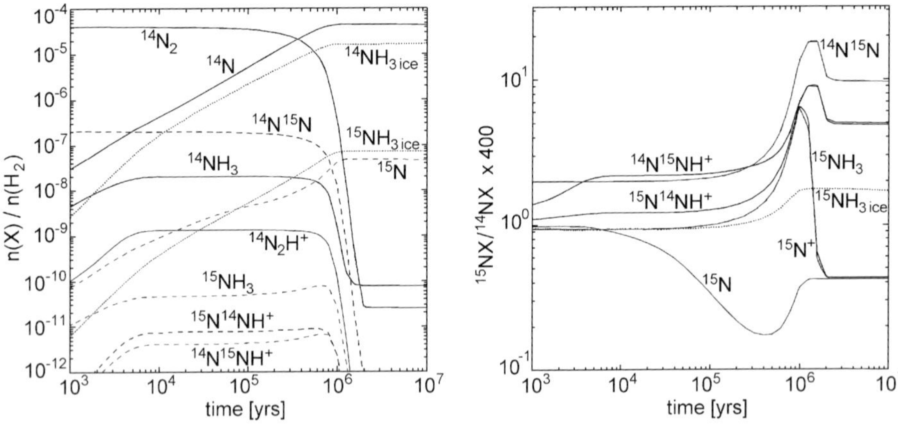

Figure 2. Left: Abundance vs. time of N-bearing molecules and their isotopomers in a cold dense cloud ($n = 10^7$ cm^{-3}, $T = 10$ K), with all species except H_2, N_2, and atoms accreting onto grains. Decline in the N_2 abundance is due to reactions with He^+. *Right:* Relative fractionation of ^{15}N in different species. Initial $^{14}N/^{15}N$ ratio is assumed to be 400. From Charnley and Rodgers (2002).

are decisive for the $^{14}N/^{15}N$ ratio: atomic N is enriched in ^{14}N, whereas molecular N_2 and ammonia (NH_3) are enriched in ^{15}N (Figure 2). Variations in $^{14}N/^{15}N$ in the (outer) solar system are much larger than those in $^{12}C/^{13}C$ and $^{16}O/^{17,18}O$ because N is diverted in both the solid and the gas phase, with the gaseous component of N fractionated with respect to the adsorbed or solid component, whereas C and O entirely rest in the solid phase. Fractionation is particularly strong, if a lack of hydroxyl (OH) molecules prevents the recycling of N atoms into N_2. Then, the gas phase is strongly depleted in ^{15}N. On the other hand, ammonia ices are formed rapidly so that a significant fraction of N is supplied to the solar system in the form of NH_3 enriched by up to 80% in ^{15}N over the cosmic $^{15}N/^{14}N$ ratio. This model accounts for both the enrichments of ^{15}N in IDPs and the depletion of N_2 in comets.

Nitrogen isotopic abundances in the solar system still leave open questions. In any case, nitrogen isotopes appear to undergo the strongest fractionation among the isotopes of heavy elements. Therefore, nitrogen isotopes may in fact be the best tracers to study the evolution of the solar system. For example, in N isotopic signatures evidence may be found that a molecular cloud or protoplanetary accretion shock had affected the spatial or temporal distribution of molecular abundances in gas or in ice.

Fractionation processes may not only occur *within* different components of molecular cloud or post-molecular cloud material. The question arises as to whether fractionation may have occurred during the collapse of the parent molecular cloud core in the protostellar and protoplanetary disk phases. The argument of researchers for believing that there was not much fractionation is that the timescales ($\approx 10^5$ y) for the episodes of strong infall were too short to allow for fractionation (e.g.,

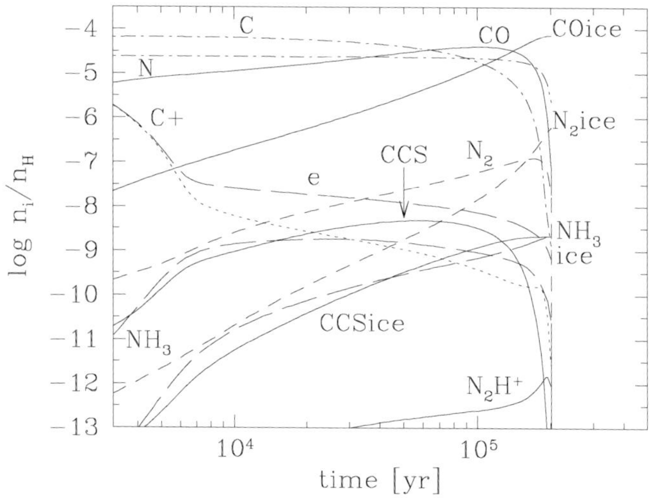

Figure 3. Evolution of molecular abundances in a fluid element that migrates from 8200 to 1000 AU in 2×10^5 y, while the gas density n_H varies from 1.7×10^4 to 5.8×10^6 cm^3. From Aikawa *et al.* (2001).

Hartmann, 2000). Such arguments can be tested with detailed chemical models of fractionation.

Recently, Aikawa *et al.* (2001) modeled the evolution of molecular abundances during the earliest stages of the collapse of a molecular cloud core (Figure 3), the so-called pre-stellar core stage, in which a core is being formed but not yet being heated above 10 K. The gas-phase deuterium fractionation is found to increase as the core evolves because of the passive effect of the depletion of gas-phase species onto the dust particles. The results are in reasonable agreement with observed abundance ratios for the molecular ions HCO^+ and HN_2^+.

A more speculative fractionation scenario could be imagined as follows: When the pre-solar molecular cloud collapsed, the neutral fraction slipped against the charged fraction, which was supported by the magnetic field, and, thus, the neutral fraction may have dominated the composition of the cold core. The infalling part could mainly be neutral dust or grains with low charge-to-mass ratio, typically enriched in ^{15}N, whereas the magnetically supported charged fraction may mainly consist of ionized atoms depleted in ^{15}N. This scenario, however, is unlikely to be efficient since very little N can be ionized in the interstellar medium. The ionization potential of N is higher than that of H, so that the UV photons dominating the spectrum are not capable to ionize N. Cosmic rays can ionize N but the amount produced is very small, N^+/N is less than 10^{-6}.

To consider deuterium fractionation at a time closer to the PSN stage, Aikawa and Herbst (1999) studied how deuterium fractionation in an interstellar cloud core is changed by the gas-phase chemistry that occurs as the core collapses into the midplane of a protoplanetary disk. Much of the material eventually ends up in the condensed phase. The condensed-phase material that reaches 30 AU from the central star was emphasized since this material can be compared with primitive objects that formed at such distances; e.g. short-period comets. Indeed, the authors found that the dominant ices in their model were those that constitute comets. On the topic of deuterium fractionation, they found that the D/H ratios in species such as HCN and water are somewhat less than found in cold molecular clouds, and in reasonable agreement with what has been observed for comets. Here the passive effect of the grains is countered by the somewhat higher temperatures (20-30 K).

The conclusion from these two papers is that chemical and physical processes during the collapse of a molecular cloud core into the PSN do affect deuterium fractionation although D/H values for trace species do remain large.

4. Embedded Protostars: Outflow Phase

Figure 1c shows that low-mass star formation generally includes a phase in which the embedded protostar drives a bipolar outflow that eventually disperses the molecular cloud and surrounding disk. The nature of the outflows themselves, usually atomic, hot and fast-moving, means that the standard fractionation processes, which are essentially low-temperature effects, do not operate. Nonetheless, observations of protostars are essential if one is to understand fractionation processes, particularly those involving deuterium, in molecular clouds. Furthermore, such observations can be used to probe the interstellar-solar system connection.

Systematic studies of low-mass protostars are only just underway. Shah and Wootten (2001) carried out a survey of NH_2D in dark cores and Class 0 sources, which are cold and possess highly collimated outflows. Some Class 0 sources do show large abundances of hydrogenated molecules, usually interpreted as a result of grain surface chemistry, as opposed to the dark cores where gas-phase chemistry must dominate. Shah and Wootten find that fractionation ratios are independent of kinetic temperature above 20 K, whereas theory would suggest that fractionation decreases with increasing temperature. This indicates that the Class 0 sources have fractionation ratios set earlier in their evolution. This is also indicated by relatively low fractionation in DCO^+.

Roberts *et al.* (2002) have surveyed both HDCO and DCN in Class 0 and Class I protostars – the latter have bolometric temperatures between 70 and 250 K – but again fractionation ratios are essentially constant with bolometric temperature. Fractionation in DCN is a factor of about 1.5 larger than that predicted by the pure gas-phase models between 10 and 20 K, but both it and fractionation in HDCO can be explained if accretion onto grains has occured for a period of around 5×10^4 y. One sees immediately that there is a discrepancy between the bolometric temperatures and the kinetic temperatures associated with the fractionation process. There are a number of possibilities for this. One is that the bolometric (dust) and kinetic (gas) temperatures refer to the same region but are uncoupled. This appears unlikely at the high densities in these sources. A second is that they refer to different spatial regions in the source, and a third is that the temperature derived from fractionation is a 'fossil' temperature. The latter two possibilities can be investigated by multi-transition, high spatial resolution observations of deuterated species. Unfortunately, this remains to be done in any systematic way. Multi-transition studies are difficult for small, linear molecules such as DCN and DCO^+ because their successive rotational transitions are far apart in frequency but can be done for species such as NH_2D and HDCO. Even in these cases, one must be careful that emission from both deuterated and non-deuterated molecules comes from the same volume of gas. Turner (2002) has argued that this is not generally the case; he suggests a scenario in which emission from deuterated molecules comes from smaller, denser regions than that from their hydrogenated parents. If true, this could alter derived fractionation ratios to a significant extent. For example, Turner derives fractionation ratios of N_2D^+ and NH_2D which are more than an order of magnitude less than other workers in the dust core TMC-1. Clearly differences as large as this needs to be resolved.

Single dish mapping of deuterated molecules is becoming available and has shown some interesting results. Caselli *et al.* (1999) have compared DCO^+ maps in L1544 with CO, HCO^+ and dust continuum emission maps and find that the DCO^+ follows the dust distribution whereas CO and HCO^+ do not. Their observations give support to the view that the largest fractionation ratios should be detected in cold cores with CO depletion. Ceccarelli *et al.* (2001) used the IRAM 30 m telescope to map D_2CO emission around the Class 0 protostar IRAS 16293-2422 which also contains ND_2H. This low-luminosity source is complex: it is a proto-binary star, contains two warm circumstellar disks, a colder circumbinary disk, multiple outflows, infalling, rotating material and ambient gas which contains dense clumps. They found that the emission from H_2CO could be interpreted with three phases: (i) a small (< 150 AU) hot (>100 K) core, (ii) a larger (700 AU), warmer (50-100 K) envelope, and (ii) an extended colder (<50 K) envelope. They were able to derive fractionation ratios and showed that the fractionation ratios in D_2CO decrease from 0.16 at small distances to 0.003 at large distances. Ceccarelli *et al.* argue that this is consistent with: (i) the thermal evaporation of onion-like ice mantles which have (ii) larger fractionation in D_2CO at greater depths. Again we note that the kinetic temperature of the gas where D_2CO is detected is too large to have efficient fractionation, that is, D_2CO is a 'fossil' species. Ceccarelli et al. argue that the D_2CO is in fact made on the cold dust grains and subsequently evaporated by thermal effects. They argue against gas-phase fractionation modified by accretion because, in their view, by the time accretion can enhance fractionation, so little material remains in the gas phase that it has no effect on the total number of grain molecules when it is finally accreted. Thus any large fractionation which occurs in the gas phase in the final stages of accretion is effectively diluted when the grain mantles evaporate.

This will be the case if large depletions are needed to enhance fractionation. However, Rodgers and Charnley (2001) noted that very large fractionation can be achieved when CO is depleted by only a factor of two. In this case, the large fractionation ratios evident in the depleted gas would be diluted by only a factor of two.

Ceccarelli *et al.* prefer a synthesis in which D_2CO is formed on the grain surface by conversion of a trace amount of CO, in a process which also produces HDCO and H_2CO. Unfortunately, models of ice chemistry in interstellar clouds are very rudimentary and there are many major uncertainties, including binding energies, diffusion barriers, activation energy barriers, and so on. We note that any chemistry which allows CO to be hydrogenated and deuterated should also predict the formation of HDO and D_2O from the hydrogenation and deuteration of atomic oxygen. The former is seen only in a few low-mass sources while the latter has yet to be detected despite sensitive searches, although for both of these species physical conditions may prevent efficient excitation.

Loinard *et al.* (2002) have searched about 20 low-mass protostars for D_2CO finding it in all Class 0 and Class I sources with an abundance ratio of D_2CO/H_2CO of 0.02-0.40. They did not detect it in three Class II sources, two of which did not contain H_2CO either, nor in six high-mass protostars with an upper limit to the fractionation about an order of magnitude less than those observed in the Class 0 and I sources.

Finally, one might expect that shocks associated with the outflows would desorb fractionated species from grains. Models show that the fractionation ratios set in the grains will survive for at least 10^4 years upon desorption (Rodgers and Millar, 1996). Outflows in IRAS 16293 are traced by strong SiO emission but do not show any enhanced D_2CO emission. Conversely, where D_2CO has its strongest emission, SiO is absent. Loinard *et al.* (2002) study also targeted shocked regions in low-mass protostellar outflows but failed to detect D_2CO in them. Buckle and Fuller (2002) surveyed 18 Class 0 and 12 Class I sources in DCN, finding it toward 40 of the 60 positions observed, and CH_3OH which was detected at all positions. Although they find enhanced abundances and broad linewidths of CH_3OH, which they associate with shock evaporation of grain mantles by the outflows, DCN shows a different velocity structure, one which picks out lower velocities but which may be associated with the interaction between the outflow and the ambient molecular cloud. This general lack of fractionation in outflows and in shocks may simply be due to the fact that the outflows are interacting with gas which was never cold enough to fractionate efficiently in the first place, or that ambient densities were so low that accretion, and hence the formation of grain mantles, took place on a longer time-scale than star formation. We recall that outflows are normally orthogonal to the highest density material which has a disk-like structure around low-mass protostars and will propagate preferentially in low density material. We note too that shock waves compress gas so that shocked clumps seen in SiO, for example, may not have been at high density for very long. Again, the intrinsic density of the pre-shock gas might have been quite low.

More detailed comparison of fractionation at the smaller spatial scales important in star formation require interferometric studies. Shah and Wootten (2002) have made the first interferometric maps of DCO^+ using the BIMA array in the S68N molecular cloud. They find radically different spatial distributions in HCO^+, $H^{13}CO^+$ and DCO^+, with DCO^+ providing the best tracer of the cold dust clumps – and thus potential sites of star formation – in the region. The fact that the hydrogenated ion is less useful is partly due to large optical depths, which leads to self-absorbed line profiles, and to the enhanced fractionation which occurs in cold, dense clouds in which accretion is dominating.

It is clear that several careful interferometric studies are needed to discriminate between the theories. Species to be studied include CH_3OH, which appears to be enhanced in grain mantles because of surface chemistry, N_2H^+, which is deuterated only by D and H_2D^+, and is enhanced in regions of depleted CO, NH_3 which probes temperature, and to a lesser extent, grain surface chemistry, and SiO which traces shocked gas. Studies of line profiles, and not just abundances, will help correlate the chemical compositions while observations of singly and multiply-deuterated molecules can be used to delineate between gas and grain chemistry.

5. Isotopic Fractionation in the Protosolar Nebula

The word "protosolar nebula" has very different meanings for different authors or different "schools of thought". We will start with the standard description of the protosolar nebula by Cameron (1995) to subsequently trying to define the terms "protosolar cloud" (PSC), "protosolar nebula" (PSN), and "accretion disk" or "circumstellar disk" (CD) as separate astrophysical objects, and to discuss possible isotopic fractionation processes in these objects.

According to Cameron (1995), the history of the PSN has four epochs: In a first stage (Figure 4a, and 1b), the nebular disk is assembled by infalling material from the collapsing molecular cloud core. The collapsing fraction of this cloud will be called "protosolar cloud" (PSC) in the context of isotopic abundance ratios. In a typical interstellar cloud, about half of the heavy elements reside in icy silicate dust grains. Except He, most volatile elements, such as C, N, and O, are locked up in solid particles. This stage lasts for a few times 10^5 y. Most (if not all) of the matter in the disk during this stage ultimately goes into the proto-Sun. Therefore, the present Sun is the most authentic witness of the composition of the PSC. Possible isotopic fractionation processes during the collapse phase are reviewed in Section 3.2. Assuming that fractionation during this stage is weak, the Sun represents the composition of the PSC. The abundances of the PSC are referred to as protosolar abundances (Section 5.1).

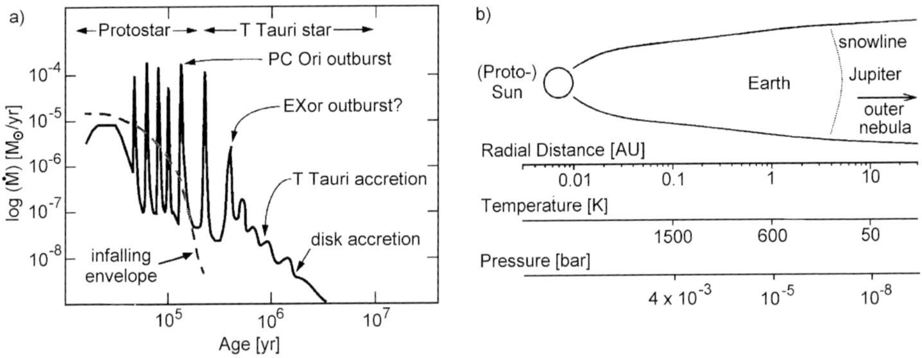

Figure 4. a) Temporal evolution of accretion to the protosolar nebula and from the protosolar nebula to the proto-Sun. The Sun acquires most of its mass, while there is still infall from the interstellar cloud (adapted from Hartmann, 2000). *b)* Temperature and pressure profiles in the protosolar nebula (adapted from Fegley, 2000). Temperature and pressure values are those of Willacy *et al.* (1998), based on the original work of Urey (1952).

In a second stage (Figure 1c), the Sun accretes at a rate of $\approx 2 \times 10^{-5}\ M_{\odot}\ \mathrm{y}^{-1}$. Within about 50'000 y most mass falling onto the accretion disk is transported through the disk into the proto-Sun. The nebula embedding the proto-Sun at this stage is called "protosolar nebula" (PSN). Although not explicitly written in the theoretical papers describing the evolution of the medium surrounding the early Sun, the existence of a PSN remains controversial. The concept of PSN refers here to a thick, dense, opaque and turbulent gas which should in principle not be confounded to the optically thick disk observed around young stars whose dimensions exceed 500 AU (Waelkens *et al.*, 1996). Part of this confusion comes from the fact that, frequently in the literature, the words "protosolar nebula" or "solar nebula" are used without specific definitions; for example a static, or a diffuse PSN is sometimes advocated to interpret the results obtained in meteorites and in this respect, is in contradiction with the available models of PSN *stricto sensu*. The PSN is supposed to be located at the center of the disks with dimensions – from 10 to 100 AU – depending on its age. Therefore, on one hand the existence of disks around young stars is now widely accepted while, on the other hand, the existence of a PSN remains an open question.

In most papers, the empirical distinction between the "circumstellar disk" (CD) of the T-Tauri Sun and the PSN is not mentioned. This deficiency is linked to two issues. First, the PSN is a theoretical object too small to be observed by the present day astronomical probes. Second, no theoretical attempts have been made to describe the transition in time and space between the low density medium dominated by the interstellar chemistry and the dense medium of the PSN dominated by homogeneous and heterogeneous reactions in a neutral gas. In the framework of the PSN model, the regions of an active interstellar chemistry are imagined to be located at the periphery of the nebula but are not described by the model. Deciding if such a PSN has ever existed around the young Sun has profound consequences on the actual isotopic distributions in solar system objects. Indeed, in the PSN, two processes must have erased in part the signatures of an interstellar heritage: (1) high temperatures (up to 2000 K) through which the interstellar solids – pre-existing the formation of the PSN – are destroyed and (2) turbulence through which gas and grains are mixed on distances up to 50 AU.

The strength of fractionation processes in the PSN (including the CD) depends on the temperature and pressure distribution due to turbulence. The proto-Sun dissipates angular momentum by spiral density waves (see Adams and Laughlin, 2000, for a review), disk-driven bipolar outflows (see Section 4 for a discussion of possible isotopic fractionation in the outflows), and magnetic instabilities (see Balbus and Hawley, 2000, for a review). The temperature gradient in the PSN and its temporal evolution results in a condensation sequence. The first abundant materials to condense at high temperatures of about 1600 K are oxides of Al, Ti, Ca, and Fe and Ni grains at 1400 K. These form the cores of planets and iron meteoroids. At 1300 K, Si condenses as silicate minerals found in the stony-iron meteorites. Apparently, condensation of disk material has not fractionated isotopes very much, as can be concluded from data on $^{41}K/^{39}K$ ratios in CI chondrites (Wimmer-Schweingruber, 2002). The fact, that some original disk material has survived this second stage and has been preserved in meteorites, is witnessed, for instance, by ^{26}Al-bearing minerals. Meteorites must have formed from different types of dust, including the refractory dust from the inner solar system, interstellar dust that has survived in the outer solar system, as well as volatile ices from the outer solar system. Between 300 K and 150 K, the elements H, C, N, and O condense in icy particles, preserved to the present day in comets and the outer planets formed outside the "snow line" (Figure 4b; Hueso and Guillot, 2003; Owen and Encrenaz, 2003). The amount of material transported outward and "contaminating" the outer nebula with products from thermochemical processing in the innermost few AU of the PSN remains controversial and uncertain, though (Fegley, 2000).

Because of the lack of direct observations of any PSN around Young Stellar Objects (YSOs), the validity of theoretical models on the PSN is established by comparing the predictions of the model with some relevant parameters which can be measured in the present day solar system or in T-Tauri stars (Morfill and Wood, 1989; Drouart *et al.*, 1999). Although the quality of the fitting between predictions and observations is mostly a matter of taste, it should be emphasized that the PSN model is the only one that fulfills the constraints of several of these parameters. The behavior of the PSN is described in Section 5.2, followed by a discussion on its ability to account for various observations.

In the third stage (Figure 1d, e) in the scenario of Cameron (1995), lasting about 1–2 My, the Sun completes its accumulation at rates of $\approx 10^{-7} - 10^{-8} M_\odot$ y^{-1}. The proto-Sun becomes a classical T Tauri star. The newest model to describe the processes in the close vicinity of the star during this stage has been developed by Shu *et al.* (1997) and is reviewed in Section 5.4.1. Jupiter and Saturn have almost completely formed during this stage, while the accretion of other planets took longer.

The fourth stage lasts 3–30 My. The PSN expands in space, which depletes the gas in the disk. The Sun does no longer accrete material and becomes a weak line T Tauri star. The properties of T Tauri and post-T-Tauri winds and their implications for isotopic fractionation processes in the early solar system are discussed in Section 6. According to Fegley (2000), the Sun's T Tauri wind also removes gas in the inner nebula – in addition to the depletion that occurs by the PSN expansion – and its UV radiation evaporates may remove gas in the outer nebula. While ultraviolet photochemistry and ion-molecule chemistry may be important disequilibrating processes in the inner nebula during the third stage, they may act in the outer nebula during the fourth stage. The implications of ion and UV chemistry on the evolution of grains during the late stage of the PSN have not yet been addressed theoretically. Without such an approach, no qualitative description of these implications can be proposed because they are a complex function of the evolution in time and space of the density of the nebula. During the fourth stage terrestrial planets form. Isotopic fractionation processes in the bulk and in the atmospheres of terrestrial planets are discussed by several authors in this volume, in particular in the accompanying working group report by Becker *et al.* (2003).

5.1. Protosolar Isotopic Abundances

Protosolar abundances are the basis for interpreting isotopic data obtained in the solar wind and in planets, satellites and comets. Also, the theory of nucleosynthesis is largely based on elemental and isotopic measurements in the solar system. The isotopic abundances in the Protosolar Cloud (PSC) are readily derived from meteorite data for the majority of the refractory elements. The situation, however, is different for the most volatile elements, because their strong depletion in all types of meteorites has caused changes in isotopic abundances that cannot be easily corrected. In recent years, essentially based on the discovery that most solar system solids (terrestrial planets, comets, meteorites, IDPs) are enriched in ^{15}N and in D relative to the protosolar gas, a new idea has emerged: in the PSN, gas and grains may have had different isotopic compositions and, in most cases, never equilibrate. For these elements, therefore, representative isotopic ratios in the PSC are derived from data taken in the solar wind and in the atmospheres of giant planets.

Abundance measurements of D and ^{3}He are of particular importance for modeling the chemical evolution of the Galaxy. Deuterium is almost entirely produced in the Big Bang Nucleosynthesis (BBN), also called primordial nucleosynthesis. Only small contributions of order $< 10^{-4}$ may originate from spallation processes in stellar flares (Section 7.4). Therefore, it can be assumed that the Galactic D abundance decreases monotonically, at least in a closed system. ^{3}He, on the other hand, has increased from its original Big Bang abundance, due to D burning in stars, and to a lesser extend, due to incomplete H burning. In the Galaxy the decrease in D is approximately offset by the increase in ^{3}He, so that (D+^{3}He)/H has remained relatively constant throughout galactic history. Indeed, within the limits of error there is no difference between (D+^{3}He)/H in the PSC and in the present-day interstellar medium. Even in the time interval from the Big Bang to the PSC this ratio has only decreased by $\approx 10\%$.

In this volume, Geiss and Gloeckler (2003) review the data and give preferred values for the isotopic abundances of hydrogen, helium and neon in the PSC. The value $^3He/^4He = (1.66 \pm 0.06) \times 10^{-4}$ obtained with the Galileo Probe Mass Spectrometer by Mahaffy *et al.* (1998) is adopted for the PSC. The ratio $D/H = (2.0 \pm 0.35) \times 10^{-5}$ given by Geiss and Gloeckler (2003) is the average of two independent determinations. One is derived from infrared spectroscopy of the Jovian atmosphere (Lellouch *et al.*, 2001), the other from solar wind ^{3}He/^{4}He results obtained by Gloeckler and Geiss (2000) with SWICS Ulysses.

Among the elements with $Z < 12$, only O and Ne have three stable isotopes. The noble gases are particularly interesting because there is no good reason to invoke non-mass dependent isotopic fractionation by physical or chemical processes in the PSN and in planets. In the case of Ne, ^{20}Ne, ^{21}Ne and ^{22}Ne abundance measurements permit to distinguish between differences in their abundances caused by fractionation from differences caused by nuclear reactions. For instance, it has been shown that in the Earth atmosphere there is a small excess of ^{21}Ne, which – at least in part – is caused by nuclear reactions in the interior of the Earth, as predicted by Wetherill (1954).

On the other hand, the Ne isotopic ratios of other components found in solar system samples lie exactly on the fractionation line of solar wind Ne. From this, one would conclude that the "SEP" component (Section 6.2) of the gases trapped in lunar surface material is of solar system origin, as well as the "Q" component of noble gases in primitive meteorites (Section 5.4.2).

5.2. The Concept of the "Protosolar Nebula"

Starting out with the protosolar isotopic abundances, the first fractionation processes may occur in the PSN. The qualitative description of the PSN presented here is based on the results of analytical models (see the pioneering work of Lynden-Bell and Pringle, 1974, and Cassen, 1994, for a thorough review of the models; Dubrulle, 1992, 1993; Dubrulle and Valdettaro, 1992; Dubrulle *et al.*, 1995; Stepinski, 1998). These types of models allow straightforward estimates of the parameters dictating the evolution of the system. Note that several numerical – but not analytical – models have been proposed in the literature which yields qualitatively to similar results (Cameron and Pine, 1972; Cameron, 1978; Morfill *et al.*, 1985; Ruden and Lin, 1986; Boss *et al.*, 1989; Cassen, 1994; Ruden and Pollack, 1991). Recent models are in 2D (Hersant *et al.*, 2001) while the original Dubrulle's model was constructed in 1D, hence its analytical form. As a general assumption, the calculation starts (at time $t = 0$) when the infall from the parent molecular cloud has stopped and when the Sun, in its T-Tauri phase, has reached $\approx 90\%$ of its present day mass. In this respect, PSN models should be regarded as "closed models" i.e. that no input of matter (as gas or grains) from the interstellar surrounding regions takes place during the evolution of the disk.

5.2.1. *The Parameters of Present PSN Models*

Among the numerous parameters calculated in the model, only three have to be assumed in order to entirely dictate the evolution of the disk: (1) the coefficient of turbulent viscosity α, (2) the initial mass of the solar nebula $M_{\mathrm{D}}(t = 0)$, and (3) its initial radius $R_{\mathrm{D}}(t = 0)$ (see also Hueso and Guillot, 2003).

1. Because it is impossible to describe the turbulent movements of the gas at all scales, Shakura and Sunyaev (1973) have defined α, the phenomenological coefficient of turbulent viscosity through which the relations between the turbulent movements and the physical parameters of the gas can be calculated. In this respect PSN models are sometimes referred in the literature to as "α disk". Of course, the aim of the theoreticians is to replace this somewhat arbitrary approach of the turbulence by a more rigorous physical treatment of the problem where the evolution with time of turbulent vortex will be calculated at all scales. In such a turbulent solar nebula, the transfer rate of the angular momentum from the Sun outwards is dictated by the value of α. Although the origin of the turbulence is unknown, its presence was postulated to account for the observed large excess of energy radiated by young stars embedded in disks. So far, this hypothesis has not been seriously challenged ...
2. M_{D} and R_{D} expressed in solar mass and astronomical units, respectively, are defined at a time $t(0)$. The gas and the grains – when < 1 cm, grains are coupled with the gas – have two distinct movements: (i) an inward fall which increases the mass of the Sun (accretion), (ii) an outward movement due to the expansion in space of the nebula because of its high turbulence. These opposite movements result in an efficient mixing between the internal and the external zones of the PSN (Cassen, 1994). As a function of time, R_{D} increases due to the outward expansion while M_{D} decreases because of accretion. When the initial conditions are defined i.e. when $M_{\mathrm{D}}(t = 0)$, $R_{\mathrm{D}}(t = 0)$, and α are fixed, the evolution in time and space of the geometrical and thermodynamical parameters of the PSN are determined. The stochasticity of the turbulence has not been modeled yet.

At time t and at the heliocentric distance R, all relevant physical quantities X can be written under the form:

$$X(R, t) = f(M_{\mathrm{D}}, R_{\mathrm{D}}, \alpha) \,. \tag{1}$$

X can be the temperature T, the pressure P or the surface density Σ. As the PSN evolves with time, or as R increases, T, P and Σ decreases. The surface density Σ is the usual density of the gas (g cm^{-3}) normalized to the thickness of the disk – hence expressed per surface unit; thus Σ stands for the total amount of matter present at a given distance from the Sun. The turbulence is the dominant force that dictates the temperature of the disk.

5.2.2. *The PSN Model in Front of Observable Parameters*

The adjustment of the three parameters M_D, R_D, and α are able to reproduce 6 observational constraints. Four constraints are derived from giant planet observations (Pollack and Bodenheimer, 1989; Pollack *et al.*, 1996):

1. The life time of the PSN is in agreement with the modeled formation time of the giant planets ($t = 2 \times 10^7$ y) and also with the observed lifetime of the disk surrounding the Young Stellar Objects.
2. The condensation temperature of water ice should be reached at 5 AU at the time of formation of Jupiter.
3. The minimum initial surface density (Σ) of the solar nebula is $> 15\,\mathrm{g\,cm^{-2}}$ at 20 AU allowing the formation of a planet like Uranus within $\approx 10^7$ y.
4. The angular momentum is transported outwards to all the regions where giant planets can be formed. This constraint is quantitative i.e. the observed angular momentum of the planets is in agreement with the calculations.
5. The last two constraints belong to chemical observations: One constraint is derived from the distribution of the D/H ratio in the solar system (Drouart *et al.*, 1999; Mousis *et al.*, 2000; Hersant *et al.*, 2001). The PSN model accounts for the fact that objects formed far away from the center of the disk should have higher D/H ratio than those formed in the inner regions. This difference is qualitatively observed between meteorites and comets (Bockelée-Morvan *et al.*, 1998; Deloule and Robert, 1995; Robert *et al.*, 2000).
6. Another constraint based on chemical observations is derived from the presence of crystalline silicates in comets (Gail, 2001; Wehrstedt and Gail, 2002; Bockelée-Morvan *et al.*, 2002). Silicates formed in interstellar conditions are amorphous and the temperature of the outer solar nebula is too cold to provoke their crystallization. Contrary to this conjecture, crystalline silicates are observed in disks around pre-main sequence stars and in cometary grains. It has been shown that, in a PSN, the turbulent diffusion is able to carry crystalline silicates, formed in the hot and inner regions of the disk, to the outer disk region within timescales of a few 10^4 y. Thus, the observations of crystalline silicates distributed over large distances from the central star (Chiang *et al.*, 2001) is accounted for the PSN model.

As a whole, judging from the remarkable predictive power of the PSN models, it can be concluded that the theoretical PSN should have some physical reality.

5.3. Consequences From PSN Models for Isotope Systematics in the Present Day Solar System Objects

Besides the constraints (5) and (6) (the D/H ratio and the crystalline silicates), little chemistry has been addressed through the PSN model. In cosmochemistry, the thermodynamical calculations accounting for the condensation of oxides and silicates at temperatures between 1000 and 2000 K are performed in a static nebula (Grossman and Larimer, 1974). Similarly, the molecular evolution of the disk is rarely performed (Prinn and Fegley, 1989) within the framework of the PSN model (see the pioneer approach of this issue by Gail, 1998). Thus judging from the possible isotopic preservation of interstellar solids in the PSN is qualitatively impossible. A complete computation of each isotopic system would have to be performed to answer to the following questions: As a function of time and heliocentric distances (1) for each specific mineralogical composition of the grains, what is the gas to grains ratio? (2) What is the temperature of the grains since the temperature dictates the isotopic exchange between the gas and the grains? (3) What is the evolution of the isotopic composition of the gas caused by the evaporation of interstellar grains? (4) How fast is the isotopic composition of the gas homogeneized in the PSN?

Qualitative answers to these questions are impossible to propose for the following reason. In the PSN model, the diffusion term depends upon the pressure and the surface density which is proportional to the parameter α, expressed in the Prandtl number. The diffusion acts as a "stirring force": the higher α, the higher the tendency of the solar nebula to be isotopically mixed. For example, for low α values ($\alpha < 10^{-6}$), the solar nebula is isotopically zoned: at the periphery of the disk where the temperature is low, no evaporation of the grains nor isotopic exchanges between gas and grains can take place, while evaporation and isotopic reactions are rapid in the hot and inner zones of the solar nebula. On the contrary, for high α values, the whole solar nebula has a tendency to be rapidly isotopically homogeneized and no grains can survive in such a nebula. Therefore only intermediate α values can preserve the observed isotopic heterogeneities of the solar system. Such an issue has not been quantitatively addressed and therefore it is difficult to decide theoretically which interstellar isotopic signatures have survived the formation of the solar system and in which objects they are likely to be found. In this respect, the isotopic anomalies found in meteorites and ascribed to a heritage of the ISM are interesting probes for testing the PSN models.

5.4. Limitations of the Standard PSN Models

The PSN model has several limitations, arising from very recent but also from well-known observations. Five are selected here, although many more are advocated in the literature.

1. First the observation of Fu Orionis stars. The possible relation between the observed different stages of star evolution and the initial conditions defined at $t = 0$ for the PSN model, has been discussed by Bell *et al.* (2000). Young stars show every 10^4-10^5 years, outbursts during which the luminosity emitted by their inner disks (not by the central star itself) increases by a factor of 100. These so-called "Fu-Orionis" objects have a life time of 10^5 y that may corresponds to the early epoch of the PSN. The PSN model may indeed cover

the two stages of the observed evolution of a star i.e. the Fu-Orionis followed by the T-Tauri stage. The thermodynamical parameters – and especially the surface temperature and density of the PSN – should be markedly different during the Fu-Orionis outbursts compared with those calculated in the framework of the classical PSN model. According to Bell *et al.* (2000), the midplane temperatures of the PSN are not affected by the outbursts whose sources are restricted within less than 0.3 AU. The situation is nevertheless different at the surface of the disk. For example, the classical PSN model predicts that water can exist as ice at the surface of the disk down to 3 AU while, when taking into account the luminosity associated with the outbursts, it is totally vaporized during the outbursts at distances up to 5 AU.

2. The Fu-Orionis is followed by the T-Tauri stage. As modeled by Shu *et al.* (1996) in the T-Tauri stage, during the first $\approx 10^5$ y and when close to the Sun (<0.02 AU), the matter drifting to the Sun is exposed to X-rays and nuclei at MeV-energies associated with solar type flares (André and Montmerle, 1994; for details see Section 5.4.1). In these regions the temperature is so high that pre-existing solids should evaporate and recondensed to form – according to Shu – chondrules and CAIs. These newly formed solids are launched above the mid-plane and water the surface of the nebula at distances up to 3 AU. This model has not been coupled with PSN models and the consequences of such an activity in the center of the disk have not been taken into account in PSN models. Consequently, the role that the turbulence may play on the distribution over the whole disk of the solids formed at such close distance from the Sun is unknown.

3. During the T-Tauri phase, X-rays water the disk and it seems likely that an interstellar type of chemistry can take place at the very surface of the solar nebula (Glassgold *et al.*, 2000). Since available PSN models are "closed", the contribution of the grains and molecules formed in these conditions cannot be calculated (see also a discussion of this problem in Lunine *et al.*, 1991). Intuitively these interstellar products – interstellar in the sense that they are formed by an interstellar type of chemistry but not in cold interstellar clouds (see Section 3.2) – should be reprocessed when settling in the neutral regions of the central plane of the disk. Without such a description it is difficult to estimate when and where interstellar products can survive in the PSN to be finally embedded in planetary objects.

4. At the end of the T-Tauri stage ($t > 10^6$ y), when giant planets such as Jupiter and Saturn are formed, it has been shown that a gap appears in the disk. This gap can be viewed as a ring where the surface density drops by several orders of magnitudes. The consequence of this gap on the turbulence (and temperature and pressure) has not been calculated in the framework of PSN models. Empirically it seems likely that the validity of the physical description of the PSN vanishes with the formation of such gaps.

5. In a pioneering study, Lewis *et al.* (1975) have discovered a special component of noble gases in "phase Q" of primitive meteorites. The elemental and isotopic composition of these "Q" noble gases strongly deviates from the mean solar composition. "Despite the many efforts, the nature of phase Q is still not known" (Busemann *et al.*, 2000).

These five examples show that the available models describing the PSN are too limited to provide reliable constraints on the behavior of the isotopic compositions of volatile elements when solar system objects were formed. Present PSN models may be rectified due to new observations on the physical conditions near young stars. For the purpose of chemical and isotopic modeling of the solar system, there is also the need to study some microscopic processes in the laboratory to understand how volatiles have been incorporated into solids of the PSN.

In the following, we mention recent approaches to explain volatile isotopic abundances in solar system samples that go beyond the considerations of present PSN models, but are not incorporated yet into these models. These are the models on (1) self-shielding (Clayton, 2003) in the X-region of the PSN postulated by Shu *et al.* (1996) and (2) a number of processes that could account for the isotopic composition of the noble gases in "phase Q".

5.4.1. *The X-model: Isotopic Volatile Fractionation Near the X-region?*

When the accretion disk has formed as part of the PSN, and when the infall from the parent molecular cloud has stopped, the Sun is assumed to be in its T Tauri phase (Figure 1d, e). Chemical reactions in the outer solar system may occur at low temperatures of 10–30 K (Charnley and Rodgers, 2002). Temperatures may also be high at the outer boundary of the solar nebula, if it is heated by a nearby giant star (Johnstone *et al.*, 1998). In the inner solar system, temperatures may be as high as 1000 K in the so-called X-region close to the Sun (Figure 5; Shu *et al.*, 1996). There, gas, dust, and precursor rocks of the accretion disk are exposed to X-rays and energetic particles from flares, and a fraction of this disk material is re-launched as an outflowing X-wind. Shu *et al.* (1996) suggest, that chondrites have been reprocessed near the X-region.

The strong magnetic flaring is assumed to be a consequence of the interaction of the magnetosphere of the early Sun with the magnetized accretion disk, from which material falls in at a rate of perhaps $10^{-8} M_{\odot}\ y^{-1}$. According to a survey with the Chandra observatory of the X-ray luminosity of 43 stars in the Orion nebula with ages between 0.3 and 10 My, the pre-main sequence Sun may have emitted 10^5 times more energetic protons than the contemporary Sun (Feigelson *et al.*, 2002). This supports the view of Shu *et al.* (1997) that the now extinct short-lived radio-activities ^{26}Al, ^{41}Ca, and ^{53}Mn observed in meteorites may have been produced inside the solar system. The recent observation of extinct ^{10}Be and the large variations in the ^{11}B/^{10}B ratio have reinforced this interpretation, since Be and B are only produced by spallation reactions and, contrary to the other extinct radionucleides such as ^{26}Al, ^{41}Ca and ^{53}Mn, cannot be produced in stars. Impulsive

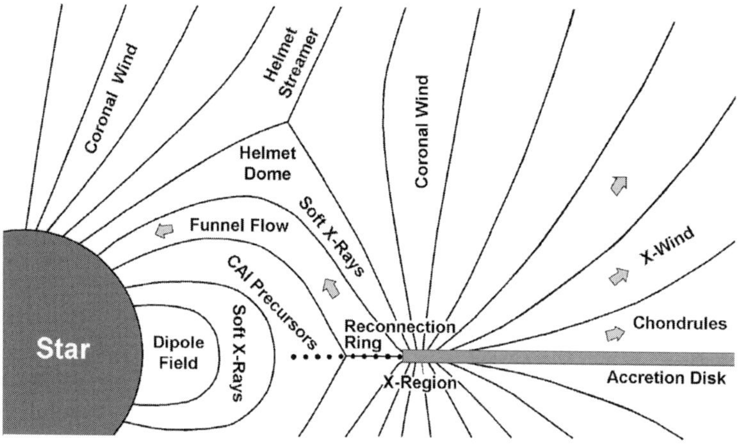

Figure 5. Illustration of the X-model (Shu *et al.*, 1997).

flares of the present Sun are known to cause isotopic fractionation in energetic particles (Kallenbach, 2003, and references therein), in particular they are very rich in ^{3}He, which can initiate spallation reactions in addition to the reactions with energetic protons (Shang *et al.*, 2000).

As the processes near the early Sun may have been quite violent, some kind of isotopic fractionation could have resulted. The strength of large-scale fractionation in the accretion disk by hydrodynamic escape from the X-region is quite small. For efficient fractionation by hydrodynamic escape to occur, either the critical mass m_c (Pepin, 2003) should not be much larger than the atomic mass m_2 of the isotope, for which fractionation is evaluated, or the disk material must circulate through the X-region very often. The critical mass is given by

$$m_c = m_1 + \frac{kTF_1}{bgX_1}, \tag{2}$$

where k = Boltzmann's constant, T = gas temperature, g = gravitational acceleration by the Sun at the X-point, X_1 is the mole fraction of hydrogen (which remains near unity throughout the escape episode since the hydrogen inventory dominates in a solar-composition gas mixture), F_1 is the outflow flux of hydrogen, m_1 the proton mass ($m_1 = m_p$), and b is the diffusion parameter (the product of diffusion coefficient and total number density) of a particular gas in hydrogen. Bochsler *et al.* (2001) have evaluated this process for expected physical conditions at the X-point and suggest that it may account for the ^{52}Cr/^{53}Cr gradient in solar system samples (Shukolyukov and Lugmair, 2000), but not for a substantial abundance variation of light purely volatile isotopes.

Recently, however, it has been suggested that the photodissociation by solar UV radiation of CO and N_2 molecules may have very different depths of self-shielding near the X-region when comparing molecules of different isotopic compositions. The accretion disk is optically thinner for the transitions of molecules with rare isotopes so that photodissociation is more efficient than for the molecules with the abundant isotopes. The process only disequilibrates isotopic compositions, if the dissociated atoms are trapped in some stable chemical compound before isotope exchange with the non-dissociated molecules occurs. This appears to be possible for O and N, but not for C (Clayton, 2002; Clayton, 2003; Becker *et al.*, 2003). The process had already been proposed by Navon and Wasserburg (1985), but only in the new scenario of the X-model by Shu *et al.* (1997), where CO and N_2 molecules are circulated through the X-region very near to the Sun, it may operate efficiently to cause the variations in O isotopic compositions in meteorites.

5.4.2. *Origin of the Noble Gases in "Phase Q" of Primitive Meteorites?*

Different classes of primitive as well as differentiated meteorites contain a component of noble gases that is strongly fractionated with respect to solar composition. This component is retained in a residue named "phase Q" (after "quintessence"; Lewis *et al.*, 1975) that remains after demineralization of the original bulk sample by HF and HCl. Phase Q only comprises 0.02% – 0.04% of the bulk meteorite but dominates the Kr and Xe inventory of most meteorites. Determining the chemical nature of the Q phase is still a pending question. The HF-HCl insoluble residues are essentially an assemblage of organic macromolecules (Gardinier *et al.*, 2000). Since the Q phase is released by a nitric acid treatment of the HF-HCl residues, it can be postulated (but this is in no way certain) that the noble gases are trapped in the organic network of the macromolecular structures which are oxidized by this nitric acid treatment. Laser microprobe analyses (Nakamura *et al.*, 1999) of phase Q in CM chondrites have revealed that phase Q also resides in the dust rims accreted on chondrules. Phase Q is well mixed with grains that were present in the solar nebula, such as nanodiamonds, interstellar grains containing Ne-E, and grains of solar composition.

It was originally proposed that the Q noble gases were incorporated into solids inside the pre-solar molecular cloud (e.g., Huss and Alexander, 1987), while the He isotopic composition indicates that He in phase Q is post-deuterium burning He and, thus, must have been incorporated in the solar nebula (e.g., Ozima *et al.*, 1998). Therefore, most efforts of studies on Q gases concentrate on trying to relate their composition to the solar composition.

The average fractionation of the 3 lightest Xe-Q isotopes (Busemann *et al.*, 2000) relative to solar wind Xe isotopes (Table I of Pepin, 2003) is ≈0.45%/amu, while the 4 heaviest Xe isotopes give ≈1.5%/amu. Pepin (2003) suggests that this fractionation pattern arises from a mass fractionation of ≈ 0.45%/amu and a higher proportion of Xe-H in Q-Xe compared to SW-Xe. The $^{20}Ne/^{22}Ne$ ratio varies between ≈ 10.1 and ≈ 10.7 in phase Q of different chondrites (Busemann *et al.*, 2000). The latter indicates a large deviation from the solar ratio $^{20}Ne/^{22}Ne \approx 13.8$ (Geiss *et al.*, 1972; Benkert *et al.*, 1993; Kallenbach *et al.*, 1997; Palma *et al.*, 2002). The depletion of He-Q relative to solar He is $\approx 10^{-7}$, while Xe-Q is about as abundant as solar Xe. The elemental fractionation among the Q noble gases is not linearly mass-dependent, particularly there is a discontinuity of two orders of magnitude between Ne and Ar (Hohenberg *et al.*, 2002; Figure 1 of Pepin, 2003).

Unlike C, N, and O, the noble gases neither participate in the gas-phase nor in the grain-surface chemistry, even at the low temperatures of dense molecular clouds or of the outer solar nebula. Terrestrial noble gases are believed to be enriched in their heavier isotopes by hydrodynamic escape from the Earth's atmosphere early in the history of the solar system (Hunten *et al.*, 1987).

In this volume, Pepin (2003) explores the capability of hydrodynamic escape to generate the Q noble gas isotopic (and elemental) composition from the solar isotopic composition. Calculated enrichments of ^{22}Ne over ^{20}Ne in the H outflow from the outer solar nebula region heated by UV radiation of a nearby giant star (Johnstone *et al.*, 1998) can be of order 10% under favorable (and probably unrealistic physical) conditions and, thus, still falls short by a factor 2 to account for the difference between the ^{20}Ne/^{22}Ne ratios in Ne-Q and the solar wind. Calculated Xe fractionations, moreover, are comparable to those in Ne and so do not replicate the Q-gas isotopic fractionation pattern noted above for Xe and Ne. If, as appears more likely, fractionating hydrodynamic outflow is established in the body of the nebula rather than just in its UV-heated outer boundary, fractionations are governed by Rayleigh distillation of the entire remaining nebula and are negligible (below 1‰) at the time when accretion disk erosion is halted by the gravitational potential of the young Sun embedded in the disk.

Pepin (2003) also explores the possibility that Q-gases could have been produced from solar gases during fractionating hydrodynamic blowoff of transient methane-rich atmospheres degassed from large primitive planetesimals accreted in an early mass-rich asteroid belt. The process can generate fractionated Xe, Kr, Ar and Ne compositions in the residual atmosphere that closely match observed meteoritic isotopic distributions, and Q-gas elemental ratios are approximated by adsorption of fractionated gases on planetesimal surface grains using plausible values of relative Henry Law constants. The variability of the Ne-Q isotope ratios (Busemann *et al.*, 2000) can be viewed as a consequence of mixing between the actual Ne-Q component generated by escape fractionation (with ^{20}Ne/^{22}Ne ≈ 9.4) and solar Ne.

However, while relatively successful in replicating Q-gas isotopic distributions, it is not clear that the blowoff model can readily accommodate additional requirements for adsorption of sufficient absolute amounts of Q-gases on carrier grains, ejection to space and mixing in the nebula without excessive dilution by nebular dust from other sources, and dispersal into meteorite bodies and into the chondrule rims found to contain Q-gases by Nakamura *et al.* (1999). In this latter context, it is noted that the origin of chondrules is heavily debated. Most authors (e.g., Desch and Connolly, 2002) presume that chondrules are inclusions that resulted from transient melting by a shock wave or lightning inside the solar accretion disk. Before they have been incorporated into the primary matrix of meteorites they acquired some dust from the solar nebula environment. Opposing this, Shu *et al.*

(2001) propose in the frame of their X-model (Section 5.4.1) on the very early solar system that chondrules have been formed by thermal processing very close to the Sun. The two models may give very different sites and conditions for the incorporation of Q noble gases into the rims of chondrules.

Other possible processes that may bear on production of Q-gases, such as 'physisorption' within a 'labyrinth of micropores at or near the surfaces of amorphous carbon' (Wacker *et al.*, 1985), have been simulated in the laboratory. Hohenberg *et al.* (2002) have demonstrated the mechanism of active capture and anomalous adsorption of noble gases during deposition of a low-Z metal on a substrate. When new surface sites are created continuously in vacuum, they are occupied by heavy noble gases before they can be preempted by chemically more active species. At such surface sites, noble gases are not only trapped by van der Waals forces, but also by chemical forces. The creation of dangling bonds at trapping sites of organic materials deposited in layers inside the pre-solar molecular cloud or in the solar nebula is promoted by ionizing radiation, for instance by UV photons or cosmic rays (Sandford *et al.*, 1998). The anomalous adsorption explains the preferential trapping of heavy noble gases, however, it predicts stronger isotopic fractionation for Ar and Kr than is observed in Q gases.

Verchovsky *et al.* (2002) propose that the noble gases were implanted into phase Q by mechanisms similar to implantation of primordial noble gases into nanodiamonds. This way, Q noble gases may be derived from solar gases by mass fractionation during implantation (Mewaldt *et al.*, 2001). However, the source from which the implanted noble gases had originated remains unclear, as phase Q was shielded from irradiation by solar particles (Nakamura *et al.*, 1999).

6. Isotopic Fractionation in the Early Solar System

Multi-wavelength studies of solar-like G-type stars at several stages of their main sequence evolution investigated under the "Sun in time" program indicate that our early Sun may have undergone a highly active phase in its particle and radiation environment 3.5–4.5 Gy ago. An overview of that problem can be found e.g. in Guinan and Ribas (2002). The "Sun in Time" program is a comprehensive project to study the magnetic evolution of the Sun using a sample of single nearby G0-V stars that have known rotation periods and well-determined physical properties and ages. The sample comprises most of the Sun's main sequence lifetime from 130 My up to 8.5 Gy. To study the evolution of the Sun's high-energy emissions, a large number of multi-wavelength (X-ray, EUV, FUV, UV, optical) observations has been collected, directed towards achieving a full description of the spectral irradiance of the targets. Full spectral irradiance tables covering from 1 Å (12 keV) to 3300 Å have already been completed for five of the stars in the sample: EK Dra (130 My), π^1 UMa (300 My), κ^1 Cet (750 My), β Com (1.6 Gy), and β Hyi (6.7 Gy). An excellent correlation between the emitted flux and stellar age is clearly observed.

The study of these solar proxies shows that the coronal X-ray-EUV emissions of the young main-sequence Sun (similar to EK Dra) were over 1000 times stronger than those of the present Sun (similar to β Hyi). The data show that while the NUV/optical flux is similar for all stars, very large differences exist in the high-energy part. The high frequency of large flares observed with EUVE by Audard *et al.* (1999) in solar proxies such as EK Dra and 47 Cas β could indicate explosive episodic releases of plasma. Wood *et al.* (2002) found that the mass loss rate per unit surface area of solar-like stars is correlated with the X-ray surface flux. Recent observations by the Hubble Space Telescope of absorption from hot H I in high-resolution Lyman-α spectra of solar-like stars resulting from collisions between stellar winds and the local interstellar medium (Wood *et al.*, 2002) can be used as a diagnostic for the stellar mass loss rate of the stars studied. By using these data one finds that the early solar wind may have been 1000 to 10000 times denser at 1.0 AU during the time when the Sun reached the main-sequence.

According to Wood *et al.* (2002), the mass loss rate of young solar-like stars scales as $\dot{M} \propto t^{-2.00\pm0.52}$ (Wood *et al.*, 2002). Based on the Sun's present-day mass loss rate of $\dot{M}_\odot = 2 \times 10^{-14} M_\odot\ \mathrm{y}^{-1}$ (e.g., Feldman *et al.*, 1977) by the solar wind, the mass loss may have been as high as $2 \times 10^{-6} M_\odot$/My or more during the first 500 My of solar system history.

However, from measurements of noble gas isotopes in lunar samples, Geiss and Bochsler (1991) concluded that the solar wind flux over the past $\approx$3 Gy was not stronger than ten times the present flux. Recently, von Steiger (2002) suggested that the record in lunar regolith grains indicates that the solar wind must have been a continuous and ubiquitous phenomenon since 4 Gy ago (von Steiger, 2002).

Sackman and Boothroyd (2002) have developed a high-precision evolutionary model of the Sun that complements the standard model by Iben (1965) (Section 7). They point out that the change of the solar mass loss over time could be constraint by helioseismology data for the present Sun. In their simulations they vary the initial mass of the Sun at 4.57 Gy ago from 1.01 to 1.07 $M_\odot$ and assume different temporal changes of the mass loss, e.g., an exponential decrease. The authors find that the sound speed in the solar interior is influenced by typically 10^{-3} or less at various depths. However, present helioseismology data are only marginally significant to identify the initial mass of the Sun and the temporal change of the solar wind flux.

6.1. Planetary Atmospheres in a Strong Solar Wind

The dissociative recombination from the radiation, ion pick-up and sputtering processes should have been important to the efficient erosion of atmospheric constituents during this early solar system period. A higher activity of the Sun during the first 500 My of solar system history could account for the enhancement in $^{15}N/^{14}N$ from the terrestrial ratio of ≈ 0.00367 or even from the protosolar ratio of ≈ 0.0023 (Owen *et al.*, 2001) to the ratio $^{15}N/^{14}N \approx 0.017$ observed on Saturn's major satellite Titan (Lammer and Bauer, 2003) by the combined processes of atmospheric sputtering and pick-up loss due to a strong solar wind.

Solar wind erosion is also a candidate process to account for changes in the atmospheres of Mars and Venus. D/H isotope anomalies imply that both planets may have lost a large amount of water to space since the end of hydrodynamic escape. Within at most 700 My after the Sun's birth, the magnetic dipole of Mars presumably ceased (Acuña *et al.*, 1999). At that time, Mars may have been exposed to a 40 times stronger solar wind. This may explain water losses on Venus and Mars, corresponding to about 2% of a terrestrial ocean on Venus and a global H_2O ocean with a depth of $\approx$ 30 m on Mars (Lammer *et al.*, 2000; Wood *et al.*, 2002).

6.2. The "SEP" Component

If the early Sun has emitted 10^5 times more energetic particles (Feigelson *et al.*, 2002) and a 1000 times stronger solar wind, one may consider that the ratio of solar energetic particle (SEP) fluence to solar wind ion fluence might have been 100 times larger than at present day. It remains open, how long such a situation might have persisted. If it did, this may shed new light on the problem of the observed "SEP" component of noble gases in lunar soils (e.g., Benkert *et al.*, 1993) and in interplanetary dust particles (Pepin *et al.*, 2000). Present-day SEP fluences fall short by one to two orders of magnitude to account for the abundance ratio of the particles implanted from the solar wind and from SEPs (Wieler *et al.*, 1986). This is confirmed on the basis of recent ACE data (Mewaldt *et al.*, 2001). On the other hand, Mewaldt *et al.* (2001) reconcile the isotopic fractionation of the "SEP" component relative to solar wind particles by implantation effects. The "SEP" component in IDPs, which may have been irradiated very recently, is typically two orders of magnitude less abundant relative to implanted solar wind (SW) particles, when compared to the typical SEP/SW abundance ratios in lunar soil samples. One may explain this discrepancy with the higher ratio of SEP/SW fluence in the past, when the lunar soil was irradiated only in the early history of the solar system. However, there is a problem with sample 71501, which has been irradiated within the past 100 My (Benkert *et al.*, 1993). At that time, the SEP/SW fluence ratio must have been close to the present-day ratio.

7. Isotopic Fractionation in the Sun and at the Solar Surface

We base our discussion on nuclear processes and mixing inside the Sun and at the solar surface on the original work of Iben (1965). In this scenario, a star of solar mass gravitationally contracts on the classical Kelvin-Helmholtz timescale. For a few ten thousand years the Sun contracts with its luminosity decreasing from $\approx 1000\,L_\odot$ to $\approx 10\,L_\odot$ and surface temperature slightly increasing from 3000 to 4000 °K. Then, the nuclear reaction $D + H \rightarrow {}^3$ He sets in, depleting the Sun's core of D after $\approx 10^5$ y. The Sun remains fairly cool and convective for the first few million years, so that actually the whole Sun is depleted of D. The contraction ceases after approximately ten million years, when the luminosity is about $0.5\,L_\odot$. Then, the core has become hot enough to replenish the solar luminosity with the thermonuclear power generated by the fusion reaction of hydrogen to helium (Ezer and Cameron, 1966). A radiative zone develops around the central core and grows monotonically until the energy flux is controlled by the interior opacity rather than the surface opacity. Within about 30 million years, the temperature of the core, where the thermonuclear reactions take place, reaches about 15 MK, and the core density levels off at 1.6×10^5 kg m^{-3}.

In the standard scenario of Iben (1965), the luminosity and the surface temperature rise together during the first 30 My of the Sun's life until the main sequence branch in the Hertzsprung-Russell diagram is reached. During the following 4.5 Gy until the present day, the solar luminosity rises slowly, and the surface temperature approaches the present-day photospheric temperature of $\approx$ 5800 K. While the scenario described by Iben (1965) appears to be sufficient to study the nuclear processes that occured over the Sun's lifetime, it does not properly predict the observed luminosities of young solar-like stars observed in the frame of the "Sun in Time" project (see the previous Section 6.1). In this section, we disregard these open questions.

7.1. Nucleosynthesis in the Sun's Core

Two processes change isotopic abundance ratios in the core of the Sun, the so-called pp-chain reactions (Ezer and Cameron, 1966) and the CNO-bicycle (Table III). The pp-chain reactions basically result in the conversion of H to He ($4\text{H} \rightarrow {}^4\text{He} + 2e^+ + 2\nu$), while the CNO-bicycle modifies the abundance of C and N isotopes. The ^{17}O/^{16}O ratio is not influenced much in the core of the Sun at its temperature of $\approx$ 15 MK (Caughlan and Fowler, 1962). The ^{18}O/^{16}O ratio remains fairly unchanged as well. Energy production by the pp-chain is still somewhat larger than that of the CNO-bicycle. In hotter stellar cores, the energy production by the CNO-bicycle dominates, and the O isotopic abundances are affected.

7.2. Mixing Inside the Sun

There is quite significant impact on volatile isotope abundances in the Sun's core as a consequence of the nuclear processes denoted pp-chain and CNO bi-cycle (Table III). Bochsler and Geiss (1973) have modeled the abundances of the H, D, ^{3}He, ^{4}He, ^{7}Li, ^{9}Be, ^{10}B, ^{11}B, ^{12}C, ^{13}C, ^{14}N, and ^{15}N isotopes in a non-mixing Sun as a function of radial distance to the solar center (Figure 6, *left*; the solar wind value of 4×10^{-4} was adopted for the ^{3}He/^{4}He ratio at the solar surface). Without mixing of material to the Outer Convective Zone from deeper regions of the Sun, these nuclear processes do not change solar surface (photospheric) abundances.

TABLE III

Thermonuclear processes in the Sun's core. The reaction rates between nuclei species 1 and 2 are calculated as $r_{12} = 3.6 \times 10^{-19} n_1 n_2 \xi^2 \exp(-\xi) S_0/(A Z_1 Z_2)$, where $\xi = 42.8(Z_1^2 Z_2^2 A/T_6)^{1/3}$, T_6 is temperature in MK, Z_1 and Z_2 number of protons in the nuclei, $A = A_1 A_2/(A_1 + A_2)$ with A_1 and A_2 atomic mass numbers, and n_1 and n_2 densities in g cm^{-3}.

Reaction	S_0 [keV barn]	ξ	τ [y]	Reaction	S_0 [keV barn]	ξ	τ [y]
pp-chain I				*CNO-cycle I*			
^{1}H(p, $\beta^+\nu$)^{2}H	3.78×10^{-22}	13.7	1.2×10^{10}	^{12}C(p, γ)^{13}N	1.4	55.4	1.4×10^6
^{2}H(p, γ)^{3}He	2.5×10^{-4}	15.1	4×10^{-8}	^{13}N(e$^+$ ν)^{13}C			2.8×10^{-5}
^{3}He(^{3}He, 2p)^{4}He	5.0×10^3	49.8	2×10^5	^{13}C(p, γ)^{14}N	5.5	55.5	4.1×10^5
^{3}He(α, γ)^{7}Be	4.7×10^{-1}	52.0	6×10^5	^{14}N(p, γ)^{15}O	2.7	61.6	3.6×10^8
pp-chain II				^{15}O(e$^+$ ν)^{15}N			5.5×10^{-6}
^{7}Be(e$^-$, ν)^{7}Li				^{15}N(p, α)^{12}C	5.3×10^4	61.7	2.0×10^4
^{7}Li(p, α)^{4}He	1.2×10^2	34.4	1×10^{-5}	*CNO-cycle II*			
pp-chain III				^{15}N(p, γ)^{16}O	2.7×10^1	61.7	3.9×10^7
^{7}Be(p, γ)^{8}B	4.0×10^{-2}	41.6	5×10^1	^{16}O(p, γ)^{17}F	1.0×10^1	67.5	3.4×10^{10}
^{8}B(β^+ γ)^{8}Be			4×10^{-8}	^{17}F(e$^+$ ν)^{17}O			3.0×10^{-6}
^{8}Be(α)^{4}He			1×10^{-23}	^{17}O(p, α)^{14}N	resonant	67.6	2.8×10^{10}

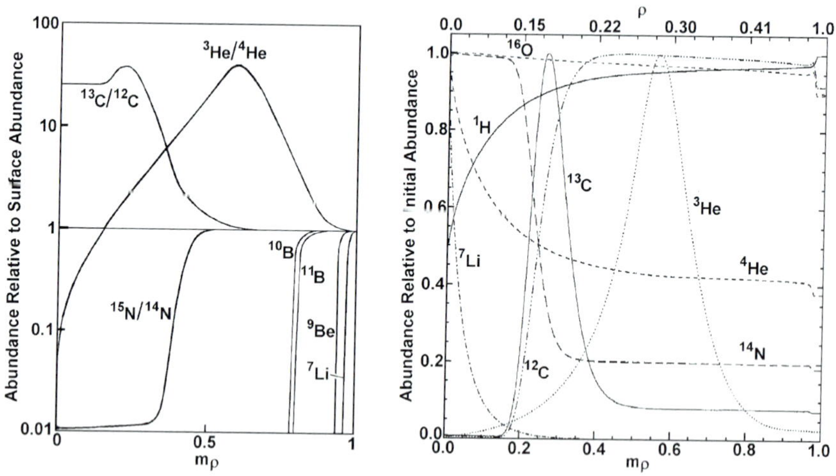

Figure 6. Left: Present abundances and abundance ratios of light nuclei in a non-mixing Sun relative to the respective unchanged surface values (model calculations by Bochsler and Geiss, 1973, as a function of fractional mass $m_\rho = M(\rho)/M_\odot$ with $\rho = r/R_\odot$ the distance from the solar center in units of the solar radius $R_\odot$). *Right:* Present-day abundances in the Sun according to a recent model (Turck-Chièze *et al.*, 2001) that reproduces the results from helioseismology within 0.2% precision. Mixing is assumed to be strong in the Sun's convective zone at $r > 0.713 R_\odot$ (Basu *et al.*, 1997).

The abundances of the isotopes of H, He, Li and Be measured in the solar wind or at the solar surface constrain the degree of secular mixing inside the Sun. From isotope measurements and from solar seismic data it was concluded (Vauclair, 1998; Wieler and Heber, 2003; Geiss and Gloeckler, 2003) that addition of ^{3}He produced by incomplete hydrogen burning at intermediate solar depth has indeed been limited: During the past several Gy ^{3}He in the OCZ was probably not increased by more than a few percent.

The helioseismology experiments on board SOHO gave the newest insights into the turbulence and mixing processes inside the Sun. Turck-Chièze *et al.* (2001) conclude that internal rotation plays an important role for mixing in the Sun. This in particular applies for the interface ('tachocline') region between the radiative zone and the base of the convection zone. On the other hand, turbulent mixing appears not to be important in the region of maximum ^{3}He abundance (as a consequence of incomplete H burning) and in the very central part of the Sun. The latter result relies on observations of acoustic modes, whereas instabilities in the core of the Sun generated by g modes are not accessible by SOHO. Based on the present knowledge on intersolar mixing, Turck-Chièze *et al.* (2001) have modeled the extent to which the abundances of some volatile isotopes inside the Sun have been changed relative to their initial abundances 4.6 Gy ago (Figure 6, *right*). This may have some minor impact on the photospheric isotopic ratios of C and N.

7.3. Gravitational Settling at the Bottom of the Convection Zone

Particularly interesting in the discussion of intersolar dynamics is the process of gravitational settling at the bottom of the convection zone. Particles settle into the radiative zone because the gravitational acceleration g is larger than the radiative acceleration g_{Rad}. This leads to a drift speed of the ions towards the solar center (Michaud and Vauclair, 2001):

$$v_{\mathrm{D}} = -D_{\mathrm{int}}\left(\frac{\partial \ln c}{\partial r} + \left[\left(A - \frac{Z}{2} - \frac{1}{2}\right) g - A g_{\mathrm{Rad}}\right]\frac{m_p}{kT} - D_T \frac{\partial \ln T}{\partial r}\right), \quad (3)$$

where D_{int} is the interdiffusion coefficient of the species with atomic mass A and nuclear charge Z, c is its concentration, and D_T is the thermal diffusion coefficient. Neglecting the thermal and concentration gradients, Bochsler (2000) has estimated the isotopic fractionation of material in the Outer Convective Zone due to gravitational settling over the lifetime of the Sun to be $f_{4,3} = 0.967$, enriching ^{3}He over ^{4}He, and $f_{12,13} \approx 1.01$, enriching ^{12}C over ^{13}C. For the volatiles N, O, Ne, and Ar, the fractionation factors are between about 0.9%/amu and 0.2%/amu. The effect is not purely mass-dependent, but atomic mass plays an important role. Future high-precision analysis of Genesis data on solar wind isotopic abundances should take into account these intersolar processes as well as possible isotopic fractionation in the solar wind (Kallenbach, 2003).

7.4. Nuclear Processes at the Solar Surface

There is observational evidence that at least two nuclei are synthesized in the solar atmosphere: 1) Deuterium is created in solar flares by the reaction $^{1}H(n,\gamma = 2.223\ MeV)D$ (Mullan and Linsky, 1998), and 2) the isotope ^{6}Li is created due to spallation of ^{16}O and ^{12}C by energetic protons (Chaussidon and Robert, 1999). Geiss and Bochsler (1982) have estimated the efficiency of these spallation reactions for the production of ^{15}N. They conclude that there should not be a significant contribution of spallation ^{15}N to ^{15}N in the solar wind, and therefore no secular variation of the solar wind $^{15}N/^{14}N$ ratio.

7.4.1. *Deuterium Synthesis in the Solar Atmosphere*

Evidence for the synthesis of D in the solar atmosphere is provided by the 2.223 MeV γ-ray line (Chupp, 1984). Protons that are accelerated in flares to energies of $E_p \geq 6.3$ MeV (Lingenfelter and Ramaty, 1967) knock out neutrons from ambient nuclei. These free neutrons thermalize within seconds by scattering off ambient protons and then, within minutes, undergo radiative capture to form D. Most of the scattering occurs in the photosphere. Some neutrons are scattered above the photosphere and detected near Earth (Chupp *et al.*, 1982; Evenson *et al.*, 1983). Non-thermal spallation reactions (Burbidge *et al.*, 1957) have been ruled out by Geiss (1993) to be important for the production of the light nuclei D and ^{3}He inside the solar system, while they are a major source of the isotopes of the light elements Li, Be, and B in the interstellar medium. Observations by Severny (1957) in the spectral range of the Balmer-α line had constrained the D/H ratio in sunspots to be about 10^{-4}. These observations do not provide information on the photosphere, where more D may be synthesized in flares. However, considering the small contribution on the order of at most 10^{-3} of flares to the present particle flux from the Sun to interplanetary space, there appears to be little impact on the solar system D/H ratio.

Hoyle and Fowler (1973) give an upper limit of $10^{-2} - 10^{-1}$ for the D/H ratio in regions of magnetic activity at the surfaces of young T Tauri stars. A very active young Sun may have produced more D than the present Sun.

7.4.2. *^{6}Li Production by Spallation of ^{16}O and ^{12}C*

Recently, Chaussidon and Robert (1999) have found evidence for spallation reactions in the solar atmosphere, significantly influencing the solar wind $^{7}Li/^{6}Li$ ratio. In the photosphere, the elemental abundance of Li is a factor of 140 lower than in meteorites, which is assumed to represent the solar system value (Anders and Grevesse, 1989; Reeves, 1993). This difference has been attributed to the destruction of Li at the base of the convection zone of the Sun (Herzog, 1965). This nuclear destruction would yield a ratio $^{7}Li/^{6}Li$ of $\approx 10^{6}$ as ^{6}Li is destroyed much more easily at a lower temperature than ^{7}Li. However, the meteoritic $^{7}Li/^{6}Li$ is 12.14 (Chaussidon and Robert, 1998), whereas $^{7}Li/^{6}Li = 31 \pm 4$ in the solar wind. The low solar wind ratio indicates spallation of ^{16}O and ^{12}C in the solar atmosphere due to energetic protons from solar flares. Assuming an energetic proton flux of 3.2×10^6 cm^{-2}, Li ions appear to accumulate for at least 100 years in the solar atmosphere.

7.4.3. *^{15}N Production by Spallation of ^{16}O and ^{12}C?*

If ^{15}N isotopes accumulated in the solar atmosphere for the whole lifetime of the Sun, the energetic proton flux would be sufficient to significantly increase the solar wind $^{15}N/^{14}N$ ratio over its protosolar value (Owen *et al.*, 2001). Geiss and Bochsler (1982) have studied this problem and suggest that this is not the case.

8. Inflow of Interstellar Matter (Gas, Dust, Micrometeorites)

The Sun moves through space with $\approx$ 16.5 parsec/My (e.g., Frisch, 2002). Consequently, the Galactic environment of the Sun has ever changed. Talbot and Newman (1977) have estimated that the Sun probably has passed through 135 interstellar molecular clouds with densities larger than 100 cm^{-3} and may have had 50 encounters with clouds that are dense enough to switch off the solar wind. Very dense ($> 10^5$ cm^{-3}) structures observed in the interstellar gas have sizes of about 1 AU (2×10^{-5} parsec) and may cause changes in the structure of the heliosphere on time scales of less than ten years. At molecular cloud densities below n = 100 cm^{-3} the number of encounters presumably was $N_e(\geq n) \approx 10\,(10^{-3}\text{cm}^3\,n)^{-0.9}$. Even at interstellar densities between 1 and 10 cm^{-3} the termination shock of the solar wind may come as close to the Sun as the orbit of Jupiter or even Earth (Scherer *et al.*, 2002).

At present, the heliosphere (Figure 7, *right*) has a size of 100 AU given by the pressure balance between the solar wind and the Local Interstellar Cloud (LIC) that has a neutral hydrogen density of $n_{\rm H} \approx 0.1$ cm^{-3}, a temperature of $\approx$ 7000 K, a degree of ionisation $n_p/n_{\rm H}$ of 0.5, a speed of 26 km/s with respect to the Sun, and an interstellar magnetic field of $\approx 1.4\,\mu$G with unknown orientation (Gloeckler *et al.*, 1997).

8.1. Neutral Gas Flow from the Interstellar Medium

The accretion of interstellar matter of the Sun during its lifetime was probably on the order of $10^{-4} M_\odot$, but there is still a 5% probability that the Sun has accreted up to $10^{-2} M_\odot$ (Talbot and Newman, 1977). The mass of the Outer Convective Zone (OCZ) of the Sun is about 10^{-2} $M_\odot$ (Geiss and Bochsler, 1982), so that some probability is given that photospheric isotopic ratios may have changed to some small degree, if the interstellar isotopic ratios have deviated significantly from the protosolar ratios. According to Geiss and Gloeckler (2003), many local interstellar isotopic ratios (e.g. $^3He/^4He$, $^{22}Ne/^{20}Ne$, $^{18}O/^{16}O$) have not changed as

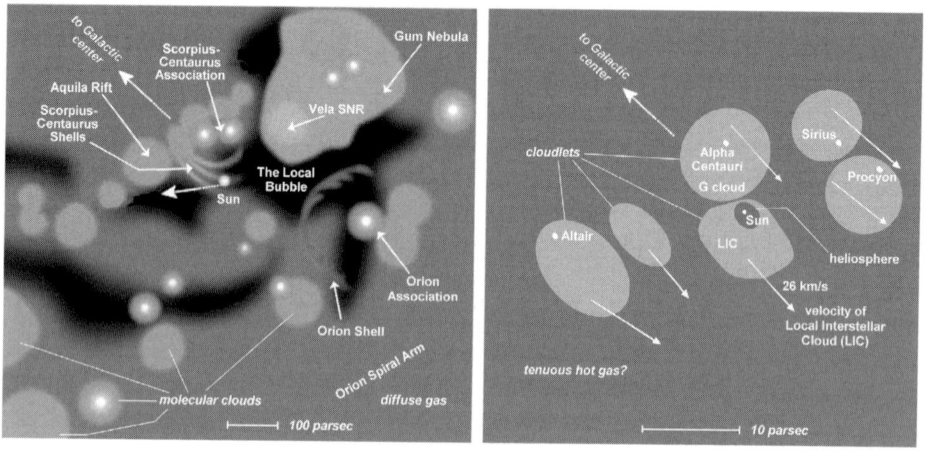

Figure 7. Journey of the Sun through the interstellar medium as approximately seen in the Galactic plane. From Frisch (2002) and Lallement (1996).

much as expected from the standard GCE (Scully *et al.*, 1996) because of infall of matter that carries the compositional signatures of dwarf galaxies with a high primary/secondary ratio.

8.2. Is the "SEP" Component of Interstellar Origin?

Although their fluence is lower than that of the bulk neutral interstellar wind, the heliospheric energetic particles (Wimmer-Schweingruber and Bochsler, 2001), ionized by solar UV radiation, picked-up by the solar wind and accelerated in turbulence and at shocks in interplanetary space, may contribute significantly to the budget of particles implanted into geologically inert planetary surfaces. Benkert *et al.* (1993) have found that "SEP" particles implanted into lunar soil at energies > 100 keV/amu have a characteristic isotopic composition and contribute a quantity of as much as 20 – 50% of the particles identified as implanted solar ions. Relative to the solar wind component, Pepin *et al.* (2000) observe about two orders of magnitude less "SEP" particles in IDPs that probably were irradiated very recently, perhaps within the past few My.

Wimmer-Schweingruber and Bochsler (2001) can reconcile the "SEP" abundance in lunar surface samples, if 1) the density of the interstellar medium was in average 1 cm^{-3} during the solar system's history, 2) the mass distribution function of the molecular clouds approximately scaled with m_{MC}^{-2}, and 3) if the pick-up ion density in the heliosphere scales with n_{MC}^2. The youngest lunar sample containing a high "SEP" abundance is ilmenite 71501 (Benkert *et al.*, 1993), which was probably irradiated within the past 100 My after it came to the lunar surface by the impact creating the Tycho crater. During the past 250'000 years, the Sun moved in the present Local Interstellar Cloud. Before, it has passed the hot Local Bubble that has basically no neutral component (Breitschwerdt, 1996). Its size is of order

200 parsec, so that the solar system may have passed it for about 10 My. If the journey of the Sun could be reconstructed for the past 100 My, this may give strong constraints on the origin of the "SEP" component in lunar soils.

8.3. Inflow of Dust and Micrometeorites

Interstellar dust grains are continuously deposited on planetary surfaces in the solar system. Those surfaces, which are geologically inert, will provide a record of contractions and expansions of the heliosphere or compositional variations in the interstellar dust due to a changing stellar environment (Frisch, 2000). Frisch (1995) suggests that the interstellar cloud complex presently flowing past the Sun is composed of molecular clouds disrupted by star formation in the Scorpius-Centaurus Association. This picture is consistent with the observed asymmetry in the micrometeorite flux. The regions of highest nearby column densities are found in the southern hemisphere in both ecliptic and Galactic coordinates. Micrometeorites may also originate in the atmospheres of evolved stars, since AGB stars are a formation site for pre-solar grains, but this would not explain the spatial asymmetry. De Geus (1992) suggests that type II supernovae are the source of micrometeorites. There were $\approx$ 10 supernova explosions in the 15 My history of the Scorpius-Centaurus Association. Large dust particles ($> 10^{-9}$ g) may originate several hundred parsecs away, because they decouple from the interstellar magnetic field.

Outer planets, comets, and Kuiper Belt objects are more likely to be exposed to pristine interstellar material than inner planets. When the interstellar clouds are denser than at present, the surfaces of outer objects are exposed to the full flux of interstellar dust grains. The Moon provides a record of encounters with interstellar clouds containing large (radii $> 0.1 \mu$m) grains. The dust cloud near the Sun, the zodiacal cloud, gives an example of dust dynamics in a cosmic plasma and radiation environment. It predominantly contains the second generation dust grains following the first generation of interstellar grains that have formed the solar system 4.6 Gy ago. These second generation dust grains are a consequence of the activity of comets and the collisional evolution of asteroids, meteorites and possibly Kuiper belt objects. Recent in situ measurements have shown the existence of interstellar dust in addition to the dust "classically" expected in the zodiacal cloud (Mann, 2001).

About 50% of IDPs in the 10-100 μm size range come from the asteroid belt and $\approx$10% of the asteroidal dust come from 3 known asteroid collision families (Dermott *et al.*, 1996). As shown from IDP ion probe studies on the one hand (Aléon *et al.*, 2001; Messenger, 2000), and from scale height of dust near the Earth on the other hand, some of the local dust is cometary in origin. A significant contribution of cometary matter is predicted from simple mass balance considerations. For example, comet P/Halley may contribute of the order of 1% of the zodiacal dust during its maximum life of 10^5 y. Estimates of the Kuiper Belt parent body population suggest that the surface density of KP dust produced by collisional event may be comparable to that of the zodiacal cloud present in the inner solar system (Backman *et al.*, 1995), depending on the efficiency of contribution of KP dust in the inner solar system. Due to electromagnetic shielding by the heliosphere, interstellar dust having a charge-to-mass ratio small enough might enter the inner solar system limiting the ISM contribution to particles $> 0.1\ \mu$m (Mann and Kimura, 2000), and theoretical considerations suggest that the fraction of dust of interstellar origin must be small. Measurements during the Ulysses experiment have detected ISP dust coming from the direction of the solar system motion in the galaxy (Grün *et al.*, 1994), and ion probe analysis of IDPs have recently shown the existence of interstellar silicates in IDPs (Messenger *et al.*, 2002). In situ measurements have allowed to establish the surface density of ISP dust to 0.1% in the inner solar system at 1 AU (Grogan *et al.*, 1996). Outside the heliosphere, Oort and Kuiper belt dust populations probably contain a larger fraction of interstellar dust, and fluctuation of the size of the heliosphere in the past is likely to have modulated to a large extent the flux of this component, although little is known at present on such variations.

Surprisingly, the total amount of matter of asteroidal origin in the zodiacal cloud is very small ($10^{18} - 10^{19}$ g) when compared to planetary masses as it would correspond to a single solid body of only 5 – 10 km in radius. Grains large enough not to be expelled from the solar system by the radiation pressure spiral towards the Sun under the influence of the Poynting-Robertson (PR) drag, with lifetime typical of $\approx 10^5$ y for particles of few tenths of mm in the inner solar system (Burns *et al.*, 1979). Dust mutual collision lifetime is longer ($10^7 - 10^8$ y) so that the PR drag controls mainly the lifetime of dust in the inner solar system. However, the time necessary to clear off completely dust after a collisional event between km-sized body is much longer than the PR drag lifetime because debris resulting from the collision will continue to collide and produce dust for long. Collisions of 10 km-sized bodies occur in the asteroidal belt every $\approx 10^7$ y, and complete dispersal of debris would take approximately three time this 10^7 y interval. Hence one should expect long-term ($10^7 - 10^8$ y) fluctuation of the flux of asteroidal dust in the inner solar system. Recently, Hashizume *et al.* (2002) (see also Marty *et al.*, 2003) have proposed that the flux of micrometeorites on the Lunar surface might have increased significantly during the last 0.4 Gy.

Since about $\approx$4.0 Gy, the flux of IDPs is likely to have been of limited impact on the terrestrial (and Venusian) atmosphere and hydrosphere. The isotopic composition of paleoatmospheric nitrogen recorded in Archean and Proterozoic sediments has been nearly constant since this 3.5 Gy (Beaumont and Robert, 1999; Pinti, 2002), whereas contribution of IDPs which are enriched in ^{15}N (Marty *et al.*, 2003) should have affected profoundly this composition if such contribution would have been significant. Likewise, most of IDPs present ^{20}Ne/^{22}Ne ratios in the range 11 – 14 (Pepin *et al.*, 2000) higher than the atmospheric value of 9.8, precluding single derivation of atmospheric Ne from such a source. From a mass balance of N contents in IDPs including micrometeorites ($\approx 10^3$ ppm) and taking a present-day IDP mass flux of $\approx$40,000 tons y^{-1} (Love and Brownlee, 1993), the IDP contribution of nitrogen to air over 4.5 Gy at constant flux would be only 5×10^{-5} times the atmospheric abundance. The situation might have been different during the Hadean since the original asteroid belt mass might have been $\approx$1000 times larger than the present-day mass of $5 \times 10^{-4} M_{\oplus}$, resulting in a zodiacal cloud 10^6 times denser than the present-day one. It is interesting to note that a contribution of about $0.3 M_{\oplus}$ of chondritic-type material is required after core differentiation to supply the amount of platinum group elements to the terrestrial mantle (Righter and Drake, 1997), and such contribution could have supplied the amount of water at the Earth's surface with the adequate D/H ratio (Dauphas *et al.*, 2000). Whether such volatile contribution was supplied by micrometeorites, or by volatile-rich bodies ejected from regions colder than the inner solar system (Morbidelli *et al.*, 2000) will deserve further studies on fluxes and detailed volatile compositions of IDPs.

8.4. Recent Pollution by Supernova Material

During the many billion years since the formation of the Galaxy until the formation of the Sun, innumerable stars, in particular millions of supernovae, have contributed to the supply of nuclei to the solar system. Only for nuclei locked-up in pre-solar grains, for which an age can be determined, or for radioactive isotopes (e.g. ^{26}Al, ^{36}Cl, ^{41}Ca, ^{44}Ti, ^{53}Mn, ^{56,57}Co, ^{60}Fe, ^{129}I, ^{146}Sm, ^{182}Hf, and ^{244}Pu) it is possible to estimate the time of their nucleosynthesis and the type of star, in which they have been produced.

In primitive meteorites, isotopic effects in the daughter nuclei of short-lived now extinct radioactivities evidence the last direct contribution (pollution) of supernova material to solar system matter shortly before and after the formation of the solar system (Meyer and Clayton, 2000). The observed strong excesses of ^{3}He in IDPs from several stratospheric collectors (Pepin *et al.*, 2000) also may indicate a late external source of nuclides contributing to solar system matter. These excesses are by a factor of up to 40 above the levels expected for implantation of solar corpuscular radiation, and model calculations on the plausible free-space or regolith burial lifetimes of IDPs from either asteroidal or Kuiper Belt sources demonstrate that these excesses are almost certainly not due to cosmic-ray spallation reactions. They may well be a primordial signature in previously unsampled solar-system matter, generated by a currently unknown process which could include, among other possibilities, irradiation of pre-cometary matter in the outer regions of the solar accretion disk. Among the possible sources that produce excess ^{3}He are supernovae and so-called planetary nebulae (Bania *et al.*, 2002). The latter have enhancements of the ^{3}He/^{4}He ratio of typically a factor 10 relative to the solar ^{3}He/^{4}He ratio.

9. Conclusions

Volatile isotope abundances are tracers for many processes during the history of the solar system. Protosolar isotopic abundances and the isotopic ratios of material introduced into the solar system after the birth of the Sun put the solar system into context with the chemical evolution of the Galaxy. Among the various types of samples present in the solar system and the processes that can be studied with help of the isotopic signatures in these samples we mention:

1. Interstellar matter that has survived the violent processes during the formation of the solar system provides the possibility to study chemical processes in the interstellar medium.
2. The same material gives insights into the nucleosynthetic processes in stars of our Galaxy that predate the formation of the Sun.
3. Interstellar material that continuously enters the solar system carries information on the chemical and nucleosynthetic processes that occur in our interstellar environment. If the age of the interstellar material or the time of its entry into the solar system can be determined, it gives insight into the history of our Galactic environment.
4. The isotopic compositions of solids formed during the few million years before and after the birth of the Sun are tracers for the complicated evolution of the protosolar cloud, the protosolar nebula, and the planetesimals from which the larger solar system bodies such as the terrestrial planets, asteroids, comets, and giant planets formed.
5. The gaseous envelopes of the giant planets, in particular those of Jupiter, are representative samples for the light volatiles present in the gas of the protosolar nebula.
6. The outer convective zone of the Sun is an authentic witness of the isotopic composition of heavy volatiles in the protosolar cloud. This composition is represented in the solar wind within typically 1% precision (Kallenbach, 2003).
7. For lighter volatiles such as H, D, ^{3}He, ^{4}He, ^{6}Li, ^{7}Li, ^{10}B, and ^{11}B the solar wind abundances reflect nuclear processes in the Sun's core and at the Sun's surface.
8. Volatile isotope abundances in meteorites and interplanetary dust particles are tracers for microscopic and/or macroscopic processes in the protosolar nebula that have established part of the volatile inventory in solids of the solar system.
9. Volatile isotope abundances in pre-solar grains found in meteorites and in pre-solar IDPs indicate the nucleosynthetic origin of the isotopes of volatile elements and possibly are tracers for the processes that have incorporated volatiles into pre-solar solids.
10. The isotopic abundances in planetary samples are tracers for the processes of planet formation, for processes in planetary atmospheres, but also represent the abundances of components implanted from the solar wind, heliospheric energetic particles, cosmic rays, and the abundances of spallogenic products.

Geologically inert planetary surfaces also record the inflow of interstellar dust and micrometeorites as mentioned above.

At present, the complicated processes during the protosolar nebula stage are still not fully understood. Topics for possible future research that address the open questions regarding the protosolar nebula include:

1. Young solar-like stellar objects need to be explored with higher spatial resolution, but also with the capability to analyze elemental or even isotopic abundances in more detail.
2. More complete 3D models on the evolution of molecular clouds, their collapse to the protosolar nebula, and on the formation of planetesimals and planets including the evolution of the proto-Sun and its magnetosphere may be developed.
3. Based on these models the chemical kinetics of the protosolar nebula could be evaluated.
4. The isotopic composition of the solar wind needs to be studied in-situ with higher precision than was achievable with the SOHO, ACE, and WIND particle instruments. The Genesis mission will presumably provide more precise data, but does not have the capability to analyze any compositional variations in different solar wind regimes.
5. The composition of outer solar-system objects needs to be studied in more detail. This provides the possibility to analyze remainders of interstellar chemistry, but also to evaluate the composition of rocky and icy planetesimals during the formation of the solar system. The Rosetta mission is aimed at these goals.
6. More laboratory studies should be performed on microscopic processes that could have occured inside the protosolar nebula during the various evolutionary stages.
7. More experimental research on interstellar dust will fortunately be continued by the ESA and the NASA space agencies.
8. The same applies for the study of planetary atmospheres in the solar system.
9. The search for extrasolar planetary systems that resemble our solar system continues.

References

Acuña, M.H., Connerney, J.E.P., Ness, N.F., Lin, R.P., Mitchell, D., Carlson, C.W., McFadden, J., Anderson, K.A., Rème, H., Mazelle, C., Vignes, D., Wasilewski, P., and Cloutier, P.: 1999, 'Global Distribution of Crustal Magnetization Discovered by the Mars Global Surveyor MAG/ER Experiment', *Science* **284**, 790–793.

Adams, F.C., and Laughlin, G.: 2000, 'Protostellar Disk Formation and Early Evolution', *Space Sci. Rev.* **92**, 23–38.

Aikawa, Y., and Herbst, E.: 1999, 'Deuterium Fractionation in Protoplanetary Disks', *Astrophys. J.* **526**, 314–326.

Aikawa, Y., Ohashi, N., Inutsuka, S.-I., Herbst, E., and Takakuwa, S.: 2001, 'Molecular Evolution in Collapsing Prestellar Cores', *Astrophys. J.* **552**, 639–653.

Aléon, J., Robert, F., Chaussidon, M., and Engrand, C.: 2002, '^{15}N Excesses in Deuterated Organics from two Interplanetary Dust Particles', *Lunar Planet Sci.* **XXXIII**, #1397.

Alexander, C.M.O'D.: 1993, 'Pre-solar SiC in Chondrites: How Variable and How Many Sources?', *Geochim. Cosmochim. Acta* **57**, 2869–2888.

Alloin, D., Collin-Souffrin, S., Joly, M., and Vigroux, L.: 1979, 'Nitrogen and Oxygen Abundances in Galaxies', *Astron. Astrophys.* **78**, 200–216.

Amari, S., Lewis, R.S., and Anders, E.: 1995, 'Interstellar Grains in Meteorites: III. Graphite and its Noble Gases', *Geochim. Cosmochim. Acta* **59**, 1411–1426.

Amari, S., Nittler, L.R., Zinner, E., Lodders, K., Lewis, R.S.: 2001a, 'Pre-solar SiC Grains of Type A and B: Their Isotopic Compositions and Stellar Origins', *Astrophys. J.* **559**, 463–483.

Amari, S., Gao, X., Nittler, L.R., Zinner, E., José, J., Hernanz, M., and Lewis, R.S.: 2001b, 'Pre-solar Grains from Novae', *Astrophys. J.* **551**, 1065–1072.

Anders, E., and Grevesse, N.: 1989, 'Abundances of the Elements – Meteoritic and Solar', *Geochim. Cosmochim. Acta* **53**, 197–214.

Anders, E., and Zinner, E.: 1993, 'Interstellar Grains in Primitive Meteorites: Diamond, Silicon Carbide, and Graphite', *Meteoritics* **28**, 490–514.

André, P., and Montmerle, T.: 1994, 'From T Tauri Stars to Protostars: Circumstellar Material and Young Stellar Objects in the p Ophiuchi Cloud', *Astrophys. J.* **420**, 837–862.

Audard, M., Güdel, M., and Guinan, E.F.: 1999, 'Implications From Extreme-Ultraviolet Observations for Coronal Heating of Active Stars', *Astrophys. J.* **513**, L53–L56.

Backman, D.E., Dasgupta, A., and Stencel, R.E.: 1995, 'Model of a Kuiper Belt Small Grain Population and Resulting Far-infrared Emission', *Astrophys. J.* **450**, L35–L38.

Balbus, S.A., and Hawley, J.F.: 2000, 'Solar Nebula Magnetohydrodynamics' *Space Sci. Rev.* **92**, 39–54.

Bania, T.M., Rood, R.T., and Balser, D.S.: 2002, 'The Cosmological Density of Baryons from Observations of ^{3}He$^+$ in the Milky Way', *Nature* **415**, 54–57.

Basu, S., Chaplin, W.J., Christensen-Dalsgaard, J., Elsworth, Y., Isaak, G.R., New, R., Schou, J., Thompson, M.J., and Tomczyk, S.: 1997, 'Solar Internal Sound Speed as Inferred from Combined BiSON and LOWL Oscillation Frequencies', *Month. Not. Roy. Astron. Soc.* **291**, 243–251.

Beaumont, V., Robert, F.: 1999, 'Nitrogen Isotope Ratios of Kerogens in Precambrian Cherts: A Record for the Evolution of Atmosphere Chemistry?', *Precambrian Res.* **96**, 63–82.

Becker, R.H., Clayton, R.N., Galimov, E.M., Lammer, H., Marty, B., Pepin, R.O., and Wieler, R.: 2003, 'Isotopic Signatures of Volatiles in Terrestrial Planets', *Space Sci. Rev.*, this volume.

Bell, K.R., Cassen, P.M., Wasson, J.T., and Woolum, D.S.: 2000, 'The Fu Orionis Phenomenon and Solar Nebula Material', *in* V. Mannings, A.P. Boss, and S.S. Russel (eds.), *Protostars and Planets IV*, Univ. Arizona Press, Tucson, pp. 897–926.

Benkert, J.-P., Baur, H., Signer, P., and Wieler, R.: 1993, 'He, Ne, and Ar from the Solar Wind and Solar Energetic Particles in Lunar Ilmenites and Pyroxenes', *J. Geophys. Res.* **98**, 13147–13162.

Bensch, F., Pak, I., Wouterloot, J.G.A., Klapper, G., and Winnewisser, G.: 2001, 'Detection of ^{13}C^{17}O and Observations of Rare CO Isotopomers toward the ρ Ophiuchi Molecular Cloud', *Astrophys. J.* **562**, L185–L188.

Benz, W., Kallenbach, R., and Lugmair, G.W. (eds.): 2000, *From Dust to Terrestrial Planets*, reprinted from *Space Sci. Rev.* **92**, Kluwer Academic Publishers, Dordrecht.

Bernatowicz, T.J., and Zinner, E. (eds.): 1997, *Astrophysical Implications of the Laboratory Study of Pre-solar Materials*, American Institute of Physics, Woodbury, N.Y.

Black, D.C., and Pepin, R.O.: 1969, 'Trapped Neon in Meteorites', *Earth Planet. Sci. Lett.* **6**, 395–405.

Bochsler, P.: 2000, 'Abundances and Charge States of Particles in the Solar Wind', *Rev. Geophys.* **38**, 247–266.

Bochsler, P., and Geiss, J.: 1973, 'Solar Abundances of Light Nuclei and Mixing of the Sun', *Sol. Phys.* **32**, 3–11.

Bochsler, P., Wimmer-Schweingruber, R.F., and Wurz, P.: 2001, 'Sun, Solar Wind and Interstellar Medium: What are the Compositional Relations?', *in* R.F. Wimmer-Schweingruber (ed.), *Proc. Joint SOHO/ACE Workshop "Solar and Galactic Composition", Bern, Switzerland, 6–9 March, 2001, AIP Conf. Proc.* **598**, 381–386.

Bockelée-Morvan, D., Gautier, D., Lis, D.C., Young, K., Keene, J., Phillips, T., Owen T., Crovisier, J., Goldsmith, P.F., Bergin, E.A., Despois, D., and Wootten, A.: 1998, 'Deuterated Water in Comet C/1996 B2 (Hyakutake) and its Implications for the Origin of Comets', *Icarus* **193** 147–162.

Bockelée-Morvan, D., Gautier, D., Hersant, F., Huré, J.-M., and Robert, F.: 2002, 'Turbulent Radial Mixing in the Solar Nebula as the Source of Crystalline Silicates in Comets', *Astron. Astrophys.* **384**, 1107–1118.

Boss, A.P., and Vanhala, H.A.T.: 2000, 'Triggering Protostellar Collapse, Injection, and Disk Formation', *Space Sci. Rev.* **92**, 13–22.

Boss, A.P., and Vanhala, H.A.T.: 2001, 'Injection of Newly Synthesized Elements into the Protosolar Cloud', *Phil. Trans. Roy. Soc. Lond.* **A359**, 2005–2017.

Boss, A.P., Morfill, G.E., and Tscharnuter, W.M.: 1989, 'Models of the Formation and Evolution of the Solar Nebula', *in* S.K. Atreya *et al.* (eds.), *Origin and Evolution of Planetary and Satellite Atmospheres*, Univ. Arizona Press, Tucson, pp. 35–77.

Breitschwerdt, D.: 1996, 'The Local Bubble: Current State of Observations and Models', *Space Sci. Rev.* **78**, 173–182.

Breneman, H.H., and Stone, E.C.: 1985, 'Solar Coronal and Photospheric Abundances from Solar Energetic Particle Measurements', *Astrophys. J.* **299**, L57–L61.

Buckle, J.V., and Fuller, G.A.: 2002, 'The Interaction between Protostars and their Environment: Carbon-bearing Species towards Low-mass Protostars', *Astron. Astrophys.* **381**, 77–97.

Burbidge, E.M., *et al.*: 1957, *Rev. Mod. Phys.* **29**, 547.

Burns, J., Lamy, P.L., and Soter, S.: 1979, 'Radiation Pressure on Small Particles in the Solar System', *Icarus* **40**, 1–48.

Busemann, H., Baur, H., and Wieler, R.: 2000, 'Primordial Noble Gases in "Phase Q" in Carbonaceaous and Ordinary Chondrites Studied by Closed-system Stepped Etching', *Met. Planet. Sci.* **35**, 949–973.

Busemann, H., *et al.*: 2001, 'Applications of Abundance Data and Requirements for Cosmochemical Modeling', *in* R.F. Wimmer-Schweingruber (ed.), *Proc. Joint SOHO/ACE Workshop "Solar and Galactic Composition", Bern, Switzerland, 6–9 March 2001, AIP Conf. Proc.* **598**, 357–379.

Cameron, A.G.W.: 1978, 'Physics of the Primitive Solar Accretion Disk', *Moon and Planets* **18**, 5–40.

Cameron, A.G.W.: 1995, 'The First Ten Million Years in the Solar Nebula', *Meteoritics* **30**, 133–161.

Cameron, A.G.W., and Pine, M.R.: 1972, 'Numerical Models of the Primitive Solar Nebula', *Icarus* **18**, 377–406.

Caughlan, G.R., and Fowler, W.A.: 1962, 'The Mean Lifetimes of Carbon, Nitrogen, and Oxygen Nuclei in the CNO Bicycle', *Astrophys. J.* **136**, 453–464.

Caselli, P., Walmsley, C.M., Tafalla, M., Dore, L., and Myers, P.C.: 1999, 'CO Depletion in the Starless Cloud Core L1544', *Astrophys. J.* **523**, L165–L169.

Cassen, P.: 1994, 'Utilitarian Models of the Solar Nebula', *Icarus* **112**, 405–429.

Ceccarelli, C., Loinard, L., Castets, A., Tielens, A.G.G.M., Caux, E., Lefloch, B., and Vastel, C.: 2001, 'Extended D_2CO Emission: The Smoking Gun of Grain Surface Chemistry', *Astron. Astrophys.* **372**, 998–1004.

Charnley, S.B., and Rodgers, S.D.: 2002, 'The End of Interstellar Chemistry as the Origin of Nitrogen in Comets and Meteorites', *Astrophys. J.* **569**, L133–L137.

Charnley, S.B., Tielens, A.G.G.M., and Rodgers, S.D.: 1997, 'Deuterated Methanol in the Orion Compact Ridge', *Astrophys. J.* **482**, L203–L206.

Chaussidon, M., and Robert, F.: 1998, '^{7}Li/^{6}Li and ^{11}B/^{10}B Variations in Chondrules from the Semarkona Unequilibrated Chondrite', *Earth Planet. Sci. Lett.* **164**, 577–589.

Chaussidon, M., and Robert, F.: 1999, 'Lithium Nucleosynthesis in the Sun inferred from the Solar-wind ^{7}Li/^{6}Li-Ratio', *Nature* **402**, 270–273.

Chiang, E.I., Joung, M.K., Creech-Eakman, M.J., Qi, C., Kessler, J.E., Blake, G.A., and van Dishoeck, E.F.: 2001, 'Spectral Energy Distributions of Passive T Tauri and Herbig Ae Disks: Grain Mineralogy, Parameter Dependences and Comparison With Infrared Space Observatory LWS Observations', *Astrophys. J.* **547**, 1077–1089.

Chupp, E.L., Forrest, D.J., Ryan, J.M., Heslin, J., Reppin, C., Pinkau, K., Kanbach, G., Rieger, E., and Share, H.: 1982, 'A Direct Observation of Solar Neutrons Following the 01:18 UT Flare on 1980 June 21', *Astrophys. J.* **263**, L95–L99.

Chupp, E.L.: 1984, 'High-Energy Neutral Radiations from the Sun', *Ann. Rev. Astron. Astrophys.* **22**, 359–387.

Clayton, D.D.: 1983, 'Principles of Stellar Evolution and Nucleosynthesis', University of Chicago Press, Chicago.

Clayton, D.D.: 1989, 'Origin of Heavy Xenon in Meteoritic Diamonds', *Astrophys. J.* **340**, 613–619.

Clayton, D.D., and Timmes, F.X.: 1997, 'Placing the Sun in Galactic Chemical Evolution: Mainstream SiC Particles', *Astrophys. J.* **483**, 220–227.

Clayton, D.D., Meyer, B.S., Sanderson, C.I., Russell, S.S., and Pillinger, C.T.: 1995, 'Carbon and Nitrogen Isotopes in Type II Supernova Diamonds', *Astrophys. J.* **447**, 894–905.

Clayton, R.N.: 2002, 'Self-shielding in the Solar Nebula', *Nature* **415**, 860–861.

Clayton, R.N.: 2003, 'Oxygen Isotopes in the Solar System', *Space Sci. Rev.*, this volume.

Clayton, R.N., Onuma, N., and Mayeda, T.K.: 1976, 'A Classification of Meteorites Based on Oxygen Isotopes', *Earth Planet. Sci. Lett.* **30**, 10–18.

Dalgarno, A.: 1987, *in: Physical Processes in Interstellar Clouds; Proc. NATO Adv. Study Inst.*, Reidel, Dordrecht, pp. 219–239.

Dauphas, N., Robert, F., and Marty, B.: 2000, 'The Late Asteroidal and Cometary Bombardment of the Earth as Recorded in Water Deuterium to Protium Ratio', *Icarus* **148**, 508–512.

De Geus, E.J.: 1992, 'Interaction of Stars and Interstellar Matter in Scorpio Centaurus', *Astron. Astrophys.* **262**, 258–270.

Deloule, E., and Robert, F.: 1995, 'Interstellar Water in Meteorites?', *Geochim. Cosmochim. Acta* **59**, 4695–4706.

Dermott, S.F., Grogan, K., Gustafson, B.A.S., Jayaraman, S., Kortenkamp S., and Xu, Y.L.: 1996, 'Size Distributions of Asteroidal Dust: Possible Constraints on Impact Strengths', *in* B.A.S. Gustafson and M.S. Hanner (eds.), *Physics, Chemistry, and Dynamics of Interplanetary Dust*, p. 143.

Desch, S.J., and Connolly, H.C., Jr.: 2002, 'A Model of the Thermal Processing of Particles in Solar Nebula Shocks: Application to the Cooling Rates of Chondrules', *Met. Planet. Sci.* **37**, 183–207.

Diaz, A.I., and Tosi, M.: 1986, 'The Origin of Nitrogen and the Chemical Evolution of Spiral Galaxies', *Astron. Astrophys.* **158**, 60–66.

Drouart, A., Dubrulle, B., Gautier, D., and Robert, F.: 1999, 'Structure and Transport in the Solar Nebula From Constraints on Deuterium Enrichment and Giant Planet Formation', *Icarus* **140**, 129–155.

Dubrulle, B.: 1992, 'A Turbulent Closure Model for Thin Accretion Disks', *Astron. Astrophys.* **266**, 592–604.

Dubrulle, B., and Valdettaro, L.: 1992, 'Consequences of Rotation on Energetics of Accretion Disks', *Astron. Astrophys.* **263**, 387–400.

Dubrulle, B.: 1993, 'Differential Rotation as a Source of Angular Momentum Transfer in the Solar Nebula', *Icarus* **106**, 59–76.

Dubrulle, B., Morfill, G., and Sterzik, M: 1995, 'The Dust Subdisk in the Protoplanetary Nebula', *Icarus* **104**, 237–246.

Edmunds, M.G., and Pagel, B.E.J.: 1978, 'Nitrogen Synthesis and the 'Age' of Galaxies', *Month. Not. Roy. Astron. Soc.* **185**, 77–80.
Ehrenfreund, P., and Charnley, S.B.: 2000, 'Organic Molecules in the Interstellar Medium, Comets, and Meteorites: A Voyage from Dark Clouds to the Early Earth', *Ann. Rev. Astron. Astrophys.* **38**, 427–483.
Evans, N.J., II: 1999, 'Physical Conditions in Regions of Star Formation', *Ann. Rev. Astron. Astrophys.* **37**, 311–362.
Evenson, P., Meyer, P., and Pyle, K.R.: 1983, 'Protons from the Decay of Solar Flare Neutrons', *Astrophys. J.* **274**, 875–882.
Ezer, D., and Cameron, A.G.W.: 1966, 'The Contraction Phase of Stellar Evolution', *in* R.F. Stein and A.G.W. Cameron (eds.), *Stellar Evolution*, Plenum Press, New York.
Fegley, B., Jr.: 'Kinetics of Gas-grain Reactions in the Solar Nebula' *Space Sci. Rev.* **92**, 177–200.
Feigelson, E.D., Garmire, G.P., and Pravdo, S.H.: 2002, 'Magnetic Flaring in the Pre-main Sequence Sun and Implications for the Early Solar System', *Astrophys. J.* **572**, 335–349.
Feldman, W.C., Asbridge, J.R., Bame, S.J., and Gosling, J.T.: 1977, *in* O.R. White (ed.), *The Solar Output and its Variation*, Colorado Associated Univ. Press, Boulder, CO, USA, p. 351.
Frisch, P.C.: 1995, 'Characteristics of Nearby Interstellar Matter', *Space Sci. Rev.* **72**, 499–592.
Frisch, P.C.: 2000, 'The Galactic Environment of the Sun', *J. Geophys. Res.* **105**, 10279–10290.
Frisch, P.C.: 2002, 'The Journey of the Sun', submitted.
Gail, H.-P.: 1998, 'Chemical Reactions in Protoplanetary Accretion Disks: IV. Multi-component Dust Mixture', *Astron. Astrophys.* **332**, 1099–1122.
Gail, H.-P.: 2001, 'Radial Mixing in Protoplanetary Accretion Disks: I. Stationary Disk Models with Annealing and Carbon Combustion', *Astron. Astrophys.* **378**, 192–213.
Gallino, R., Busso, M., Picchio, G., and Raiteri, C.M.: 1990, 'On the Astrophysical Interpretation of Isotope Anomalies in Meteoritic SiC Grains', *Nature* **348**, 298–302.
Gardinier, A., Derenne, S., Robert, F., Behar, F., Largeau, C., and Maquet, J.: 2000, 'Solid State CP/MAS ^{13}C NMR of the Insoluble Organic Matter of the Orgueil and Murchison Meteorites: Quantitative Study', *Earth Planet. Sci. Lett.* **184**, 9–21.
Geiss, J.: 1993, 'Primordial Abundances of Hydrogen and Helium Isotopes', *in* N. Prantzos, E. Vangioni-Flam, and M. Cassée (eds.), *Origin and Evolution of the Elements*, Cambridge University Press, Cambridge, UK.
Geiss, J., and Bochsler, P.: 1982, 'Nitrogen Isotopes in the Solar System', *Geochim. Cosmochim. Acta* **46**, 529–548.
Geiss, J., and Bochsler, P.: 1991, 'Long Time Variations in Solar Wind Properties – Possible Causes versus Observations', *in* C. Sonnett, M. Giampapa, and M. Matthews (eds.), *The Sun in Time*, Univ. Arizona Press, Tucson, pp. 98–117.
Geiss, J., and Gloeckler, G.: 2003, 'Isotopic Composition of H, He and Ne in the Protosolar Cloud', *Space Sci. Rev.*, this volume.
Geiss, J., Bühler, F., Cerutti, H., Eberhardt, P., and Filleux, C.: 1972, 'Solar Wind Composition Experiment', *NASA SP* **315**, 14.1.
Geiss, J., Gloeckler, G., and Charbonnel, C.: 2002, 'Chemical Evolution in our Galaxy During the Last 5 Gyr', *Astrophys. J.* **578**, 862–867.
Glassgold, A., Feigelson, E.D., and Montmerle, T.: 2000, 'Effects on Energetic Radiation in Young Stellar Objects', *in* V. Mannings, A.P. Boss, and S.S. Russel (eds.), *Protostars and Planets IV*, Univ. Arizona Press, Tucson, pp. 429–455.
Gloeckler, G., and Geiss, J.: 2000, 'Deuterium and Helium-3 in the Protosolar Cloud' *in* L. da Silva (ed.), *The Light Elements and their Evolution, Proc. IAUS* **198**, 224–233.
Gloeckler, G., Fisk, L.A., and Geiss, J.: 1997, 'Anomalously Small Magnetic Field in the Local Interstellar Cloud', *Nature* **386**, 374–377.
Grogan, K., Dermott, S.F., and Gustafson, B.A.S.: 1996, 'An Estimation of the Interstellar Contribution to the Zodiacal Thermal Emission', *Astrophys. J.* **472**, 812–817.

Grossman, L., and Larimer, J.W.: 1974, 'Early Chemical History of the Solar System', *Rev. Geophys. Space Phys.* **12-1**, 71–101.

Grün, E., Gustafson B., Mann, I., Baguhl, M., Morfill, G.E., Staubach, P., Taylor, A., and Zook, H.A.: 1994, 'Interstellar Dust in the Heliosphere', *Astron. Astrophys.* **286**, 915–924.

Guinan, E.F., and Ribas, I.: 2002, 'Our Changing Sun: The Role of Solar Nuclear Evolution and Magnetic Activity on Earth's Atmosphere and Climate', *in* B. Montesinos, A. Giménez, and E.F. Guinan (eds.), *The Evolving Sun and its Influcence on Planetary Environments*, ASP Conf. Ser. **269**, 85–106.

Hartmann, L.: 2000, 'Observational Constraints on Transport (and Mixing) in Pre-Main Sequence Disks', *Space Sci. Rev.* **92**, 55–68.

Hasegawa, T.I., and Herbst, E.: 1993, 'New Gas-grain Chemical Models of Quiescent Dense Interstellar Clouds: The Effects of H_2 Tunnelling Reactions and Cosmic Ray Induced Desorption', *Month. Not. Roy. Astron. Soc.* **261**, 83–102.

Hashizume, K., Chaussidon, M., Marty, B., and Robert, F.: 2000, 'Solar Wind Nitrogen on the Moon: Deciphering Pre-solar From Planetary Nitrogen', *Science* **290**, 1142–1145.

Hashizume, K., Marty, B., and Wieler, R.: 2002, 'Analyses of Nitrogen and Argon in Single Lunar Grains: Towards a Quantification of the Asteroidal Contribution to Planetary Surfaces', *Earth Planet. Sci. Lett.* **202**, 201–216.

Henning, Th., and Salama, F: 1998, 'Carbon in the Universe', *Science* **282**, 2204–2210.

Herbst, E.: 1995, 'Large Molecules in the Interstellar Medium', *in* A.G.G.M. Tielens and T.P. Snow (eds.), *The Diffuse Interstellar Bands*, Astrophysics and Space Science Library **202**, Kluwer Academic Publishers, Dordrecht, p. 307.

Herbst, E.: 2001, 'The Chemistry of Interstellar Space', *Chem. Soc. Rev.* **30**, 168–176.

Herbst, E., and Klemperer, W.: 1973, 'The Formation and Depletion of Molecules in Dense Interstellar Clouds', *Astrophys. J.* **185**, 505–534.

Hersant, F., Gautier, D., and Huré, J.-M.: 2001, 'A Two-dimensional Model for the Primordial Nebula Constrained by D/H Measurments in the Solar System: Implications for the Formation of Giant Planets', *Astrophys. J.* **554**, 391–407.

Herzog, G.H.: 1965, 'Lithium Abundances in F5-G8 Dwarfs', *Astrophys. J.* **141**, 588–609.

Higdon, J.C., Lingenfelter, R.E., and Ramaty, R.: 1998, 'Cosmic-Ray Acceleration from Supernova Ejecta in Superbubbles', *Astrophys. J.* **509**, L33–L36.

Hohenberg, C.M., Thonnard, N., and Meshik, A.: 2002, 'Active Capture and Anomalous Adsorption: New Mechanisms for the Incorporation of Heavy Noble Gases', *Met. Planet. Sci.* **37**, 257–267.

Hoppe, P., and Zinner, E.: 2000, 'Pre-solar Dust Grains from Meteorites and their Stellar Sources', *J. Geophys. Res.* **105**, 10371–10385.

Hoyle, F., and Fowler, W.A.: 1973, *Nature* **241**, 384.

Hueso, R., and Guillot, T.: 2003, 'Evolution of the Protosolar Nebula and the Formation of the Giant Planets', *Space Sci. Rev.*, this volume.

Hunten, D.M., Pepin, R.O., and Walker, J.C.G.: 1987, 'Mass Fractionation in Hydrodynamic Escape', *Icarus* **69**, 532–549.

Huss, G.R.: 1997, 'Pre-solar Grains as Tracers of Nebular and Parent Body Processing of Chondritic Material', *Proc. Workshop on Parent-Body and Nebular Modification of Chondritic Materials, Maui, Hawaii, July 17-19, 1997.*

Huss, G.R., and Alexander, E.C., Jr.: 1987, 'On the Pre-solar Origin of the 'Normal Planetary' Noble Gas Component in Meteorites', *J. Geophys. Res.* **92**, 710–716.

Huss, G.R., and Lewis, R.S.: 1995, 'Pre-solar Diamond, SiC, and Graphite in Primitive Chondrites: Abundances as a Function of Meteorite Class and Petrologic Type', *Geochim. Cosmochim. Acta* **59**, 115–160.

Huss, G.R., Fahey, A.J., Gallino, R., and Wasserburg, G.J.: 1994, 'Oxygen Isotopes in Circumstellar Al_2O_3 Grains from Meteorites and Stellar Nucleosynthesis', *Astrophys. J.* **430**, L81–L84.

Huss, G.R., Hutcheon, I.D., and Wasserburg, G.J.: 1997, 'Isotopic Systematics of Pre-solar Silicon Carbide from the Orgueil (CI) Chondrite – Implications for Solar System Formation and Stellar Nucleosynthesis', *Geochim. Cosmochim. Acta* **61**, 5117–5148.

Iben, I., Jr.: 1965, 'Stellar Evolution. I. The Approach to the Main Sequence', *Astrophys. J.* **141**, 993–1018.

Johnstone, D., Hollenbach, D., and Bally, J.: 1998, 'Photoevaporation of Disks and Clumps by Nearby Massive Stars: Application to Disk Destruction in the Orion Nebula', *Astrophys. J.* **499**, 758–776.

Kallenbach, R., *et al.*: 1997, 'Isotopic Composition of Solar Wind Neon Measured by CELIAS/MTOF on Board SOHO', *J. Geophys. Res.* **102**, 26895–26904.

Kallenbach, R.: 2003, 'Isotopic Fractionation by Plasma Processes', *Space Sci. Rev.*, this volume.

Lallement, R.: 1996, 'Relations Between ISM Inside and Outside the Heliosphere', *Space Sci. Rev.* **78**, 361–374.

Lambert, D.L., Gustafsson, B., Eriksson, K., and Hinkle, K.H.: 1986, 'The Chemical Composition of Carbon Stars. I – Carbon, Nitrogen, and Oxygen in 30 Cool Carbon Stars in the Galactic Disk', *Astrophys. J. Suppl.* **62**, 373–425.

Lammer, H., and Bauer, S.J.: 2003, 'Isotopic Fractionation by Gravitational Escape', *Space Sci. Rev.*, this volume.

Lammer, H., Stumptner, W., Molina-Cuberos, G.J., Bauer, S.J., and Owen, T.: 2000, 'Nitrogen Isotope Fractionation and its Consequence for Titan's Atmospheric Evolution', *Planet. Space Sci.* **48**, 529–543.

Lellouch, E., Bézard, B., Fouchet, T., Feuchtgruber, H., Encrenaz, T., and de Graauw, T.: 2001, 'The Deuterium Abundance in Jupiter and Saturn from ISO-SWS Observations', *Astron. Astrophys.* **370**, 610–622.

Lester, D.F., Dinerstein, H.L., Werner, M.W., Watson, D.M., Genzel, R., and Storey, J.W.V.: 1987, 'Far-infrared Measurements of N/O in H II Regions – Evidence for Enhanced CN Process Nucleosynthesis in the Inner Galaxy', *Astrophys. J.* **320**, 573–585.

Lewis, R.S., Srinivasan, B., and Anders, E.: 1975, 'Host Phase of a Strange Xenon Component in Allende', *Science* **190**, 1251–1262.

Lewis, R.S., Tang, M., Wacker, J.F., Anders, E., and Steel, E.: 1987, 'Interstellar Diamonds in Meteorites', *Nature* **326**, 160–162.

Lewis, R.S., Amari, S., and Anders, E.: 1994, 'Interstellar Grains in Meteorites: II. SiC and its Noble Gases', *Geochim. Cosmochim. Acta* **58**, 471–494.

Lingenfelter, R.E., and Ramaty, R.: 1967, *in* B. Shen (ed.), *High Energy Nuclear Reactions in Astrophysics*, Benjamin, New York, p. 99.

Lingenfelter, R.E., Ramaty, R., and Kozlovsky, B.: 1998, 'Supernova Grains: The Source of Cosmic-Ray Metals', *Astrophys. J.* **500**, L153–L156.

Little-Marenin, I.R.: 1986, 'Carbon Stars with Silicate Dust in their Circumstellar Shells', *Astrophys. J.* **307**, L15–L19.

Loinard, L., Castets, A., Ceccarelli, C., Lefloch, B., Benayoun, J.-J., Caux, E., Vastel, C., Dartois, E., and Tielens, A.G.G.M.: 2002, 'Doubly Deuterated Formaldehyde in Star-forming Regions: An Observational Approach', *Plan. Space Sci.*, in press.

Love, S., and Brownlee, D.E.: 1993, 'A Direct Measurement of the Terrestrial Mass Accretion Rate of Cosmic Dust', *Science* **262**, 550–553.

Lunine, J.I., Engel, S., Rizk, B., and Horanyi, M.: 1991, 'Sublimation and Reformation of Icy Grains in the Primitive Solar Nebula', *Icarus* **94**, 333–344.

Lynden-Bell, D., and Pringle, J.E.: 1974, 'The Evolution of Viscous Disks and the Origin of the Nebula Variables', *Month. Not. Roy. Acad. Sci.* **168**, 603–638.

Mahaffy, P.R., Donahue, T.M., Atreya, S.K., Owen, T.C., and Niemann, H.B.: 1998, 'Galileo Probe Measurements of D/H and ^{3}He/^{4}He in Jupiter's Atmosphere', *Space Sci. Rev.* **84**, 251–263.

Mann, I.: 2001, 'Dust in the Solar System and in Other Planetary Systems', *in* I.P. Williams and N. Thomas (eds.), *Solar and Extra-Solar Planetary Systems*, Lecture Notes in Physics **577**, 218.

Mann, I., and Kimura, H.: 2000, 'Interstellar Dust Properties Derived from Mass Density, Mass Distribution and Flux Rates in the Heliosphere', *J. Geophys. Res.* **105**, 10317–10328.

Marty, B., Hashizume, K., Chaussidon, M., and Wieler, R.: 2003, 'Nitrogen Isotopes on the Moon: Archives of the Solar and Planetary Contributions to the Inner Solar System', *Space Sci. Rev.*, this volume.

Mason, G.M.: 2001, 'Heliospheric Lessons for Galactic Cosmic-ray Acceleration', *Space Sci. Rev.* **99**, 119–133.

McKeegan, K.D., Chaussidon, M., and Robert, F.: 2000, 'Incorporation of Short-lived ^{10}Be in Calcium-aluminium-rich Inclusion from the Allende Meteorite', *Science* **289**, 1334–1336.

Messenger, S.: 2000, 'Identification of Molecular-cloud Material in Interplanetary Dust Particles', *Nature* **404**, 968–971.

Messenger, S., Keller, L.P., and Walker, R.M.: 2002, 'Discovery of Abundant Interstellar Silicates in Cluster IDPs', *Lunar Planet Sci.* **XXXIII**, abstract # 1887 (CD-ROM).

Messenger, S., Stadermann, F.J., Floss, C., Nittler, L.R., and Mukhopadhyay, S.: 2003, 'Isotopic Signatures of Presolar Materials in Interplanetary Dust', *Space Sci. Rev.*, this volume.

Mewaldt, R.A.: 1989, 'The Abundances of Isotopes in the Cosmic Radiation', *in: Cosmic Abundances of Matter*, AIP Conf. Proc. **183**, New York, pp. 124–146.

Mewaldt, R.A., Ogliore, R.C., Gloeckler, G., and Mason, G.M.: 2001, 'A New Look at Neon-C and SEP-Neon', *in* R.F. Wimmer-Schweingruber (ed.), *Proc. Joint SOHO/ACE Workshop "Solar and Galactic Composition" Bern, Switzerland, 6–9 March 2001, AIP Conf. Proc.* **598**, 393–398.

Meyer, J.-P., Drury, L.O'C., and Ellison, D.C.: 1997, 'Galactic Cosmic Rays from Supernovae Remnants I. A Cosmic-Ray Composition Controlled by Volatility and Mass-to-Charge Ratio', *Astrophys. J.* **487**, 182–196.

Meyer, B.S., and Clayton, D.D.: 2000, 'Short-lived Radioactivities and the Birth of the Sun', *Space Sci. Rev.* **92**, 133–152.

Meyer, B.S., Clayton, D.D, and The, L.-S.: 2000, 'Molybdenum Isotopes from a Neutron Burst', *Lunar Planet. Sci.* **XXXI**, abstract # 1458 (CD-ROM).

Meynet, G., Arnould, M., Paulus, G., and Maeder, A.: 2001, 'Wolf-Rayet Star Nucleosynthesis and the Isotopic Composition of the Galactic Cosmic Rays', *Space Sci. Rev.* **99**, 73–84.

Michaud, G., and Vauclair, S.: 1991, 'Element Separation by Atomic Diffusion', *in* A.N. Cox, W.C. Livingston, and M.S. Matthews (eds.), *Solar Interior and Atmosphere*, University of Arizona Press, Tucson, pp. 304–325.

Millar, T.J.: 2003, 'Deuterium Fractionation in Interstellar Clouds', *Space Sci. Rev.*, this volume.

Millar, T.J., Bennett, A., and Herbst, E.: 1989, 'Deuterium Fractionation in Dense Interstellar Clouds', *Astrophys. J.* **340**, 906–920.

Morbidelli, A., Chambers, J., Lunine, J.I., Petit, J.M., Robert, F., Valsecchi, G.B., and Cyr, K.E.: 2000, 'Source Regions and Timescales for the Delivery of Water on Earth' *Met. Planet. Sci.* **35**, 1309–1320.

Morfill, G.E., and Wood, J.A.: 1989, 'Protoplanetary Accretion Disk Models: The Effects of Several Meteoritic, Astronomical and Physical Constraints', *Icarus* **82**, 225–243.

Morfill, G.E., Tscharnuter, W., and Völk, H.J.: 1985, 'Dynamical and Chemical Evolution of the Protoplanetary Nebula', *in* D.C. Black and M.S. Matthews (eds.), *Protostars and Planets II*, Univ. Arizona Press, Tucson, pp. 493–533.

Mousis, O., Gautier, D., Bockelée-Morvan, D., Robert, F., Dubrulle, B., and Drouart, A.: 2000, 'Constraints on the Formation of Comets From D/H Ratios Measured in H_2O and HCN', *Icarus* **148**, 513–525.

Mullan, D.J., and Linsky, J.L.: 1998, 'Nonprimordial Deuterium in the Interstellar Medium', *Astrophys. J.* **511**, 502–512.

Nakamura, T., Nagao, K., and Takaoka, N.: 1999, 'Microdistribution of Primordial Noble Gases in CM Chondrites Determined by in Situ Laser Microprobe Analysis: Decipherment of Nebular Processes', *Geochim. Cosmochim. Acta* **63**, 241–255.

Navon, O., and Wasserburg, G.J.: 1985, 'Self-shielding in O_2 – A Possible Explanation for Oxygen Isotopic Anomalies in Meteorites?', *Earth Planet. Sci. Lett.* **73**, 1985, 1–16.

Nicolussi, G.K., Pellin, M.J., Lewis, R.S., Davis, A.M., Amari, S., and Clayton, R.N.: 1998, 'Molybdenum Isotopic Composition of Individual Pre-solar Silicon Carbide Grains from the Murchison Meteorite', *Geochim. Cosmochim. Acta* **62**, 1093–1104.

Nittler, L.R., Hoppe, P., Alexander, C.M.O'D., Amari, S., Eberhardt, P., Gao, X., and Lewis, R.S., Strebel, R., Walker, R.M., and Zinner, E.: 1995, 'Silicon Nitride from Supernovae', *Astrophys. J.* **453**, L25–L28.

Nittler, L.R., and Cowsik, R.: 1997, 'Galactic Age Estimates from O-rich Stardust in Meteorites', *Phys. Rev. Lett.* **78**, 175–178.

Nittler, L.R., Alexander, C.M.O'D., Gao, X., Walker, R.M., and Zinner, E.: 1997, 'Stellar Sapphires: the Properties and Origins of Pre-solar Al_2O_3 in Meteorites', *Astrophys. J.* **483**, 475–495.

Ott, U.: 1996, 'Interstellar Diamond Xenon and Timescales of Supernova Ejecta', *Astrophys. J.* **463**, 344–348.

Ott, U.: 2002, 'Meteoritic Anomalies, Rapid Neutron Capture, and Gamma-rays', *New Astron. Rev.* **46**, 513–518.

Ott, U.: 2003a, 'The Most Primitive Material in Meteorites', *in:* Th. Henning (ed.), *Astromineralogy*, Springer, in press.

Ott, U.: 2003b, 'Isotopes of Volatiles in Pre-solar Grains', *Space Sci. Rev.*, this volume.

Owen, T., and Encrenaz, T.: 2003, 'Element Abundances and Isotope Ratios in the Giant Planets and Titan', *Space Sci. Rev.*, this volume.

Owen, T., Mahaffy, P.R., Niemann, H.B., Atreya, S., and Wong, M.: 2001, 'Protosolar Nitrogen', *Astrophys. J.* **553**, L77–L79.

Ozima, M., Wieler, R., Marty, B., and Podosek, F.A.: 1998, 'Comparative Studies of Solar, Q-gases and Terrestrial Noble Gases, and Implications on the Evolution of the Solar Nebula', *Geochim. Cosmochim. Acta* **62**, 301–3124.

Palma, R.L., Becker, R.H., Pepin, R.O., and Schlutter, D.J.: 2002, 'Irradiation Records in Regolith Materials, II: Solar-wind and Solar-energetic-particle Components in Helium, Neon, and Argon Extracted from Single Lunar Mineral Grains and from the Kapoeta Howardite by Stepwise Pulse-heating', *Geochim. Cosmochim. Acta* **66**, 2929–2958.

Parizot, E.: 2001, 'Galactic Cosmic Rays and the Light Elements', *Space Sci. Rev.* **99**, 61–71.

Pellin, M.J., Davis, A.M., Lewis, R.S., Amari, S., and Clayton, R.N.: 1999, 'Molybdenum Isotopic Composition of Single Silicon Carbide Grains from Supernovae', *Lunar Planet. Sci.* **XXX**, abstract # 1969 (CD-ROM).

Pepin, R.O.: 2000, 'On the Isotopic Composition of Primordial Xenon in Terrestrial Planet Atmospheres', *Space Sci. Rev.* **92**, 371–395.

Pepin, R.O.: 2003, 'On Noble Gas Processing in the Solar Accretion Disk', *Space Sci. Rev.*, this volume.

Pepin, R.O., Palma, R.L., and Schlutter, D.J.: 2000, 'Noble Gases in Interplanetary Dust Particles, I: The Excess Helium-3 Problem and Estimates of the Relative Fluxes of Solar Wind and Solar Energetic Particles in Interplanetary Space', *Met. Planet. Sci.* **35**, 495–504.

Pinti, D.L.: 2002, 'The Isotopic Record of Archean Nitrogen and the Evolution of the Early Earth', *Trends Geochem.* **2**, 1–17.

Pollack, J.B., and Bodenheimer, P.: 1989, 'Theories of the Origin and Evolution of the Giant Planets', *in* S.K. Atreya, J.B. Pollack, and M.S. Matthews (eds.), *Origin and Evolution of Planetary and Satellite Atmospheres*, Univ. Arizona Press, Tucson, pp. 564–602.

Pollack, J.B., Hubicky, O., Bodenheimer, P., Lissauer, J.L., Podolak, M., and Greenzweig, Y.: 1996, 'Formation of the Giant Planets by Concurrent Accretion of Solids and Gas', *Icarus* **124**, 62–85.

Prinn, R.G., and Fegley, B.J.: 1989, 'Solar Nebula Chemistry: Origin of Planetary, Satellite, and Cometary Volatiles', *in* S.K. Atreya *et al.* (eds.), *Origin and Evolution of Planetary and Satellite Atmospheres*, Univ. Arizona Press, Tucson, pp. 78–136.

Reeves, H.: 1993, *in:* N. Prantzos, E. Vangioni-Flam, and M. Cassé (eds.), *Origin and Evolution of the Elements*, Cambridge Univ. Press, pp. 168–197.

Renzini, A., and Voli, M.: 1981, 'Advanced Evolutionary Stages of Intermediate-mass Stars. I - Evolution of Surface Compositions', *Astron. Astrophys.* **94**, 175–193.

Reynolds, J.H., and Turner, G.: 1964, 'Rare Gases in the Chondrite Renazzo', *J. Geophys. Res.* **69**, 3263–3281.

Richter, S., Ott, U., and Begemann, F.: 1998, 'Tellurium in Pre-solar Diamonds as an Indicator for Rapid Separation of Supernova Ejecta', *Nature* **391**, 261–263.

Righter, K., and Drake, M.J.: 1997, 'Metal/silicate Equilibrium in a Homogeneous Accreting Earth: New Results for Re', *Earth Planet. Sci. Lett.* **146**, 541–553.

Robert, F., Gautier, D., and Dubrulle, B.: 2000, 'The Solar System D/H Ratio: Observations and Theories', *in* W. Benz, R. Kallenbach, and G.W. Lugmair (eds.), *From Dust to Terrestrial Planets*, reprinted in *Space Sci. Rev.* **92**, 201–224.

Roberts, H., Fuller, G.A., Millar, T.J., Hatchell, J., and Buckle, J.V.: 2002, 'A Survey of [HDCO]/[H_2CO] and [DCN]/[HCN] Ratios Towards Low-mass Protostellar Cores', *Astron. Astrophys.* **381**, 1026–1038.

Rodgers, S.D., and Millar, T.J.: 1996, 'The Chemistry of Deuterium in Hot Molecular Cores', *Month. Not. Roy. Astron. Soc.* **280**, 1046–1054.

Rodgers, S.D., and Charnley, S.B.: 2001, 'Gas-phase Production of NHD_2 in L134N', *Astrophys. J.* **553**, 613–617.

Ruden, S.P., and Lin, D.N.C.: 1986, 'The Global Evolution of the Primordial Solar Nebula', *Astrophys. J.* **308**, 883–901.

Ruden, S.P., and Pollack, J.B.: 1991, 'The Dynamical Evolution of the Protosolar Nebula', *Astrophys. J.* **375**, 740–760.

Sackmann, I.-J., and Boothroyd, A.I.: 2002, 'Our Sun. V. A Bright Young Sun Consistent with Helioseismology and Warm Temperatures on Ancient Earth and Mars', *Astrophys. J.*, in press.

Sandford, S.A., Bernstein, M.P., and Swindle, T.D.: 1998, 'The Trapping of Noble Gases by the Irradiation and Warming of Interstellar Ice Analogs', *Met. Planet. Sci.* **32**, A135.

Scherer, K., Fichtner, H., and Stawicki, O.: 2002, 'Shielded by the Wind: The Influence of the Interstellar Medium on the Environment of Earth', *J. Atmos. Solar-Terr. Phys.* **64**, 795–804.

Scully, S.T., Cassé, M., Olive, K.A., Schramm, D.N., Truran, J., and Vangioni-Flam, E.: 1996, 'The Local Abundance of ^3He: A Confrontation between Theory and Observation', *Astrophys. J.* **462**, 960–968.

Serrano, A., and Peimbert, M.: 1983, 'Chemical Evolution of Galaxies – Part Three – the N/O versus O/H Relationship', *Revista Mexicana de Astron. y Astrof.* **8:2**, 117.

Severny, A.B.: 1957, 'A Spectroscopic Investigation of the Deuterium D_α Line in Active Regions on the Sun', *Soviet Astron.* **1**, 324.

Shah, R.Y., and Wootten, A.: 2001, 'Deuterated Ammonia in Galactic Protostellar Cores', *Astrophys. J.* **554**, 933–947.

Shah, R.Y., and Wootten, A.: 2002, 'Identifying Cold Clumps in Star Forming Regions with Deuterated Molecules', *Planet. Space Sci.*, in press.

Shakura, N.L., and Sunyaev, R.A.: 1973, 'Black Holes in Binary Systems. Observational Appearance', *Astron. Astrophys.* **24**, 337–355.

Shang, H., Shu, F.H., Lee, T., and Glassgold, A.E.: 2000, 'Protostellar Winds and Chondritic Meteorites', *Space Sci. Rev.* **92**, 153–176.

Shu, F.H., Adams, F.C., and Lizano, S.: 1987, 'Star Formation in Molecular Clouds – Observation and Theory', *Ann. Rev. Astron. Astrophys.* **25**, 23–81.

Shu, F.H., Shang, H., and Lee, T.: 1997, 'Towards an Astrophysical Theory of Chondrites', *Science* **271**, 1545–1552.

Shu, F.H., Shang, H., Glassgold, A.E., and Lee, T.: 1997, 'X-rays and Fluctuating X-winds from Protostars', *Science* **277**, 1475–1479.

Shu, F.H., Shang, H., Gounelle, M., Glassgold, A.E., and Lee, T.: 2001, 'The Origin of Chondrules and Refractory Inclusions in Chondritic Meteorites', *Astrophys. J.* **548**, 1029–1050.

Shukolyukov, A., and Lugmair, G.W.: 2000, 'On the ^{53}Mn Heterogeneity in the Early Solar System', *Space Sci. Rev.* **92**, 225–236.

Stepinski, T.F.: 1998, 'The Solar Nebula as a Process – An Analytical Model', *Icarus* **132**, 100–112.

Swart, P.K., Grady, M.M., Pillinger, C.T., Lewis, R.S., and Anders, E.: 1983, 'Interstellar Carbon in Meteorites', *Science* **220**, 406–410.

Talbot, R.J., and Newman, M.J.: 1997, 'Encounters Between Stars and Dense Interstellar Clouds', *Astrophys. J.* **34**, 295–308.

Terzieva, R., and Herbst, E.: 2000, 'The Possibility of Nitrogen Isotopic Fractionation in Interstellar Clouds', *Mon. Not. R. Astron. Soc.* **317**, 563–568.

Timmes, F.X., and Clayton, D.D.: 1996, 'Galactic Evolution of Silicon Isotopes: Application to Presolar SiC Grains from Meteorites', *Astrophys. J.* **472**, 723–741.

Torres-Peimbert, S., Peimbert, M., and Fierro, J.: 1989, 'Physical Conditions of H II Regions in M101 and the Pregalactic Helium Abundance', *Astrophys. J.* **345**, 186–195.

Turck-Chièze, S., Nghiem, P., Couvidat, S., and Turcotte, S.: 2001, 'Solar Internal Composition and Nuclear Reaction Rates in the Light of Helioseismology', *Sol. Phys.* **200**, 323–342.

Turner, B.E.: 2002, 'Deuterated Molecules in Translucent and Dark Clouds', *Astrophys. J., Suppl.* **136**, 579–629.

Urey, H.C.: 1952, *The Planets*, Yale University Press, New Haven, Cincinnati.

Van Dishoeck, E.F., and Blake, G.A.: 1998, 'Chemical Evolution of Star-Forming Regions', *Ann. Rev. Astron. Astrophys.* **36**, 317–368.

Vauclair, S.: 1998, 'Evidence of Element Diffusion inside the Sun and the Stars and its Consequences on the Lithium Primordial Abundance', *Space Sci. Rev.* **84**, 265–271.

Verchovsky, A.B., Sephton, M.A., Wright, I.P., and Pillinger, C.T.: 2002, 'Separation of Planetary Noble Gas Carrier from Bulk Carbon in Enstatite Chondrites During Stepped Combustion', *Earth Planet. Sci. Lett.* **99**, 243–255.

von Steiger, R.: 2002, 'History of the Solar Wind', *in* W. Schröder (ed.), *Solar Variability and Geomagnetism*, Lectures from the IAGA Assembly in Hanoi, Vietnam, 2001.

Wacker, J.F., Zadnik, M.G., and Anders, E.: 1985, 'Laboratory Simulation of Meteoritic Noble Gases. I - Sorption of Xe on Carbon: Trapping Experiments', *Geochim. Cosmochim. Acta* **49**, 1035–1048.

Waelkens, C., Waters, L.B.F.M., de Graauw, M.S., Huygen, E., Malfait, K., Plets, H., Vandenbussche, B., Beintema, D.A., Boxhorn, D.R., Habing, H.J., Heras, A.M., Certer, D.J.M., Lahuis, F., Morris, P.W., Roelfsema, P.R., Salama, A., Siebenmorgen, R., Traus, N.R., van der Bliek, N.R., Valentjin, E.A., and Wenelius, P.R.: 1996, 'SWS Observations of Young Main-sequence Stars with Dusty Circumstellar Disks', *Astron. Astrophys.* **315**, L245–L248.

Wehrstedt, M., and Gail, H.-P.: 2002, 'Radial Mixing in Protoplanetary Accretion Disks: II. Time-dependent Disk Models with Annealing and Carbon Combustion', *Astron. Astrophys.* **385**, 204–220.

Wetherill, G.W.: 1954, 'Variations in the Isotopic Abundances of Ne and Ar Extracted from Radioactive Minerals', *Phys. Rev.* **96**, 679–683.

Wiedenbeck, M.E., *et al.*: 2001, 'The Origin of Primary Cosmic Rays: Constraints from ACE Elemental and Isotopic Composition Observations', *Space Sci. Rev.* **99**, 15–26.

Wielen, R., and Wilson, T.L.: 1997, 'The Evolution of the C, N, and O Isotope Ratios from an Improved Comparison of the Interstellar Medium with the Sun', *Astron. Astrophys.* **326**, 139–142.

Wieler, R., and Heber, V.S.: 2003, 'Noble Gas Isotopes on the Moon', *Space Sci. Rev.*, this volume.

Wieler, R., Baur, H., and Signer, P.: 1986, 'Noble Gases from Solar Energetic Particles Revealed by Closed System Stepwise Etching of Lunar Soil Minerals', *Geochim. Cosmochim. Acta* **50**, 1997–2017.

Willacy, K., Klahr, H.H., Millar, T.J., and Henning, Th.: 1998, 'Gas and Grain Chemistry in a Protoplanetary Disk', *Astron. Astrophys.* **338**, 995–1005.

Wilson, T.L., and Rood, R.: 1994, 'Abundances in the Interstellar Medium', *Ann. Rev. Astron. Astrophys.* **32**, 191–226.

Wimmer-Schweingruber, R.F.: 2002, 'The Composition of the Solar Wind', *Adv. Space Res.* **30**, 23–32.

Wimmer-Schweingruber, R.F., and Bochsler, P.: 2001, 'Lunar Soils: A Long-term Archive for the Galactic Environment of the Heliosphere?', *in* R.F. Wimmer-Schweingruber (ed.), *Proc. Joint SOHO/ACE Workshop "Solar and Galactic Composition" Bern, Switzerland, 6–9 March 2001, AIP Conf. Proc.* **598**, 399–404.

Wood, B.E., Müller, H.-R., Zank, G.P., and Linsky, J.L.: 2002, 'Measured Mass-loss Rates of Solar-like Stars as a Function of Age and Activity', *Astrophys. J.* **574**, 412–425.

Zinner, E.: 1998, 'Stellar Nucleosynthesis and the Isotopic Composition of Pre-solar Grains from Primitive Meteorites', *Ann. Rev. Earth Planet. Sci.* **26**, 147–188.

Address for Offprints: Reinald Kallenbach, International Space Science Institute, Hallerstrasse 6, CH-3012 Bern, Switzerland; kallenbach@soho.unibe.ch

ISOTOPIC SIGNATURES OF VOLATILES IN TERRESTRIAL PLANETS

Working Group Report

R.H. BECKER[1], R.N. CLAYTON[2], E.M. GALIMOV[3], H. LAMMER[4], B. MARTY[5], R.O. PEPIN[1] and R. WIELER[6]

[1] *School of Physics and Astronomy, University of Minnesota, Minneapolis, MN 55455, USA*

[2] *Dept. of the Geophysical Sciences, University of Chicago, Chicago, IL 60637, USA*

[3] *V.I. Vernadsky Institute of Geochemistry and Analytical Chemistry, Russian Academy of Sciences, 117975 Moscow, 19, Russia*

[4] *Space Research Institute, Austrian Academy of Sciences, A-8042 Graz, Austria*

[5] *Centre de Recherches Pétrographiques et Géochimiques, CNRS UPR, 9046, rue Notre-Dame des Pauvres BP 20, F-54501 Vandoevre-les Nancy-Cedex*

[6] *Isotopengeologie, ETH Zürich, 8092 Zürich, Switzerland*

Received: 30 September 2002; Accepted in final form: 20 November 2002

Abstract. Variations in the isotopic ratios of volatile elements in different reservoirs on the terrestrial planets carry information about processes that operated on the planets since their formation. Comparisons between primordial planetary compositions, to the extent they can be determined, may help us understand the planetary formation process. This working group report summarizes our knowledge of terrestrial planet volatile inventories.

1. Introduction

This report considers isotopic variations in volatile elements and what they might tell us about the origins of, and processes operating on, the terrestrial planets. For our purposes, volatile elements encompass the noble gases, carbon, nitrogen and hydrogen. Sulfur is not considered here because, while some data are available on sulfur isotopes in meteorites (Cooper *et al.*, 1997; Farquhar *et al.*, 2000a, and references therein), and the sulfur cycle on Earth has been studied in contexts such as the evolution of oxygen in the atmosphere (Canfield *et al.*, 2000; Farquhar *et al.*, 2000b), there is little currently known about sulfur isotopes on the other terrestrial planets (but see Farquhar *et al.*, 2000c). We include oxygen, although in a separate context from the other elements mentioned as it is a major rather than a trace constituent of the planets. Planetary oxygen occurs for the most part in the form of nonvolatile silicate and oxide minerals, but the large cosmic abundance of oxygen relative to magnesium, silicon and iron requires that in the solar nebula most of the oxygen was bound in gaseous molecules, such as CO and H_2O.

Among the terrestrial planets we do not consider Mercury here, because there are no isotopic data from Mercury for the elements we are considering. Sodium and potassium have been observed in Mercury's atmosphere (Hunten and Sprague, 1997; Killen and Ip, 1999) and may therefore be considered volatile elements on

Space Science Reviews **106:** 377–410, 2003.

that planet, but isotopic variations are nonexistent for Na and unavailable for K from Mercury. In contrast, while not a planet, the Earth's moon will receive some consideration in this report because volatiles trapped in its surface regolith yield significant information with respect to solar system history.

The initial planetary volatile isotopic compositions would have been established at the time of formation of the planets, reflecting an averaging of the various reservoirs that went into making up the planets. For oxygen, and probably carbon and nitrogen as well, the primary reservoirs would have been solid objects, representatives of which may still exist in the various meteorite groups (Clayton, 2003; Grady and Wright, 2003). In the case of the noble gases and hydrogen, it is possible that the initial reservoirs were dominated by nebular gases of solar composition (Pepin and Porcelli, 2002). The isotopic compositions of those reservoirs are the subject of much of the rest of this book, but we do not know how much of any specific reservoir was incorporated in any specific planet. The initial planetary compositions were then subject to processes that added further material to or removed material from the planets.

Modification processes may have occurred early in solar system history, as with hydrodynamic escape of atmospheric gases driven by high UV flux from the early sun, or they may have operated continuously, as with gravitational escape from the top of an atmosphere or infall of micrometeoritic or cometary materials, or in the case of the Moon implantation of solar wind. Radiogenic production of specific isotopes, while not strictly speaking adding to or removing material from the planets, also modified isotopic compositions and has been a major factor in helping to understand planetary histories, providing both a means of age-dating events and tracing physical processes. Radiogenic production will be discussed where appropriate in the context of the other subjects. Finally, while not affecting bulk planetary compositions, biological processes can produce isotopic effects in materials containing C, H, and N, making these elements potential markers for the presence of life on the terrestrial planets.

Since most of the processes we consider occur at the interface between a planet's atmosphere and space, an issue which must be addressed is the degree to which changes in the atmosphere may have been incorporated into or buffered by the bulk planetary composition. That is to say, does mixing occur between the atmosphere and the solid planet? This bears on the question of whether we might expect to find evidence for the initial bulk isotopic compositions of planetary volatiles preserved within the interiors of any of the planets.

What we know about present-day isotopic compositions of the planets is limited by the observations that have thus far been carried out. For Earth, we have abundant samples of crustal and upper mantle rocks to study, and a well-determined atmosphere. For Venus, we have atmospheric data only, with significant uncertainties for many of the isotopic ratios. For Mars, as for the Earth, we have data from the atmosphere as well as for some mantle-derived rocks, in the form of the Martian meteorites. The number and variety of rock samples, however, is much more limited than for Earth, and the Martian atmospheric ratios are in general less well-determined.

2. Available Data on Terrestrial Planet Volatiles

In this section we attempt to provide a summary of the currently available data base on isotopic abundances of volatile elements in the atmospheres of Venus, Earth, and Mars and also in interior reservoirs of the latter two planets. We will concentrate on recent references where relevant data are compiled, but will also highlight some crucial original publications. For a summary of the noble gas and nitrogen inventories on the Moon the reader is referred to two other papers in this book (Wieler and Heber, 2003; Marty *et al.*, 2003) as well as to a recent review (Wieler, 2002). Carbon in lunar samples has received little attention in recent years. Most of the lunar regolith carbon data can be found in Becker (1980) and Norris *et al.* (1983) and in references therein. Oxygen is discussed separately in Section 3 of this report.

2.1. Elemental Abundances in the Atmospheres of Earth, Venus, and Mars

Overall abundances of noble gases (apart from He, which is not retained in the current terrestrial and Martian atmospheres but may be on Venus), carbon, and nitrogen in planetary atmospheres and in carbonaceous chondrites are shown, e.g., in Pepin (1991) and Donahue and Russell (1997). The two most conspicuous features seen are i) a gross depletion of all noble gases relative to condensible elements in the respective solid body and relative to solar abundances and ii) the fact that the light noble gases are considerably more depleted (by several orders of magnitude) than the heavier gases. Further important observations are iii) carbon and nitrogen are less depleted than the noble gases, particularly in meteorites, iv) the ratio of nonradiogenic Ne/Ar is very similar in all three planets and in carbonaceous chondrites, v) the Venus atmosphere contains 1-2 orders of magnitude more nonradiogenic Ne and Ar than the terrestrial atmosphere and probably also several times more nonradiogenic Kr and Xe, and Mars is further depleted in all noble gases by roughly two orders of magnitude compared to the terrestrial values, vi) the Earth and Mars, and probably also Venus, have similar Kr/Xe ratios, which are, however, several times higher than the CI chondrite value.

2.2. Isotopic Ratios of Volatiles on the Earth, Venus, and Mars

Recent comprehensive reviews of the isotopic compositions of noble gases in terrestrial atmospheric and interior reservoirs are provided by two books (Ozima and Podosek, 2002; Porcelli *et al.*, 2002). Distributions of terrestrial carbon and nitrogen isotopic compositions are covered by Hoefs (1997) and Coplen *et al.* (2002).

The isotopic compositions of Ne and Ar in the atmosphere of Venus have mainly been determined by several Venera missions (Istomin *et al.*, 1983). An upper limit on $^{3}He/^{4}He$ was provided by Pioneer Venus (Hoffman *et al.*, 1980). These data are compiled by e.g. Pepin (1991), Donahue and Russell (1997), and Wieler (2002).

As noted above, data for Mars are considerably more abundant than for Venus. The Martian atmosphere has been directly analyzed during the Viking mission (Nier and McElroy, 1977), and information on the hydrogen, carbon and oxygen compositions is also provided by infrared spectroscopy from Earth (Krasnopolsky *et al.*, 1996). Furthermore, trapped atmospheric gases have been measured in some of the Martian meteorites (Bogard and Johnson, 1983; Becker and Pepin, 1984; Wiens *et al.*, 1986). Such analyses in some cases were consistent with Viking values (actually yielding the best evidence so far that these meteorites are from Mars) and in other cases provided data either not available from Viking or improving upon Viking results. Martian atmospheric data are compiled by Pepin (1991), Owen (1992), Bogard *et al.* (2001), and Swindle (2002). Carbon and nitrogen data for Martian magmatic and low-temperature mineral phases are given by Grady and Wright (2003).

Some Martian meteorites also appear to have retained noble gases from the early (as opposed to recent) Martian atmosphere (e.g. Mathew and Marti, 2001), while others very likely contain the signature of Martian mantle noble gases (e.g., Ott, 1988). This may seem surprising with a sample collection comprising just some 20 randomly provided rocks, given that decades of work have not unambiguously revealed terrestrial paleoatmosphere with an isotopic signature different from modern values and that terrestrial samples with uncontaminated mantle noble gases are rarely found. This may reflect the absence of plate tectonics, which will have inhibited mixing of atmospheric and interior noble gases, and, perhaps even more importantly, a lack of alteration by liquid water. This combination allowed the preservation of at least one > 4 Gy meteorite that suffered several early shock events but has been little altered since (Swindle, 2002).

2.3. Observations Requiring Particular Attention

Several features exhibited by the isotopic data can be singled out for special mention. For one, all three planets show indications of "solar-like" noble gas patterns, at least in some reservoirs and for some gases. Apart from nucleogenic ^{21}Ne, there is a component of neon in the terrestrial mantle similar to and possibly nearly identical to the solar wind value (Figure 11 of Graham, 2002). Earth mantle $^{3}He/^{4}He$ ratios are higher than the atmospheric value, although still considerably lower than the protosolar (pre-D burning) or solar composition (Graham, 2002). One Martian meteorite, Chassigny, contains Xe indistinguishable in composition from solar wind Xe, and presumably represents the Martian mantle (Ott, 1988; Mathew and Marti, 2001). The Venus atmosphere has $^{20}Ne/^{22}Ne$ and possibly also $^{36}Ar/^{38}Ar$ values closer to the solar wind values than the respective ratios in the terrestrial atmosphere (Istomin *et al.*, 1983, cf. Wieler, 2002). The interpretation of these data is not straightforward, but they clearly bear on our understanding of the evolutionary history of the planets.

In contrast to the above, where all three planets show affinities with solar compositions, there are also remarkable differences in some noble gas signatures among the different planets. One is the strikingly low $^{36}Ar/^{38}Ar$ ratio of Martian atmospheric argon, probably in the range of 3.0-3.6 (Bogard, 1997), much lower than those measured for Earth and Venus, and considered to imply significant loss of Ar from the Mars atmosphere (see Section 5.2.1). Another key difference between Mars and Earth is that both $^{40}Ar/^{36}Ar$ and $^{129}Xe/^{132}Xe$ ratios are more radiogenic (i.e. higher) in the Martian atmosphere than in the terrestrial one, although the $^{40}Ar/^{36}Ar$ value derived by Viking has been considerably revised downwards based on analyses of Martian meteorites (e.g., Garrison and Bogard, 1998). Also, whereas on Earth both Ar and Xe in the interior are more radiogenic than in the atmosphere, the opposite is true for Mars. These observations have obvious significance with regard to the degassing histories of the two planets.

A further difference between Mars and Earth is in the composition of their atmospheric Xe. Whereas nonradiogenic Xe on Mars is best explained as an isotopically fractionated version of solar wind (Swindle and Jones, 1997; Pepin, 2000; Swindle, 2002), this is not the case for the Earth, where a primordial component (called U-Xe), depleted in the two heaviest isotopes ^{134}Xe and ^{136}Xe relative to solar wind Xe, appears necessary (Pepin, 2000).

Among the non-noble gases, some key observations are that the D/H ratios on both Mars and Venus are considerably higher than the terrestrial ocean water value, $\approx$5 times higher on Mars and perhaps $\approx$150 times higher on Venus (Owen, 1992; Donahue and Russell, 1997; Bogard *et al.*, 2001). Another important observation is that nitrogen in the Martian atmosphere is heavier by about 60% that the terrestrial atmospheric value, as well as the Martian interior value, whereas Venus shows, within a quite substantial uncertainty, the same nitrogen isotopic composition as the Earth's atmosphere (Bogard *et al.*, 2001; Pepin, 1989). Carbon is also enhanced in the heavy isotope in the Martian atmosphere relative to Mars interior carbon, but only by about 6% (Jakosky and Jones, 1997).

3. Oxygen Isotopes

Distributions of isotopes yielded by chemical reactions, whether controlled by equilibrium thermodynamics or reaction kinetics, or by physical processes, such as evaporation or diffusion, are expected to be predominantly mass-dependent, so that, in the case of oxygen, variations in $^{17}O/^{16}O$ ratios should be one-half of the corresponding variations in $^{18}O/^{16}O$ ratios. For terrestrial samples this has generally been the case. When oxygen isotope measurements were extended from terrestrial to meteoritic samples, it was found that the isotopic variations in many meteorites did not obey the expected mass-dependent relationship, but instead reflected processes yielding $\delta^{17}O = \delta^{18}O$, referred to as non-mass-dependent fractionation. This observation presents both a challenge, to understand the processes involved, and an opportunity, to use the measured variation in understanding solar system evolution.

3.1. Proposed Sources of Oxygen Isotopic Variations in the Solar Nebula

3.1.1. *Nucleosynthetic ^{16}O*

The initial observations of mass-independent oxygen isotopic effects lent themselves to a straightforward explanation, that of addition of a pure ^{16}O component to a more-or-less standard terrestrial oxygen isotopic composition Clayton *et al.* (1973). Samples in which anomalous oxygen was found also contained anomalies in isotopic compositions of other elements, attributable to nucleosynthesis, including evidence for decayed ^{26}Al. The ^{16}O-enriched component of oxygen was therefore considered to have a similar origin. The model was later modified by postulating that the entire dust component of the nebula was enriched in ^{16}O by about 5% relative to the gas component, reflecting different proportions of supernova products in the two reservoirs Clayton *et al.* (1977). The observed mass-independent isotope effects in meteorites would then result from interactions between the two reservoirs.

3.1.2. *Mass-Independent Chemical or Photochemical Effects*

As a result of observations of mass-independent effects in the isotopic composition of ozone in the Earth's atmosphere (Mauersberger, 1981) and in the laboratory (Heidenreich and Thiemens, 1983), the idea that the ^{16}O-enriched component in meteorites had an origin other than nucleosynthetic was raised by Thiemens and coworkers (Thiemens and Heidenreich, 1983). They initially suggested self-shielding by $^{16}O_2$ in the formation of O_3, but this was shown not to be a viable mechanism in this context (see Thiemens, 1999). They later suggested the effects in ozone were the result of symmetries in the reactive state of the O_3 molecule and that similar results could have been obtained in the solar nebula as are observed in the laboratory. They proposed that such symmetries occurred in reactive states in other species and that one might therefore look for isotopic effects in other elements as well as oxygen (Bhattacharya and Thiemens, 1989).

A thorough analysis of the reaction channels for all oxygen isotope combinations leading to ozone molecules, reported by Mauersberger (2003), indicates that symmetries in the reactive state do not explain the ozone observations. Therefore, this mechanism is unlikely to be the cause of the oxygen variations seen in meteorites. However, another alternative has just been raised to the nucleosynthetic origin for the anomalous oxygen (Clayton, 2003). The model discussed there explains some of the observations from meteorites better than the nucleosynthetic model does, but is incomplete in other respects. It does have the potential of explaining variations in nitrogen isotopes in planetary and meteoritic bodies in the context of an initial nebula that began with a $^{15}N/^{14}N$ ratio such as is observed in Jupiter (Owen *et al.*, 2001). This model is based on isotopic self-shielding in the photodissociation of carbon monoxide. This is a well-known process in cold molecular clouds (van Dishoeck and Black, 1988), in which photolysis of CO is associated with absorption of ultraviolet light from stars outside the cloud. The absorbed wavelengths are isotopically separated, and the cloud becomes optically thick for wavelengths that dissociate $^{12}C^{16}O$, while remaining transparent for wavelengths that dissociate $^{12}C^{17}O$ and $^{12}C^{18}O$, thus enhancing the isotope ratios $^{17}O/^{16}O$ and $^{18}O/^{16}O$ in atoms in the cloud interior. Yurimoto and Kuramoto (2002) proposed that these atoms reacted with hydrogen to produce water ice that has heavy-isotope enrichment relative to the rock-forming oxides. Upon subsequent heating in the solar nebula, these two reservoirs could interact to give the mass-independent effects observed in meteorites. Under the conditions pertaining in these clouds, the objections raised in regard to self-shielding in ozone do not apply.

In an alternative version of the self-shielding model, Clayton (2002) proposed that the isotope-selective photolysis of CO and N_2 occurred within the solar nebula, with the young Sun as the source of ultraviolet radiation. Because of the high density (and opacity) of the solar nebula, such irradiation must occur at a distance from the Sun of much less than 1 AU, as in the X-wind model of Shu *et al.* (1997). The principal motivation for the self-shielding models is the recent ion-microprobe observation that the same ^{16}O-rich compositions, originally found in refractory inclusions, are common in many minerals that have chemical characteristics of primary condensates from a gas (Clayton, 2003, and references therein). The implication was that the nebular gas had this ^{16}O-rich composition, so that, in a sense, the refractory inclusions are isotopically 'normal' (i.e. solar), while the rest of the inner Solar system is 'anomalous'.

An obvious question that arises in the context of the self-shielding of CO (Clayton, 2002) is what happens to the ^{13}C that will be produced in the region of the nebula where it is optically thick for $^{12}C^{16}O$ but is producing dissociated ^{17}O and ^{18}O. The expected ^{13}C enrichments will in fact not be observed, as a consequence of rapid isotope exchange reaction between CO and C^+ ions. This occurs for carbon, but not for nitrogen and oxygen, because the first ionization potential for C is less than that of H, whereas for N and O, it is greater, so N^+ and O^+ are not formed. With regard to any ^{15}N-enriched atoms produced by this mechanism, the means of trapping them in some form of condensate is still an open question. One possibility is reaction with iron vapor, but this is an area that needs further investigation.

The models proposed to explain the large mass-independent oxygen isotopic variations observed in various solar system samples make specific predictions of the isotopic compositions of solar oxygen, and perhaps of nitrogen as well. As these will be subject to measurement in solar wind being collected by NASA's Genesis mission (Wiens *et al.*, 1999), it is possible that significant progress in this area will be made in the next couple of years.

3.2. Oxygen Isotopes and the Origin of Planets

The existence of non-mass related oxygen isotopic effects is observable not only in inclusions within certain meteorites, but to a lesser degree within whole-rock measurements of meteorites (Clayton, 1993). This has provided a means of grouping meteorites within classes based on oxygen isotopes using 3-isotope plots (i.e., Figure 5 of Clayton, 2003), and relating otherwise ungrouped meteorites to known meteorite classes. This process can be extended to the scale of planetary-sized bodies as well. As noted by Clayton (2003), there are conjectures that ^{16}O-enrichments in the solar nebula may have initially been radially dependent, varying in a monotonic fashion. There are however models of terrestrial planet accretion that suggest considerable mixing of material from different radial distances occurred during the formation of any given planet (Wetherill, 1994). The observation of a regular variation in the ^{16}O-enrichments of the planets would support the first of these possibilities and reduce the likelihood of the second. As of yet, the available data (Clayton, 2003) do not support a monotonic radial variation.

The use of the 3-isotope plot also figures into the question of the origin of the Moon. The Moon is indistinguishable from the terrestrial fractionation line (Clayton and Mayeda, 1996; Wiechert *et al.*, 2001), as might be expected if it were formed from material excavated from the Earth by the impact of a large body. Some caution must be exercised in this regard, however, as a difference in Δ-^{17}O (a measure of the ^{16}O-enrichment or depletion relative to the terrestrial composition) can rule out a relationship between two (differentiated) objects whereas identical Δ-^{17}O does not in itself prove that two objects must be related. One need merely look at Figure 5 of Clayton (2003) to see that many different combinations of meteorites could be assembled to arrive at an average on the terrestrial fraction line. Thus, suggestions that the Moon inherited its composition from the impactor rather than from the Earth (Taylor, 2001) cannot be excluded based on oxygen isotopes, and need to be considered on their merits.

This same consideration arises in the context of the question of whether the Earth is made up of enstatite meteoritic material. There is overlap here not only in oxygen isotopes but in nitrogen as well, strengthening the argument in the minds of some (see Section 6.3). Consider however that, since enstatite chondrites contain two orders of magnitude more nitrogen (Grady and Wright, 2003) than the Earth, it would be necessary to either dilute the enstatite chondrite component or deplete it 100-fold in N without altering O or N isotopic compositions significantly. Again, in a body which may have been formed from a large variety of asteroidal objects, some of which may be represented by present day meteorites and some not, one cannot rule out that the end result has bulk properties that in some aspects mirrors some specific class of meteorite purely by chance.

4. Initial Evolution of Planetary Atmospheres

Modern theories of atmospheric evolution on the terrestrial planets are focused on the nebula as the primary supplier of primordial planetary volatiles, either directly, i.e., in ways that do not fractionate isotopes (by adsorption on planetary embryos, "ingassing" from dense, gravitationally condensed atmospheres, or solar wind implantation onto precursor material) or by accretion of cometary ices carrying noble gases that could have been either isotopically solar or mildly fractionated during trapping from the ambient nebula (see Pepin and Porcelli, 2002). The observation that nonradiogenic Ne, Ar, Kr and Xe in Earth's current atmosphere are all isotopically heavier than their solar (i.e., solar-wind) counterparts is an important clue to the nature of the processes that subsequently acted on these primordial planetary reservoirs. This is also the case on Mars, except for Kr which is isotopically solar (and therefore an interesting challenge for evolutionary modeling), and on Venus as well, as far as one can tell from the limited data on hand (Wieler, 2002; Pepin and Porcelli, 2002).

4.1. Hydrodynamic Escape and Early Atmospheres

Recent modeling of atmospheric evolution on the terrestrial planets has focused most extensively on fractionation from primordial source compositions during thermally-driven, hydrodynamic atmospheric escape (reviewed by Pepin and Porcelli, 2002). Hydrogen-rich primordial atmospheres on partially or fully accreted planets are heated at high altitudes by intense extreme-ultraviolet (EUV) radiation from the young sun or alternatively, in the case of the Earth, by energy deposited in a large Moon-forming impact. Under these conditions hydrogen escape fluxes can be large enough to exert upward drag forces on heavier atmospheric constituents sufficient to lift them out of the atmosphere. Lighter species are entrained with the outflowing hydrogen and lost more readily than heavier ones, leading to mass fractionation of the residual atmosphere.

4.1.1. *Earth*

Models in which a planetary atmosphere was driven from isotopically solar-like progenitors to its present compositional state by energetic hydrodynamic escape of the primary atmosphere, followed by mixing of the isotopically fractionated atmospheric residue with solar gases evolved from the interior, are able to account for many of the details of contemporary noble gas isotopic distributions on Earth (Pepin, 1991, 1997). The one glaring exception is Xe. Isotope ratios of terrestrial nonradiogenic Xe cannot be generated by hydrodynamic-escape fractionation of the measured solar-wind Xe composition. An initial composition called U-Xe, which appears to be isotopically identical to solar-wind Xe except for sharply lower abundances of the two heaviest isotopes, is required (Pepin, 2000). There is some evidence that a U-Xe-like composition may have been present in the presolar molecular cloud and at some stage in the history of the solar nebula (Pepin, 2000, 2003). But if so, we are still left with the puzzling question of why it differs isotopically from the Xe composition in the solar wind and presumably also in the Sun.

4.1.2. *Venus*

Turning to Venus, high noble gas abundances and, within their substantial uncertainties, solar-like elemental ratios (except for Ne/Ar) suggest that at least the heavier noble gases in this atmosphere are not greatly evolved from their primordial states (Cameron, 1983; Pepin, 1991, 1997). Neon and Ar isotope ratios also appear to be displaced toward solar values compared to their terrestrial counterparts. The Venus atmosphere therefore may have been altered from its initial composition by planet-specific fractionating loss mechanisms to a much smaller extent than those of Earth and Mars. Applied to Venus, hydrodynamic escape modeling generates Ne and Ar isotope ratios close to values observed in its contemporary atmosphere, and noble gas elemental ratios in its primordial atmosphere very similar to those derived for Earth's initial atmosphere (Pepin, 1997). The model also makes specific predictions for the currently unknown Kr and Xe isotopic distributions; this opportunity for an observational test of the modeling assumptions is one of the primary scientific drivers in current proposals for a suitably instrumented Discovery mission to the planet.

4.1.3. *Mars*

Mars, as usual, is interestingly different. Atmospheric Kr is isotopically identical to solar-wind Kr, as noted above, and so is evidently unfractionated. Severe fractionation, however, is needed to establish present Xe isotope ratios, and here the composition of its primordial progenitor is completely consistent with solar-wind Xe, not U-Xe. Current Ar and Ne isotopic compositions appear to result from interplay of planetary degassing and fractionating sputtering loss from the exosphere over the past several Gy.

Current modeling divides Martian atmospheric history into early and late evolutionary periods (e.g. Haberle *et al.*, 1994; Jakosky *et al.*, 1994; Pepin, 1994; Hutchins and Jakosky, 1996; Hutchins *et al.*, 1997; Carr, 1999). The first is characterized by an initial episode of hydrodynamic escape, followed by high CO_2 pressures and a possible greenhouse, although it appears that greenhouse gases other than CO_2 are needed to raise the surface temperature above the freezing point (Kasting, 1991). The second is characterized by either a sudden transition to a low pressure environment perhaps initiated by abrupt polar CO_2 condensation $\approx$3.7 Gy ago (Haberle *et al.*, 1994) or by a more uniform decline to present pressure. Early in the pre-3.7 Gy epoch, nonradiogenic Xe isotopes are assumed to be hydrodynamically fractionated to their present composition, with corresponding depletions and fractionations of lighter primordial atmospheric species (Pepin, 1994). Subsequent CO_2 pressure history and isotopic evolution during this early period are dictated by the interplay of losses to impact erosion, sputtering, and carbonate precipitation, additions by outgassing and carbonate recycling, and perhaps also by feedback stabilization under greenhouse conditions (Pepin, 1994; Carr, 1999). It should be stressed, however, that little is actually known about the values of the parameters governing these various processes, and so models of this early period are no more than qualitative illustrations of how they might have driven atmospheric behavior in the first few hundred My of Martian history.

4.2. Inferences from Short-Lived Radioactivities

Abundances of radiogenic ^{129}Xe and fissiogenic $^{131-136}$Xe in contemporary planetary atmospheres, generated respectively by decay of the extinct radionuclides ^{129}I and ^{244}Pu, are important clues to the provenance of accretionary materials, the timing of planetary growth, and the subsequent geochemical and outgassing histories of the planets (reviewed in Pepin and Porcelli, 2002). Earth atmosphere modeling points to the presence of an atmospheric Xe component due principally to ^{244}Pu fission, superimposed on escape-fractionated U-Xe and amounting to $\approx$4.6% of the present ^{136}Xe inventory; in contrast, Martian atmospheric Xe is almost perfectly consistent (except for ^{129}Xe) with a composition generated entirely by fractionation of solar-wind Xe by hydrodynamic escape in the early evolutionary epoch (Pepin, 2000). Additional contributions to the Martian Xe inventory from degassed fission Xe are thus either very minor or absent altogether. And yet the present atmosphere is clearly heavily enriched in ^{129}Xe from ^{129}I decay. The presence of radiogenic ^{129}Xe but apparently little if any ^{244}Pu fission Xe, a situation quite unlike that on Earth, has been awkward to reconcile with models of Martian geochemical evolution and degassing history. A less severe fractionation of SW-Xe would allow the presence of an additional component with Pu-Xe fission yield ratios, amounting to $\approx$5% of total ^{136}Xe (Swindle and Jones, 1997) similar to the fraction on Earth. At the moment the central question of the presence or absence of Pu-Xe on Mars is plagued by apparent coincidences. The excellent match of fractionated SW-Xe alone to the atmospheric data is presumably fortuitous if Pu-Xe is present; and if it is absent, the fact that a weaker fractionation generates a pattern in good accord with ^{244}Pu fission yields must likewise be accidental.

4.3. Further Questions to Consider

Escape fractionation of the Earth's primordial atmosphere generates the abundance and nonradiogenic isotope ratios of present-day atmospheric Xe, and also leaves behind highly fractionated residual abundances of the lighter noble gases. Mixing of these residuals with solar-composition Kr, Ar and Ne degassed from the interior replicates their present elemental and isotopic compositions. But Xe is again a problem, on both Earth and Mars only minor amounts of degassed Xe could have been added to their post-escape atmospheres without perturbing the isotopic distributions generated by hydrodynamic escape. Either initial Xe inventories were preferentially outgassed very early by fractional degassing, well before the bulk of the lighter noble gases, and therefore most of it was already present in the primary atmospheres prior to hydrodynamic escape, or most of it was stabilized and segregated permanently in some way deep in the planetary interiors.

The modeling approach considered above implies that noble gases in residual primordial volatile reservoirs in planetary interiors should display solar isotopic compositions. This seems to be the case on Mars, at least for Xe; nonradiogenic Xe isotope ratios in Chassigny appear to be indistinguishable from solar values (Mathew and Marti, 2001). A central question for terrestrial atmospheric formation is whether solar-like noble gases are present in the Earth's mantle. There are strong indications in the mid-ocean-ridge and ocean-island basalt data base that this is indeed the case for He and Ne (reviewed by Graham, 2002). But the observational problem has been that solar isotopic signatures have appeared only subtly if at all in analyses of the heavier noble gases in mantle-derived samples; their nonradiogenic isotope ratios have generally been found to be indistinguishable or only slightly offset from those in air – a consequence, one suspects, of massive air-derived contamination by post-collection adsorption, seawater-magma interaction, subduction, or combinations of all of these (Patterson *et al.*, 1990; Pepin, 1998; Ballentine and Barfod, 2000). There is some isotopic evidence for a solar-like Ar component in measurements on two ocean-ridge and ocean-island sample suites (Pepin, 1998), but both the data and their interpretation are controversial (e.g., Kunz, 1999).

5. Long-term Alteration Processes

After the establishment of atmospheric and internal volatile reservoirs during the accretionary and early post-accretionary phases of planet formation, further modifications to isotopic ratios may occur over the long term (see, for example, Figure 6 of Jakosky and Jones, 1997), such as by infall of material, exchange between atmosphere and the crust or interior, and as a result of gravitational escape-related processes (Fox and Hać, 1997; Lammer and Bauer, 2003). In the context of escape processes, fractionation in planetary atmospheres results mainly from the diffusive separation by mass of isotopic (and elemental) species and occurs between the homopause, the level where diffusion rather than turbulent mixing becomes the controlling process, and the exobase, where collisions become rare. The lighter particles are more abundant at the exobase and exosphere than the heavier species. When particles are removed from a planetary exosphere by atmospheric loss processes, the lighter isotopes are preferentially lost and the heavier ones become enriched in the residual gas.

Diffusive separation leads to an enrichment of the lighter isotope in the exosphere, depending on the homopause altitude (Lammer and Bauer, 2003). This effect enhances the importance of all atmospheric escape processes that act at the exobase level. Atmospheric escape mechanisms which can lead to isotope fractionation in a planetary atmosphere are dissociative recombination, impact dissociation of molecules via energetic electrons, charge exchange and atmospheric sputtering (Johnson, 1990; Lammer and Bauer, 1993).

The fractionation of an atmospheric species due to loss to space of a component of that species is described by the Rayleigh fractionation law

$$R(t) = R(0)\,f^{(r-1)}, \qquad (1)$$

where $R(t)$ is the ratio of isotopic abundances in the bulk atmosphere at time t, $R(0)$ is the initial isotope ratio in the atmosphere, f is the fraction which has been lost and r is the fractionation between the homopause and the the exobase. Note that Rayleigh fractionation strictly applies only if the atmospheric species under consideration is not buffered by isotopic exchange with a reservoir in the planet's crust or mantle or by addition of fresh, unfractionated material to the atmosphere. Either of those conditions will result in observed isotopic effects being reduced, possibly to a large degree, relative to those predicted by loss of an atmospheric species to space.

5.1. Specific Examples – Venus

Measurements of the D/H ratio in the atmosphere of Venus showed an enrichment in D of two orders of magnitude relative to the terrestrial oceanic ratio (Donahue *et al.*, 1982). There have been two schools of thought regarding the implications of this observation. The first is that the high D/H ratio can be associated with a large amount of missing H_2O on Venus; i.e., Venus once had an abundance of water similar to the Earth and Mars and that this water, or at least the hydrogen, was mostly lost to space, with a resulting large D/H fractionation (Dayhoff *et al.*, 1967; Walker *et al.*, 1970). Because of the high surface temperature, H_2O on Venus must have always been in the vapor phase in the atmosphere. Hydrogen could have escaped via one of several possible mechanisms. Donahue *et al.* (1982) suggested that dynamic escape from an H_2-dominated thermosphere of high exospheric temperature would have depleted H_2 and HD together until the H_2 mixing ratio dropped to 2%, at which point D loss would become negligible and enrichment of D/H would have occurred. McElroy *et al.* (1982a) argued that collisions of hot O atoms with hydrogen would cause loss of H atoms but not D. For both mechanisms the D/H enrichment, assuming a starting D/H ratio equivalent to that of Earth, requires a minimum loss of at least 0.3% of the terrestrial ocean mass, but losses prior to the onset of D/H enrichment cannot be determined. Bauer (1983) argued that an early loss of H_2, H and O from the Venus atmosphere corresponding to as much as 2% of a terrestrial ocean would have been possible, if the early Sun underwent a highly active phase.

The current loss rate of O atoms from the Venus atmosphere to space matches that of H_2, but in the past losses of hydrogen would have been greater (McElroy *et al.*, 1982b). Preferential loss of hydrogen relative to oxygen requires a sink for the excess oxygen on the planet. Lewis (1970) presented geochemical arguments against the existence of such a sink for oxygen on Venus. Working from a model of the nebula with a thermal gradient that predicted an absence of hydrated minerals at the formation distance of Venus (Lewis, 1974a), and having no sink for excess oxygen, Lewis argued that the planet formed from nebular condensates containing little or no H_2, that it never had large amounts of H_2O, and that the present H_2O content of the Venus atmosphere is due to infall of cometary material (e.g. Lewis, 1970; Lewis, 1974b). The D/H ratios for Venus and Earth could have been inherently different from each other when they formed. Since D/H ratios in solar system materials containing hydrogen, other than ices, tend to be significantly higher than ocean water, a dry Venus would have started out with a higher D/H ratio than most people assume, and therefore need not have lost as much hydrogen as is calculated. Of course, the solar system models that Lewis was working from have been superseded by ones in which the source materials for the terrestrial planets need not be strongly dependent on radial distance from the Sun (for example, Wetherill, 1994), and Lewis' view of the state of water on Venus may no longer be viable.

5.2. Specific Examples – Mars

As in the case of Venus, an elevated D/H ratio in the Martian atmosphere suggests that there has been loss of water to space accompanied by isotopic fractionation. The situation is, however, even more complicated than Venus. The D/H ratio in the present Mars atmospheric H_2O vapor is $(8.1 \pm 0.3) \times 10^{-4}$ (Owen *et al.*, 1988; Yung *et al.*, 1988; Krasnopolsky *et al.*, 1997), greater than terrestrial ocean water by a factor of 5.2 ± 0.2. Unlike Venus, most of the water on Mars is not in the vapor phase at any given time. Enrichment in D/H as a result of loss of hydrogen to space is therefore buffered by exchange with condensed water reservoirs, either in the polar caps or within the Martian regolith or both. Determining how much water currently exists on Mars and how much has been lost over time using fractionation of D/H is thus very dependent on how one models the reservoirs and the exchange behavior over time.

5.2.1. *Mars and Hydrogen*

Modeling by Yung *et al.* (1988) indicated that Mars currently has the equivalent of a global layer of H_2O of about 0.2 m thick, the residual of an initial reservoir of $\approx$3.6 m. In a revised version of that model, using an estimate from the O loss rate to space over the Martian history that the planet has lost a global 50 m layer of H_2O, Kass and Yung (1999) derived an 8 m thick residual global reservoir exchanging with the Martian atmosphere. However, Lammer *et al.* (1996, 2002) showed that H and O are not lost from Mars in the stoichiometric water ratio, and thus raised doubts about whether one can calculate water loss based on oxygen. Their study implies a loss of H_2O to space corresponding to an equivalent depth of $\leq$17 m over 3.5 Gy. Recently, the Far Ultraviolet Spectroscopic Explorer (FUSE) spacecraft detected H_2 for the first time in the upper atmosphere of Mars. Krasnopolsky and Feldman (2001) use this observation as a starting point for a photochemical model of the upper atmosphere from which they calculate that the amount of H_2O loss required to derive the D/H ratio in the current Martian H_2O supply from a D/H ratio considered to represent Mars after the end of hydrodynamic escape (1.9 times the terrestrial ocean value, observed by Leshin, 2000, in a Martian meteorite sample) is equivalent to a global ocean of about 30 m deep.

5.2.2. *Mars and Argon*

Of the mechanisms for long-term loss to space given above, the only one capable of explaining the observed $^{36}Ar/^{38}Ar$ depletion in the Mars atmosphere (Bogard, 1997) appears to be sputtering. The probable existence of a sputtering loss mechanism on Mars was demonstrated by Luhmann *et al.* (1992). Oxygen atoms in the Martian exosphere, ionized by solar EUV radiation and accelerated in the motional $\mathbf{v} \times \mathbf{B}$ electric field of the solar wind, can impact species near the top of the atmosphere with enough energy to eject them from the planet's gravitational field. The process generates large elemental and isotopic fractionations of Ne and Ar in the residual atmosphere (Jakosky *et al.*, 1994; Pepin, 1994; Pepin and Porcelli, 2002). Loss rates of the dominant atmospheric constituent (CO_2 on Mars) in this sputtering process depend on the magnitudes of the EUV flux and the solar wind velocity, and so estimates of how both electromagnetic and corpuscular radiation have evolved over solar history are needed in order to calculate sputtering losses in past epochs. Note that this same requirement for estimates of flux history applies as well to early atmospheric evolution by EUV-driven hydrodynamic escape.

CO_2 pressure collapse on Mars near 3.7 Gy would have triggered the rapid sputtering removal of preexisting Ne and Ar from the atmosphere (Kr and Xe are too heavy to be affected). This has the interesting consequence that no isotopic memory of their earlier processing survives (Pepin, 1994). Current abundances and isotopic compositions are entirely determined by the action of sputtering on gases supplied by outgassing during the late evolutionary epoch, and final distributions of the light noble gases are therefore decoupled from whatever their inventories might have been in the pre-3.7 Gy atmosphere. The issue of how much Ne and Ar degassing was required to prevent their complete sputtering annihilation from the atmosphere during this epoch is unresolved. A conclusion of one study was that the Ne-Ar outgassing flux attributable to degassing during major episodes of volcanic activity would have been $\approx$1-3 orders of magnitude too low to balance sputtering losses, and thus another major source of juvenile volatiles must have contributed to the atmosphere over geologic time, perhaps via input from gas-enriched hydrothermal systems (Hutchins and Jakosky, 1996).

Sputtering losses are greatly attenuated by the presence of a planetary magnetic field, most importantly because it deflects the solar wind around the planet and shields atoms photoionized in the outer atmosphere from the solar wind electric field that would otherwise accelerate some of them downward toward the exobase

(Hutchins *et al.*, 1997). For this reason the process would not have been important on Earth for as long as the core dynamo has existed, and it seems unlikely that the composition of Ne and heavier noble gases in the massive Venusian atmosphere could have been substantially affected by sputtering loss with or without the protection of a paleofield. Escape of sputtered species is also impeded by the higher gravity of these two planets. In contrast, a thin, magnetically unshielded, and more weakly bound Martian atmosphere is particularly vulnerable to sputtering erosion. Efficient operation of this fractionating loss mechanism over time on Mars is therefore centrally linked both to atmospheric pressure history and to the timing of the disappearance of the Martian paleomagnetic field (Hutchins *et al.*, 1997; Connerney *et al.*, 1999).

5.2.3. *Mars and Nitrogen*

Nitrogen accounts for about 2.5% of the present Martian atmosphere. Since ^{15}N is enriched in the atmosphere by about $(62 \pm 16)\%$ relative to the terrestrial standard (Nier and McElroy, 1977) and the presumed magmatic source composition (Grady and Wright, 2003), the concentration of nitrogen on Mars must have been significantly higher in the past. The anomalous $^{15}N/^{14}N$ ratio therefore provides yet another means of modeling the history of the Martian atmosphere. Monte Carlo simulations by Fox and Hać (1997) have shown that the escape of nitrogen by dissociative recombination would have been much too efficient given the current atmospheric density. The escape of N had to have been inhibited by some means, most likely by the presence of a dense early Martian atmosphere at the start of the late evolutionary period mentioned in Section 4.1.3. This early atmosphere was then lost over time. Fox and Hać (1997) found that the current ratio could be produced if Mars had an initial CO_2 pressure above at least 114 mbar.

Note that the calculation of Fox and Hać (1997) considered only one loss mechanism for nitrogen. We know that other mechanisms, particularly sputter loss, must have been operating as well. Fox and Hać (1997) neglected the sputter contribution to N isotopic evolution because of the uncertainty of parameters involved; this process, however, would certainly have affected nitrogen to some degree potentially increasing the expected ^{15}N enhancement. Johnson and Luhmann (1998) and Leblanc and Johnson (2001) argued that feedback processes could limit the sputter loss rates during the earlier epochs. More research, including studies of the early solar radiation and particle environment, is needed to get a better idea of how efficiently the sputtering mechanism would have affected the nitrogen isotope fractionation process. Other effects, such as buffering of nitrogen to some degree by an as yet unrecognized reservoir, might be worth considering as a means of explaining the lower than expected ^{15}N enhancement.

5.3. Mars and Carbon

As noted in Section 2.3, carbon in the Martian atmosphere is also enhanced in the heavy isotope, but by an order of magnitude less than nitrogen. It is tempting to attribute the ^{13}C enhancement to the same processes that act on nitrogen, with the smaller degree of enhancement having to do with a lower scale height for CO_2 than N_2 and a lower efficiency for the dissociation reactions involved. However, Galimov (2000) has argued that the loss of carbon to space would have been insufficient to produce a 6% enrichment in the atmospheric $^{13}C/^{12}C$ ratio, at least in the early history of Mars before it lost its magnetic field and atmospheric sputtering become important. He points out that the carbonates in ALH 84001, dated at about 4 Gy (Borg *et al.*, 1999) and thought by many to have an atmospheric source for their carbon (Treiman and Romanek, 1998), have similar $^{13}C/^{12}C$ to the recent Mars atmosphere, and that it is therefore necessary to have a mechanism for producing ^{13}C-enriched CO_2 in the early Mars atmosphere. Galimov suggests that the initial Martian atmosphere was reduced, with minor amounts of CO_2 present in isotopic equilibrium with the dominant carbon species at temperatures which would yield the appropriate carbon isotopic composition in the CO_2. The CO_2 is sequestered as carbonates in the Martian crust, and, after the reduced carbon is lost, the carbonates become the source of ^{13}C-enriched CO_2.

Since we know that atmospheric loss processes have been operating in the time period since 4 Gy ago, and yet the present-day atmospheric carbon remains similar to that found in ALH 84001 carbonates, the model presented by Galimov (2000) would appear to require that atmospheric CO_2 is strongly buffered by crustal carbonates. It should also be noted that Grady *et al.* (1998) find nitrogen in ALH 84001 which is enriched by at least 20% in ^{15}N, suggesting either that atmospheric loss processes were already efficiently operating at the time alteration of this rock occurred, or that at least some atmospheric species were introduced much more recently than 4 Gy ago. These possibilities obviously need further investigation.

6. Planetary Processes

In this section we address differences in the composition of volatile elements between those retained in planetary atmospheres and those trapped in planetary interiors. The only two planets for which this exercise is possible are Earth and Mars.

6.1. Earth

It had long been assumed that atmospheric gases originated from the terrestrial mantle, mainly through early catastrophic degassing, possibly linked to terrestrial differentiation (Allègre *et al.*, 1983; Ozima, 1975), and later on through residual mantle degassing as exemplified by helium. This was based mostly on differences between the atmosphere and mantle in the amounts of radiogenic rare gas isotopes produced by long-lived (^{40}K-^{40}Ar, leading to ^{40}Ar/^{36}Ar = 295.5 in air versus 40,000 in the upper mantle, Moreira *et al.*, 1998) and short lived (^{129}I-^{129}Xe, ^{129}Xe/^{130}Xe = 6.50 in air versus 8.2 in the upper mantle, Staudacher *et al.*, 1989) radioactive elements. This view has dramatically changed with the discovery of a neon isotopic composition in the mantle different from that of the atmosphere, the mantle presenting a close to solar signature (e.g., ^{20}Ne/^{22}Ne = 13.8) whereas the atmosphere is characterized by a ^{20}Ne/^{22}Ne ratio of 9.80 (Hiyagon *et al.*, 1992; Honda *et al.*, 1991; Marty, 1989; Sarda *et al.*, 1988). Such a difference would not preclude a mantle origin for atmospheric gases, if the latter were isotopically fractionated through atmospheric processes after degassing. However, a careful examination of the ^{129}I-^{129}Xe short-lived decay ($T_{1/2}$ = 16 My) and ^{244}Pu-$^{131-136}$Xe short-lived fission ($T_{1/2}$ = 80 My; after correction for the contribution of ^{238}U-$^{131-136}$Xe long-lived fission, $T_{1/2}$ = 4.46 Gy) revealed that mantle gases could not straightforwardly be associated with atmospheric gases (Ozima *et al.*, 1985), requiring two different sources of volatile elements for the Earth (Marty, 1989).

The terrestrial mantle indeed contains volatile elements that were never degassed to the Earth's surface nor recycled back into the mantle through plate subduction. The best evidence for this is the isotopic composition of helium measured in mantle-derived rocks and minerals and in volcanic and geothermal gases, which indicates the occurrence of ^{3}He trapped during Earth's accretion (Clarke *et al.*, 1969; Mamyrin *et al.*, 1969). Helium degassing is still active and its recycling is negligible because helium is lost from the terrestrial atmosphere and because, as a light rare gas, it can hardly be recycled back into the mantle (see Hilton *et al.*, 2002, for a discussion of noble gases and mantle recycling on Earth).

The carbon isotopic composition of the surface inventory (organic plus inorganic) is about −5‰ (Galimov, 1968), not significantly different from the isotopic composition of mantle carbon of −5.4‰ to −5.4‰ (Deines, 1992). This similarity has been taken as evidence that carbon is exchanged between the mantle and the surface through magma generation along divergent plate boundaries on one hand, and through subduction of sediments and altered oceanic crust on another hand, at rates that are sufficiently fast on a geological time scale to insure efficient stirring of the concerned reservoirs (e.g., Javoy *et al.*, 1986). Note, however, that there is actually considerable variability in carbon isotopes in mantle phases (Deines, 1992) and that this complicates any models that attempt to explain carbon in terms of well-stirred reservoirs.

The isotopic composition of atmospheric xenon, when radiogenic and fissiogenic contributions are discarded, is fractionated by ≈4% per amu relative to solar, or Q, xenon (see Figure 2 of Pepin, 2000). Such fractionation could be due either to a "pre-terrestrial" process like gravitational fractionation in planetesimals (Ozima and Nakasawa, 1980), in which case the mantle should have a Xe isotope composition similar to that of air, or it could have been produced during early atmospheric processing (see Section 4.1.1), so that the mantle could have preserved non-fractionated Xe. This problem is not yet fully settled, mainly because mantle-derived samples contain extremely small amounts of trapped Xe which are difficult to analyze with precision, and which are also prone to surface contamination during magmatic evolution, lava eruption, or laboratory processing. So far, all samples analyzed with sufficient analytical precision to address this problem present an air-like isotopic composition (Moreira *et al.*, 1998; Staudacher *et al.*, 1989), with the notable exception of Xe present in some natural gases rich in CO_2 of mantle derivation, for which the high abundance of Xe allowed extremely precise analysis of Xe isotopes. These analyses have revealed that CO_2-well gas Xe contains a small (10%) component of planetary-like (or solar-like) Xe (Caffee *et al.*, 1988). This discovery suggests that the solid Earth contains non-atmospheric Xe which was diluted by atmospheric Xe recycled into the mantle through subduction processes. Recently, Pepin (1998) proposed that the mantle contains solar-like argon with 38Ar/36Ar lower than the atmospheric value, based on data from mid-ocean ridge basalts published by Niedermann *et al.* (1997). To date, analyses of argon in mantle-derived samples having sufficiently high precision have failed to confirm this claim (Marty *et al.*, 1998; Moreira *et al.*, 1998). Krypton in the mantle is also air-like, but no strong isotopic difference is expected since air Kr more closely resembles extraterrestrial krypton.

The nitrogen isotopic composition of mantle-derived rocks and minerals shows that the mantle is likely to be isotopically heterogeneous for this element. Eclogitic, peridotitic, and fibrous diamonds show a distribution of $\delta^{15}N$ values, with a strong mode around −5‰ (Boyd and Pillinger, 1994; Cartigny *et al.*, 1997; Javoy *et al.*, 1984). The distribution may be due to mixing between a light nitrogen end-member that could represent the primordial signature of the silicate Earth (Javoy, 1997) and a heavy nitrogen component originally residing at the Earth's surface and progressively recycled into the mantle. The distribution of $\delta^{15}N$ values in diamonds may also be due to isotope fractionation occurring during diamond growth. Indeed, it is similar in extent to the range of carbon isotopic variations observed in diamonds which, together with extremely variable nitrogen contents, has been attributed to fractionation processes (Cartigny *et al.*, 1998, 1991, 2001).

The mantle source of fluid-forming diamonds probably acquired this specific light nitrogen isotope signature (−5‰) as far back the Archean. Basalts erupted on the sea floor under considerable hydrostatic pressure contain CO_2-rich inclusions representing quenched gas bubbles that concentrated magmatic volatiles before eruption and have been preserved from surface-derived contamination. Nitrogen and argon extracted by crushing of Mid-Ocean Ridge Basalt (MORB) glasses are correlated, and permit the identification of a mantle source nitrogen end-member depleted in ^{15}N by 3-5‰ on average (Javoy and Pineau, 1991; Marty and Humbert, 1997). The overall resemblance of the $\delta^{15}N$ distribution for MORB and diamond data as well as comparable C/N ratios for MORB and diamonds strongly suggest that both sets of mantle-derived materials have sampled a common mantle nitrogen component. In contrast, $\delta^{15}N$ values of mantle-derived magmatic rocks associated with continental or oceanic plumes, as attested by their rare gas signatures, are positive within the range from 0‰ to +6‰ (Dauphas and Marty, 1999; Marty and Dauphas, 2002). The $\delta^{15}N$ values of plume-derived samples are comparable to those of oceanic nitrates, modern sediments, and high grade metamorphic minerals from paleosubduction zones and, together with trace element concentrations, radiogenic isotopic compositions, and stable isotopic ratios in plume-related lavas, are best explained by addition of sediments or altered oceanic crust to the source of plumes.

The origin of the light N signature in the MORB and diamond mantle could be related to early incorporation of light N from enstatite chondrites [$\delta^{15}N = (-30 \pm 10)$ ‰], which is consistent with a common oxygen isotope composition for the Earth and this class of meteorites (Javoy *et al.*, 1986). However, it could also represent a surface N component that was recycled long ago since (i) nitrogen metabolized during the anoxic Archean and incorporated later on in the crust and sediments might have been light (Beaumont and Robert, 1999; Pinti *et al.*, 2001), and (ii) there is petrologic evidence that nitrogen is indeed pro parte recycled into the mantle (Andersen *et al.*, 1995). Comparable N/K ratios between the surface (sediments + crust) and the mantle constitute a strong argument for a recycled origin of nitrogen in the mantle (Marty *et al.*, 2002), and nitrogen isotopes may therefore provide a sensitive tool for investigating crust-mantle interaction through time.

6.2. Mars

Stable isotopic compositions of Martian reservoirs indicate that the atmosphere-hydrosphere has been extensively fractionated and efficiently isolated from the Martian interior. The Martian atmosphere has a δD value of +4200‰ (Bjoraker *et al.*, 1989). Martian crustal rocks have been altered by a fluid that has exchanged with the Martian atmosphere and/or hydrosphere with δD values up to +2100‰ (e.g., Leshin *et al.*, 1996, and references therein). Ion probe investigations of magmatic minerals in Martian meteorites, by contrast, have identified a light hydrogen end-member having δD values down to −200‰, interpreted as representing a Martian mantle component of hydrogen sharing similarities with that of the Earth (Deloule, 2002). This interpretation needs to be reconciled with that of Leshin (2000), who derived a δD value for Martian magmatic water of (+900 ± 250)‰. Additional measurements of D/H in Martian minerals will be required to sort this difference out.

The $\delta^{15}N$ value of the Martian atmosphere is (+620 ± 160)‰ (Nier and McElroy, 1977). In contrast, the isotopic composition of nitrogen trapped in Martian meteorites is much lighter, with $\delta^{15}N$ values down to −30‰ in some of the low to medium temperature stepwise-heating extractions (Mathew *et al.*, 1998; Murty *et al.*, 1996; Mathew and Marti, 2001). The close association of light N and solar-like Xe, thought to represent Martian interior xenon (see below), has been taken as evidence that the light N component represents Martian mantle nitrogen (Mathew and Marti, 2001). An intriguing problem with this claim is that this light N component is extracted at relatively low temperature, 300 °C for Chassigny. It seems unlikely that a nitrogen mantle component trapped in a meteorite having been shocked to such an extent as to allow its extraction from Mars could still contain such a labile N component in its original siting. This requires an adequate process and phase to allow the hosting of mantle volatiles in such a low energy site, which has not yet been identified.

The isotopic composition of xenon in the Martian atmosphere (as recorded in SNC impact glasses) is air-like (Swindle *et al.*, 1986), whereas that of Martian mantle Xe is solar-like (Mathew *et al.*, 1998; Ott, 1988), contrary to the case of terrestrial mantle Xe. The strong difference for Xe between the Martian surface and mantle is fully consistent with the D and N isotope records, which indicate efficient insulation between the surface and interior of Mars. The isotope geochemistry of Xe allows the setting of some time constraints on this insulation.

SNCs contain radiogenic ^{129}Xe from the decay of ^{129}I and fissiogenic Xe isotopes from ^{244}Pu fission. Modeling these systems gives results that are consistent with early differentiation of Mars within the first several tens of My, possible late stage degassing during several hundreds of My, and then efficient insulation between the atmosphere and mantle during most of Martian history (Marty and Marti, 2002). Independent evidence for a low exchange rate of volatiles is provided by the ^{40}Ar/^{36}Ar ratio, which is much higher in the atmosphere ($\approx$1900, inferred from SNC meteorites by Garrison and Bogard, 1998) than in the interior (not well determined, but taken to be $\approx$300 by Garrison and Bogard, 1998). Exactly the reverse occurs in the Earth (^{40}Ar/^{36}Ar = 295.5 in air and 40,000 in the upper mantle). The Mars interior ^{40}Ar/^{36}Ar ratio, when compared to the much higher Earth mantle ratio (both planets having comparable K contents as far as orders of magnitude are concerned), indicates that Mars is still volatile-rich, implying low rates of exchange for volatile elements between the Mars interior and the surface.

6.3. Implications for the Origin of Terrestrial Planet Building Materials and for Tectonics on the Earth and Mars

Comparing terrestrial and Martian geochemical characteristics is for the moment hampered by the lack of samples of atmosphere, sediments, crust, and mantle returned from Mars. Such returned samples would allow the documentation with precision of the isotopic compositions of volatile elements and would erase ambiguities concerning the origin of SNCs and the processes experienced by these meteorites. If nitrogen with δ^{15}N down to -30‰ characterized the terrestrial mantle before dilution by N recycled from the surface (which is not a certainty given the available mantle-derived sample data), then Mars and Earth might have shared a common ancestor material, despite difference in their respective oxygen isotopic compositions. Indeed for the Earth, enstatite chondrite-like material would fit both the O isotope constraints as well as a light N ancestor (but see the caution in Section 3.2), whereas in the case of Mars, the O isotope composition would prohibit a direct link with this class of meteorites. Therefore, future investigations should be devoted to better documenting the N isotopic composition together with that of rare gases of the terrestrial mantle, particularly the deep mantle. The phase(s) hosting the light N and solar-like Xe released at low temperature from SNCs will also need to be characterized. Furthermore, we are lacking data concerning light rare gases and, given the tremendous amount of information that helium and neon have given us on the structure of the terrestrial mantle and on its convection, there is no doubt that these two elements would aid in the investigation of past and present fluxes of volatiles between the interior and the atmosphere of Mars.

Taken at face value, the isotope data on H, N, O and rare gases for Mars and Earth outline unambiguously major differences in tectonic regimes between the two planets. The Earth appears much more outgassed than Mars, although both planets experienced major episodes of differentiation early in their history which might have led, either during planetary accretion or during magma ocean episodes, to efficient degassing. However, the Martian mantle still contains a high content of rare gases compared to Earth, so that our planet must have subsequently lost gases. Magmatism at divergent plate margins may provide the required mechanism, but one should not forget that major episodes of mantle degassing, particularly the upper mantle sampled by MORBs, must have taken place early in the Earth's history to allow subsequent radiogenic in-growth of isotopes produced by long-lived radioactivities such that of ^{40}K. Thus the early Earth (subsequent to magma ocean episodes) probably had a tectonic regime much more active than that of Mars.

On Earth, subduction of oceanic plates allowed the ongoing erasure of isotopic differences between the surface and the mantle. In some cases (S, C and possibly Xe), no difference in the stable isotope compositions exists anymore. In other cases (e.g., Ar, N, H), there is some ambiguity as to the role of recycling in fixing the isotopic compositions of the two reservoirs. Neon provides unambiguous evidence that there exist deep terrestrial reservoir(s) not homogenized with the surface inventory. For Mars, as far as the available data are representative of the interior and surface reservoirs, all available tracers (H, N, Xe) indicate that the two reservoirs have been isolated very efficiently, probably after a few tens to a few hundreds of Ma as recorded in the Xe isotopes produced by extinct radioactivities. Besides gaining reliable data for the Martian reservoirs, particular effort in the future will have to be focused on integrating information gained from different tracers in a global Martian model of interior-atmosphere evolution.

7. Biological Processes

It is well known that biological processes on Earth can lead to isotopic fractionation effects in C, H and N. Isotopic fractionation effects may arise through thermodynamic or kinetic processes (Galimov, 2003), the latter being the ones of interest in this context. Organisms in the course of using CO_2, H_2O or small organic molecules process these through reaction pathways where isotopically light species tend to react faster than heavier ones. As a result organisms may concentrate isotopically light species in themselves or their byproducts. How this manifests itself for carbon is shown in Figure 1 of Galimov (2003). The size of the effect for a single stage of fractionation may be of the order of 10‰ for nitrogen, 2-3 times that for carbon, and another few times that for hydrogen.

As a result of the effects seen in carbon on Earth, the belief exists that isotopically light organic carbon "provides a distinct biomarker when compared to the inorganic carbon (p. 102, Report from the ESA Exobiology Team Study, 1997-1998). One would like to have available an unambiguous marker for biology when investigating samples such as ALH 84001, the Martian meteorite which has generated considerable controversy over claims that it contains evidence of life (Andersen *et al.*, 1995; Galimov, 1997). However this type of isotope fractionation is not a peculiar property of biological processes. It is caused by kinetic isotope effects, which accompany the formation or breaking of chemical bonds and which therefore can occur both in biological and non-biological processes.

In the biosphere the effect is related to photosynthetic fixation of CO_2. However, if we turn to the mantle and the lower crustal environment, the domain of non-biological processes, we see the same type of isotope effects in carbon. Complex organic material (so-called bitumen) occurring in magmatic rocks can be synthesized from simple precursors (CO, CH_4, etc.). Isotopically light bitumen, presumably produced in this way, has been found as fluid inclusions in minerals of alkaline igneous rocks and in olivine and garnet from kimberlites (Galimov and Petersilie, 1967; Galimov, 1991). Its carbon is depleted in ^{13}C to almost the same degree as biogenic carbon from sedimentary rocks. This is a consequence of the kinetic isotope effect in the formation of C-C bonds. Carbon compounds, found in mineral inclusions, show consecutive enrichment in the light isotope from CH_4 to C_2, C_3 and eventually bitumen (Galimov, 2000). The same type of isotope distribution has been observed in spark discharge experiments (Des Marais *et al.*, 1981). Thus, enrichment in light carbon isotopes would not by itself be conclusive evidence for a life-related origin of organic material. Similar ambiguities hold for nitrogen, oxygen and hydrogen. In order to positively distinguish a substance as being of biological origin we have to rely on properties which are unique for the chemistry of life (Galimov, 2003). The following recommendations for the future study may be made, based on that paper's analysis:

1. Emphasize the determination of isotope distributions between compounds or structures of organic materials, rather than the isotopic composition of total organic carbon.
2. Study the isotope fractionations that arise during the synthesis of such molecules of prebiotic significance as ATP (adenosine triphosphate), amino acids, nucleic acids, etc.

Suggestions for future science missions would be:

1. Seeking out brine seepages on the Martian surface as they are speculated to be favorable sites to search for extinct or even extant life. If a landing mission is designed with this objective, organic substances eluted from evaporite samples collected by a Mars lander should be tested for the presence of ATP, the compositions of isomers of amino acids, and intermolecular carbon isotope distributions.
2. A Phobos sample return mission, which can provide valuable information on prebiotic organic background on a planetary surface, since it is likely that materials present on this Martian satellite would be representative of those materials which might have come to the surface of planets during the period of the giant bombardment.

8. The Moon as a Special Case

For an object that is not actually a terrestrial planet, contains little or no internal volatile reservoirs, and has almost no atmosphere, the Moon receives a great deal of attention in attempting to understand solar system volatiles. The reasons for this are its lack of atmosphere and indigenous volatiles, which allow direct access of solar wind to its surface and provide a low background against which to measure trace amounts of implanted species. Unlike measurements of the atmospheres of Jupiter or Saturn, or satellite measurements of present-day solar wind, which provide snapshots of compositions of solar system volatiles at particular times, the lunar regolith potentially contains the entire history of the composition of solar wind, to the extent that we can decipher it.

8.1. Lunar Interior Volatiles

Geochemically the Moon is depleted in volatile elements relative to the Earth, whether directly because of its formation process or because it inherited that state from the material that formed it (Taylor, 2001). One way in which this manifests itself is in an apparent absence of indigenous noble gases (Wieler and Heber, 2003). Nitrogen and carbon have been measured in some lunar rock samples, and lunar meteorites at ppm levels (Des Marais, 1998; Grady and Wright, 2003; Marty *et al.*, 2003). It is difficult, given the ubiquity of N_2 and organic carbon on the Earth, to exclude the possibility that these are contaminants derived from sample handling and storage, but since these elements can form chemical bonds in silicate solids and liquids whereas the noble gases cannot (Marty *et al.*, 2003), it is also possible that the Moon does contain small amounts of indigenous C and N that survived the formation event. Hydrogen in lunar rocks would be hard to determine in the presence of terrestrial water contamination and the D/H ratio would be subject to some degree of alteration by cosmic-ray spallation in surface-exposed rocks.

The lunar interior does contain some volatiles however, namely those produced by radioactive decay. Radiogenic isotopes, such as ^{40}Ar, may be used in the usual way to date lunar samples (references in Wieler and Heber, 2003). When released into the lunar atmosphere and then implanted into the lunar regolith (Manka and Michel, 1971), they provide a possible means of determining the time at which a regolith sample was exposed to the solar wind, its so-called "antiquity."

8.2. Lunar Regolith Volatiles

We know that the lunar regolith contains a historical record of solar wind implanted species. In order to use the information in the record to evaluate changes in solar wind over time, we need to know when the solar wind in a given sample was implanted, and whether it has been modified by addition of nonsolar species or fractionated by loss. These issues are addressed by Wieler and Heber (2003) and by Marty *et al.* (2003).

8.2.1. *Lunar Antiquity*

Radiogenic gases released from the lunar interior are implanted in lunar grains along with solar wind, although nominally at lower energies. If the amount of radiogenic gas implanted varies in a predictable way with respect to the solar flux, the ratio of the radiogenic to solar isotopes in the grains yields secular information. As outlined by Wieler and Heber (2003), it is assumed that ^{40}Ar is released from the lunar interior at a rate proportional to its production so that its release declines with time. Absolute dates for a few samples with known ^{40}Ar/^{36}Ar then allow a calibration of the ^{40}Ar/^{36}Ar antiquity indicator (Eugster *et al.*, 2001).

The real world is not necessarily as simple as just stated. As discussed by Wieler and Heber (2003), it is not necessarily expected that ^{40}Ar release from the lunar interior follow the production rate. Even if it does today, it may not have in the past. Large impacts in K-rich crustal rocks might yield temporary transients in the atmospheric ^{40}Ar density. Lunar grains may lose ^{40}Ar preferentially over time relative to ^{36}Ar because of differing implantation depths of the two species. (Note that most analyses that suggest similar release conditions, i.e., similar sitings, for ^{40}Ar and ^{36}Ar have been done on bulk soils, where the vast majority of the trapped gases reside in grain surfaces buried within glassy particles, not pristine surfaces, and most other analyses exclude early steps with high ^{40}Ar/^{36}Ar as residual terrestrial argon.) Mixtures of soil grains may give apparent antiquities that do not correspond to any of the individual grains contained within them. Although at some level one may use the $^{40}Ar/^{36}Ar$ ratio as a measure of the relative time of exposure of a lunar regolith sample, one must exercise caution not to over-interpret the $^{40}Ar/^{36}Ar$ ratios.

8.2.2. *Nonsolar Components in the Regolith*

As discussed by Kerridge (1993) and references therein, when noble gas, N, and C contents of bulk lunar soils are compared to one another or to measures of solar wind exposure such as the ferromagnetic resonance parameter Is/FeO (Morris, 1976), very good linear correlations are obtained. The absence of large amounts of scatter was taken to indicate that there could not be random additions of nonsolar components, and any nonsolar element ratios obtained in these correlations reflected fractional loss of the depleted element. For example, the $^{14}N/^{36}Ar$ ratio in soils, roughly 10 times solar, implied a 90% loss of solar wind argon from lunar soils. Since Ar/Kr/Xe ratios in lunar soils were fractionated relative to "solar" ratios, one could make a consistent argument, and the variation in regolith $\delta^{15}N$ ratios would be dealt with separately.

In light of work by Wieler *et al.* (1999) and Wieler *et al.* (1996) on single soil grains, this view is no longer tenable. The former paper shows that 36Ar can vary independently of N in individual grains, the latter that Ar/Kr/Xe ratios in single grains are relatively constant in a soil independent of the amount of these gases in the grains and that the fractionated elemental ratios mentioned above were probably due to ionization potential effects on the solar surface. Together these papers indicate that there is almost without question significant nonsolar N in lunar soils, amounting to 90% on average of the total N present in the individual soils, there does not need to be massive Ar diffusional loss from lunar soils, and there does not appear to be such loss. Although single grain measurements have not been done on carbon, bulk soil data indicate excess carbon relative to solar Ar in the same way that they indicate excess N. It can be assumed therefore that there is a nonsolar carbon contribution to the lunar regolith in addition to the nonsolar nitrogen. The extent to which they may be related is a matter for future investigation.

The Wieler *et al.* (1999) paper suggests an explanation for the tight correlations observed in bulk soils. Individual lunar grains vary widely in inter-element ratios, large sets of individual grains (bulk soils) average towards the mean of the range of possible values, and therefore yield similar ratios from one soil to another. A clear inference is that future studies of lunar volatiles should concentrate on single grains or small subsets of grains (preferably of a single mineral) rather than bulk soils.

8.2.3. *Multiple Components in the Regolith and The Solar Wind $\delta^{15}N$ Value*

Marty *et al.* (2003) conclude that solar wind nitrogen in lunar soils has a $\delta^{15}N$ value of the order of −250‰ or below. This value is based on measured data obtained by ion probe depth profiling (Hashizume *et al.*, 2000) of one or a few grains from a soil whose $^{40}Ar/^{36}Ar$ antiquity (and cosmic-ray age) indicates an old surface exposure (no Ar data were reported on the specific grain(s) profiled) and total grain analyses for about 3 dozen grains for which ^{36}Ar was measured along with nitrogen (Figure 4 of Marty *et al.*, 2003). While Marty *et al.* may well be correct in their conclusion, and the consistency of that value with other primitive solar system N isotope ratios lends support to it (Figure 8 of Marty *et al.*, 2003), some of the authors of this chapter believe that their data allow alternative interpretations, which will be discussed below.

The vast majority of the 60-70 grains for which both N and ^{36}Ar were determined have N/Ar ratios at least 5 time solar, i.e., > 80% of the N could be considered nonsolar. Marty *et al.* (2003) suggest that the grains could have lost half their Ar, but this appears to be inconsistent with the result of Wieler *et al.* (1996) on the noble gas ratios. It is probable that the data reported by Hashizume *et al.* (2000) are from a grain dominated by nonsolar N, and it could therefore be argued that the light N in this grain, and indeed in the set of grains in Figure 4 of Marty *et al.* (2003) belong to the nonsolar rather than the solar component. Bochsler (1994) discussed a means of transferring ^{14}N-enriched nitrogen from the Earth's upper atmosphere to the lunar regolith early in lunar history which could have provided the light end-member for a variable mixture of two nonsolar nitrogen components, the other end-member perhaps being the micrometeorites discussed by Marty *et al.* (2003). The data in Figure 4 of Marty *et al.* do not allow one to unequivocally establish actual end-members in the 3-component system suggested by that figure. Some caution should perhaps be exercised in drawing the conclusions put forward there.

The particular question of the solar wind delta ^{15}N value, while still considered open to debate by some of the authors of this chapter, is expected to be settled in the next couple of years with the return of the Genesis mission mentioned in Section 3.1.2. Once that value is known, questions about the nonsolar contributions to the lunar regolith will become easier to answer.

References

Allègre, C.J., Staudacher, T., Sarda, P., and Kurz, M.: 1983, 'Constraints on Evolution of Earth's Mantle from Rare Gas Systematics', *Nature* **303**, 762–766.

Anders, E.: 1996, 'Evaluating the Evidence for Past Life on Mars', *Science* **274**, 2119–2121.

Andersen, T., Burke, E.A.J., and Neumann, E.-R.: 1995, 'Nitrogen-rich Fluid in the Upper Mantle: Fluid Inclusions in Spinel Dunite from Lanzarote, Canary Islands', *Contrib. Mineral. Petrol.* **120**, 20–28.

Ballentine, C.J., and Barfod, D.N.: 2000, 'The Origin of Air-like Noble Gases in MORB and OIB', *Earth Planet. Sci. Lett.* **180**, 39–48.

Bauer, S.J.: 1983, 'Water on Venus: Lack or Loss?', *Annales Geophysicae* **1**, 477–480.

Beaumont, V., and Robert, F.: 1999, 'Nitrogen Isotope Ratios of Kerogens in Precambrian Cherts: A Record of the Evolution of Atmosphere Chemistry?', *Precamb. Res.* **96**, 63–82.

Becker, R.H.: 1980, 'Evidence for a Secular Variation in the $^{13}C/^{12}C$ Ratio of Carbon Implanted in Lunar Soils', *Earth Planet. Sci. Lett.* **50**, 189–196.

Becker, R.H., and Pepin, R.O.: 1984, 'The Case for a Martian Origin of the Shergottites: Nitrogen and Noble Gases in EETA 79001', *Earth Planet. Sci. Lett.* **69**, 225–242.
Bhattacharya, S.K., and Thiemens, M.H.: 1989, 'New Evidence for Symmetry Dependent Isotope Effects: O + CO Reaction', *Z. Naturforsch.* **44a**, 435–444.
Bjoraker, G.L., Mumma, M.J., and Larson, H.P.: 1989, 'Isotopic Abundance Ratios for Hydrogen and Oxygen in the Martian Atmosphere', *Bull. Amer. Astron. Soc.* **21**, 991.
Bochsler, P.: 1994, 'Solar Wind Composition from the Moon', *Adv. Space Res.* **14(6)**, 161–173.
Bogard, D.D.: 1997, 'A Reappraisal of the Martian $^{36}Ar/^{38}Ar$ Ratio', *J. Geophys. Res.* **102**, 1653–1661.
Bogard, D.D., and Johnson, P.: 1983, 'Martian Gases in an Antarctic Meteorite?', *Science* **221**, 651–654.
Bogard, D.D., Clayton, R.N., Marti, K., Owen, T., and Turner, G.: 2001, 'Martian Volatiles: Isotopic Composition, Origin, and Evolution', *Space Sci. Rev.* **96**, 425–458.
Borg, L.E., Connelly, J.N., Nyquist, L.E., Shih, C.-Y., Wiesmann, H., and Reese, Y.: 1999, 'The Age of the Carbonates in Martian Meteorite ALH 84001', *Science* **286**, 90–94.
Boyd, S.R., and Pillinger, C.T.: 1994, 'A Preliminary Study of $^{15}N/^{14}N$ in Octahedral Growth Form Diamonds', *Chem. Geol.* **116**, 43–59.
Caffee, M.W., Hudson, G.B., Velsko, C., Alexander, E.C., Jr., Huss, G.R., and Chivas, A.R.: 1988, 'Non-atmospheric Noble Gases from CO_2 Well Gases', *Lunar Planet. Sci.* **XIX**, 154–155.
Cameron, A.G.W.: 1983, 'Origin of the Atmospheres of the Terrestrial Planets', *Icarus* **56**, 195–201.
Canfield, D.E., Habicht, K.S., and Thamdrup, B.: 2000, 'The Archean Sulfur Cycle and the Early History of Atmospheric Oxygen', *Science* **288**, 658–661.
Carr, M.H.: 1999, 'Retention of an Atmosphere on Early Mars', *J. Geophys. Res.* **104**, 21897–21909.
Cartigny, P., Boyd, S.R., Harris, J.W., and Javoy, M.: 1997, 'Nitrogen Isotopes in Peridotitic Diamonds from Fuxian, China: The Mantle Signature', *Terra Nova* **9**, 175–179.
Cartigny, P., Harris, J.W., Phillips, D., Girard, M., and Javoy, M.: 1998, 'Subduction-related Diamonds? – The Evidence for a Mantle-derived Origin from Coupled $\delta^{13}C$ - $\delta^{15}N$ Determinations', *Chem. Geol.* **147**, 147–159.
Cartigny, P., Harris, J.W., and Javoy, M.: 2001, 'Diamond Genesis, Mantle Fractionations and Mantle Nitrogen Content: A Study of $\delta^{13}C$-N Concentrations in Diamonds', *Earth Planet. Sci. Lett.* **185**, 85–98.
Clarke, W.B., Beg, M.A., and Craig, H.: 1969, 'Excess ^{3}He in the Sea: Evidence for Terrestrial Primordial Helium', *Earth Planet. Sci. Lett.* **6**, 213–220.
Clayton, R.N.: 1993, 'Oxygen Isotopes in Meteorites', *Ann. Rev. Earth Planet. Sci.* **21**, 115–149.
Clayton, R.N.: 2002, 'Self-shielding in the Solar Nebula', *Nature* **415**, 860–861.
Clayton, R.N.: 2003, 'Oxygen Isotopes in the Solar System', *Space Sci. Rev.*, this volume.
Clayton, R.N., and Mayeda, T.K.: 1996, 'Oxygen Isotope Studies of Achondrites', *Geochim. Cosmochim. Acta* **60**, 1999–2017.
Clayton, R.N., Grossman, L., and Mayeda, T.K.: 1973, 'A Component of Primitive Nuclear Composition in Carbonaceous Meteorites', *Science* **182**, 485–488.
Clayton, R.N., Onuma, N., Grossman, L., and Mayeda, T.K.: 1977, 'Distribution of the Pre-solar Component in Allende and Other Carbonaceous Chondrites', *Earth Planet. Sci. Lett.* **34**, 209–224.
Connerney, J.E.P., Acuña, M.H., Wasilewski, P.J., Ness, N.F., Rème, H., Mazelle, C., Vignes, D., Lin, R.P., Mitchell, D.L., and Cloutier, P.A.: 1999, 'Magnetic Lineations in the Ancient Crust of Mars', *Science* **284**, 794–798.
Cooper, G.W., Thiemens, M.H., Jackson, T.L., and Chang, S.: 1997, 'Sulfur and Hydrogen Isotope Anomalies in Meteorite Sulfonic Acids', *Science* **277**, 1072–1074.
Coplen, T.B., *et al.*: 2002, 'Compilation of Minimum and Maximum Isotope Ratios of Selected Elements in Naturally Occurring Terrestrial Materials and Reagents', *USGS Water-Resources Investigations Report* **01-4222**, U.S. Geological Survey, 98 pp.

Dauphas, N., and Marty, B.: 1999, 'Heavy Nitrogen in Carbonatites of the Kola Peninsula: A Possible Signature of the Deep Mantle', *Science* **286**, 2488–2490.

Dayhoff, M.O., Eck, R.V., Lippincott, F.R., and Sagan, C.: 1967, 'Venus: Atmospheric Evolution', *Science* **155**, 556–558.

Deines, P.: 1992, 'Mantle Carbon: Concentration, Mode of Occurrence, and Isotopic Composition', *in* M. Schidlowski, S. Golbic, M.M. Kimberly, D.M. McKirdy, and P.A. Trudinger (eds.), *Early Organic Evolution: Implications for Mineral and Energy Resources*, Springer-Verlag, Berlin Heidelberg, pp. 133–146.

Deloule, E.: 2002, 'D/H Ratio Ion Probe Measurements on Magmatic Minerals from Martian Meteorites: Implications for Degassing of the Martian Mantle', *Lunar Planet. Sci.* **XXXIII**, abstract #1607, Lunar and Planetary Institute, Houston (CD ROM).

Des Marais, D.J.: 1978, 'Carbon, Nitrogen and Sulfur in Apollo 15, 16 and 17 Rocks', *Proc. Lunar Planet. Sci. Conf. 9th*, 2451–2467.

Des Marais, D.J., Donchin, J.H., Nehring, N.L., and Truesdell, A.H.: 1981, 'Molecular Carbon Isotopic Evidence for the Origin of Geothermal Hydrocarbons', *Nature* **292**, 826–828.

Donahue, T.M., and Russell, C.T.: 1997, 'The Venus Atmosphere and Ionosphere and their Interaction with the Solar Wind: An Overview', *in* S.W. Bougher, D.M. Hunten, and R.J. Phillips (eds.), *Venus II*, Univ. Arizona Press, Tucson, pp. 3–31.

Donahue, T.M., Hoffman, J.H., Hodges, R.R., Jr., and Watson, A.J.: 1982, 'Venus was Wet: A Measurement of the Ratio of Deuterium to Hydrogen', *Science* **216**, 630–633.

Eugster, O., Terribilini, D., Polnau, E., and Kramers, J.: 2001, 'The Antiquity Indicator Argon-40/Argon-36 for Lunar Surface Samples Calibrated by Uranium-235–Xenon-136 Dating', *Meteorit. Planet. Sci.* **36**, 1097–1115.

ESA Exobiology Team Study Report (1997-1998). Exobiology in the Solar System and the Search for life on Mars. ESA Publication Division, 188 pp.

Farquhar, J., Jackson, T.L., and Thiemens, M.H.: 2000a, 'A ^{33}S Enrichment in Ureilite Meteorites: Evidence for a Nebular Sulfur Component', *Geochim. Cosmochim. Acta* **64**, 1819–1825.

Farquhar, J., Bao, H., and Thiemens, M.H.: 2000b, 'Atmospheric Influence of Earth's Earliest Sulfur Cycle', *Science* **289**, 756–758.

Farquhar, J., Savarino, J., Jackson, T.L., and Thiemens, M.H.: 2000c, 'Evidence of Atmospheric Sulphur in the Martian Regolith from Sulphur Isotopes in Meteorites', *Nature* **404**, 50–52.

Fox, J.L., and Hać, A.: 1997, 'The ^{15}N/^{14}N Isotope Fractionation in Dissociative Recombination of N_2^+', *J. Geophys. Res.* **102**, 9191–9204.

Galimov, E.M.: 1968, *Geochemistry of Carbon Stable Isotopes*, Nedra-Press, Moscow, 224 pp.

Galimov, E.M.: 1991, 'Isotope Fractionation Related to Kimberlite Magmatism and Diamond Formation', *Geochim. Cosmochim. Acta* **55**, 1697–1708.

Galimov, E.M.: 1997, 'Concerning the Existence of Life on Mars', *Solar System Res.* **31**, 183–190.

Galimov, E.M.: 2000, 'On the Phenomenon of Enrichment of Mars in ^{13}C: A Suggestion on the Reduced Initial Atmosphere', *Icarus* **147**, 472–476.

Galimov, E.M.: 2003, 'Isotopic Criteria of Origin of Organic Carbon on Earth and Meteorites', *Space Sci. Rev.*, this volume.

Galimov, E.M., and Petersilie, I.A.: 1967, 'Carbon Isotope Composition of Methane Isolated in Pores of Some Igneous Minerals', *Dokl. Akad. Nauk SSSR.* **176**, 914–917 (In Russian).

Garrison, D.H., and Bogard, D.D.: 1998, 'Isotopic Composition of Trapped and Cosmogenic Noble Gases in Several Martian Meteorites', *Meteorit. Planet. Sci.* **33**, 721–736.

Grady, M.M., and Wright, I.P.: 2003, 'Elemental and Isotopic Abundances of Carbon and Nitrogen in Meteorites', *Space Sci. Rev.*, this volume.

Grady, M.M., Wright, I.P., and Pillinger, C.T.: 1998, 'A Nitrogen and Argon Stable Isotope Study of Allan Hills 84001: Implications for the Evolution of the Martian Atmosphere', *Meteorit. Planet. Sci.* **33**, 795–802.

Graham, D.W.: 2002, 'Noble Gas Isotope Geochemistry of Mid-Ocean Ridge and Ocean Island Basalts: Characterization of Mantle Source Reservoirs', *Rev. Mineral. Geochem.* **47**, 247–317.
Haberle, R.M., Tyler, D., McKay, C.P., and Davis, W.L.: 1994, 'A Model for the Evolution of CO_2 on Mars', *Icarus* **109**, 102–120.
Hashizume, K., Chaussidon, M., Marty, B., and Robert, F.: 2000, 'Solar Wind Record on the Moon: Deciphering Presolar from Planetary Nitrogen', *Science* **290**, 1142–1145.
Heidenreich, J.E., III, and Thiemens, M.H.: 1983, 'A Non-mass-dependent Isotope Effect in the Production of Ozone from Molecular Oxygen', *J. Chem. Phys.* **78**, 892–895.
Hilton, D.R., Fischer, T.P., and Marty, B.: 2002, 'Noble Gases and Volatile Recycling at Subduction Zones', *Rev. Mineral. Geochem.* **47**, 319–370.
Hiyagon, H., Ozima, M., Marty, B., Zashu, S., and Sakai, H.: 1992, 'Noble Gases in Submarine Glasses from Mid-oceanic Ridge and Loihi Seamount: Constraints on Early History of the Earth', *Geochim. Cosmochim. Acta* **56**, 1301–1316.
Hoefs, J.:1997, *Stable Isotope Geochemistry*, Springer-Verlag, Berlin, Heidelberg, 201 pp.
Hoffman, J.H., Oyama, V.I., and von Zahn, U.: 1980, 'Measurements of the Venus Lower Atmosphere Composition: A Comparison of Results', *J. Geophys. Res.* **85**, 7871–7881.
Honda, M., McDougall, I., Patterson, D.B., Doulgeris, A., and Clague, D.A.: 1991, 'Possible Solar Noble-gas Component in Hawaiian Basalts', *Nature* **349**, 149–151.
Hunten, D.M., and Sprague, A.L.: 1997, 'Origin and Character of the Lunar and Mercurian Atmospheres', *Adv. Space Res.* **19**, 1551–1560.
Hutchins, K.S., and Jakosky, B.M.: 1996, 'Evolution of Martian Atmospheric Argon: Implications for Sources of Volatiles', *J. Geophys. Res.* **101**, 14933–14949.
Hutchins, K.S., Jakosky, B.M., and Luhmann J.G.: 1997, 'Impact of a Paleomagnetic Field on Sputtering Loss of Martian Atmospheric Argon and Neon', *J. Geophys. Res.* **102**, 9183–9189.
Istomin, V.G., Grechnev, K.V., and Kochnev, V.A.: 1983, 'Mass-spectrometry of the Atmosphere by Venera 13 and Venera 14', *Cosmic Res.* **21**, 329–338.
Jakosky, B.M., and Jones, J.H.: 1997, 'The History of Martian Volatiles', *Rev. Geophys.* **35**, 1–16.
Jakosky, B.M., Pepin, R.O., Johnson, R.E., and Fox, J.L.: 1994, 'Mars Atmospheric Loss and Isotopic Fractionation by Solar-Wind-Induced Sputtering and Photochemical Escape', *Icarus* **111**, 271–288.
Javoy, M.: 1997, 'The Major Volatile Elements of the Earth: Their Origin, Behavior, and Fate', *Geophys. Res. Lett.* **24**, 177–180.
Javoy, M., and Pineau, F.: 1991, 'The Volatiles Record of a "Popping" Rock from the Mid-Atlantic Ridge at 14° N: Chemical and Isotopic Composition of Gas Trapped in the Vesicles', *Earth Planet. Sci. Lett.* **107**, 598–611.
Javoy, M., Pineau, F., and Demaiffe, D.: 1984, 'Nitrogen and Carbon Isotopic Composition in the Diamonds of Mbuji Mayi (Zaïre)', *Earth Planet. Sci. Lett.* **68**, 399–412.
Javoy, M., Pineau, F., and Delorme, H.: 1986, 'Carbon and Nitrogen Isotopes in the Mantle', *Chem. Geol.* **57**, 41–62.
Johnson, R.E.: 1990, *Energetic Charged-Particle Interactions with Atmospheres and Surfaces*, Springer-Verlag, Berlin, Heidelberg, New York, 232 pp.
Johnson, R.E., and Luhmann, J.G.: 1998, 'Sputter Contribution to the Atmospheric Corona on Mars', *J. Geophys. Res.* **103**, 3649–3653.
Kass, D.M., and Yung, Y.L.: 1999, 'Water on Mars: Isotopic Constraints on Exchange Between the Atmosphere and Surface', *Geophys. Res. Lett.* **26**, 3653–3656.
Kasting, J.F.: 1991, 'CO_2 Condensation and the Climate of Early Mars', *Icarus* **94**, 1–13.
Kerridge, J.F.: 1993, 'Long-term Compositional Variation in Solar Corpuscular Radiation: Evidence from Nitrogen Isotopes in the Lunar Regolith', *Rev. Geophys.* **31**, 423–437.
Killen, R.M., and Ip, W.-H.: 1999, 'The Surface-bounded Atmospheres of Mercury and the Moon', *Rev. Geophys.* **37**, 361–406.

Krasnopolsky, V.A., and Feldman, P.D.: 2001, 'High-Resolution Detection of HDO of Molecular Hydrogen in the Atmosphere of Mars', *Science* **294**, 1914–1917.

Krasnopolsky, V.A., Mumma, M.J., Bjoraker, G.L., and Jennings, D.E.: 1996, 'Oxygen and Carbon Isotope Ratios in Martian Carbon Dioxide: Measurements and Implications for Atmospheric Evolution', *Icarus* **124**, 553–568.

Krasnopolsky, V.A., Bjoraker, G.L., Mumma, M.J., and Jennings, D.E.: 1997, 'Spectroscopy of Mars at 3.7 and 8 μm: A Sensitive Search for H_2O_2, H_2CO, HCL, and CH_4, and Detection of HDO', *J. Geophys. Res.* **102**, 6525–6534.

Kunz, J.: 1999, 'Is there Solar Argon in the Earth's Mantle?', *Nature* **399**, 649–650.

Lammer, H., and Bauer, S.J.: 1993, 'Atmospheric Mass Loss from Titan by Sputtering', *Planet. Space Sci.* **41**, 657–663.

Lammer, H., and Bauer, S.J.: 2003, 'Isotopic Fractionation by Gravitational Escape', *Space Sci. Rev.*, this volume.

Lammer, H., Stumptner, W., and Bauer, S.J.: 1996, 'Loss of H and O from Mars: Implications for the Planetary Water Inventory', *Geophys. Res. Lett.* **23**, 3353–3356.

Lammer, H., Lichtenegger, H.I.M., Kolb, C., Ribas, I., Guinan, E.F., and Bauer, S.J.: 2002, 'Loss of Water from Mars: Implications for the Oxidation of the Soil', *Icarus*, submitted.

Leblanc, F., and Johnson, R.E.: 2001, 'Sputtering of the Martian Atmosphere by Solar Wind Pick-up Ions', *Planet. Space Sci.* **49**, 645–656.

Leshin, L.A.: 2000, 'Insights into Martian Water Reservoirs from Analyses of Martian Meteorite QUE94201', *Geophys. Res. Lett.* **27**, 2017–2020.

Leshin, L.A., Epstein, S., and Stolper, E.M.: 1996, 'Hydrogen Isotope Geochemistry of SNC Meteorites', *Geochim. Cosmochim. Acta* **60**, 2635–2650.

Lewis, J.S.: 1970, 'Venus: Atmospheric and Lithospheric Composition', *Earth Planet. Sci. Lett.* **10**, 73–80.

Lewis, J.S.: 1974a, 'The Temperature Gradient in the Solar Nebula', *Science* **186**, 440–443.

Lewis, J.S.: 1974b, 'Volatile Element Influx on Venus from Cometary Impacts', *Earth Planet. Sci. Lett.* **22**, 239–244.

Luhmann, J.G., Johnson, R.E., and Zhang, M.H.G.: 1992, 'Evolutionary Impact of Sputtering of the Martian Atmosphere by O^+ Pickup Ions', *Geophys. Res. Lett.* **19**, 2151–2154.

Mamyrin, B.A., Tolstikhin, I.N., Anufriev, G.S., and Kamensky, I.L.: 1969, 'Anomalous Isotopic Composition of Helium in Volcanic Gases', *Dokl. Akad. Nauk SSSR* **184**, 1197–1199 (in Russian).

Manka, R.H., and Michel, F.C.: 1971, 'Lunar Atmosphere as a Source of Lunar Surface Elements', *Proc. 2nd Lunar Sci. Conf.*, 1717–1728.

Marty, B.: 1989, 'Neon and Xenon Isotopes in MORB: Implications for the Earth-atmosphere Evolution', *Earth Planet. Sci. Lett.* **94**, 45–56.

Marty, B., and Dauphas, N.: 2002, 'The Nitrogen Record of Crust-mantle Interaction from Archean to Present', *Earth Planet. Sci. Lett.*, submitted.

Marty, B., and Humbert, F.: 1997, 'Nitrogen and Argon Isotopes in Oceanic Basalts', *Earth Planet. Sci. Lett.* **152**, 101–112.

Marty, B., and Marti, K.: 2002, 'Signatures of Early Differentiation of Mars', *Earth Planet. Sci. Lett.* **196**, 251–263.

Marty, B., Tolstikhin, I., Kamensky, I.L., Nivin, V., Balaganskaya, E., and Zimmermann, J.-L.: 1998, 'Plume-derived Rare Gases in 380 Ma Carbonatites from the Kola Region (Russia) and the Argon Isotopic Composition in the Deep Mantle', *Earth Planet. Sci. Lett.* **164**, 179–192.

Marty, B., Matrajt, G., Zimmermann, L., Engrand, C., and Duprat, J.: 2002, 'Nitrogen and Noble Gas Isotopes in Antarctic Micrometeorites', *Lunar Planet. Sci.* **XXXIII**, abstract #1578, Lunar and Planetary Institute, Houston (CD ROM).

Marty, B., Hashizume, K., Chaussidon, M., and Wieler, R.: 2003, 'Nitrogen Isotopes on the Moon: Archives of the Solar and Planetary Contributions to the Inner Solar System', *Space Sci. Rev.*, this volume.

Mathew, K.J., and Marti, K.: 2001, 'Early Evolution of Martian Volatiles: Nitrogen and Noble Gas Components in ALH84001 and Chassigny', *J. Geophys. Res.* **106**, 1401–1422.

Mathew, K.J., Kim, J.S., and Marti, K.: 1998, 'Martian Atmospheric and Indigenous Components of Xenon and Nitrogen in the Shergotty, Nakhla, and Chassigny Group Meteorites', *Meteorit. Planet. Sci.* **33**, 655–664.

Mauersberger, K.: 1981, 'Measurement of Heavy Ozone in the Stratosphere', *Geophys. Res. Lett.* **8**, 935–937.

Mauersberger, K.: 2003, 'Oxygen Isotope Processes and Transfer Reactions', *Space Sci. Rev.*, this volume.

McElroy, M.B., Prather, M.J., and Rodriguez, J.M.: 1982a, 'Escape of Hydrogen from Venus', *Science* **215**, 1614–1615.

McElroy, M.B., Prather, M.J., and Rodriguez, J.M.: 1982b, 'Loss of Oxygen on Venus', *Geophys. Res. Lett.* **9**, 649–651.

Moreira, M., Kunz, J., and Allègre, C.J.: 1998, 'Rare Gas Systematics in Popping Rock: Isotopic and Elemental Compositions in the Upper Mantle', *Science* **279**, 1178–1181.

Morris, R.V.: 1976, 'Surface Exposure Indices of Lunar Soils: A Comparative FMR Study', *Proc. Lunar Sci. Conf. 7th*, pp. 315–335.

Murty, S.V.S., Mohapatra, R.K., and Clement, C.J.: 1996, 'Nitrogen and Noble Gas Components in the Martian Orthopyroxenite ALH 84001', *Lunar Planet. Sci.* **XXVI**, 1019–1020.

Niedermann, S., Bach, W., and Erzinger, J.: 1997, 'Noble Gas Evidence for a Lower Mantle Component in MORBs from the Southern East Pacific Rise: Decoupling of Helium and Neon Isotope Systematics', *Geochim. Cosmochim. Acta* **61**, 2697–2715.

Nier, A.O., and McElroy, M.B.: 1977, 'Composition and Structure of Mars' Upper Atmosphere: Results from the Neutral Mass Spectrometers on Viking 1 and 2', *J. Geophys. Res.* **82**, 4341–4349.

Norris, S.J., Swart, P.K., Wright, I.P., Grady, M.M., and Pillinger, C.T.: 1983, 'A Search for Correlatable, Isotopically Light Carbon and Nitrogen Components in Lunar Soils and Breccias', *Proc. Lunar Planet. Sci. Conf. 14th, Part 1, J. Geophys. Res.* **88** (suppl.), B200–B210.

Ott, U.: 1988, 'Noble Gases in SNC Meteorites: Shergotty, Nakhla, Chassigny', *Geochim. Cosmochim. Acta* **52**, 1937–1948.

Owen, T.: 1992, 'The Composition and Early History of the Atmosphere of Mars', *in* H.H. Kieffer, B.M. Jakosky, C.W. Snyder, and M.S Matthews (eds.), *Mars*, Univ. Arizona Press, Tucson, pp. 818–834.

Owen, T., Maillard, J.P., de Bergh, C., and Lutz, B.L.: 1988, 'Deuterium on Mars: The Abundance of HDO and the Value of D/H', *Science* **240**, 1767–1770.

Owen, T., Mahaffy, P.R., Niemann, H.B., Atreya, S., and Wong, M.: 2001, 'Protosolar Nitrogen', *Astrophys. J.* **553**, L77–L79.

Ozima, M.: 1975, 'Ar Isotopes and Earth-atmosphere Evolution Models', *Geochim. Cosmochim. Acta* **39**, 1127–1134.

Ozima, M., and Nakazawa, K.: 1980, 'Origin of Rare Gases in the Earth', *Nature* **284**, 313–316.

Ozima, M., and Podosek, F.A.: 2002, *Noble Gas Geochemistry*, Cambridge Univ. Press, 300 pp.

Ozima, M., Podosek, F.A., and Igarashi, G.: 1985, 'Terrestrial Xenon Isotope Constraints on the Early History of the Earth', *Nature* **315**, 471–474.

Patterson, D.B., Honda, M., and McDougall, I.: 1990, 'Atmospheric Contamination: A Possible Source for heavy noble gases in basalts from Loihi seamount, Hawaii', *Geophys. Res. Lett.* **17**, 705–708.

Pepin, R.O.: 1989, 'Atmospheric Compositions: Key Similarities and Differences', *in* S.K. Atreya, J.B. Pollack, M.S. Matthews (eds.), *Origin and Evolution of Planetary and Satellite Atmospheres*, Univ. Arizona Press, Tucson, pp. 291–305.

Pepin, R.O.: 1991, 'On the Origin and Early Evolution of Terrestrial Planet Atmospheres and Meteoritic Volatiles', *Icarus* **92**, 2–79.

Pepin, R.O.: 1994, 'Evolution of the Martian Atmosphere', *Icarus* **111**, 289–304.

Pepin, R.O.: 1997, 'Evolution of Earth's Noble Gases: Consequences of Assuming Hydrodynamic Loss Driven by Giant Impact', *Icarus* **126**, 148–156.

Pepin, R.O.: 1998, 'Isotopic Evidence for a Solar Argon Component in the Earth's Mantle', *Nature* **394**, 664–667.

Pepin, R.O.: 2000, 'On the Isotopic Composition of Primordial Xenon in Terrestrial Planet Atmospheres', *Space Sci. Rev.* **92**, 371–395.

Pepin, R.O.: 2003, 'On Noble Gas Processing in the Solar Accretion Disk', *Space Sci. Rev.*, this volume.

Pepin, R.O., and Porcelli, D.: 2002, 'Origin of Noble Gases in the Terrestrial Planets', *Rev. Mineral. Geochem.* **47**, 191–246.

Pinti, D.L., Hashizume, K., and Matsuda, J.-I.: 2001, 'Nitrogen and Argon Signatures in 3.8 to 2.8 Ga Metasediments: Clues on the Chemical State of the Archean Ocean and the Deep Biosphere', *Geochim. Cosmochim. Acta* **65**, 2301–2316.

Porcelli, D., Ballentine, C. J., and Wieler, R. (eds.): 2002, *Noble Gases in Geochemistry and Cosmochemistry, Rev. Mineral. Geochem.* **47**, Mineralogical Society of America, pp. 844.

Sarda, P., Staudacher, T., and Allègre, C.J.: 1988, 'Neon Isotopes in Submarine Basalts', *Earth Planet. Sci. Lett.* **91**, 73–88.

Shu, F.H., Shang, H., Glassgold, A.E., and Lee, T.: 1997, 'X-rays and Fluctuating X-winds from Protostars', *Science* **277**, 1475–1479.

Staudacher, T., Sarda, P., Richardson, S.H., Allègre, C.J., Sagna, I., and Dmitriev, L.V.: 1989, 'Noble Gases in Basalt Glasses from a Mid-Atlantic Ridge Topographic High at 14° N: Geodynamic Consequences', *Earth Planet. Sci. Lett.* **96**, 119–133.

Swindle, T.D.: 2002, 'Martian Noble Gases', *Rev. Mineral. Geochem.* **47**, 171–190.

Swindle, T.D., and Jones, J.H.: 1997, 'The Xenon Isotopic Composition of the Primordial Martian Atmosphere: Contributions from Solar and Fission Components', *J. Geophys. Res.* **102**, 1671–1678.

Swindle, T.D., Caffee, M.W., and Hohenberg, C.M.: 1986, 'Xenon and Other Noble Gases in Shergottites', *Geochim. Cosmochim. Acta* **50**, 1001–1015.

Taylor, S.R.: 2001, 'Does the Lunar Composition Retain a Memory from the Early Solar Nebula?', *Meteorit. Planet. Sci.* **36**, 1567–1569.

Thiemens, M.H.: 1999, 'Mass-Independent Isotope Effects in Planetary Atmospheres and the Early Solar System', *Science* **283**, 341–345.

Thiemens, M.H., and Heidenreich, J.E.: 1983, 'The Mass-independent Fractionation of Oxygen: A Novel Isotope Effect and its Possible Cosmochemical Implications', *Science* **219**, 1073–1075.

Treiman, A.H., and Romanek, C.S.: 1998, 'Bulk and Stable Isotopic Compositions of Carbonate Minerals in Martian Meteorite Allan Hills 84001: No Proof of High Formation Temperature', *Meteorit. Planet. Sci.* **33**, 737–742.

van Dishoeck, E.F., and Black, J.H.: 1988, 'The Photodissociation and Chemistry of Interstellar CO', *Astrophys. J.* **334**, 771–802.

Walker, J.C.G., Turekian, K.K., and Hunten, D.M.: 1970, 'An Estimate of the Present-day Deep-mantle Degassing Rate from Data on the Atmosphere of Venus', *J. Geophys. Res.* **75**, 3558–3561.

Wetherill, G.W.: 1994, 'Provenance of the Terrestrial Planets', *Geochim. Cosmochim. Acta* **58**, 4513–4520.

Wiechert, U., Halliday, A.N., Lee, D.-C., Snyder, G.A., Taylor, L.A., and Rumble, D.: 2001, 'Oxygen Isotopes and the Moon-forming Giant Impact', *Science* **294**, 345–348.

Wieler, R.: 2002, 'Noble Gases in the Solar System', *Rev. Mineral. Geochem.* **47**, 21–70.
Wieler, R., and Heber, V.S.: 2003, 'Noble Gas Isotopes on the Moon', *Space Sci. Rev.*, this volume.
Wieler, R., Kehm, K., Meshik, A.P., and Hohenberg, C.M.: 1996, 'Secular Changes in the Xenon and Krypton Abundances in the Solar Wind Recorded in Single Lunar Grains', *Nature* **384**, 46–49.
Wieler, R., Humbert, F., and Marty, B.: 1999, 'Evidence for a Predominantly Non-solar Origin of Nitrogen in the Lunar Regolith Revealed by Single Grain Analyses', *Earth Planet. Sci. Lett.* **167**, 47–60.
Wiens, R.C., Becker, R.H., and Pepin, R.O.: 1986, 'The Case for a Martian Origin of the Shergottites, II. Trapped and Indigenous gas Components in EETA 79001 Glass', *Earth Planet. Sci. Lett.* **77**, 149–158.
Wiens, R.C., Huss, G.R., and Burnett, D.S.: 1999, 'The Solar Oxygen-isotopic Composition: Predictions and Implications for Solar Nebula Processes', *Meteorit. Planet. Sci.* **34**, 99–107.
Yung, Y.L., Wen, J.-S., Pinto, J.P., Allen, M., Pierce, K.K., and Paulson, S.: 1988, 'HDO in the Martian Atmosphere: Implications for the Abundance of Crustal Water', *Icarus* **76**, 146–159.
Yurimoto, H., and Kuramoto, K.: 2002, 'A Possible Scenario Introducing Heterogeneous Oxygen Isotopic Distribution in Protoplanetary Disks', *Meteorit. Planet. Sci.* **37**, A153.

Address for Offprints: Richard H. Becker, School of Physics and Astronomy, University of Minnesota, 116 Church Street S.E., Minneapolis, MN 55455, USA; `rhbecker@maroon.tc.umn.edu`

EPILOGUE

GLOSSARY

R. KALLENBACH[1] and U. OTT[2]
[1] *International Space Science Institute, CH-3012 Bern, Switzerland*
[2] *Max-Planck-Institut für Chemie, Postfach 3060, D-55020 Mainz, Germany*

1. Astrophysical Terms

Accretion Disk: rotating disk of gas and dust falling into a newborn star.
Carbon Star: rare class of *red giant* stars rich in carbon or carbon compounds.
Dwarf Galaxy: a relatively small galaxy, typically containing some 10^7 stars. The Large and Small Magellanic Clouds, visible in the Southern Hemisphere at a distance of ≈50 kpc, are two irregular dwarf galaxies that are gravitationally bound to the Milky Way. The latter contains on the order of 10^{11} stars
Galactic Halo: spherical regions around spiral galaxies that contain dim stars and globular clusters. The radius of the halo surrounding the Milky Way extends some 50 kpc from the Galactic center (Earth is at ≈28 kpc in the Galactic disk).
Heliosphere: region of space containing plasma and magnetic fields of solar origin: a cavity carved in the *interstellar medium* by the flow of the *solar wind.*
Interstellar Medium (ISM): material found between the stars, consisting of gas, dust particles, and cosmic rays. It comprises about 10% of visible matter in the disk of our Galaxy (Milky Way).
Local Interstellar Cloud (LIC): interstellar medium near the Sun with a number density of ≈ 0.5 cm^{-3}, a temperature of ≈ 7000 K, and a size of several parsecs. The Sun presumably leaves the LIC in about 10'000 years.
Local Bubble (LB): a cavity of ≈ 100 pc radius that may have been formed by a supernova explosion; probably filled with hot, tenuous plasma of $\approx 10^6$ K and a density of $\approx 0.005 - 0.05$ cm^{-3}. The LIC resides in the LB.
Main Sequence: stage of evolution in which stars spend the major part of their lifetimes. Main-sequence stars produce energy in their cores by hydrogen fusion. Their luminosity and surface temperatures are well correlated.
Molecular Cloud: interstellar cloud with relatively high density and very low temperature of gas (roughly 75% molecular hydrogen, 21-24% helium, and a large variety of heavier molecules), dust, and mineral grains. Our solar system formed out of a molecular cloud 4.57 billion years ago.
Nova: star that, over a period of a few days, becomes 10^3–10^4 times brighter than it was; observed ≈10-15 times per year in the Milky Way. Presumably, novae are binary systems, in which a *red giant* transfers matter to its white dwarf companion. This eventually leads to a thermonuclear explosion on the white dwarf's surface.
OB Star: rare short-lived star of spectral type O or B – that is, hot and blue.
Primordial Nucleosynthesis: the process of formation of nuclei in the early universe. Five stable species are resulting: ^{1}H, ^{2}H, ^{3}He, ^{4}He, ^{7}Li.

Space Science Reviews **106:** 413–422, 2003.

Protostar: star in the process of formation, which has not yet become hot enough in the core to initiate the process of nuclear fusion ($\approx 10^7$ K) to halt its gravitational collapse. It is luminous owing to the release of gravitational potential energy from the infall of nebula material from its *accretion disk*.
Red Giant: giant star in the later stages of stellar evolution after the *main sequence*. It has a surface temperature of $\approx$2000–3000 K and a diameter of 10–100 $R_\odot$.
Solar Nebula: cloud of gas and dust from which our Sun, planets, and other bodies in our solar system formed. The gas and dust in the *accretion disk* of the young Sun is often denoted as solar nebula.
Solar Wind: a supersonic flow of a fully ionized plasma produced by the expansion of the solar corona into interplanetary space.
Supernova: explosive brightening of a star in which the energy radiated by it increases by a factor of 10^{10}; occurs when the star has burnt up all its available nuclear fuel and the core collapses catastrophically. A supernova takes several years to fade. Six of them have been observed in the Milky Way in the past 1000 y.
T Tauri Star: class of very young, flaring stars on the verge of becoming a *main-sequence* star fueled by nuclear fusion; their very variable spectra show broad and very intense emission lines.

2. Planetary Science Terms

Asteroids: rocky, metallic, 1 cm – 1000 km-sized objects, orbiting the Sun, mostly in the Asteroid Main Belt between the orbits of Mars and Jupiter; consist of pristine solar system material; most likely bodies that never coalesced into a planet.
Atmosphere: mixture of gases and traces of dust, ices, and droplets gravitationally bound to a planet. Within the first $\approx$ 90 km above the Earth's surface the composition is uniform (homosphere) except for the presence of some condensable and trace gases such as H_2O and O_3. In the troposphere, the lowest layer, atmospheric temperatures decrease with height, and they rise again in the stratosphere. The third layer in the homosphere is the mesosphere where again very low temperatures prevail. Above about 95 km the atmosphere changes considerably (heterosphere): Molecular diffusion will dominate transport of mass and energy and gases start to separate according to their molecular mass. In the thermosphere temperatures rise rapidly until, at the highest part, the exosphere, isothermal conditions under high temperatures are found (around 500 km). Light gases such as hydrogen can escape.
Coma: the spherical envelope of gas and dust surrounding the nucleus of an active *comet*, created when the ambient heat causes the vaporization of comet material.
Cometesimals: (icy) *planetesimals* that presumably formed the *comets*.
Differentiated Planet: a *planet* that has separated into regions of different composition, for instance core, mantle, and crust; heavier material sinks to the core and lighter materials rise toward the surface.

Hydrosphere: the portion of Earth, or other planet, that is water, including liquid water, ice, and water vapor on the surface, underground, or in the *atmosphere*.
Ionosphere: region of charged particles in a planet's upper *atmosphere*; the Earth's ionosphere is at an altitude of $\approx 40 - 400$ km.
Kuiper Belt: flat region of space in the radial zone extending outwards from the orbit of Neptune (at 30 AU) to 50 AU, containing at least 70000 objects with diameters larger than 100 km on low-eccentricity, low-inclination orbits.
Lithosphere: rigid outer layer of a planet.
Meteorites: fragment of small interplanetary debris from *asteroids*, comets, Mars, or the Moon, that survives atmospheric entry (at speeds of $15 - 70$ km/s) and can be recovered on the ground; categorized as *stony, iron,* and *stony-iron.*
Oort Cloud: spherical distribution of some 10^{12} to 10^{13} long-period *comets* orbiting the solar system with semimajor axes between 1'000 and 50'000 AU (0.25 pc), typically with low orbital eccentricity. The inner Oort Cloud (<15000 AU) is affected by Galactic tidal forces and the outer Oort Cloud (>15000 AU) by stellar perturbations, respectively. These perturbations bring the comets into the inner solar system.
Perihelion: point at which a body's orbital motion takes it closest to the Sun.
Planetesimals: $\approx$1–100 km-sized bodies formed in the early solar system by accretion of dust (rock) and ice (if present) in the central plane of the solar nebula. Most planetesimals accreted to planets, but many – such as the *asteroids* – never combined to form large bodies.
Terrestrial Planets: Mercury, Venus, Earth, and Mars.
Troposphere: lowermost region of a planetary *atmosphere* where convection keeps the gas mixed and maintains a steady decrease of temperature with height. Most clouds are in the troposphere.

3. Meteorite Classes

Achondrites: *meteorites* in which *chondrules* are not visible; formed from the complete melting or partial melting of the original chondritic material, which had occured due to the *differentiation* of the parent body (*planet* or *asteroid*), the heat of radioactive decay, or due to impacts or any combination of these events. Achondrites are sometimes difficult to identify because they resemble terrestrial *igneous* rocks. They are found in several rare types: evolved or *differentiated* achondrites such as the HED meteorites (Howardites, Eucrites, Diogenites) presumably from the *asteroid* Vesta, *Angrites*, *Aubrites*, *SNC meteorites* from Mars, and *primitive* achondrites.
Angrites: ancient *achondrite*, chiefly consisting of clino*pyroxene*, *anorthitic plagioclase*, and kirschsteinite (*olivine* with chemical formula $CaFeSiO_4$).
Aubrites: evolved *achondrite* of *igneous* origin, consisting primarily of *enstatite*.

Carbonaceous Chondrites (CC): rare and *primitive* – though complex – *meteorites* containing organic compounds and water-bearing *minerals* that are evidence for the inclusion of water not long after formation. Chemical and *mineralogical* differences are named for the type specimens Ivuna (CI), Mighei (CM), Vigarano (CV), Ornans (CO), Karoonda (CK), and Renazzo (CR). The CI, CM, and CR meteorites are richer in carbon than the CO, CV, and CK meteorites. The petrologic grade designations of C chondrites indicate increasing *metamorphism* of the *meteorite* by heat (grades 3 to 6) or by water, with grade 2 indicating less *metamorphism* than grade 1.

Chondrites: class of primitive *stony meteorites* containing *chondrules* inside their primary *matrix*; formed shortly after the Sun's birth ($\approx$ 4.57 Gyr ago); sub-classes are *ordinary*, *carbonaceous*, *enstatite*, and Rumuruti chondrites.

Chondrules: small glassy spherical inclusions in *meteorites*.

Differentiated Meteorites: exhibit the type of chemical fractionation one would expect to occur on a *differentiated planet*.

Diogenites: *igneous achondrites* from the deeper crust of the *asteroid* Vesta; composed of Mg-rich ortho-*pyroxene* and minor amounts of *olivine* and *plagioclase*.

Enstatites (E Chondrites): rare *chondrites* rich in *enstatite*; most of their Fe is in the form of metal or sulfide, rather than oxides in *silicates*, implying that they originate from a oxygen-poor region of the *solar nebula*, possibly inside the present orbit of Mars; further classified by petrographic grades 3 to 6 and according to their Fe content (EH indicates approximately 30% Fe and EL about 25% Fe).

Eucrites: *igneous achondrites* presumably representing the *crust* of the *asteroid* Vesta; primarily composed of Ca-poor *pyroxene* (pigeonite), and Ca-rich *plagioclase* (anorthite), but, additionally often contain accessory minerals such as *silica, chromite, troilite,* and nickel-iron metal.

Howardites: *polymict breccias* consisting of *diogenic* and *eucritic* compounds.

Iron Meteorites: represent only $\approx$ 5% of *meteorite* falls; some are thought to be pieces of the shattered *cores* of *differentiated asteroids*; structure variations result from the ratio of the two Ni-Fe alloys, *kamacite* and *taenite*, that have crystallized forming the core of the parent *asteroids*.

Mesosiderites: a broad class of *stony-iron meteorites*, surface *regolith*, fragment mixture (*breccia*) of mantle rock and Ni-Fe alloy; much more uniform metal compositions than *iron meteorites*; stirred up and fused by repeated impacts.

Ordinary Chondrites (OC): the most common type of *meteorite* fall on Earth; further classified by their Fe content: H denotes an Fe content of $\approx$ 27% by weight, L corresponds to $\approx$ 23% of Fe, and LL means "low iron" and "low metal" ($\approx$ 20% Fe); numbers 3 – 7 following the H, L, and LL classifications are petrographic grades indicating the degree of *chondrule* alteration by heating.

Pallasites: a class of *stony-iron meteorites*; their metallic *matrix* contains large green *olivine* crystals; probably formed at *asteroids*' *core-mantle* boundary.

Primitive Meteorites: contain chemically unmodified or relatively unaltered material representative of the *solar nebula*; opposite of *differentiated meteorites*.

SNC Meteorites: include the 3 *meteorite* classes named after the location of the fall of a prominent member of each class: Shergottites, Nakhlites, Chassignites; identified to originate from Mars by their young ages, *basaltic* composition, and inclusion of Martian atmospheric gas. (Another Martian meteorite is ALH 84001.)
Stony-iron Meteorites: have both stone and metal fractions; two main groups: *mesosiderites* and *pallasites.*
Ureilites: coarse grained *primitive achondrites* which are made up of *olivine* and *pyroxene* with graphite or diamond present in the areas between grains; unresolved origin, but the diamond formation has been attributed to shock.

4. Trapped Noble Gases in Meteorites

Trapped noble gases in meteorites are noble gases which **neither** originate as *in-situ* products from radioactive decay ("radiogenic") including fission ("fissiogenic") **nor** from nuclear interactions with meteoritic matter ("spallogenic", "cosmogenic", "nucleogenic"). The latter are commonly marked by subscript r (e.g. radiogenic $^{40}Ar_r$), f (e.g. fissiogenic $^{136}Xe_f$), and c or s (e.g. cosmogenic $^{21}Ne_c$, spallogenic $^{21}Ne_s$). Subscript t usually designates the trapped portion of a noble gas (e.g. $^{36}Ar_t$).

4.1. Modern Nomenclature of Major Trapped Noble Gas Components in Meteorites

The following modern nomenclature on major trapped noble gas components in meteorites is based on cf. Ott (2002). The terms apply for all noble gases from He to Xe, except for the components named R and U, which only apply for Ne (R-Ne) and Xe (U-Xe), respectively.

SW (solar wind) noble gases: noble gases implanted by the solar wind into mineral grain surfaces that have been part of an *asteroidal regolith*; present in "gas-rich" *meteorites*, but also in lunar soils from the lunar *regolith.* Elemental abundances in most minerals are moderately fractionated compared to solar, favoring the heavy elements.
SEP noble gases: more deeply implanted gases than *solar wind* particles; probably of solar origin (Solar Energetic Particles), however, particles of interstellar origin may contribute as well.
Q gases: major part of "planetary" Ar, Kr, and Xe. Ar-Q, Kr-Q, Xe-Q usually dominate the inventory of these gases in *primitive meteorites.* The Q gases show strong elemental fractionation with regard to solar. Most likely, but not necessarily of "local" origin, thought to have been trapped in the *solar nebula* by the ill-defined carbonaceous carrier "phase Q." Isotopically heavier than *solar wind* gases, with differences ill understood.

HL gases: isotopically anomalous gases carried by pre-solar diamonds, from which they are released at intermediate temperatures (max. release at $\approx$1300-1400°C). Most distinctive is the isotopic Xe-HL signature with overabundances in the light (Xe-L) and heavy (Xe-H) isotopes. Xe-H and Xe-L are assumed to have a different origin, but efforts of separating them so far have not been successful. In the absence of solar gases, He-HL and Ne-HL dominate the He and Ne inventory of *primitive meteorites* ("planetary" He and Ne, He-A, Ne-A).

P3 gases: low-temperature noble gas component carried by pre-solar diamonds. Isotopically "normal", similar to but not identical to Q-gases.

P6 gases: less well-characterized component released from pre-solar diamonds at highest extraction temperatures mixed with HL gases. It is not clear whether P6 is isotopically "normal", similar to P3, or an "exotic" component.

G component: noble gas component found in pre-solar SiC grains and owing its origin to nucleosynthesis in the He-burning shell of AGB stars. Distinctive are the compositions of Kr and Xe, which show the signatures of the slow neutron capture process (s-process), and of Ne, which is dominated by ^{22}Ne. Other names: s-process-Kr, -Xe, Ne-E(H). S-process-Kr and Xe occur also in a sub-fraction of pre-solar graphite grains.

N component: isotopically approximately normal component from the envelopes of AGB stars, not influenced by nucleosynthesis in these stars. Found in pre-solar SiC grains together with the G component. Most likely G gases from the He burning shell and N gases were mixed in the envelopes of their parent stars before being trapped by SiC grains.

R component (Ne only): mono-isotopic ^{22}Ne found in a fraction of pre-solar graphite grains; presumably from ^{22}Na decay in the grains; corresponds to Ne-E(L).

U-Xe (Xe only): Xenon component with a composition derived from multidimensional correlations in stepwise release of bulk *carbonaceous chondrites.* Maybe most primitive Xe in the Solar System, but has never been observed directly. Relative to U-Xe, solar wind Xe shows overabundances of ^{134,136}Xe, but otherwise the compositions are identical.

4.2. Alternative and Historic Names

"Planetary Gases" or **"Planetary Component"**: inventory of all trapped gases in meteorites **not** of solar composition; in general dominated in Ar, Kr, Xe by Q-gases, in He and Ne by the HL component in pre-solar diamond.

AVCC-Kr, -Xe: Kr and Xe as found in *primitive (carbonaceous) chondrites* (Average Carbonaceous Chondrite). Dominated by the Q component, but including contributions from gases in pre-solar grains, with gases in diamond (HL, P3, P6) contributing most. AVCC is a planetary gas.

P1: alternative name for Q component.

OC-Xe: same as Xe-Q and Xe-P1.

He-A, Ne-A, (Ar-A): "planetary" He, Ne (and Ar) as observed in bulk meteorites; He and Ne are dominated by the HL component in pre-solar diamonds,

with contributions from the P3 (=A1) component in pre-solar diamonds, the ^{22}Ne-rich components in pre-solar silicon carbide and graphite (Ne-G, Ne-R), and minor contributions from the Q component. "Planetary" Ar is mostly Ar-Q.
Ne-A1: same as Ne-P3.
Ne-A2: same as Ne-HL (including unresolved but probably negligible Ne-P6).
Ne-B (He-B, Ar-B): most clearly idendified for Ne. Ne (He, Ar) component found in "gas-rich" meteorites. Mixture of solar wind and SEP-Ne (-He, -Ar).
Ne-C (He-C, Ar-C): component identified – most clearly for Ne – in stepwise heating of bulk gas-rich meteorites. Probably related to SEP-Ne (-He, -Ar).
Ne-E: pre-solar ^{22}Ne-rich Ne in *primitive meteorites.* Includes both Ne-G (=Ne-E(H)) and Ne-R (=Ne-E(L)).
Ne-E(H): pre-solar ^{22}Ne-rich component found in pre-solar silicon carbide grains. Named Ne-E(H) because it is hosted by a phase of high density and release temperature. Ne-E(H) is identical with Ne-G.
Ne-E(L): pre-solar ^{22}Ne-rich component found in pre-solar graphite grains. Named Ne-E(L) because it is hosted by a phase of low density and release temperature. Ne-E(L) is identical with Ne-R.
S-process-Kr, -Xe: Kr and Xe produced in the He burning shell of AGB stars; found in grains of pre-solar silicon carbide (=Kr-, Xe-G) and graphite.
CCF-Xe (CCFX-Xe): formerly used for Xe-HL. CCF stands for "*carbonaceous chondrite* fission", because the heavy part (Xe-H) was originally thought to be a fission product.

4.3. Some Less Wide-spread and/or Less-well Characterized Components

Ureilite Gases: gases as found in the carbon- and noble-gas-rich *achondritic ureilites.* Mostly contained in diamonds. In composition similar to the Q component in *primitive meteorites.*
Kenna-type Xe: Xe composition as found in *ureilites* (after the ureilite Kenna).
Subsolar Noble Gases: component with rather high Ar/Kr/Xe abundance ratios, intermediate between solar and "typical planetary" values, and isotopic compositions similar to solar. Found in *enstatite chondrites* of higher petrographic type.
Sub-Q Gases: component found in *enstatite chondrites* of lower petrographic type. Elementally even more fractionated than Q gases. May constitute an extension of the range of compositions observed for Q rather than an independent component.

5. Mineralogical and Petrographic Terms

Anorthite: calcium end member of the *plagioclase feldspar* group; $Ca(Al_2Si_2O_8)$.
Anorthosite: lunar crustal rock chiefly consisting of *anorthite.*
Amphibole: group of ferromagnesian *silicates* with a double chain (single chain: see *pyroxene*) of silicon-oxygen tetrahedra.

Basalt: fine-grained, dark-colored *igneous* rock; contains 48 – 52% silica (SiO_2); chiefly composed of *plagioclase, feldspar, pyroxene*, but also of other *minerals*, e.g. *olivine* and *ilmenite*; most common volcanic rock on the terrestrial planets.
Breccia: rock consisting of broken fragments of rock (clasts) cemented together by a fine-grained *matrix*; formed from regolith during subsequent impacts.
Breunnerite: $(Mg,Fe)CO_3$.
Calcite: calcium carbonate, $CaCO_3$; major constituent of limestone.
Clay Minerals: finely crystalline, hydrous *silicates*; *weathering* or hydrothermal alteration product of, e.g., *feldspar, pyroxene*, and *amphibole*.
Chromite: brownish-black cubic *mineral* of the *spinel* group: $(Cr,Al)_2[Fe,Mg)O_4]$.
Enstatite: belonging to the *pyroxene* group; $MgSiO_3$.
Feldspar: alumino*silicate mineral*; positive ions of K, Na, and Ca fill interstices of negatively charged framework of silicon-oxygen and aluminum-oxygen tetrahedra; depending on composition monoclinic or triclinic.
Fullerene: C_{60} molecule; "soccer-ball"-like structure.
Garnet: high-pressure, deep-seated *mineral*; $A_3B_2(SiO_4)_3$ where A = Ca, Mg, Fe^{+2}, and Mn^{+2}, and B = Al, Fe^{+3}, Mn^{+3}, V^{+3} and Cr.
Igneous: involving the solidification of hot, molten magma or lava.
Ilmenite: $FeTiO_3$; black; hexagonal crystal; found in *basalts*.
Kamacite: Ni-Fe alloy with $\approx$ 27–65% Ni.
Mafic: of or pertaining to rocks rich in dark, ferromagnesian *minerals*.
Magnetite: Fe_3O_4; member of the *spinel* group.
Matrix: solid matter in which a fossil, crystal, or rock fragment is embedded.
Metamorphic: pertaining to rocks that have recrystallized in a solid state as a result of changes in temperature, pressure, and chemical environment.
Mica: hydrous *silicates* of Al with various bases (K, Mg, Fe, and Li).
Mineral: inorganic substance of definite chemical composition and usually of definite crystal structure; most rocks contain several minerals.
Olivine: rock-forming *igneous silicate mineral*; $(Mg,Fe)_2SiO_4$.
Phyllosilicates: minerals such as *clay* or *micas* which contain structural OH and whose Si atoms are arranged in a sheet structure.
Plagioclase: light-colored *feldspar* ranging from $NaAlSi_3O_8$ to $CaAl_2\ Si_2O_8$.
Polymict: pertaining to multiple rock-types combined in a single rock.
Pyroxene: group of ferromagnesian *silicates* with a single chain (double chain: see *amphibole*) of silicon-oxygen tetrahedra; $(Fe,Mg,Ca)SiO_3$; crystal structure is monoclinic (Clinopyroxene) and orthorhombic (Orthopyroxene, varying from *enstatite* to ferrosilite); common in *basalt*.
Siderite: $FeCO_3$; named in 1845 from the Greek sideros, for iron.
Silicate: variety of *minerals* that contain Si, O, and one or more metals.
Spinel: *mafic* rock; (magnesium-iron) aluminum *silicate* material, e.g. $MgAl_2O_4$; has great hardness; forms octahedral crystals.
Taenite: Ni-Fe alloy with $\lesssim$7.5% Ni.
Troilite: *mineral* with chemical formula FeS.

6. Chemical Terms

The following explanations should serve the reader as a link to the literature of chemistry, in particular of organic chemistry.

Adenine: one of the four nitrogen-containing bases occurring in *nucleotides*; consists of two combined five-fold symmetry and six-fold symmetry rings of C and N bonds, the two rings sharing two C atoms; an NH_2 group is bonded to the six-fold symmetry ring.
Adenine Nucleotide: *ribose* sugar with an *adenine* base bonded – by one of the N atoms in the five-fold symmetry ring of *adenine* – to one of the C atoms in the *ribose* ring – substituting an OH group – , and a *phosphate* group attached at the CH_2OH group of *ribose* (substituting one H).
Adenosine Triphosphate (ATP): transfers energy from chemical bonds to endergonic (energy absorbing) reactions within the cell; consists of the *adenine nucleotide* plus two other phosphate groups. The energy is stored in the bond of the third phosphate attached to ADP (adenosine diphosphate).
Aldehyde: O=RCH, i.e. containing a carbonyl group C=O; R is, e.g., CH_3 for acetaldehyde.
Calvin cycle: second stage in the process of photosynthesis fixing carbon molecules from CO_2 by the scheme C-5 + C-1 $\rightarrow$ 2 C-3. The Calvin Cycle is a dark reaction and uses the free energy of cleavage of *ATP*, which is created by photophosphorylation, and the reducing power of NADPH.
C-3 plants: examples are wheat, rice, and soybeans; created by the *Calvin cycle*; respond readily to an increase in atmospheric CO_2 with increased productivity.
C-4 plants: examples include maize (corn), sorghum, millet, and sugarcane; created by the *Hatch-Slack cycle*; less efficient photosynthesizers in a carbon-enriched atmosphere than *C-3 plants*;
Cyanohydrin: HO−RCR−C≡N; formed by nucleophylic addition of HCN to the carbonyl group of an *aldehyde* or a *ketone*: HCN + O=CRR' $\rightarrow$ NC−RCR'−OH.
Deoxyribonucleic Acid (DNA): *nucleic acid* composed of two polynucleotide strands wound around a central axis to form a double helix; the repository of genetic information; *nucleic acid* that functions as the physical carrier of inheritance for 99% of all species. The basic unit, the *nucleotide*, consists of a molecule of *deoxyribose* sugar, a *phosphate* group, and one of four nitrogenous bases such as *adenine*.
Deoxyribose: *ribose* sugar with H instead of OH at one C atom in its ring.
Fugacity of Oxygen: a measure of the availability of oxygen to react.
Hatch-Slack cycle: addition of one molecule of CO_2 to phosphoenolpyruvate (PEP, $C_3H_2O_2$–O–PO_4), resulting in oxalacetate ($C_4H_2O_5$), the first product in the photosynthesis of *C-4 plants*.
Ice: used by planetary scientists to refer to water, methane, and ammonia which usually occur as solids in the outer solar system.

Iron-Wustite (IW) Buffer: reference buffer to which the oxygen fugacity of a system is compared (Wustite is FeO).
Ketone: O=RCR.
Nucleic Acids: *polymers* composed of *nucleotides*; e.g., *DNA* and RNA.
Nucleotides: subunits of *nucleic acids*; fundamental structural unit of the *nucleic acid* group of *organic* macromolecules; composed of a *phosphate*, a sugar such as *ribose*, and a nitrogen-containing base such as *adenine*. Some nucleotides are involved in information storage (as nucleotides in *DNA*), protein synthesis (as nucleotides in RNA), and energy transfers (e.g. as single nucleotide *ATP*).
Organic: pertaining to carbon-containing compounds. Organic compounds can be formed by both biological and non-biological processes.
Phosphate: compound with PO_4^{2-} anion.
Polycyclic Aromatic Hydrocarbons (PAHs): class of very stable organic molecules made up of only carbon and hydrogen (e.g., coronene, $C_{24}H_{12}$); flat molecules, with each C having three neighboring C atoms like graphite; carcinogenic.
Polymer: *organic* molecule composed of smaller units known as monomers.
Pyrolysis: thermal degradation of complex material in an inert atmosphere or a vacuum; causes molecules to cleave at their weakest points to produce smaller, volatile fragments called pyrolysate.
Reduced Partition Function Ratio: see Bigeleisen and Mayer (1947): $(s/s') = f^{-1}\prod_1^f u_i \exp(-u_i/2)\times[1-\exp(-u_i')]/\exp(-u_i'/2)/[1-\exp(-u_i)]/u_i'$; $u_i = h\nu_i/(kT)$; $u_i' = h\nu_i'/(kT)$; f: degree of freedom of vibrational motion; h: Planck's constant; ν_i and ν_i': frequency of the ith molecular vibration of the heavier and lighter isotopic species.
Ribose: five-carbon sugar (ring formed of 4 C atoms with one O atom; OH group and H bonded to each of three of the C atoms in the ring, while at the fourth C atom a CH_2OH group is attached instead of an OH group).
Strecker Synthesis: chemical reaction needed for the synthesis of amino acids, peptides, and proteines; nucleophilic addition then nucleophilic acyl substitution: $O{=}RCH + NH_4Cl + NaCN \rightarrow H_2N{-}RCH{-}CN + \ldots \rightarrow H_3N^+{-}RCH{-}CO_2^-$, where the last step occurs under heat and addition of H_3O^+.

References

Bigeleisen, J., and Mayer, M.G.: 1947, *J. Chem. Phys.* **15**, 261.
Ott, U.: 2002, 'Noble Gases in Meteorites: Trapped Components', *in* D. Porcelli, C.J. Ballentine, and R. Wieler (eds.), *Noble Gases and Cosmochemistry*, pp. 71–100. Reviews in Mineralogy and Geochemistry, Vol. 47, Mineralogical Society of America, Washington.

LIST OF ACRONYMS

ACE	Advanced Composition Explorer (NASA mission)
AGB	Asymptotic Giant Branch star
ALMA	Atacama Large Millimeter Array
AU	Astronomical Unit, 1.496×10^{11} m
AVCC	Average Carbonaceous Chondrite
BB	Big Bang
BBN	Big Bang Nucleosynthesis (Primordial Nucleosynthesis)
BIMA	Berkeley Illinois Maryland Association (millimeter-wave radio interferometer array)
CAI	Ca-Al-rich Inclusions
CC	Carbonaceous Chondrites
CELIAS	Charge, Element, and Isotope Analysis System on board SOHO
CHON	stands for the combination of the elements C, H, O, and N
CIRS	Cassini Infrared Spectrometer
CME	Coronal Mass Ejection
CNES	Centre National d'Etudes Spatiales
CNRS	Centre National de la Recherche Scientifique
CONTOUR	Comet Nucleus Tour (NASA discovery mission)
COROT	COnvection ROtation and planetary Transits (ESA mission)
COSIMA	Cometary Secondary Ion Mass Analyzer on board Rosetta
CP	Cyanopolyyne
CRE	Cosmic Ray Exposure age
CVODE	stiff/non-stiff ODE code in C
ESA	European Space Agency
ESO	European Southern Observatory
DFMS	Double Focusing Mass Spectrometer (ROSINA)
DNA	Deoxyribonucleic Acid
EUV	Extreme Ultraviolet
EUVE	Extreme Ultraviolet Explorer (NASA mission)
FIP	First Ionization Potential
FIRST	Far Infrared and Submillimeter Telescope (ESA mission now named Herschel)
FUSE	Far Ultraviolet Spectroscopic Explorer (NASA mission)
FUV	Far Ultraviolet
GCMS	Gas Chromatograph and Mass Spectrometer on board the Huygens probe (ESA mission combined with NASA's Cassini)
GHRS	Goddard High Resolution Spectrograph
GPMS	Galileo Probe Mass Spectrometer
HED	Howardite, Eucrite, and Diogenite meteorites
HEF	Hydrodynamic Escape Fractionation

Space Science Reviews **106:** 423–425, 2003.

HIFI	Heterodyne Instrument for the FIRST mission
HIP	High Ionisation Phase
HST	Hubble Space Telescope
ICE	International Cometary Explorer (see ISEE 3)
ICI	Ion Composition Instrument on board ISEE 3
IDP	Interplanetary Dust Particle
IF	Ionization Front
IMAPS	Interstellar Medium Absorption Profile Spectrograph
IMP	Interplanetary Monitoring Platform (NASA mission)
INSU	Institut National des Sciences de l'Univers
IR	Infrared
IRAS	Infrared Astronomical Satellite
IRIS	Infrared Spectrometer on the Voyager Spacecraft
ISAS	Institute of Space and Astronautical Science; Japan's national space research center
ISEE 3	International Sun Earth Explorer 3 (third mission by ESA/NASA later renamed ICE)
ISM	Interstellar Medium
ISO	Infrared Space Observatory (ESA/ISAS/NASA mission)
IUE	International Ultraviolet Explorer (ESA/NASA/UK mission)
JASSI	Jupiter: Atmospheric Sounding and Sensing of the Interior
JCMT	James Clerk Maxwell Telescope
LB	Local Bubble
LIC	Local Interstellar Cloud
LIP	Low Ionisation Phase
LISM	Local Interstellar Medium
LMT	Large Millimeter Telescope
LTE	Local Thermal Equilibrium
LWS	Long-wavelength Spectrometer
MIF	Mass-independent chemical Fractionation
MORB	Mid Ocean Ridge Basalt
MTOF	Mass Time-of-Flight Spectrometer (SOHO/CELIAS)
NASA	National Aeronautics and Space Administration (USA)
NGIMS	Neutral Gas and Ion Mass Spectrometer (ROSINA)
NUV	Near Ultraviolet
OC	Ordinary Chondrites
OCZ	Outer Convective Zone of the Sun
ODE	Ordinary Differential Equations
OM	Organic Molecules
PACS	Photodetector Array and Camera Spectrometer; instrument on the planned Herschel (FIRST) mission
PAH	Polyaromatic Hydrocarbons
PDR	Photon-dominated Region or Photo Dissociation Region

QSO	Quasi Stellar Object
Rosetta	ESA mission for a rendezvous with comet 46 P/Wirtanen
ROSINA	Rosetta Orbiter Spectrometer for Ion and Neutral Analysis
SCIP	Solar Composition Icy Planetesimal
SEP	Solar Energetic Particle
SIMS	Secondary Ion Mass Spectrometry
SMOW	Standard Mean Ocean Water
SNC	Shergotty, Nakhla, Chassigny (Martian meteorite classes)
SOHO	Solar and Heliospheric Observatory (ESA/NASA mission)
STIS	Space Telescope Imaging Spectrograph
SW	Solar Wind
SWC	Apollo Solar Wind Composition experiment
SWICS	Solar Wind Ion Composition Spectrometer on board Ulysses
SWS	Short-wavelength Spectrometer on board ISO
TEM	Transmission Electron Microscope
Ulysses	ESA mission exploring the heliosphere in the ecliptic plane and over the solar poles
UMIST	University of Manchester Institute of Science and Technology
UOC	Unequilibrated Ordinary Chondrites
UV	Ultraviolet
UVCS	Ultraviolet Coronal Spectrograph on board SOHO
VLT	Very Large Telescope of ESO
VLTI	Very Large Telescope Interferometer
VUV	Vacuum Ultraviolet
WIND	NASA mission to explore the solar wind
YSO	Young Stellar Object
ZPE	Zero Point Energy

Space Science Series of ISSI

1. R. von Steiger, R. Lallement and M.A. Lee (eds.): *The Heliosphere in the Local Insterstellar Medium.* 1996 ISBN 0-7923-4320-4
2. B. Hultqvist and M. Øieroset (eds.): *Transport Across the Boundaries of the Magnetosphere.* 1997 ISBN 0-7923-4788-9
3. L.A. Fisk, J.R. Jokipii, G.M. Simnett, R. von Steiger and K.-P. Wenzel (eds.): *Cosmic Rays in the Heliospehre.* 1998 ISBN 0-7923-5069-3
4. N. Prantzos, M. Tosi and R. von Steiger (eds.): *Primordial Nuclei and Their Galactic Evolution.* 1998 ISBN 0-7923-5114-2
5. C. Fröhlich, M.C.E. Huber, S.K. Solanki and R. von Steiger (eds.): *Solar Composition and its Evolution – From Core to Corona.* 1998 ISBN 0-7923-5496-6
6. B. Hultqvist, M. Øieroset, Goetz Paschmann and R. Treumann (eds.): *Magnetospheric Plasma Sources and Losses.* 1999 ISBN 0-7923-5846-5
7. A. Balogh, J.T. Gosling, J.R. Jokipii, R. Kallenbach and H. Kunow (eds.): *Co-rotating Interaction Regions.* 1999 ISBN 0-7923-6080-X
8. K. Altwegg, P. Ehrenfreund, J. Geiss and W. Huebner (eds.): *Composition and Origin of Cometary Materials.* 1999 ISBN 0-7923-6154-7
9. W. Benz, R. Kallenbach and G.W. Lugmair (eds.): *From Dust to Terrestrial Planets.* 2000 ISBN 0-7923-6467-8
10. J.W. Bieber, E. Eroshenko, P. Evenson, E.O. Flückiger and R. Kallenbach (eds.): *Cosmic Rays and Earth.* 2000 ISBN 0-7923-6712-X
11. E. Friis-Christensen, C. Fröhlich, J.D. Haigh, M. Schüssler and R. von Steiger (eds.): *Solar Variability and Climate.* 2000 ISBN 0-7923-6741-3
12. R. Kallenbach, J. Geiss and W.K. Hartmann (eds.): *Chronology and Evolution of Mars.* 2001 ISBN -7923-7051-1
13. R. Diehl, E. Parizot, R. Kallenbach and R. von Steiger (eds.): *The Astrophysics of Galactic Cosmic Rays.* 2001 ISBN 0-7923-7051-1
14. Ph. Jetzer, K. Pretzl and R. von Steiger (eds.): *Matter in the Universe.* 2001 ISBN 1-4020-0666-7
15. G. Paschmann, S. Haaland and R. Treumann (eds.): *Auroral Plasma Physics.* 2002 ISBN 1-4020-0963-1
16. R. Kallenbach, T. Encrenaz, J. Geiss, K. Mauersberger, T.C. Owen and F. Robert (eds.): *Solar System History from Isotopic Signatures of Volatile Elements.* 2003 ISBN 1-4020-1177-6

Kluwer Academic Publishers – Dordrecht / Boston / London